Metallurgy and Materials Science

Metallurgy and Materials Science

Edited by **Ricky Peyret**

NY RESEARCH
P R E S S

New York

Published by NY Research Press,
23 West, 55th Street, Suite 816,
New York, NY 10019, USA
www.nyresearchpress.com

Metallurgy and Materials Science
Edited by Ricky Peyret

International Standard Book Number: 978-1-63238-494-2 (Hardback)

Printed in the United States of America.

Contents

Preface

This book has been a concerted effort by a group of academicians, researchers and scientists, who have contributed their research works for the realization of the book. This book has materialized in the wake of emerging advancements and innovations in this field. Therefore, the need of the hour was to compile all the required researches and disseminate the knowledge to a broad spectrum of people comprising of students, researchers and specialists of the field.

Metallurgy is a subfield of materials science. It is generally applied to the purification and production of metals from their ores. Materials science on the other hand is a broader field which encompasses the discovery and design of new materials. It also requires the knowledge of engineering, physics and chemistry. There has been rapid progress in this field and its applications are finding their way across multiple industries. This book is a valuable compilation of topics, ranging from the basic to the most complex advancements in metallurgy and materials science. It attempts to understand the multiple branches that fall under the discipline of materials science and how such concepts have practical applications. It will help the readers in keeping pace with the rapid changes in this field.

At the end of the preface, I would like to thank the authors for their brilliant chapters and the publisher for guiding us all-through the making of the book till its final stage. Also, I would like to thank my family for providing the support and encouragement throughout my academic career and research projects.

Editor

A Numerical Formula for General Prediction of Interface Bonding between Alumina and Aluminum-Containing Alloys

Michiko Yoshitake,[1] Shinjiro Yagyu,[1] and Toyohiro Chikyow[2]

[1]National Institute for Materials Science, 3-13 Sakura, Tsukuba 305-0003, Japan
[2]National Institute for Materials Science, 1-1 Namiki, Tsukuba 305-0044, Japan

Correspondence should be addressed to Michiko Yoshitake; yoshitake.michiko@nims.go.jp

Academic Editor: Dina V. Dudina

Interface termination between alumina and aluminum-containing alloys is discussed from a viewpoint of thermodynamics by extending the authors' previous discussion on the interface termination between alumina and pure metal. A numerical formula to predict interface bonding at alumina-aluminum-containing alloys is proposed. The effectiveness of the formula is examined by extracting information on interface termination from experimental results and first-principle calculations in references. It is revealed that the prediction by the formula agrees quite well with the results reported in the references. According to the formula, a terminating species can be switched from oxygen to aluminum, which had been actually demonstrated experimentally. The formula uses only basic quantities of pure elements and the formation enthalpy of oxides. Therefore it can be applied for most of aluminum-containing alloys in the periodic table and is useful for material screening in developing interfaces with particular functions.

1. Introduction

Interface bonding between oxides and metals is one of the crucial factors that determine properties of materials such as bonding strength, Schottky barrier height, sensitivity of sensors, catalytic activity, and overpotential in batteries. Metal oxides, which are composed of metals and oxygen, can have differently terminated surfaces, for example, the topmost surface being occupied only by oxygen atoms or by the metal atoms that compose the oxide. When such differently terminated surfaces form the interface with metals, bonding strength and wetting properties at the interfaces depend on surface termination species [1–5]. Electron energy level alignment between the Fermi level and oxides' valence bands (band alignment) also varies largely with surface terminating species [6–10]. Regarding alumina/metal interface, which is one of the most extensively studied systems among various oxide/metal interfaces, we have studied the thermodynamics of interface termination and proposed a numerical formula to predict a terminating species at the interface [11]. A software program that gives predicted results according to the formula has been released [12].

Under conditions where a stable interface termination is determined by a metal in contact, alloying (mixing two or more metals) is one of the most frequently used techniques for modifying interfaces, especially for electric device applications where an electrode metal works only as an electronic conductor but is not chemically functioning. Because the choice of oxides is based on specific properties of the oxides, modification should be made on electrode metals, not on oxides. In this paper, therefore, the discussion on the thermodynamics of interface termination in [11] is extended to the interface between alumina and alloys that contain aluminum, and then a numerical formula for predicting a stable interface terminating species at alumina/aluminum-containing alloy interfaces is proposed.

This work aims to offer a new tool for predicting whether alloying with aluminum is effective or not for interface modification so that time-consuming trials and errors on each system would not be necessary. We believe that the formula proposed in this work would be of great use in material development.

2. Formula for Prediction

2.1. Varieties in Termination. The most stable phase of alumina is alpha, which has a corundum structure with hexagonal symmetry. The planes parallel to the *c*-axis (*c*-plane) have differently terminated surfaces and could be O-terminated, Al-terminated, or Al-double-layer-terminated. Because experiments on the *c*-plane have been mostly conducted using metal/alumina interfaces for both solid-state bonding and film growth experiments, the *c*-plane is considered in this study. Therefore, the interface bonding species can be either Al (M–Al-alumina, including Al-double layer) or O (M–O-alumina) when an interface is formed with a metal (M).

The interface between pure metal (M) and alumina will be terminated by either (1) M–Al-alumina (Al-termination) or (2) M–O-alumina (O-termination). Here, the influence of the coverage is neglected as discussed in the previous paper [11]. When the discussion is extended to the interface with aluminum-containing alloy (MAl), M–Al′-alumina (Al-termination, Al′ denotes terminating Al) in pure M system is replaced by (MAl)–Al′-alumina, where the atom that binds to Al′ can be either M* or Al* in MAl (here, M* and Al* denote atoms from alloys). Likewise, M–O′-alumina (O-termination, O′ denotes terminating O) is replaced by (MAl)–O′-alumina, where the atom that binds to O′ can be either M* or Al* in MAl. Then, there would be four different types of termination at the interface:

(alloy) | interface | alumina

(A) (MAl)—$\underline{M^*}$–Al—alumina   (Al-termination)

(B) (MAl)—$\underline{Al^*}$–Al—alumina   (Al-termination)

(C) (MAl)—$\underline{M^*}$–O—alumina   (O-termination)

(D) (MAl)—$\underline{Al^*}$–O—alumina

Although (D) is initially derived from the O-terminated interface between (MAl) alloy and alumina, it can be regarded as Al-termination because this is the same as either (A) or (B).

2.2. Procedure for Prediction. Thermodynamic equilibrium among the four types of termination (A)–(D) has been considered. The equilibrium is determined by Gibbs energy difference among the terminations [11]. In our previous paper on the interface between alumina and pure metal, we use M–Al and M–O bonding energy to obtain Gibbs energy difference among different terminations for simplicity [11]. Here, for the interface between alumina and aluminum-containing alloy (MAl), we use M–Al, Al–Al, M–O, and Al–O bonding energy.

As in the previous paper, M–Al bonding energy is estimated either by the adsorption energy of Al on M (Approx-1) or by subtracting the adsorption energy of M on M from

that of Al on M (Approx-2). The subtraction is considered because the values of adsorption energy include not only the influence of chemical interaction between Al and M but also that of cohesion energy. M–O bonding energy is estimated either by the adsorption energy of oxygen on M (Approx-1) or by subtracting the dissociation energy of molecular oxygen from the adsorption energy of oxygen on M (Approx-2). The reason of adopting adsorption energy of metals to estimate M–Al or Al–Al bonding energy and that of oxygen for M–O bonding energy is as follows: values of formation enthalpy of oxides and intermetallic compounds or of mixing enthalpy include terms not only from chemical interaction but also from structural change. At the interface structural relaxation occurs more easily than in bulk and the structural term would be smaller and can be neglected for rough estimation.

The adsorption energy of Al on M (=Al on Al when M = Al) and M on M were calculated using (1), which is based on Meadima's formula [13]:

$$\Delta H_{ad}(A \text{ on } B) = -F \times \gamma_B \times S_A + (1 - F) \times \gamma_A \times S_A + F \times \Delta H_{sol}(A \text{ in } B) - \Delta H_{vap}(A), \quad (1)$$

where $\Delta H_{ad}(A \text{ on } B)$ is the adsorption energy of A on B, γ_A, and S_A, and γ_B and S_B are surface energy and surface area of A and B, respectively. $\Delta H_{sol}(A \text{ in } B)$ is the heat of mixing of A in B; ΔH_{vap} is vaporization enthalpy. F is the portion of the area of A in contact with B, which is typically around 0.4. Here, the energy is described per mol. $\Delta H_{sol}(A \text{ in } B)$ is calculated by the following equation:

$$\Delta H_{sol}(A \text{ in } B) = 2V(A)^{2/3} \times \left(n(A)^{-1/3} + n(B)^{-1/3}\right)^{-1} \times N_0 \times P \times \left\{-e(\Delta \phi)^2 + \frac{Q}{P(\Delta n^{1/3})^2} - \frac{R}{P}\right\}, \quad (2)$$

where $V(A)$ is the molar volume of metal A, $n(A)$, and $n(B)$ are the electron density of A and B at the boundary of the Wigner-Seitz cell, $\Delta \phi$ is the work function difference between A and B, and P, Q, and R are parameters. N_0 is the Avogadro's number. The detail of the calculation and values for γ_A and γ_B, ΔH_{vap}, $V(A)$, $n(A)$, and $n(B)$, $\Delta \phi$, and three parameters P, Q, and R are described in [14]. The values of the calculated adsorption energy for each M–Al combination were obtained using the software [15] released by one of the authors and listed in Table 1.

The adsorption energy of oxygen on M is estimated in the following way. It has been reported [16] that the initial heat of adsorption of oxygen on some metals [17] has linear dependence on the standard enthalpy of formation [18] of the corresponding oxides with the highest oxidation state. The values of the formation enthalpy and the valence of the corresponding oxides were reexamined using other references [19, 20]. We decided to use the values from [20] to correlate the initial heat of adsorption of oxygen (Hads) with the formation enthalpy of the corresponding oxide with the highest oxidation state (Hform) except Cr. The following

TABLE 1: Adsorption energy of Al and other metals (M) on M and their subtracted values.

Metal-M	Al on M Adsorption energy (kJ/mol)	M on M Adsorption energy (kJ/mol)	(Al on M) – (M on M) Energy difference (kJ/mol)
Al	270	270	0
Si	277	359	−82
Ti	384	363	21
V	400	401	−1
Cr	377	303	74
Fe	392	316	76
CO	408	335	73
Ni	407	340	67
Cu	332	265	67
Zn	299	113	186
Ga	243	227	16
Ge	258	297	−39
Zr	389	490	−101
Nb	409	582	−173
Mo	410	523	−113
Ru	454	521	−67
Rh	447	435	12
Pd	413	283	130
Ag	280	222	58
In	228	198	30
Sn	230	254	−24
La	312	358	−46
Hf	399	492	−93
Ta	433	627	−194
W	434	695	−261
Re	496	612	−116
Os	487	636	−149
Ir	474	535	−61
Pt	450	448	2
Au	325	293	32
Hg	262	60	202
Pb	215	153	62
Bi	211	169	42

numerical relationship has been obtained with the correlation coefficient of 0.977:

$$Hads \ (kJ/mol\text{-}O) = 0.719 \times Hform \ (kJ/mol\text{-}M) \\ + 230 \ (kJ/mol\text{-}O). \quad (3)$$

We use values of Hads calculated by (3) as the adsorption energy of oxygen on M.

In Table 2, formation enthalpy of various oxides to be used for the calculation (after [20] except those in italics which are from [19]) and the calculated adsorption energy values are listed. For readers' convenience, values for metals

that were not reported in references we discuss later are also given.

In the previous paper, the following two expressions were used in order to predict whether the interface termination is either M–Al-alumina or M–O-alumina between pure metal and alumina.

Approx-1 is

$$(AlonM) - (OonM). \quad (4)$$

Approx-2 is

$$\{(AlonM) - (MonM)\} - \left\{(OonM) \\ - \frac{1}{2}(O_2 \text{ dissociation energy})\right\}, \quad (5)$$

where O_2 dissociation energy = 493.07 kJ/mol [21]. Approx-2 has been proposed because the value of (AlonM) includes both chemical interaction between Al and M and cohesive energy of atomic Al, and therefore, in order to extract the term caused by only chemical interaction, subtraction of cohesive energy is necessary. The situation is the same for (OonM), where both chemical interaction between O and M and cohesive energy of atomic O are included and the subtraction of cohesive energy (=the half of O_2 dissociation energy) is needed. Prediction with Approx-1 is that if (4) is positive, that is, (AlonM) > (OonM), M–Al bonding is preferred to M–O bonding and if (AlonM) < (OonM), M–O bonding is preferred. With Approx-2, prediction goes as follows: if (5) is positive, that is, {(AlonM) – (MonM)} > {(OonM) – 1/2(O_2 dissociation energy)}, M–Al bonding is preferred, and vice versa.

Four types of terminations (A)–(D) for aluminum-containing alloy, as described in Section 2.1, are derived from either M–Al-alumina or M–O-alumina interface in pure metal, where (A) and (B) are derived from M–Al-alumina and (C) and (D) are derived from M–O-alumina. Therefore, in order to predict which one of terminations is realized at the interface with aluminum-containing alloy, we first predict whether M–Al-alumina or M–O-alumina is realized at the interface without aluminum in the alloy using either Approx-1 or Approx-2. Once M–Al-alumina is predicted, the second step is to predict whether the interface is (A) or (B) at the interface with aluminum-containing alloy. When M–O-alumina is predicted, the second step is to predict whether the interface is (C) or (D). Whether (A) or (B) is realized is determined by comparing the value of (AlonAl) with that of (AlonM). If (AlonAl) > (AlonM), Al–Al bonding is preferred to M–Al bonding and the termination becomes (B). Similarly, (C) or (D) is determined by the value of (OonAl) with respect to (OonM). Here, the value of (OonAl) is obtained by calculating (OonM) with M = Al and is 833.06 kJ/mol. If (OonAl) > (OonM), Al–O bonding is preferred to M–O bonding and the termination becomes (D), which is regarded as Al-termination. Here, comparison between (AlonAl) and (AlonM) or between (OonAl) and (OonM) does not need subtraction like in (5), because the

TABLE 2: Values of oxide formation enthalpy and calculated adsorption energy of oxygen on various metals (M) and related values.

		kJ/mol	kJ/mol-M	kJ/mol-O	kJ/mol-O	Energy difference (kJ/mol)
Mg	MgO	601.6	601.6	601.6	664.14	419.61
Al	Al_2O_3	1675.7	837.85	558.5667	833.06	588.53
Si	SiO_2	910.7	910.7	455.35	885.15	640.62
Ti	TiO_2	944	944	472	908.96	664.43
V	V_2O_5	1550.6	775.3	310.12	788.34	543.81
Cr	Cr_2O_3	1139.7	569.85	379.9	641.44	396.91
Mn	MnO_2	520	520	260	605.80	361.27
Fe	Fe_2O_3	824.2	412.1	274.7333	528.65	284.12
CO	CO_3O_4	891	297	222.75	446.36	201.83
Ni	Ni_2O_3	489.5	244.75	163.1667	409.00	164.47
Cu	CuO	157.3	157.3	157.3	346.47	101.94
Zn	ZnO	350.5	350.5	350.5	484.61	240.08
Ga	Ga_2O_3	1089.1	544.55	363.0333	623.35	378.82
Ge	GeO_2	580	580	290	648.70	404.17
Zr	ZrO_2	*1094.3*	*1094.324*	*547.162*	1016.44	771.91
Nb	Nb_2O_5	1899.5	949.75	379.9	913.07	668.54
Mo	MoO_3	745.1	745.1	248.3667	766.75	522.22
Ru	RuO_4	239.3	239.3	59.825	405.10	160.57
Rh	Rh_2O_3	343.0	171.5	114.3333	356.62	112.09
Pd	PdO	85.4	85.4	85.4	295.06	50.53
Ag	Ag_2O_2	24.3	12.15	12.15	242.69	−1.84
In	In_2O_3	925.8	462.9	308.6	564.97	320.44
Sn	SnO_2	577.63	577.63	288.815	647.01	402.48
La	La_2O_3	1793.7	896.85	597.9	875.25	630.72
Hf	HfO_2	1144.7	1144.7	572.35	1052.46	807.93
Ta	Ta_2O_5	2046	1023	409.2	965.45	720.92
W	WO_3	842.9	842.9	280.9667	836.67	592.14
Re	Re_2O_7	1240.1	620.05	177.1571	677.34	432.81
Os	OsO_4	*391.248*	*391.248*	*97.812*	513.74	269.21
Ir	IrO_2	274.1	274.1	137.05	429.98	185.45
Pt	PtO_2	133.3	133.3	66.65	329.31	84.78
Au	AuO_x	<0	<0	<0	<0	<0
Hg	HgO	90.79	90.79	90.79	298.91	54.38
Pb	PbO	218	218	218	389.87	145.34
Bi	Bi_2O_3	573.9	191.3	286.95	370.78	126.25

cohesive energy is canceled when (AlonAl) and (AlonM) are compared [(AlonAl) − (AlonM) = {(AlonAl) − (AlonAl)} − {(AlonM) − (AlonAl)}], as well as for (OonAl) and (OonM) [(OonAl) − (OonM) = {(OonAl) − 1/2(O_2 dissociation energy)} − {(OonM) − 1/2(O_2 dissociation energy)}]. Therefore, the expressions for each termination for aluminum-containing alloy (Approx-1) are as follows, where the flow chart for finding an expression is shown in Figure 1(a):

(a) (OonM) < (AlonM) > (AlonAl)

(b) (OonM) < (AlonM) < (AlonAl)

(c) (OonAl) < (OonM) > (AlonM)

(d) (OonAl) > (OonM) > (AlonM)

When Approx-2 is used in the first step to predict whether M–Al-alumina or M–O-alumina is realized at the interface

without aluminum in the alloy, the comparison between (OonM) and (AlonM) should be made by replacing (OonM) by {(OonM) − 1/2(O_2 dissociation energy)} and (AlonM) by {(AlonM) − (MonM)}. Then the corresponding expressions to (a)–(d) become as follows, where the flow chart for finding an expression is shown in Figure 1(b):

(a′) (OonM) − 1/2(O_2 dissociation energy) < (AlonM) − (MonM) > (AlonAl) − (MonM)

(b′) (OonM) − 1/2(O_2 dissociation energy) < (AlonM) − (MonM) < (AlonAl) − (MonM)

(c′) (OonAl) − 1/2(O_2 dissociation energy) < (OonM) − 1/2(O_2 dissociation energy) > (AlonM) − (MonM)

(d′) (OonAl) − 1/2(O_2 dissociation energy) > (OonM) − 1/2(O_2 dissociation energy) > (AlonM) − (MonM).

(a) Approx-1

(b) Approx-2

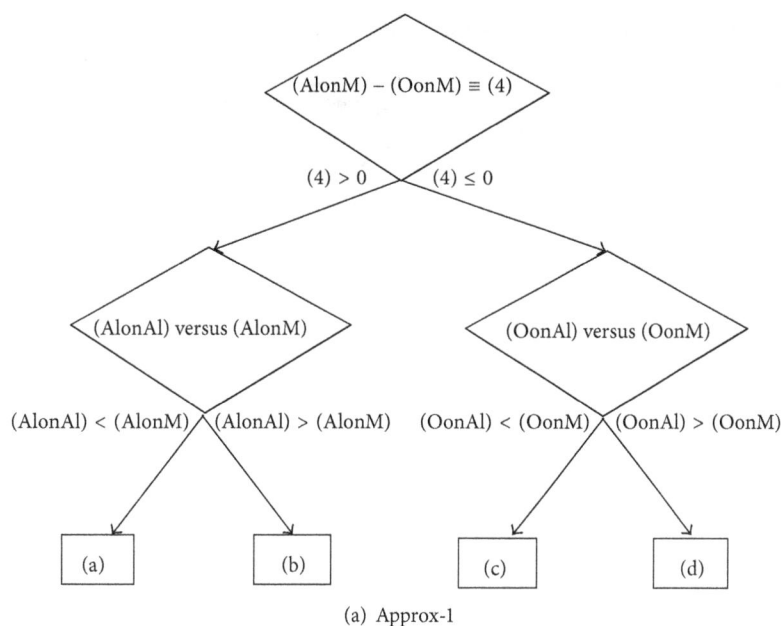

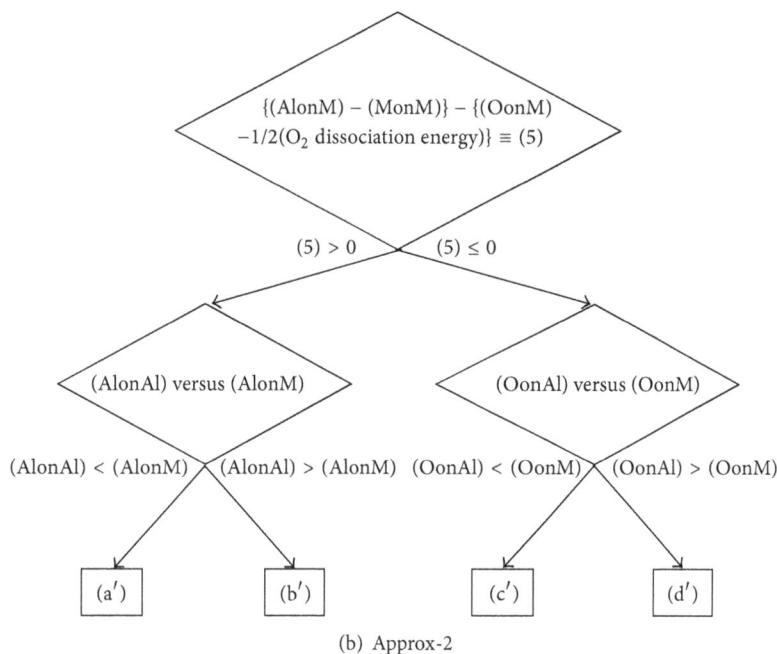

FIGURE 1: Flow chart for finding an expression that predicts termination.

If (d) or (d′) is satisfied, the interface between alumina and pure metal M is O-terminated, whereas the interface with M alloyed with Al is Al-terminated. This is the key to switch interface termination species from oxygen to aluminum by adding Al to metals that satisfy (d) or (d′). Whether this switching of interface termination species occurs or not is governed by the adsorption energy of oxygen on M and Al. As in the expressions (d) and (d′), when the adsorption energy of oxygen on Al is larger than that on M, Al–O bonding is preferred at the interface and type (D) termination, that is, Al-termination, is formed. On the other hand, it is clear that switching Al-terminated interface to O-terminated one

by alloying with Al is impossible from the expressions (a)–(d′). It should be noted that the systems where different termination is predicted in the first step using by either Approx-1 or Approx-2 are limited only to Ru, Rh, Ir, Pt, and Hg (predictions for these systems are M–Al-alumina with Approx-1 and M–O-alumina with Approx-2). Furthermore, aluminum-containing alloys of these metals are predicted to be Al-terminated whether interface termination is predicted as M–Al-alumina or M–O-alumina in the first step (termination (A) or (D)). Therefore, we may be able to use Approx-1 in the first step for predicting the interface in aluminum-containing alloy.

TABLE 3: Binding energies of Al 2p in Ti-Al compounds prepared in various conditions compared to those in alumina on Cu and Ni related metals as references.

	Preparation conditions	Binding energy (eV)			Termination	Reference
		Al 2p3/2 or Al 2p (ox)	Al 2p3/2 or Al 2p (MAl)	Al 2p3/2 or Al 2p (ox) − (MAl)		
Cu9Al(111)		74.90	72.58	2.32	Al	[24]
Cu(111)		74.11		1.53	O	[10]
NiAl(110)		74.91	72.47	2.44	Al	[10]
Ni(111)		74.11		1.64	O	[10]
TiAl(111)	1×10^{-5} Pa, 923 K	75.5	72.2	3.3		[25]
Ti45-Al55	1×10^{-5} Pa, 923 K	75.5	72.3	3.2		[26]
Ti55-Al45	$<5 \times 10^{-8}$ Pa, ~873 K	74.9	71.6	3.3		[27]
Ti_3Al	1×10^{-5} Pa, 923 K	75.5	72.3	3.2		[26]
TiAl	1.3×10^{-5} Pa, 673–873 K	74.4	71.7	2.7		[28]
Ti45-Al55	10 Pa, 423 K & 623 K	75.2	72.57	2.63		[29]
TiAl	air, 573–673 K	74.9	72.3	2.6		[30]

In summary, to find a type of interface bonding in aluminum-containing alloy (MAl), interface termination at alumina-corresponding pure metal (without aluminum in the alloy) should be first examined. Then, if the interface with pure metal is Al-terminated, values of (AlonM) and (AlonAl) are to be compared. For O-terminated interface with pure metal, values of (OonM) and (OonAl) should be compared. This procedure gives the type of interface bonding in aluminum-containing alloy from expressions (A)–(D). After looking for M that satisfies (AlonAl) > (AlonM) (M = Ga, Ge, In, Sn, Hg, Pb, Bi in Table 1), we found that there is no M with (OonM) < (AlonM). Therefore, there is no M that satisfies expression (B).

It should be noted that the values of (OonM) are derived from the standard formation enthalpy of corresponding oxide, which is defined at 1 bar pressure. Because both oxidation and reduction of metal can occur in the same system at different oxygen pressure, a terminating species would vary with oxygen pressure, especially for metals with relatively small standard formation enthalpy values.

There is one more thing to be noted. In all the above discussion, the possibility of alumina reduction is excluded. However, if the oxide formation enthalpy of metal M per mol-O is larger than that of alumina, formation of oxide with M and reduction of alumina should occur, which is expected for M = La, Hf in Table 2.

3. Termination in Aluminum-Containing Alloys in References

3.1. Experimental Results. There are only a limited number of references that handle interface termination between alumina and aluminum-containing alloys. We have investigated interface termination using NiAl(110) and Cu-9Al(111) [10] and showed that Al 2p XPS peak is a good measure to judge a type of termination. If Al 2p peak has a component between that for Al_2O_3 and metallic Al, the component is attributed to the interface and the interface is Al-terminated. For NiAl(110),

the shoulder in Al 2p peak in oxidized substrate has been known [22], which were attributed to Al that binds the substrate and alumina film using calculation and STM [23].

Although other references did not discuss interface termination, by examining the reported Al 2p XPS spectra, a type of interface termination can be estimated in the above way. For FeAl, a similar shoulder in Al 2p peak as in NiAl and Cu-9Al was reported [31], which indicates that the interface between FeAl and alumina formed by the oxidation of FeAl was Al-terminated though the authors of the paper did not mention it. In addition to NiAl(110), when NiAl(111), a different orientation of the same intermetallic, was oxidized, a similar shoulder in Al 2p was reported [32].

There are XPS studies on the oxidation of TiAl and Ti_3Al, but well resolved Al 2p spectra were not reported. However, we are able to estimate the interface termination difference by examining the reported Al 2p binding energies in the following way. In Table 3, Al 2p binding energy values of alumina and of intermetallics taken from references [10, 24–30] are listed. In the case of Cu and Ni systems, where all the data come from our laboratory under the same energy calibration conditions, differences of Al $2p_{3/2}$ values in alumina (Al $2p_{3/2}$ (ox)) with respect to the ones in M–Al (Al $2p_{3/2}$ (MAl)) for Al-terminated samples (2.3-2.4 eV) are clearly different from those for O-terminated ones (1.5-1.6 eV). For Ti systems, the energy difference between Al 2p (ox) and Al 2p (MAl) falls in two categories, one 3.2-3.3 eV and the other 2.6-2.7 (in the references, Al $2p_{3/2}$ and Al $2p_{1/2}$ were not resolved). The smaller energy difference appears to suggest O-termination, while the larger one is for Al-termination. If we examine the preparation conditions for all these experiments in Table 3, it seems that the suggested terminating species is dependent on the oxidation potential during the interface formation. In the case of lower oxygen pressure and/or higher temperature (=lower oxidation potential), larger energy difference, that is, Al-termination, appears to be realized. The interface termination deduced from reported experiments is schematically summarized in Table 4.

TABLE 4: Schematic representation of interface terminating species at interfaces with alumina reported in experiments.

Mg										
Ca	Sc	Ti*	V*	Cr*	Mn	Fe*	CO*	Ni*	Cu*	Zn
		TiAl‡				FeAl†	COAl	NiAl†	Cu(Al)†	
Sr	Y	Zr	Nb*	Mo	Tc	Ru	Rh	Pd	Ag‡	Cd
Ba	La	Hf	Ta	W	Re	Os	Ir	Pt	Au	Hg

*O-term.
†Al-term.
‡Oxygen pressure dependent.

3.2. Theoretical Results. To the authors' knowledge, there is no reference that calculates the stability of interface termination at alumina/aluminum-containing alloy (including intermetallics) by first-principle calculations. The references that discuss the chemical potential of Al, $\Delta\mu_{Al}$, as a parameter in the thermodynamic study of interface termination for alumina-Ni, Cu, Ag, and Au [33–35] interfaces handle Al-containing intermetallics. Their conclusion is that an interface terminating species changes from oxygen to aluminum according to the increase of the chemical potential of Al in metals as schematically shown in Figure 2. This $\Delta\mu_{Al}$ is a function of both oxygen partial pressure and aluminum activity a_{Al} in metals. The figure assumes that alumina is more stable than oxide of metal M, MO. Here, using Figure 2, we discuss the influence of aluminum activity under constant oxygen partial pressure, where alumina is stable. From the right to the left in the figure, metal composition changes from pure metal to Al intermetallics. On the border B at the right side in Figure 2, the interface is not Al_2O_3/M (pure metal) but Al_2O_3/MO_x (metal oxide). When a_{Al} is larger than that at the border B, the interface is O-terminated. If mixed oxide phase MAlO exists, the border C appears in the figure and the interface with alumina would be Al_2O_3/$MAlO_x$ instead of Al_2O_3/MO_x. The interface Al_2O_3/$MAlO_x$ is regarded as O-terminated from a bonding point of view, because M–O bonding, not M–Al, exists at the interface. On the border A at the left side in Figure 2, Al_2O_3 reduces to Al metal that forms an intermetallic compound MAl, such as Cu_3Al and Ni_3Al, and the interface is Al_2O_3/MAl, not Al_2O_3/M. Therefore, the reference tells us that the interface between MAl (M = Ni, Cu, Ag and Au) and alumina is Al-terminated.

4. Comparison between Prediction and Results in References

Here, we examine the prediction derived from the proposed expressions for each system and compare with the results deduced from the reported results.

In Table 5, the prediction for pure metal (M), the examination of the expression described in Section 2.2 (both Approx-1 and Approx-2), and the resulting prediction for aluminum-containing alloy (MAl) are listed for various metals. The results from the experimental references are also shown in the table. For M on which experimental results both for M and MAl are available (M = Fe, Ni, Cu), the interface is terminated by oxygen for pure M. Therefore, the termination for MAl should be either (C) or (D). Because (OonM) is

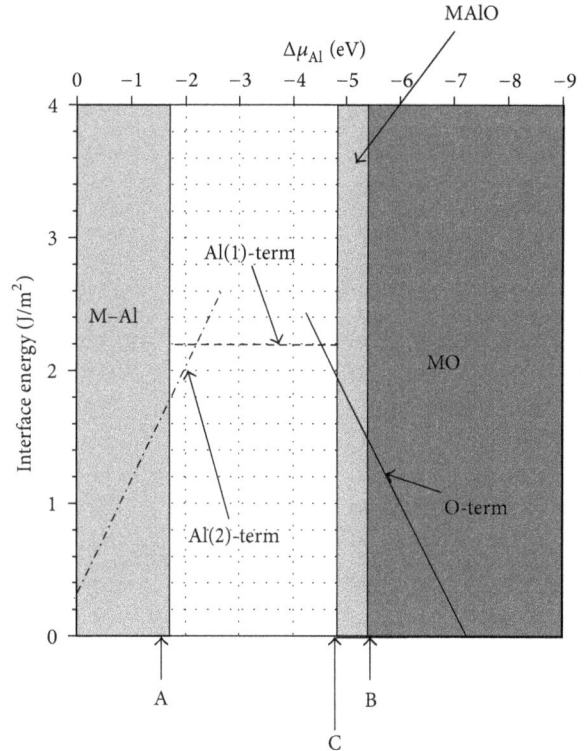

FIGURE 2: Schematic diagram of interface energy and preferred interface termination as a function of $\Delta\mu_{Al}$ (chemical potential of Al) for M with intermetallic compound (MAl) and oxides (MO) formation taken into account. On the left side of the border A, Al_2O_3 reduces to Al metal. On the right side of the border B, the interface is not Al_2O_3/M (pure metal) but Al_2O_3/MO_x (metal oxide). When a_{Al} is larger than that at the border B, the interface is O-terminated. If mixed oxide phase MAlO exists, the border C appears in the figure and the interface with alumina would be Al_2O_3/$MAlO_x$ instead of Al_2O_3/MO_x.

smaller than (OonAl), expression (d) or (d′) applies. This means that our formula predicts Al-termination for MAl (M = Fe, Ni, Cu). The experimental results agree with this prediction. In Section 2.2 it is noted that switching O-terminated interface with pure M to Al-termination by adding Al in M should be possible. Our experiments on alumina/Ni, NiAl, Cu, and Cu-9Al interfaces [10] actually demonstrated the above idea of termination switching. The experiment clearly showed that O-terminated interface with

TABLE 5: The interface prediction for pure metal (M), examination of interface type and interface prediction for aluminum-containing alloy, and experimental results on interface termination in references.

M	Predicted interface termination for pure metal (M)	Predicted interface type for MAl		Predicted interface termination for alloy (MAl)	Experimental results from references
		Approx-1	Approx-2		
Si	O	C	C	O	
Ti	O	C	C	O	Al, O
V	O	D	D	Al	
Cr	O	D	D	Al	
Fe	O	D	D	Al	Al
CO	O	D	D	Al	
Ni	O	D	D	Al	Al
Cu	O	D	D	Al	Al
Zn	O	D	D	Al	
Ga	O	D	D	Al	
Ge	O	D	D	Al	
Zr	O	C	C	O	
Nb	O	C	C	O	
Mo	O	D	D	Al	
Ru	Al, O	A	D	Al	
Rh	Al, O	A	D	Al	
Pd	Al	A	A	Al	
Ag	Al	A	A	Al	
In	O	D	D	Al	
Sn	O	D	D	Al	
La	O	—	—	Al$_2$O$_3$ reduction	
Hf	O	—	—	Al$_2$O$_3$ reduction	
Ta	O	C	C	O	
W	O	C	C	O	
Re	O	D	D	Al	
Os	O	D	D	Al	
Ir	Al, O	A	D	Al	
Pt	Al, O	A	D	Al	
Au	Al	A	A	Al	
Hg	Al, O	A	D	Al	
Pb	O	D	D	Al	
Bi	O	D	D	Al	

pure Ni and pure Cu changed to Al-terminated with NiAl and Cu-9Al.

For M = Ti, where the expression (c) or (c′) is satisfied, our formula predicts O-termination. In the experimental reports, both O-termination and Al-termination appear to be obtained depending on the conditions (oxygen potential) for the interface formation as discussed in Section 3.1. Al-termination, which is in disagreement with the prediction, was obtained under low oxygen pressure at high temperature. Under such condition, adsorbed oxygen is known to dissolve into bulk Ti [36]. For Ti, although (OonAl) < (OonM), it appears that dissolution of oxygen at the interface into Ti occurs, resulting in Al-termination. The dissolution of oxygen into metal or alloy is highly dependent on a kind

of metals or alloy and is not taken into account in the prediction formula. Among Si, Ti, Zr, Nb, Ta, and W, which satisfy expression (c) or (c′), similar behavior as for Ti is expected for Zr, Nb, and Ta, because these three metals dissolve considerable amount of oxygen according to the phase diagrams.

One more thing to be noted is that all the metals that satisfy expression (c) or (c′) have a mixed oxide phase described as MAlO in Figure 2, Al$_2$O$_3$·SiO$_2$, Al$_2$O$_3$·TiO$_2$, 2Al$_2$O$_3$·ZrO$_2$, 1/2(Al$_2$O$_3$·Nb$_2$O$_5$) (=AlNbO$_4$), 1/2(Al$_2$O$_3$·Ta$_2$O$_5$) (=AlTaO$_4$), 2Al$_2$O$_3$·6WO$_3$ (=Al$_2$(WO$_4$)$_3$).

As mentioned in Section 3.2, interface for these metals could be Al$_2$O$_3$/MAlO (border C in Figure 2), which contains Al–O–M–O– bonding at the interface and hence is regarded

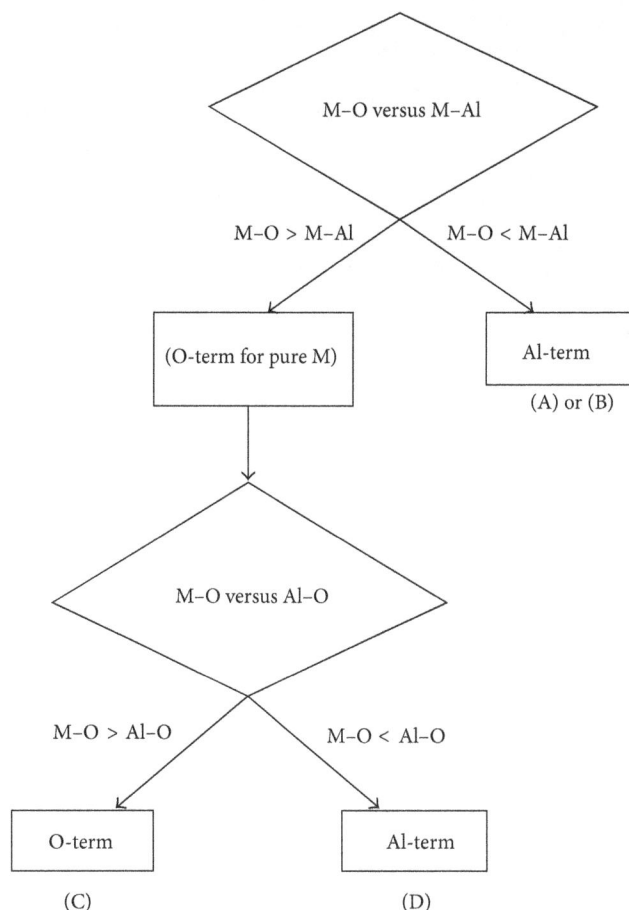

FIGURE 3: Flow chart for predicting a termination type among (A)~ (D) in the text.

as O-termination in this paper. When the formation enthalpy of metal oxide increases, approaching that of alumina, the position of borders B and C shifts toward left side in Figure 2, while the position of border A is not influenced by oxide formation enthalpy but mainly by the strength of M–Al bonding (if the M-Al bonding is stronger, the position of border A moves toward right side).

The features shown in Figure 2 in the theoretical study are based on thermodynamics and are common among all the systems calculated. Therefore, we can expect the features to be universal for any metal. Then, when the interface is Al-terminated, alumina is expected to be in equilibrium with aluminum-containing alloys or intermetallics under oxygen partial pressure that metal is not oxidized. This guides the practically useful technique to switch interface bonding from an originally O-terminated interface to Al-termination by adding Al in a metal M. The exception of the application of this technique is the case where (OonAl) < (OonM) (expression (c) or (c′) is satisfied). Interface bonding for metals that satisfy expression (c) or (c′) would be process-dependent in practice, because the thermodynamic stability of Al-terminated and O-terminated interface is very close for metals that satisfy both (OonAl) < (OonM) and (formation enthalpy of oxide of metal M) < (formation enthalpy of

Al_2O_3). For such case, prediction is possible by observing which one of metal elements is preferentially oxidized in Al-containing alloys or intermetallics. If Al is preferentially oxidized, the interface would be Al-terminated, whereas the preferential oxidation of M would results in O-terminated.

It should be noted that if MO is more stable than Al_2O_3, interface Al/MO instead of Al_2O_3/M should be formed under thermodynamic equilibrium. Among metals we consider in Table 2, Mg, La, and Hf correspond to the case.

By incorporating the discussion on Figure 2 into the expressions presented here, we can make an algorithm to find interface termination in Al-containing alloy as in Figure 3. This algorithm guides a novel method to control interface termination; for a metal with (OonM) < (OonAl), an interface that exhibits O-termination in pure metal can be switched to Al-termination by alloying the metal with Al. It also concludes that a stable Al-terminated interface cannot be formed for metals with (OonM) > (OonAl) under equilibrium conditions if oxygen partial pressure is not low enough to reduce Al_2O_3. Therefore, for such metals, utilizing a quenching process is necessary to obtain Al-terminated interfaces. One example of a quenching process is depositing Al on metals followed by oxidation without sufficient annealing, which avoids atomic diffusion needed to reach thermodynamically stable O-termination.

The influence of oxygen partial pressure is not taken into account in (1) and (3), which are used to calculate the adsorption energies of Al on M and oxygen on M, respectively. On the other hand, the strengths of M–O and Al–O bonds should depend on the oxygen partial pressure. Therefore, prediction by this method is not accurate, especially for easily reduced metals. However, it provides a guide for termination, which we believe is quite useful for material development.

5. Conclusions

Interface bonding between alumina and aluminum-containing alloy (MAl) has been investigated. A method to predict an interface terminating species is proposed by extending the prediction method already proposed for the interface between alumina and pure metals. In the method, to find the most stable interface termination, the interface bonding energies of differently terminated interfaces, which are estimated using the adsorption energy of Al on base-metal M and that of M on M, and the adsorption energy of oxygen on M and Al are compared. In the algorithm for prediction, interface termination at alumina-pure metal interface should be first examined. Then, if the interface with pure metal is Al-terminated, values of (AlonM) and (AlonAl) are compared. For O-terminated interface with pure metal, values of (OonM) and (OonAl) should be compared. This procedure gives the type of interface bonding in aluminum-containing alloy according to the expressions (a)–(d) or (a′)–(d′) in the text. Based on the algorithm, it is also revealed that O-terminated interface can be switched to Al-terminated one by adding Al to pure metal M.

The predicted results are compared with those deduced from experimental studies. The agreement is very good. For aluminum-containing alloys where there is little difference between (OonM) and (OonAl), termination type would be dependent on temperature and oxygen partial pressure and these influences should be taken into account for more accurate and precise prediction. However, for most of metals, the formula for prediction proposed here should be very effective and useful for material screening in developing interfaces because the method is based on thermodynamics and uses only basic parameters of metals and oxides.

Conflict of Interests

The authors declare that there is no conflict of interests regarding the publication of this paper.

Acknowledgments

Michiko Yoshitake greatly appreciates partial support by Grants-in-Aid for Scientific Research from the Japan Society for the Promotion of Science (no. 20560027) and from The Mitsubishi Foundation.

References

[1] U. Alber, H. Müllejans, and M. Rühle, "Wetting of copper on α-Al$_2$O$_3$ surfaces depending on the orientation and oxygen partial pressure," *Micron*, vol. 30, no. 2, pp. 101–108, 1999.

[2] V. Merlin and N. Eustathopoulos, "Wetting and adhesion of Ni-Al alloys on α-Al$_2$O$_3$ single crystals," *Journal of Materials Science*, vol. 30, no. 14, pp. 3619–3624, 1995.

[3] D. Chatain, F. Chabert, V. Ghetta, and J. Fouletier, "New experimental setup for wettability characterization under monitored oxygen activity: II, wettability of sapphire by silver-oxygen melts," *Journal of the American Ceramic Society*, vol. 77, no. 1, pp. 197–201, 1994.

[4] S. Shi, S. Tanaka, and M. Kohyama, "First-principles study of the tensile strength and failure of α-Al$_2$O$_3$ (0001)/Ni(111) interfaces," *Physical Review B*, vol. 76, Article ID 075431, 2007.

[5] S. Shi, S. Tanaka, and M. Kohyama, "First-principles investigation of the atomic and electronic structures of α-Al$_2$O$_3$(0001)/Ni(111) Interfaces," *Journal of the American Ceramic Society*, vol. 90, no. 8, pp. 2429–2440, 2007.

[6] S. Shi, S. Tanaka, and M. Kohyama, "Influence of interface structure on Schottky barrier heights of α-Al$_2$O$_3$(0001)/Ni(111) interfaces: a first-principles study," *Materials Transactions*, vol. 47, no. 11, pp. 2696–2700, 2006.

[7] K. Shiraishi, T. Nakayama, T. Nakaoka, A. Ohta, and S. Miyazaki, "Theoretical investigation of metal/dielectric interfaces? Breakdown of Schottky barrier limits?" *ECS Transactions*, vol. 13, no. 2, pp. 21–27, 2008.

[8] T. Nagata, P. Ahmet, Y. Z. Yoo et al., "Schottky metal library for ZNO-based UV photodiode fabricated by the combinatorial ion beam-assisted deposition," *Applied Surface Science*, vol. 252, no. 7, pp. 2503–2506, 2006.

[9] A. Asthagiri, C. Niederberger, A. J. Francis, L. M. Porter, P. A. Salvador, and D. S. Sholl, "Thin Pt films on the polar SrTiO$_3$(1 1 1) surface: an experimental and theoretical study," *Surface Science*, vol. 537, no. 1–3, pp. 134–152, 2003.

[10] M. Yoshitake, S. Nemšák, T. Skála et al., "Modification of terminating species and band alignment at the interface between alumina films and metal single crystals," *Surface Science*, vol. 604, no. 23-24, pp. 2150–2156, 2010.

[11] M. Yoshitake, S. Yagyu, and T. Chikyow, "Novel method for the prediction of an interface bonding species at alumina/metal interfaces," *Journal of Vacuum Science and Technology A*, vol. 32, no. 2, Article ID 021102, 8 pages, 2014.

[12] http://interchembond.nims.go.jp/.

[13] A. R. Miedema and J. W. F. Dorleijn, "Quantitative predictions of the heat of adsorption of metals on metallic substrates," *Surface Science*, vol. 95, no. 2-3, pp. 447–464, 1980.

[14] M. Yoshitake, Y.-R. Aparna, and K. Yoshihara, "General rule for predicting surface segregation of substrate metal on film surface," *Journal of Vacuum Science & Technology A*, vol. 19, no. 4, pp. 1432–1437, 2001.

[15] http://surfseg.nims.go.jp/SurfSeg/menu.html.

[16] K. Tanaka and K. Tamaru, "A general rule in chemisorption of gases on metals," *Journal of Catalysis*, vol. 2, no. 5, pp. 366–370, 1963.

[17] D. Brennan, D. O. Hayward, and B. M. Trapnell, "The calorimetric determination of the heats of adsorption of oxygen on evaporated metal films," *Proceedings of the Royal Society of London A*, vol. 256, pp. 81–105, 1960.

[18] N. A. Lange, *Handbook of Chemistry*, McGraw-Hill, 1956.

[19] L. Brewer, "The thermodynamic properties of the oxides and their vaporization processes," *Chemical Reviews*, vol. 52, no. 1, pp. 1–75, 1953.

[20] D. R. Lide, Ed., *CRC Handbook of Chemistry and Physics*, CRC Press, 74th edition, 1993-1994.

[21] P. Brix and G. Herzberg, "Fine structure of the Schumann-Runge bands near the convergence limit and the dissociation energy of the oxygen molecule," *Canadian Journal of Physics*, vol. 32, pp. 110–135, 1954.

[22] R. M. Jaeger, H. Kuhlenbeck, H.-J. Freund et al., "Formation of a well-ordered aluminium oxide overlayer by oxidation of NiAl(110)," *Surface Science*, vol. 259, no. 3, pp. 235–252, 1991.

[23] G. Kresse, M. Schmid, E. Napetschnig, M. Shishkin, L. Köhler, and P. Varga, "Materials science: structure of the ultrathin aluminum oxide film on NiAl(110)," *Science*, vol. 308, no. 5727, pp. 1440–1442, 2005.

[24] M. Yoshitake, W. Song, J. Libra et al., "Interface termination and band alignment of epitaxially grown alumina films on Cu-Al alloy," *Journal of Applied Physics*, vol. 103, Article ID 033707, 2008.

[25] V. Maurice, G. Despert, S. Zanna, P. Josso, M.-P. Bacos, and P. Marcus, "The growth of protective ultra-thin alumina layers on γ-TiAl(1 1 1) intermetallic single-crystal surfaces," *Surface Science*, vol. 596, no. 1–3, pp. 61–73, 2005.

[26] V. Maurice, G. Despert, S. Zanna, P. Josso, M.-P. Bacos, and P. Marcus, "XPS study of the initial stages of oxidation of α$_2$-Ti$_3$Al and γ-TiAl intermetallic alloys," *Acta Materialia*, vol. 55, no. 10, pp. 3315–3325, 2007.

[27] K. Kovács, I. V. Perczel, V. K. Josepovits, G. Kiss, F. Réti, and P. Deák, "In situ surface analytical investigation of the thermal oxidation of Ti-Al intermetallics up to 1000°C," *Applied Surface Science*, vol. 200, no. 1–4, pp. 185–195, 2002.

[28] M. Schmiedgen, P. C. J. Graat, B. Baretzky, and E. J. Mittemeijer, "The initial stages of oxidation of γ-TiAl: an X-ray photoelectron study," *Thin Solid Films*, vol. 415, no. 1-2, pp. 114–122, 2002.

[29] J. F. Silvain, J. E. Barbier, Y. Lepetitcorps, M. Alnot, and J. J. Ehrhardt, "Chemical and structural analysis of TiAl thin films

sputter deposited on carbon substrates," *Surface and Coatings Technology*, vol. 61, no. 1–3, pp. 245–250, 1993.

[30] J. Xia, H. Dong, and T. Bell, "Surface properties of a γ-based titanium aluminide at elevated temperatures," *Intermetallics*, vol. 10, no. 7, pp. 723–729, 2002.

[31] H. Graupner, L. Hammer, K. Heinz, and D. M. Zehner, "Oxidation of low-index FeAl surfaces," *Surface Science*, vol. 380, no. 2-3, pp. 335–351, 1997.

[32] E. Loginova, F. Cosandey, and T. E. Madey, "Nanoscopic nickel aluminate spinel ($NiAl_2O_4$) formation during NiAl(1 1 1) oxidation," *Surface Science*, vol. 601, no. 3, pp. L11–L14, 2007.

[33] W. Zhang, J. R. Smith, and A. G. Evans, "The connection between ab initio calculations and interface adhesion measurements on metal/oxide systems: Ni/Al_2O_3 and Cu/Al_2O_3," *Acta Materialia*, vol. 50, no. 15, pp. 3803–3816, 2002.

[34] W. Zhang and J. R. Smith, "Nonstoichiometric interfaces and Al_2O_3 adhesion with Al and Ag," *Physical Review Letters*, vol. 85, no. 15, pp. 3225–3228, 2000.

[35] J. Feng, W. Zhang, and W. Jiang, "Ab initio study of Ag/Al_2O_3 and Au/Al_2O_3 interfaces," *Physical Review B*, vol. 72, no. 11, Article ID 115423, 11 pages, 2005.

[36] M. Yoshitake and K. Yoshihara, "The surface segregation of Ti-Nb composite film and its application to a smart getter material," *Vacuum*, vol. 51, no. 3, pp. 369–376, 1998.

Effects of Heat Treatment on the Mechanical Properties of Al-4% Ti Alloy

Segun Isaac Talabi,[1] Samson Oluropo Adeosun,[2] Abdulganiyu Funsho Alabi,[1] Ishaq Na'Allah Aremu,[1] and Sulaiman Abdulkareem[3]

[1] *Department of Materials and Metallurgical Engineering, University of Ilorin, Ilorin, Nigeria*
[2] *Department of Metallurgical and Materials Engineering, University of Lagos, Nigeria*
[3] *Department of Mechanical Engineering, University of Ilorin, Ilorin, Nigeria*

Correspondence should be addressed to Segun Isaac Talabi; isaacton@yahoo.com

Academic Editor: Koppoju Suresh

This paper examines the effects of heat treatment processes on the mechanical properties of as-cast Al-4% Ti alloy for structural applications. Heat treatment processes, namely, annealing, normalizing, quenching, and tempering, are carried out on the alloy samples. The mechanical tests of the heat treated samples are carried out and the results obtained are related to their optical microscopy morphologies. The results show that the heat treatment processes have no significant effect on the tensile strength of the as-cast Al-4% Ti alloy but produce significant effect on the rigidity and strain characteristic of the alloy. With respect to the strain characteristics, significant improvement in the ductility of the samples is recorded in the tempered sample. Thus, for application requiring strength and ductility such as in aerospace industries, this tempered heat treated alloy could be used. In addition, the quenched sample shows significant improvement in hardness.

1. Introduction

Aluminium like all pure metals has low strength and cannot be readily used in applications where resistance to deformation and fracture is essential. Therefore, other elements are added to aluminium primarily to improve strength. The low density with high strength has made aluminium alloys attractive in applications where specific strength (strength-to-weight ratio) is a major design consideration. For structural use, the strongest alloy which meets minimum requirements for other properties such as corrosion resistance, ductility, and toughness is usually selected if it is cost effective. Hence, composition is the first consideration for strength [1]. Structural application of aluminium alloy at high or moderate temperature requires a fine, homogeneous, and stable distribution of crystal hardening up to the temperature of use. High melting point intermetallics phases are good candidate for that. Al_3Ti is very attractive among all intermetallics, because of its high melting point ($1350°C$) and relatively low

density ($3.3\,g/cm^3$) [2]. Recently, aluminium based alloys, especially with titanium, are becoming more useful for high temperature applications due to their excellent properties [3]. Previous researchers have looked mainly at the effect of titanium as a grain refiner in aluminium alloy as grain refinement plays an important role in determining the ultimate properties of aluminium alloy products. It improves tensile intensities and plasticity, increases feeding complex castings, and reduces the tendency of hot tearing and porosity [4]. In reality grain refinement of aluminium by titanium is due to the occurrence of a peritectic reaction at the aluminium-rich end of the aluminium-titanium phase diagram [5, 6]. A combination of titanium addition to aluminum alloy and other processing is thought as possible means of further improving the mechanical properties of this alloy especially for high temperature usage.

Thus, in this study, the effect of heat treatment is employed as a means of improving the mechanical properties of aluminium-4% titanium alloy.

TABLE 1: Chemical composition of as-cast Al-4% Ti alloy.

Element	Fe	Si	Mn	Cu	Zn	Ti	Mg	Pb	Sn	Al	Wa	B
Weight percentage	0.738	0.308	0.377	0.027	0.748	4.061	0.092	0.343	1.207	88.73	0.027	2.807

2. Experimental Methods

The as-cast aluminium-4% titanium alloy rod with chemical composition shown in Table 1 was cut and machined at ambient temperature into standard tensile and hardness samples. The samples were then subjected to annealing, normalising, quenching, and tempering processes before tensile and hardness tests were conducted on them. During annealing, the sample was heated to 500°C, held for one hour, and allowed to cool in the furnace. For the normalised and quenched samples, the samples were heated to 500°C, held for one hour, and allowed to cool in air and water, respectively. The tempered sample was heated to 500°C, held for one hour, quenched rapidly in water, reheated to 100°C, held at this temperature for one hour, and then allowed to cool in air. The corresponding heat treated samples were designated as "as-cast sample (AC)," "annealed sample (AS)," "normalised sample (NS)," quenched sample (QS)," and "tempered sample (TS)."

Tensile test was carried out in accordance with ASTM E8 on standard samples using Instron Universal Tester, model 3369. Hardness test was done using a Vickers microhardness tester model "Deco" 2005 with a test load of 490 kN and a dwell time of 10 s. A minimum of 3 indentations were made on each of the samples. Standard microstructural test pieces were prepared and ground using emery paper with grit 220 to 600 microns in succession. The ground surfaces of the pieces are polished using a mixture of alumina and diamond paste and then etched in a solution containing 5 g of sodium hydroxide (NaOH) dissolved in 100 mL of water. The etched surfaces were left for 20 seconds before being rinsed with water and dried. The samples' crystals morphologies were viewed under a Digital Metallurgical Microscope at ×200 magnification and the photomicrographs are shown in Plates 1–5.

3. Results and Discussion

The tensile and yield properties of the as-cast rod samples with 4% titanium after heat treatment are shown in Figure 1. The heat treatment programmes have no significant influence on the tensile strength of the samples. When strength and ductility are important, the quenched sample possesses considerable strength (126 MPa) with 42% elongation while its hardness of 60 HV makes it a better candidate for engineering applications requiring a combination of these properties compared to the as-cast sample.

The rigidity of the as-cast sample is significantly affected by the heat treatment processes (Figure 2), with the tempered sample having lowest rigidity (1219 MPa). With the exception of the quenched sample in which significant increase in hardness is attained (35%, 60 HV), other heat treatment processes have little effect on the hardness characteristic of

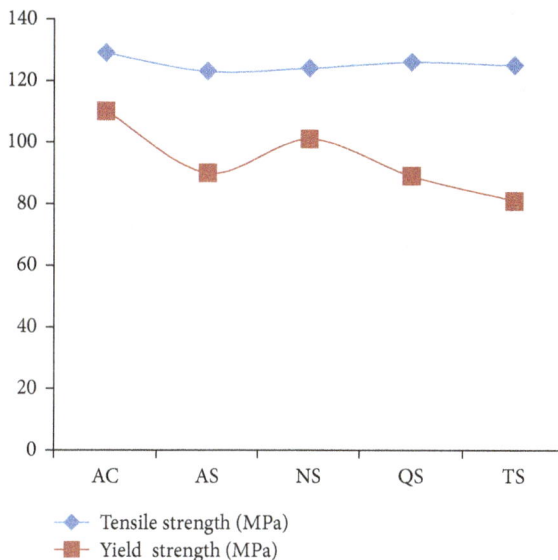

FIGURE 1: Ultimate tensile strength and yield strength of as-cast and heat treated Al-4% Ti.

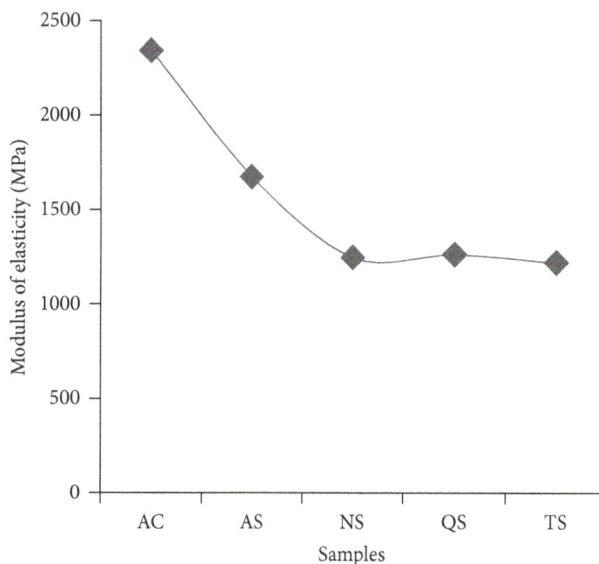

FIGURE 2: Young modulus of as-cast and heat treated Al-4% Ti.

the alloy (Figure 3). For the quenched sample, its toughness is slightly sacrificed for improved strength. The heat treatment processes employed improve the tensile elongation characteristic of the material with the tempered sample exhibiting superior (48%) elongation (see Figure 4). This result agreed with modulus of elasticity results obtained for the tested samples (see Figure 2). Thus, quenching process is preferred

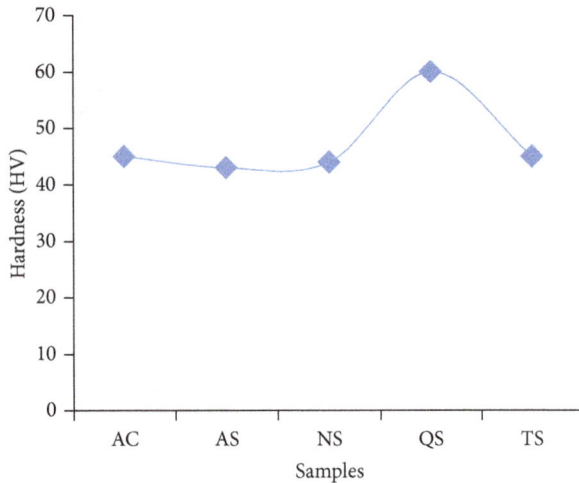

FIGURE 3: Hardness of as-cast and heat treated Al-4% Ti.

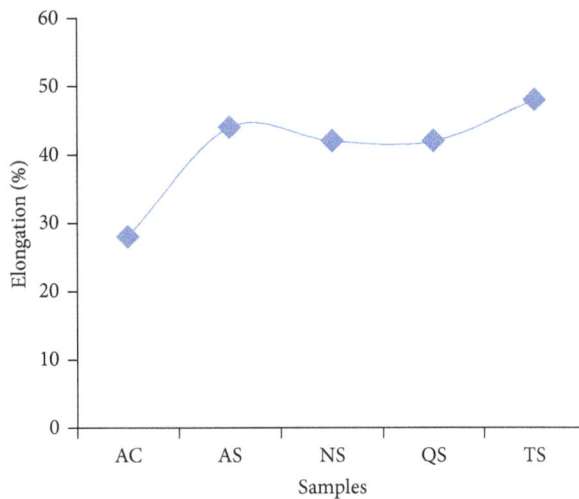

FIGURE 5: Micrographs of as-cast Al-4% Ti alloy.

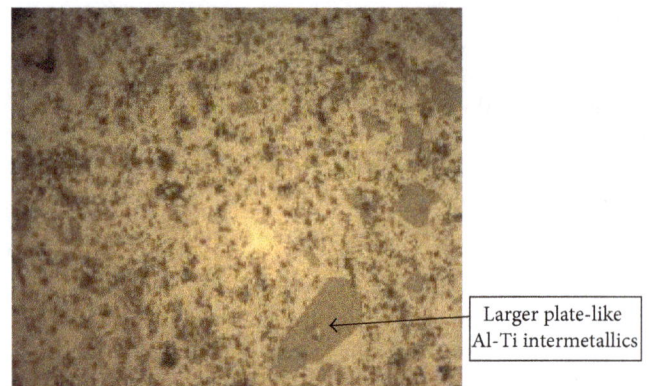

FIGURE 6: Micrographs of annealed Al-4% Ti alloy.

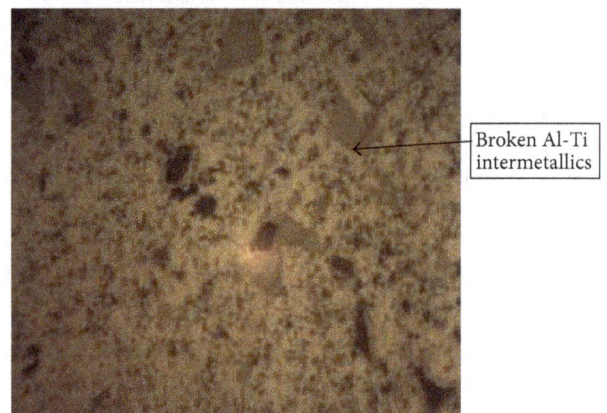

FIGURE 4: Elongation of as-cast and heat treated Al-4% Ti.

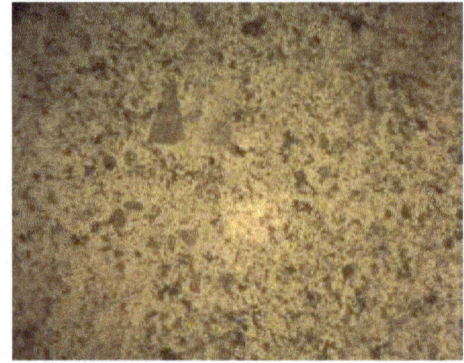

FIGURE 7: Micrographs of normalised Al-4% Ti alloy.

to other heat treatment processes as it confers optimum mechanical properties on Al-4 % Ti alloy.

The microstructure analysis results of the heat treated samples together with the as-cast samples are shown in Plates from 1 to 5. The as-cast alloy morphology has some plate-like structure and evidence of formation of an intermetallic-like phase within the matrix of the alloy (Figure 5). This plate-like structure has been identified as aluminium-titanium inter-metallics with transmission electron microscope [7] while the precipitate-like phase represents TiB_2 [8]. Previous work has shown that TiB_2 reinforcement is both thermodynamically and microstructurally stable within the aluminide matrices [9]. The mechanical properties of the samples are found to depend on the formation of aluminium-titanium inter-metallics and TiB_2 precipitates within the matrix structure. During annealing, the plate-like features within the matrix grew in size, becoming larger and soft than that of the as-cast sample (Figure 6). During tensile testing, dislocation

can glide easily because of the increased size of aluminium-titanium intermetallics which does not resist dislocation motion. This explains the slight decrease in strength as can be observed in Figure 1. The plate-like structure in the normalised sample when compared with the annealed sample appears to have been broken down (Figure 7) because of the fast cooling introduced. This is because diffusion process (crystal nucleation and growth) is more pronounced during annealing operation compared to the normalised heat treatment programme. In the quenched sample (Figure 8), an increase in volume fraction of the second TiB_2 precipitate

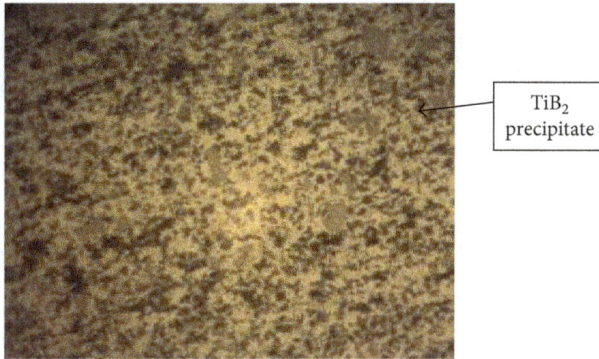

FIGURE 8: Micrographs of quenched Al-4% Ti alloy.

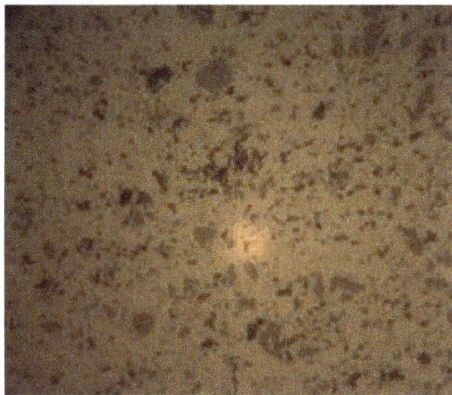

FIGURE 9: Micrographs of tempered Al-4% Ti alloy.

is observed. This increase is responsible for the observed increase in the hardness of the quenched sample (60 HV). Among the various reinforcing phases, TiB_2 is particularly attractive because it possesses many desirable properties, such as high hardness, low density, high melting temperature, high modulus, and high corrosion resistance [10, 11]. As reported by earlier researchers, evidence of agglomeration of TiB_2 precipitate was noted as being responsible for the observed results [8, 12]. The size of the plate-like crystals within the matrix of the alloy significantly declines with the crystals being more uniformly distributed within the matrix of the quenched sample. The improvement in hardness may also be attributed to this occurrence. The tempered sample matrix shows reduction in volume fraction of TiB_2 precipitate as well as that of the plate-like crystals in the matrix resulting in significant reduction in the hardness (44 HV) of the alloy (see Figure 9).

4. Conclusion

In this study tensile elongation of aluminum-4% titanium alloy is found to improve significantly with respect to the heat treatment processes. The rigidity of the as-cast sample is affected by the heat treatment processes, with the tempered sample having the lowest Young modulus value. When strength, ductility, and hardness are important, the quenched sample possesses considerable strength (126 MPa),

elongation (42%), and hardness (60 HV) which make it a better candidate than the as-cast sample with high strength but low ductility. The microstructure shows that the heat treatment programmes affect both the size and distribution of the aluminium-titanium crystals as well as the volume fraction of the secondary phase TiB_2 precipitates. However, the heat treatment processes do not significantly improve the alloy tensile strength.

References

[1] T. Murat and T. James, "Physical metallurgy and the effects of alloying additions in Aluminium alloy," in *Handbook of Aluminium*, vol. 1, pp. 81–209, Marcel Dekker, 2003.

[2] K. R. Cardoso, D. N. Travessa, A. G. Escorial, and M. Lieblich, "Effect of mechanical alloying and Ti addition on solution and ageing treatment of an AA7050 Aluminium alloy," *Materials Research*, vol. 10, no. 2, pp. 199–203, 2007.

[3] L. A. Dorbzanski, K. Labisz, and A. Olsen, "Microstructure and mechanical property of the Al-Ti alloy with Calcium addition," *Journal of Achievement in Materials and Manufacturing Engineering*, vol. 26, pp. 183–186, 2008.

[4] Z.-Y. Liu, M.-X. Wang, Y.-G. Weng, T.-F. Song, J.-P. Xie, and Y.-P. Huo, "Grain refinement effects of Al based alloys with low titanium content produced by electrolysis," *Transactions of Nonferrous Metals Society of China*, vol. 12, no. 6, pp. 1121–1126, 2002.

[5] F. A. Crossley and L. F. Mondolfo, "Mechanism of grain refinement of Aluminum alloys," *Transactions AIME*, vol. 191, pp. 1143–1148, 1951.

[6] R. O. Kaibyshev, I. A. Mazurina, and D. A. Gromov, "Mechanisms of grain refinement in aluminum alloys in the process of severe plastic deformation," *Metal Science and Heat Treatment*, vol. 48, no. 1-2, pp. 57–62, 2006.

[7] K. I. Moon and K. S. Lee, "Study of the microstructure of nanocrystalline Al-Ti alloys synthesized by ball milling in a hydrogen atmosphere and hot extrusion," *Journal of Alloys and Compounds*, vol. 291, no. 1-2, pp. 312–321, 1999.

[8] T. V. Christy, N. Murugan, and S. Kumar, "A Comparative Study on the Microstructures and Mechanical Properties of Al 6061 alloy and the MMC Al 6061/TiB_2/12$_P$," *Journal of Minerals & Materials Characterization & Engineering*, vol. 9, no. 1, pp. 57–65, 2010.

[9] S. L. Kampe, J. D. Bryant, and L. Christodoulou, "Creep deformation of TiB_2-reinforced near-γ titanium aluminides," *Metallurgical Transactions A*, vol. 22, no. 2, pp. 447–454, 1991.

[10] K. L. Tee, L. Lu, and M. O. Lai, "In situ processing of Al-TiB_2 composite by the stir-casting technique," *Journal of Materials Processing Technology*, vol. 89-90, pp. 513–519, 1999.

[11] K. L. Tee, L. Lü, and M. O. Lai, "Improvement in mechanical properties of in-situ Al-TiB_2 composite by incorporation of carbon," *Materials Science and Engineering A*, vol. 339, no. 1-2, pp. 227–231, 2003.

[12] E.-M. Uşurelu, P. Moldovan, M. Buţu, I. Ciucă, and V. Drăguţ, "On the mechanism and thermodynamics of the precipitation of TiB_2 particles in 6063 matrix aluminum alloy," *UPB Scientific Bulletin B*, vol. 73, no. 3, pp. 205–216, 2011.

Spectroscopic Study of Al^{3+}-Substituted Strontium Hexaferrite

M. Y. Salunkhe,[1] D. S. Choudhary,[2] and S. B. Kondawar[3]

[1] *Department of Physics, Institute of Science, R. T. Road, Nagpur, Maharashtra 440001, India*
[2] *Dhote Bandhu Science College, Gondia, Maharashtra 441614, India*
[3] *Department of Physics, R. T. M. Nagpur University, Nagpur, Maharashtra 440033, India*

Correspondence should be addressed to M. Y. Salunkhe; nimahsal@yahoo.com

Academic Editor: Yuanshi Li

The substituted Y type hexaferrites $Sr_2Zn_2Al_xFe_{12-x}O_{22}$ (where $x = 2, 4, 6, 10$) are prepared by standard ceramic technique. The infrared absorption spectra of the prepared compounds are studied in the range $400\ cm^{-1}$ to $4600\ cm^{-1}$. These spectra are used to locate the vibrational ranges due to substituted cations with the nearest oxygen layers and also to understand the band positions attributed to the lattice sites. The absorption regions found around $460\ cm^{-1}$ and $600\ cm^{-1}$ are the common features of all spinel ferrites. Both X-ray and IR spectroscopic studies in these ferrites do not detect the presence of Fe^{2+}. Also with increase in aluminium substitution, the higher frequency bands start disappearing. This may be due to reduced vibrations of trivalent cations Fe–O bonds at both octahedral and tetrahedral sites which in turn affect Fe^{3+}–O^{2-}–Fe^{3+} superexchange interactions present in the structure.

1. Introduction

Infrared spectroscopy is the most powerful technique for the chemical identification of crystal structure. It provides useful information about the structure of molecules without tiresome evaluation methods, which are applied in other usual methods of structural analysis such as X-ray diffraction and neutron diffraction. With the absorption of infrared radiations, the molecules of the chemical substance vibrate at many rates of vibration, giving rise to close packed absorption bands. Thus such infrared spectra give the vibrational behavior of the crystal structure. It is interesting to note here that till date much less effort has been taken in studying the vibrational spectrum of the ferrites. In that, most of the work is concentrated on spinel ferrites [1–3]. Infrared spectral analysis has been carried out for several ferrites by Waldron in 1955, who reported two absorption bands within the wave numbers 200–$800\ cm^{-1}$, which are attributed to the tetrahedral and octahedral complexes of the spinel structure. Cr substituted Ni spinel ferrite is also studied [4] and authors attributed the existence of fine structure to the Jahn-Teller effect. Gd^{3+} substituted Cd-Ca spinel ferrite [5] and Nd^{3+} substituted Zn-Mg spinel ferrite [6] are also studied by using infrared absorption spectroscopy. Thus it is clear that as far as

the hexaferrite is concerned, no previous infrared studies of these materials were found in the literature with the exception of the reflection spectrum of Ba_2-Y type [7], Sr_2-Y with divalent substitution [8], and absorption spectra of Ba-M type ferrite [9].

In the present paper, the results regarding the infrared absorption spectral analysis of four hexagonal ferrites with general chemical formula $Sr_2Zn_2Al_xFe_{12-x}O_{22}$ ($x = 2$ to 10) are discussed. The efforts are taken to find out the effect of less massive, nonmagnetic aluminum substitution in the Sr-Y type hexagonal ferrite using spectroscopic studies.

2. Experimental

The system of polycrystalline samples $Sr_2Zn_2Al_xFe_{12-x}O_{22}$ ($x = 2$ to 10) was prepared by standard ceramic technique. The analar grade reactants $SrCO_3$, ZnO, Al_2O_3, and Fe_2O_3 are mixed in the proper molar ratio and ground in the agate mortar for about 7 hours using acetone to achieve uniform grain size and homogeneity. With polyvinyl acetate as a binder and by applying a pressure of 10 tones psi for 5 to 7 minutes, pellets were prepared. These pellets are slowly heated in the furnace at $600°C$ for about 6 hours to remove

TABLE 1: Lattice sites present in the Y type hexaferrite structure.

Position	Block	Coordinates	No. of ions per unit cell
$6c_{iv}$	S	Tetrahedral	6
$3b_{vi}$	S	Octahedral	3
$18h_{vi}$	S-T	Octahedral	18
$6c_{vi}$	T	Octahedral	6
$6c_{iv*}$	T	Tetrahedral	6
$3a_{vi}$	T	Octahedral	3

the binder [10] and fired at 1200°C, sintering temperature for 120 hours. The samples then cooled at a rate of 20°C per hour up to 1100°C and then at a rate of 60°C per hour up to 500°C. Later on, the furnace is cooled at room temperature in natural way. The pellets are then finely powered and sieved through fine sieve. To confirm the completion of the solid state reaction, the compositions were subjected to characterization by X-ray diffraction.

For infrared studies, the technique used by Mazen et al. [11] is applied to the sample preparation. Nearly 2 mg of ferrite powered is mixed with powdered KBr in the ratio 1 : 100 by weighing to ensure uniform dispersion. The mixed powder is then pressed in a die of 12 mm diameter and thickness about 1 mm by applying a pressure of 10 tones per square inch for about 5 minutes. The spectra were recorded on Infrared Spectrophotometer model FTIR-8001 with resolution $4.0\ cm^{-1}$ in the region of $400\ cm^{-1}$ to $4000\ cm^{-1}$.

3. Results and Discussion

X-ray diffractograms of powdered compounds under investigation reveal the formation of single phase hexagonal Y type phase, showing well-defined reflections of allowed planes. The results regarding X-ray analysis are reported elsewhere [12].

The infrared spectra of the prepared Al^{3+} substituted Sr-Y ferrite compounds are recorded in the range $400\ cm^{-1}$ to $4600\ cm^{-1}$. The spectra are shown in Figure 1. In the spectra, no absorption bands were present above $1250\ cm^{-1}$ except the band at $3460\ cm^{-1}$ from the OH group of the moisture present (always) in the sample, which is the major disadvantage of the pressed pellet technique used in the infrared spectroscopy [13]. The spectra of all the aluminium substituted Sr-Y type ferrite are used to locate the band positions, which are shown in Table 2.

The Y type hexagonal ferrite, namely, $Ba_2Zn_2Fe_{12}O_{22}$, has alternate stacking of the spinel S and hexagonal T block along hexagonal c axis as $(TS)''| (TS)(TS)'(TS)''| (TS)$ where the prime means that the corresponding block is rotated 120° around the c axis. The space group is $R\bar{3}m$ [14]. Here the cations occupy six different tetrahedral and octahedral sites present in the structure in different blocks S and T, respectively, as shown in Table 1. We tried to throw some light on the I. R. spectra of such compounds and showed that the vibration frequencies are below $1250\ cm^{-1}$ [8]. In the substituted Y type ferrite, the divalent Zn^{2+} cations and two

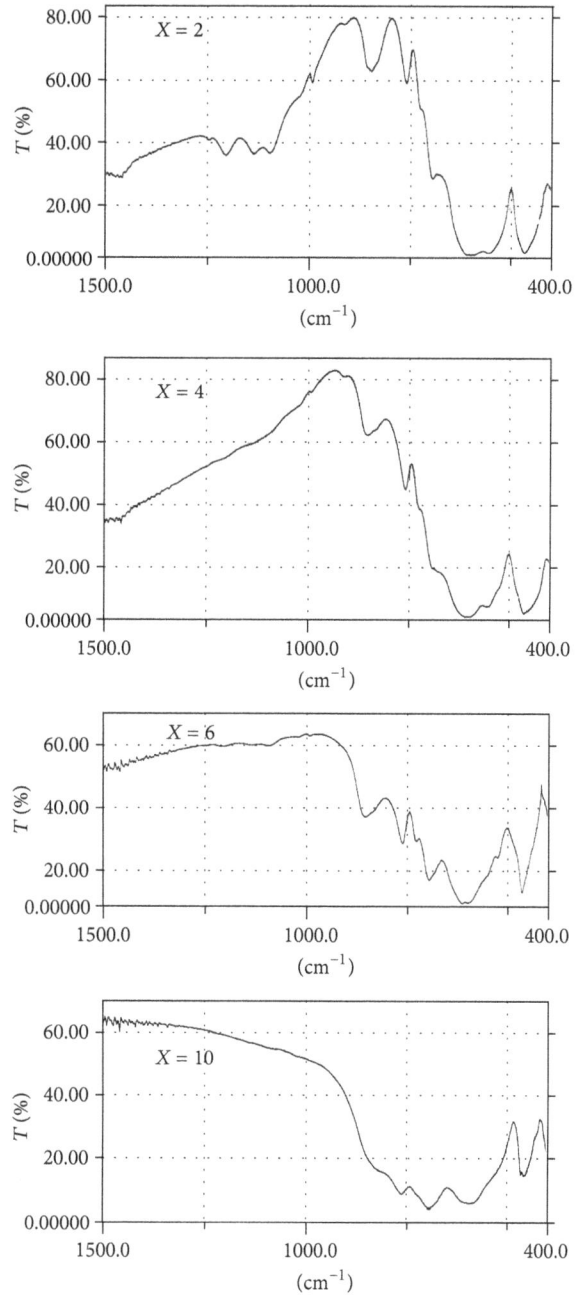

FIGURE 1: Infrared absorption spectra of $Sr_2Zn_2Al_xFe_{12-x}O_{22}$ series.

Fe^{3+} ions are statistically distributed among the tetrahedral 6c positions of the spinel and hexagonal blocks and the remaining 10 Fe^{3+} ions occupy 3a and 6c positions on the T block, 3b site in the S blocks while 18h on the T-S block boundaries [15]. Now with the replacement of Fe^{3+} ions by Al^{3+} ions, aluminium is distributed among the 3b, 18h, and 6c octahedral positions in the whole complex structure [16].

As mentioned in Introduction, very few efforts are taken for spectroscopic measurements of hexaferrites, so we tried to interpret the infrared spectra of our prepared hexaferrite samples in the shade of spinel ferrites. This is owing to the following.

TABLE 2: Infrared absorption bands in the $Sr_2Zn_2Al_xFe_{12-x}O_{22}$ system.

Compound	Infrared absorption bands (in cm^{-1})							
$Sr_2Zn_2Al_2Fe_{10}O_{22}$	1202.8	1089	994.3	848.7	759.2	694.9	601.8	464.3
$Sr_2Zn_2Al_4Fe_8O_{22}$	—	—	994.3	852.4	759.2	694.9	604.0	464.3
$Sr_2Zn_2Al_6Fe_6O_{22}$	—	—	—	852.4	759.2	964.9	612.9	464.3
$Sr_2Zn_2Al_{10}Fe_2O_{22}$	—	—	—	—	759.2	690.5	590.7	459.4

(1) Lot of work has been done on spectroscopic study of spinel ferrites.

(2) Spinel infrared study has been started with rhombohedral primitive cell by Waldron in 1955 where Y type hexaferrite is rhombohedral itself.

(3) Y type hexagonal ferrite already consists of spinel block alternately stacked with T block along c axis.

(4) In both the ferrites, only tetrahedral and octahedral sites are present.

From Figure 1, in Y type hexagonal ferrite compounds, the absorption frequency bands are in the region as follows: the low frequency bands are at $(462 \pm 3)\,cm^{-1}$, and the medium frequency bands are at $(600 \pm 12)\,cm^{-1}$, $(692 \pm 2)\,cm^{-1}$ and $759.2\,cm^{-1}$, respectively, while the higher frequency absorption bands are observed at $(850 \pm 2)\,cm^{-1}$, $1089\,cm^{-1}$ and $1202.8\,cm^{-1}$, respectively.

As far as the interpretation of the low frequency bands and the first element in the medium frequency is concerned, from the literature reported data, Waldron [1] and Hafner [17] attributed the low frequency band around $400\,cm^{-1}$ to the stretching vibrations of octahedral complexes while the band around $600\,cm^{-1}$ to the intrinsic vibrations of tetrahedral complexes in the spinel structure as these absorption ranges are the common features of all spinel ferrites. For these bands, Tarte and coworkers [3, 18] observed that this attribution is true only when tetrahedral cation is of higher valence than octahedral cation. In the present work, Tarte's assumption feels true, so here the absorption at $(462 \pm 3)\,cm^{-1}$ is assigned to the stretching vibrations of the tetrahedral complexes in spinel blocks. The intensity of this band decreases slowly with decrease in Fe^{3+} concentration, which indicates the decrease in the occupancy of the iron cations at tetrahedral sites in S blocks. Also the absorption band at $(600 \pm 12)\,cm^{-1}$ is attributed to the stretching vibrations of the octahedral group in spinel blocks. The broadness of this band decreases with decrease in magnetic substitutions, so this may be the effect of magnetic substitutions in the vibrations. Splitting is observed at higher frequency side near to $(692 \pm 2)\,cm^{-1}$. The increase in significance of this absorption band may be attributed to the presence of aluminium vibration at octahedral sites in T-S block boundaries.

In a crystal, the cation-oxygen vibration frequencies depend on the mass of the cations, the cation oxygen bonding force, distance related to cation electronic structure, ionic radius and chemical nature of the neighboring cations, and also on the unit cell parameters [19]. Thus as the aluminium has less atomic mass (atomic weight 26.982) compared to iron (atomic weight 55.847) and also less ionic radius of the Al^{3+} cation, the cell volume as well as bond strength Al–O decreases with Al^{3+} substitution. This weakening of the bond strength increases the vibrations of the octahedral aluminium cations at $759.2\,cm^{-1}$ in both S and T blocks. This band is sharp and more significant. The increase in the intensity may be due to occupancy of Al^{3+} cation on the octahedral sites in both S and T blocks of the whole complex Y structure.

Towards higher frequency side in the absorption spectra of Sr-Y type hexaferrite, it is observed that with the dilution of Fe^{3+} by Al^{3+}, the absorption bands gradually start disappearing. These higher frequency absorption bands are assigned to the vibrations of octahedral and tetrahedral substitutions of the metallic cation Fe^{3+} with oxygen layer. By replacing trivalent Fe cation by less massive trivalent Al cation, the contribution of Fe–O vibrations decreases. Thus the disappearance of the higher frequency bands is attributed to the decrease of Fe cations with decrease in Fe–O vibrations in the whole complex structure.

Conflict of Interests

The authors declare that there is no conflict of interests regarding the publication of this paper.

References

[1] R. D. Waldron, "Infrared spectra of ferrites," *Physical Review*, vol. 99, no. 6, pp. 1727–1735, 1955.

[2] P. N. Vasambekar, C. B. Kolekar, and A. S. Vaingankar, "Electrical switching in $Cd_xCo_{1-x}Fe_{2-y}Cr_yO_4$ system," *Materials Research Bulletin*, vol. 34, no. 6, pp. 863–868, 1999.

[3] P. Tarte, "Etude infra-rouge des orthosilicates et des orthogermanates-III. Structures du type spinelle," *Spectrochimica Acta*, vol. 19, no. 1, pp. 49–71, 1963.

[4] A. K. Ghatage, S. C. Choudhari, S. A. Patil, and S. K. Paranjpe, "X-ray, infrared and magnetic studies of chromium substituted nickel ferrite," *Journal of Materials Science Letters*, vol. 15, no. 17, pp. 1548–1550, 1996.

[5] C. B. Kolekar, P. N. Kamble, and A. S. Vaigankar, "X-ray and far IR characterization and susceptibility study of Gd^{3+} substituted Cu-Cd ferrites," *Indian Journal of Physics A*, vol. 68, p. 529, 1994.

[6] B. P. Ladgaonkar, C. B. Kolekar, and A. S. Vaingankar, "Infrared absorption spectroscopic study of Nd^{3+} substituted Zn-Mg ferrites," *Bulletin of Materials Science*, vol. 25, no. 4, pp. 351–354, 2002.

[7] G. Zanmarchi and P. F. Bongers, "Infrared faraday rotation in ferrites," *Journal of Applied Physics*, vol. 40, no. 3, pp. 1230–1231, 1969.

[8] M. Y. Salunkhe, D. S. Choudhary, and D. K. Kulkarni, "Infrared absorption studies of the system $Sr_2Me_2Fe_{12}O_{22}$," *Vibrational Spectroscopy*, vol. 34, no. 2, pp. 221–224, 2004.

[9] C. Sudakar, G. N. Subbanna, and T. R. N. Kutty, "Nanoparticles of barium hexaferrite by gel to crystallite conversion and their magnetic properties," *Journal of Electroceramics*, vol. 6, no. 2, pp. 123–134, 2001.

[10] M. Y. Salunkhe and D. K. Kulkarni, "Structural, magnetic and microstructural study of $Sr_2Ni_2Fe_{12}O_{22}$," *Journal of Magnetism and Magnetic Materials*, vol. 279, no. 1, pp. 64–68, 2004.

[11] S. A. Mazen, M. H. Abdallah, B. A. Sabrah, and H. A. M. Hashem, "The effect of titanium on some physical properties of $CuFe_2O_4$," *Physica Status Solidi A*, vol. 134, p. 863, 1992.

[12] M. Y. Salunkhe, D. S. Choudhary, and S. B. Kondawar, "Effect of the trivalent substitution on structural, magnetic and electrical properties of SR-Y type hexaferrite," *Der Pharma Chemica*, vol. 5, no. 2, p. 175, 2013.

[13] G. Chatwal and S. Anand, *Spectroscopy*, Himalaya, 1992.

[14] J. Smit and H. P. J. Wijn, *Ferrites*, Wiley, New York, NY, USA, 1959.

[15] P. B. Braun, *Philips Research Reports*, vol. 12, p. 491, 1957.

[16] N. N. Aganova, V. A. Sizov, and I. I. Yamazin, "Neutron diffraction study of the hexagonal ferrite $Ba_2Zn_2Al_{2.5}Fe_{9.5}O_{22}$," *Soviet Physics—Solid State*, vol. 10, no. 9, p. 2258, 1969.

[17] S. Hafner, "Ordnung/unordnung und ultrarotabsorption IV. Die absorption einiger metalloxyde mit spinellstruktur," *Zeitschrift für Kristallographie*, vol. 115, p. 331, 1961.

[18] P. Tarte and J. Preudhomme, "Infra-red spectrum and cation distribution in spinels," *Acta Crystallographica*, vol. 16, p. 227, 1963.

[19] V. R. K. Murthy, S. C. Shanker, K. V. Reddy, and J. Sobhanadri, "Mossbauer and infrared studies of some nickel zinc ferrites," *Indian Journal of Pure and Applied Physics*, vol. 16, p. 79, 1978.

4

Optical Characteristics of Polystyrene Based Solid Polymer Composites: Effect of Metallic Copper Powder

Shujahadeen B. Aziz, Sarkawt Hussein, Ahang M. Hussein, and Salah R. Saeed

Department of Physics, Faculty of Science, University of Sulaimani, Kurdistan Regional Government, Sulaimani City, Iraq

Correspondence should be addressed to Shujahadeen B. Aziz; shujaadeen78@yahoo.com

Academic Editor: Mohamed Bououdina

Solid polymer composites (SPCs) were prepared by solution cast technique. The optical properties of polystyrene doped with copper powder were performed by means of UV-Vis technique. The optical constants were calculated by using UV-Vis spectroscopy. The dispersion regions were observed in both absorption and refractive index spectra at lower wavelength. However, a plateau can be observed at high wavelengths. The small extinction coefficient compared to the refractive index reveals the transparency of the composite samples. The refractive index and optical band gap were determined from the reflectance and optical absorption coefficient data, respectively. The nature of electronic transition from valence band to conduction band was determined and the energy band gaps of the solid composite samples were estimated. It was observed that, upon the addition of Cu concentration, the refractive index increased while the energy gaps are decreased. The calculated refractive indexes (low index of refraction) of the samples reveal their availability in waveguide technology.

1. Introduction

Recent years have witnessed constant search for high permittivity materials that have wide range of technologically important applications such as microelectronic, embedded passive, and electrostrictive devices. The majority of the electronic components in microelectronic circuits are passive and occupy more than 80% of the printed wired surface area [1]. The dispersion of an electrically conductive phase within an insulating polymer matrix affects the overall performance of the heterogeneous system. It was reported that if the dispersed metallic particle is in sufficient quantity, a conductive or semiconductive composite is formed. The interesting properties of such systems make them technologically important and competitive to other alternative materials due to their cost-effectiveness [2]. Conductive polymer composites are essential for applications referring to electromagnetic interference (EMI) shielding, radio frequency interference (RFI) shielding, and electrostatic dissipation of charges (ESD). Polymer composites are used as electrical conductive adhesives and circuit elements in microelectronics and have been reported to possess anticorrosive behavior as metal

components coatings [3]. Many types of polymer composites have been studied in the pursuit to develop a system with high conductivity. These include epoxy resin matrix with iron particles [2], high-density polyethylene (HDPE) with multiwalled carbon nanotubes (MWNTs) [4], poly(p-phenylene vinylene) (PPV)-TiO_2 [5], and polyvinylidene fluoride (PVDF) with multiwalled carbon nanotube (MWCNT) [6]. Current studies reveal that the measurement and understanding of the optical properties of materials are important. The occurrence of electronic transitions in the structure of materials is directly related to the photon energy. The nature of the highest occupied molecular orbital (HOMO) and the lowest unoccupied molecular orbital (LUMO) of the linear and cyclic alkanes was understood through the study of the absorbance edge and index of refraction [7]. Nonpolar polymers have the lowest dielectric constant of any known solid polymer, and this makes them attractive for demanding electronics applications. According to Yang at al., polymers that have the lowest index of refraction are very suitable as low-index claddings for waveguide applications [8]. Optical characterization of thin films gives information about some important physical properties, such as band gap energy and band structure and role of defects, and therefore may be of permanent interest for several different applications. The widely used envelop method has been developed for transmittance measurements to evaluate the refractive index, extinction coefficient, and absorption coefficient [9]. The extensive and intensive survey of the literature reveals that there is a very little work that has been done on the optical analysis of polymers containing metallic particles. Hence, the goal of this work is to study the optical parameters of polymer composites based on polystyrene-copper system.

2. Experimental Detail

2.1. Sample Preparation. Polystyrene (procured from Sigma) and copper powders from sigma (with sizes of microns) have been used as the raw materials in this work to prepare the solid polymer composites (SPCs) using the solution cast technique. For this purpose 1 gm of polystyrene was dissolved in 25 mL of toluene solution. The mixture was stirred continuously with a magnetic stirrer for several hours at room temperature until the polystyrene has completely dissolved. While the above systems were still in the liquid state, various amounts of copper powder were added for the production of the solid composite samples. The copper powder content in the prepared samples was varying from 0 wt.% to 6 wt.0% in volume fraction and the mixtures were stirred continuously until homogeneous solutions were obtained. The solutions were then cast into different clean and dry glass Petri dish and allowed to evaporate at room temperature until solvent-free films were obtained. The films were kept in desiccators with silica gel desiccant for further drying. Table 1 shows the concentration of the prepared samples.

2.2. UV-Vis Measurement. The UV-Vis spectra of the solid polymer composites based on polystyrene have been recorded using UV-Vis spectra (model: Lambda 25) in the absorbance mode. To calculate the transmittance (T), absorption coefficient (α), extinction coefficient (k), and refractive index (n), the optical study was performed for polystyrene-xCu ($0.01 \leq x \leq 0.06$) composites by analyzing the absorption spectra as follows [7, 9–11]:

$$T_{\text{sample}} = 10^{-A},$$

$$\alpha = \frac{-1}{t} \ln(T), \quad (1)$$

$$k = \frac{\alpha \times \lambda}{4\pi t},$$

$$n = \frac{(1 + \sqrt{R})}{(1 - \sqrt{R})}, \quad (2)$$

where T_{sample} is the amount of light transmitted through the sample, A is the optical absorbance (base 10) of the sample, α is absorption coefficient, t is the sample thickness, λ is the wavelength, and k is the extinction coefficient. R is the reflectance $R = (1 - (A + T))$ of the sample which is used to estimate the refractive index. Equation (2) is valid for low loss

TABLE 1: Composition of PS$\,{:}\,x$Cu ($0.01 \leq x \leq 0.06$) solid polymer composites (SPCs).

Designation	polystyrene (g)	Cu powder (wt. %)	Cu powder (g)
SPC$_1$	1.0	0.0	0.0000
SPC$_2$	1.0	1.0	0.0101
SPC$_3$	1.0	2.0	0.0204
SPC$_4$	1.0	3.0	0.0309
SPC$_5$	1.0	5.06	0.0626

materials, that is, small extinction coefficient (k). The optical absorption coefficient was used to determine the energy band gap of the solid polymer composites, which is the most direct and simplest method, using the following relationship [12]:

$$\alpha h\nu = B(h\nu - E_g)^m, \quad (3)$$

where B is an energy-independent constant, E_g is the optical energy band gap, and m is a constant which determines the nature of the optical transition from the valence band to the conduction band (fundamental absorption). The nature of the transition can be determined by determining the value of m which is, respectively, 1/2 and 2 for allowed direct and indirect transitions.

3. Results and Discussion

The optical absorption and transmission spectra may provide some insight into the optical behavior of the samples. Figures 1 and 2 show the absorbance and transmittance spectra for all the samples, respectively. It can be noticed that upon the addition of copper powders the absorption increases very rapidly at 6 wt.%, while the transmittance decreased. It is obvious that the absorption and transmission spectra of the samples (Figures 1 and 2) with wavelength are in sharp manner, that is, nonexponential. The sharp behavior of absorbance and transmittance with wavelength for all the samples indicates the crystalline nature of the samples [13]. According to these results the behavior of absorption coefficient ($h\alpha\nu$) versus photon energy (eV) should be nonexponential.

Figures 3 and 4 show the refractive index and extinction coefficient as a function of wavelengths. It can be seen that the refractive indexes of the samples are larger than their extinction coefficients. The small extinction coefficient ($\approx 10^{-5}$) indicates that the composite samples still are very transparent [14]. Study of refractive index and the extinction coefficient for optical systems performance are important. The index of refraction is crucial for matching or optimizing the numerical aperture (NA) of the reduction optical system at all points in the optical path. The extinction coefficient is critical to determine optical losses in the system [15]. The extinction coefficient is the fraction of electromagnetic energy lost due to scattering and absorption per unit thickness in a particular medium. The variation of n and k values with wavelength reveals that some interaction takes place between photons and electrons. The increase of extinction coefficient at high wavelengths (Figure 4) is related to the higher concentration of copper powders (from 2 to 6 wt.% of Cu powder) and

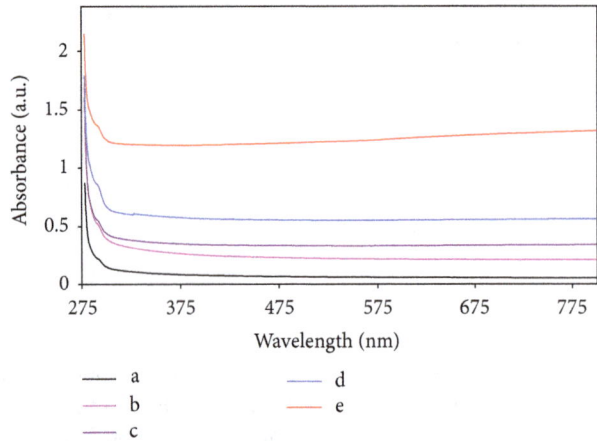

FIGURE 1: Absorption spectra as a function of wavelength for (a) SPC_1, (b) SPC_2, (c) SPC_3, (d) SPC_4, and (e) SPC_5.

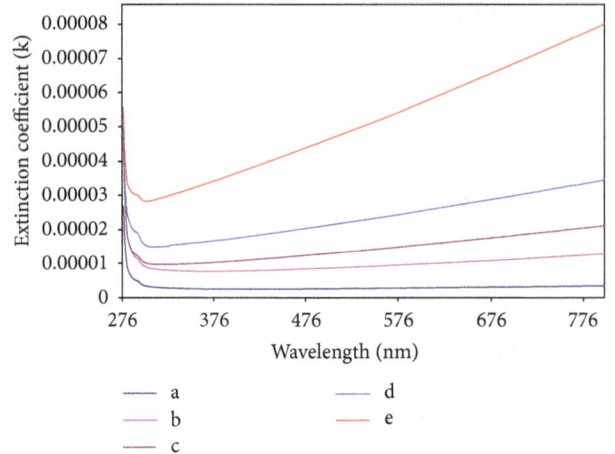

FIGURE 2: Transmittance spectra as a function of wavelength for (a) SPC_1, (b) SPC_2, (c) SPC_3, (d) SPC_4, and (e) SPC_5.

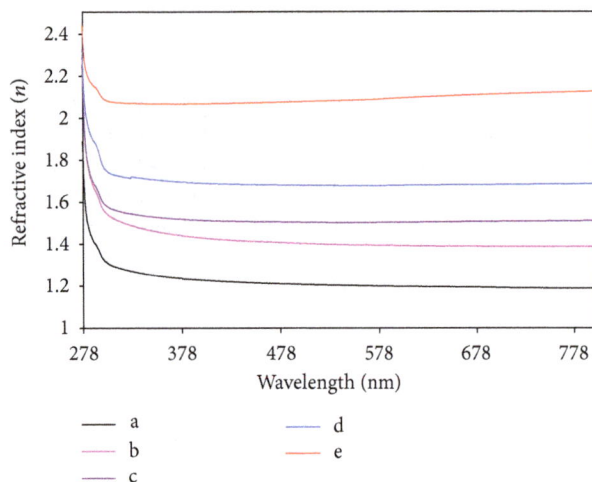

FIGURE 3: Refractive index spectra as a function of wavelength for (a) SPC_1, (b) SPC_2, (c) SPC_3, (d) SPC_4, and (e) SPC_5.

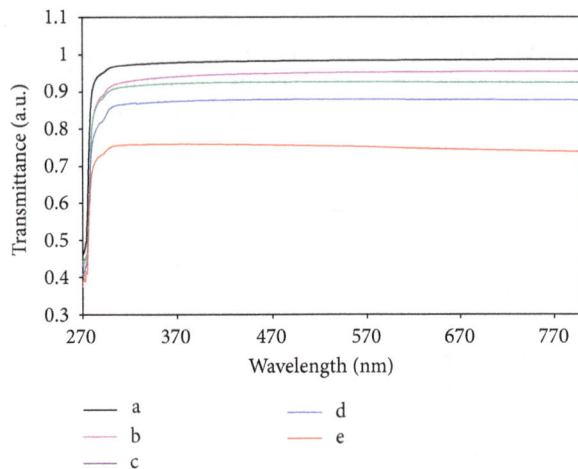

FIGURE 4: Extinction coefficient spectra as a function of wavelength for (a) SPC_1, (b) SPC_2, (c) SPC_3, (d) SPC_4, and (e) SPC_5.

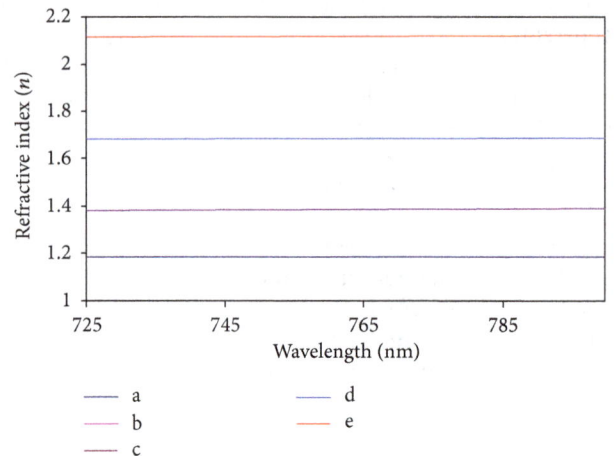

FIGURE 5: High wave length region of refractive index spectra as a function of wavelength for (a) SPC_1, (b) SPC_2, (c) SPC_3, (d) SPC_4, and (e) SPC_5.

thus more scattering of photons are occurred with the added Cu powder. The change in refractive index and extinction coefficient with wavelength of the incident light beam is due to the above mentioned interactions [16].

It can be seen that the refractive index (Figure 3) increases with increasing copper powder, that is, the refractive index of the composite samples is tunable upon the addition of copper powder concentration. The high wavelength region of refractive index represents the material (bulk) property and is almost independent on the wavelength as depicted in Figure 5. The sudden increase of refractive index at 6 wt.% may be attributable to percolation threshold phenomena. It is well known that when the content of conductive particle reaches to a critical value, that is, a percolation threshold, a continuous network can be formed by these conductive particles and thus increasing the crystallinity [17]. The sharply behavior of $((h\alpha\nu)^{\wedge 2})$ versus photon energy ($h\nu$) at higher Cu powder concentration can completely confirm the dominant

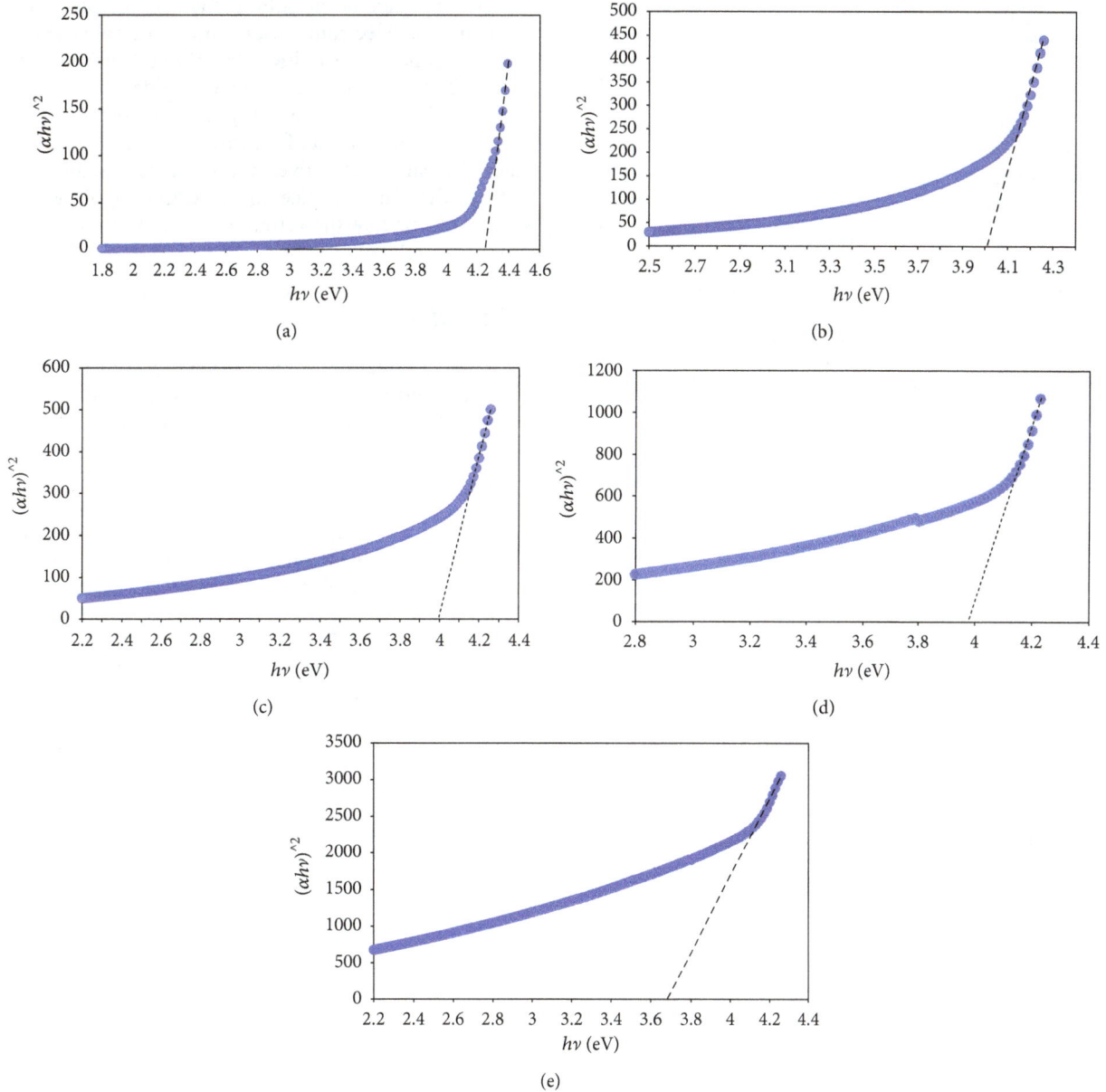

FIGURE 6: The plot of $(h\alpha\nu)^{\wedge 2}$ versus $h\nu$ for (a) SPC_1, (b) SPC_2, (c) SPC_3, (d) SPC_4, and (e) SPC_5.

of crystalline portion in the samples as can be seen in later sections.

From the refractive index (Figure 5) study in this work we conclude that the energy gap of the samples may be altered upon the addition of copper powder. This can be more understood through the study of energy band gap and refractive index as a function of copper concentration. For this purpose the plot of absorption coefficient $((h\alpha\nu)^{\wedge 2})$ as a function of photon energy $(h\nu)$ enable us to calculate the energy band gaps. In order to show the influence of copper powder on the optical band gap of the solid polymer composite films, the optical absorption edge was investigated for all the samples. Figures 6(a)–6(e) show the dependence of $(h\alpha\nu)^{\wedge 2}$ versus $h\nu$ for all the samples. The comprehensible absorption edge shown by all the samples establish the dominant of crystalline

nature of the samples [18]. It is evident that the absorption edge shifted towards the lower energy side upon the addition Cu powder especially at 6 wt.% of Cu. These results confirm the fact that the energy band gap and the refractive index are strongly correlated.

The direct energy band gaps for all samples were determined from the dotted lines intersection on the photon energy axis of Figures 6(a)–6(e) and plotted as a function of copper concentration as depicted in Figure 7. It is obvious that the energy band gap is steeply decreased at 6 wt.% of Cu powder. One possible interpretation of this experimental observation is that Cu powder introduces multiple valence states in the polystyrene structure and thus decreasing the band gap energy between the valence and conductions bands.

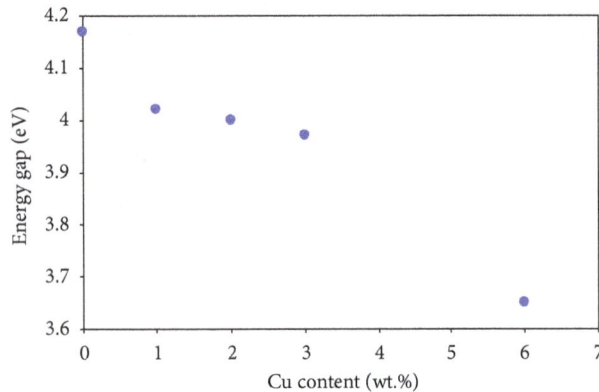

FIGURE 7: Energy gap as a function of copper concentration (wt.%).

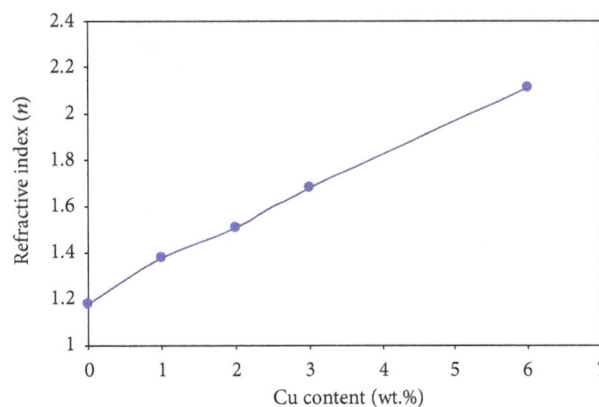

FIGURE 8: Refractive index as a function of copper concentration (wt.%).

Variation of refractive index as a function of copper concentration permits us to understand the behavior of energy gap variation with copper concentration. Figure 8 shows the copper powder dependence of refractive index which calculated from the intersection of the plateau of Figure 5 on the Y-axis. It is clear that the energy band gap (Figure 7) and the refractive index (Figure 8) follows the same trend but in a different manner. It was reported that the optical properties (index of refraction and energy gap) of a material are related to the variation in material composition and atomic arrangement [19]. The small value of the refractive indexes (1.2 to 2.18) obtained in this work reveals that PS based polymer composites are very suitable as low-index claddings for waveguide applications [8].

4. Conclusion

The dispersion region in both absorption and refractive indexes at lower wavelength can be ascribed to the sufficient time for polarization with the electric field component of the electromagnetic light. A plateau of absorption and refractive indexes at high wavelengths may be attributable to the inertia of the side groups of polystyrene which is difficult when following the electric field alternation. The calculated energy band gaps of the solid composite samples show that the nature of electronic transition from valence band to conduction band was a direct transition. Upon the addition of Cu concentration, the refractive index increased from 1.2 up to 2.1 and the energy gap decreased from 4.05 to about 3.65 eV as a result of the introduction of more multiple states. The small refractive indexes of PS based composites indicate their importance in waveguide application. This work confirms that the refractive index and energy gap are strongly correlated.

Acknowledgment

The authors gratefully acknowledge the Ministry of Higher Education and Scientific Research, Kurdistan Regional Government, University of Sulaimani, for the financial support.

References

[1] S. George and M. T. Sebastian, "Three-phase polymer-ceramic-metal composite for embedded capacitor applications," *Composites Science and Technology*, vol. 69, no. 7-8, pp. 1298–1302, 2009.

[2] G. C. Psarras, E. Manolakaki, and G. M. Tsangaris, "Dielectric dispersion and ac conductivity in—iron particles loaded: polymer composites," *Composites A*, vol. 34, no. 12, pp. 1187–1198, 2003.

[3] G. C. Psarras, "Hopping conductivity in polymer matrix-metal particles composites," *Composites A*, vol. 37, no. 10, pp. 1545–1553, 2006.

[4] F. Liu, X. Zhang, W. Li et al., "Investigation of the electrical conductivity of HDPE composites filled with bundle-like MWNTs," *Composites A*, vol. 40, no. 11, pp. 1717–1721, 2009.

[5] S. H. Yang, T. P. Nguyen, P. le Rendu, and C. S. Hsu, "Optical and electrical properties of PPV/SiO₂ and PPV/TiO₂ composite materials," *Composites A*, vol. 36, no. 4, pp. 509–513, 2005.

[6] Q. Li, Q. Xue, L. Hao, X. Gao, and Q. Zheng, "Large dielectric constant of the chemically functionalized carbon nanotube/polymer composites," *Composites Science and Technology*, vol. 68, no. 10-11, pp. 2290–2296, 2008.

[7] E. A. Costner, B. K. Long, C. Navar et al., "Fundamental optical properties of linear and cyclic alkanes: VUV absorbance and index of refraction," *Journal of Physical Chemistry A*, vol. 113, no. 33, pp. 9337–9347, 2009.

[8] M. K. Yang, R. H. French, and E. W. Tokarsky, "Optical properties of Teflon AF amorphous fluoropolymers," *Journal of Micro/ Nanolithography, MEMS, and MOEMS*, vol. 7, no. 3, Article ID 033010, pp. 1–9, 2008.

[9] M. Caglar, Y. Caglar, and S. Ilican, "The determination of the thickness and optical constants of the ZnO crystalline thin film by using envelope method," *Journal of Optoelectronics and Advanced Materials*, vol. 8, no. 4, pp. 1410–1413, 2006.

[10] A. Tadjarodi, M. Imani, and H. Kerdari, "Application of a facil solid state process to synthesize the CdO spherical nanoparticles," *International Nano Letters*, vol. 3, article 43, 2013.

[11] N. A. Hamizi and M. R. Johan, "Optical properties of CdSe quantum dots via non-TOP based route," *International Journal of Electrochemical Science*, vol. 7, pp. 8458–8467, 2012.

[12] F. Yakuphanoglu, M. Sekerci, and A. Balaban, "The effect of film thickness on the optical absorption edge and optical constants

of the Cr(III) organic thin films," *Optical Materials*, vol. 27, no. 8, pp. 1369–1372, 2005.

[13] N. Ahlawat, S. Sanghi, A. Agarwal, and S. Rani, "Effect of Li_2O on structure and optical properties of lithium bismosilicate glasses," *Journal of Alloys and Compounds*, vol. 480, no. 2, pp. 516–520, 2009.

[14] L. Bi, A. R. Taussig, H.-S. Kim et al., "Structural, magnetic, and optical properties of $BiFeO_3$ and Bi_2FeMnO_6 epitaxial thin films: an experimental and first-principles study," *Physical Review B*, vol. 78, no. 10, Article ID 104106, 2008.

[15] R. H. French, J. M. Rodríguez-Parada, M. K. Yang, R. A. Derryberry, and N. T. Pfeiffenberger, "Optical properties of polymeric materials for concentrator photovoltaic systems," *Solar Energy Materials & Solar Cells*, vol. 95, no. 8, pp. 2077–2086, 2011.

[16] T. Arumanayagam and P. Murugakoothan, "Studies on optical and mechanical properties of new organic NLO crystal: guanidinium 4-aminobenzoate (GuAB)," *Materials Letters*, vol. 65, no. 17-18, pp. 2748–2750, 2011.

[17] T. N. Zhou, X. D. Qi, and Q. Fu, "The preparation of the poly(vinyl alcohol)/grapheme nanocomposites with low percolation threshold and high electrical conductivity by using the large-area reduced grapheme oxide sheets," *Express Polymer Letters*, vol. 7, no. 9, pp. 747–755, 2013.

[18] K. C. Preetha, K. V. Murali, A. J. Ragina, K. Deepa, and T. L. Remadevi, "Effect of cationic precursor pH on optical and transport properties of SILAR deposited nano crystalline PbS thin films," *Current Applied Physics*, vol. 12, no. 1, pp. 53–59, 2012.

[19] X. Rocquefelte, S. Jobic, and M.-H. Whangbo, "Concept of optical channel as a guide for tuning the optical properties of insulating materials," *Solid State Sciences*, vol. 9, no. 7, pp. 600–603, 2007.

5

Effects of Moulding Sand Permeability and Pouring Temperatures on Properties of Cast 6061 Aluminium Alloy

Olawale Olarewaju Ajibola,[1,2] **Daniel Toyin Oloruntoba,**[1] **and Benjamin O. Adewuyi**[1]

[1]*Metallurgical and Materials Engineering Department, Federal University of Technology Akure, Akure 340252, Nigeria*
[2]*Materials and Metallurgical Engineering Department, Federal University Oye Ekiti, Oye Ekiti 371104, Nigeria*

Correspondence should be addressed to Olawale Olarewaju Ajibola; olawale.ajibola@fuoye.edu.ng

Academic Editor: Manoj Gupta

Effects of moulding sand permeabilities prepared from the combinations of four proportions of coarse and fine particle size mixtures and pouring temperatures varied from 700, 750, and 800 (±10°C) were studied on the hardness, porosity, strength, and microstructure of cast aluminium pistons used in hydraulic brake master cylinder. Three sand moulds were prepared from each of the 80 : 20, 60 : 40, 40 : 60, and 20 : 80 ratios. The surfaces and microstructures of cast samples were examined using high resolution microscopic camera, metallurgical microscope with digital camera, and scanning electron microscope with EDX facilities. The best of the metallurgical properties were obtained from the combination of 80 : 20 coarse-fine sand ratio and 750 ± 10°C pouring temperature using as MgFeSi inoculant. An 8 : 25 ratio of coarse to fine grained eutectic aluminium alloy was obtained with enhanced metallographic properties. The cast alloy poured at 750 ± 1°C has a large number of fine grain formations assuming broom-resembling structures as shown in the 100 μm size SEM image.

1. Introduction

Aluminium recycling industries are growing globally at very alarming rate. In Nigeria, the entrepreneurship challenges and opportunities that go along with this trend are also vast [1]. Hence, the research and development affect many facets such as the aluminium foundry, materials design for various fields of applications such as automobile and automotive industries [2]. The metallurgical properties of metal alloys are controlled by many factors such as the chemical composition, microstructure, processing methods such as casting [3], extrusion, and postproduction treatment such as surface deposition and heat treatment [4–6].

These mentioned factors definitely affect the microstructures of the product and consequently determine the behaviour of the cast under the stipulated service [7]. In order to improve the quality of the aluminium alloy piston used in the hydraulic brake master cylinder, a controlled casting technique involving melting and pouring temperature, selective combination of moulding sand permeability and finally the solidification and cooling process will have great deal on the microstructure, mechanical strength, hardness, and hence the wear resistance of the product. The sand particle size distribution controls the mould permeability which is the amount of air that can be trapped through the sand in the mould. It was reported that the coarse particle size results in high permeability while fine particles give low permeability of moulding sand [8].

Dissolved gases increase the chances of pores formation thereby increasing the porosity in the metal cast whereas the enclosure or inclusion of the surface energy effect makes it difficult, which likely need negative pressure to form empty spaces (voids). Dissolved gases in liquid alloy cause porosity because the solubility of gases in liquid metals usually exceeds the solubility in the solid.

It is in most cases required that the cast alloy material should be impervious to gases and liquids. The porosity is higher when more pores are contained in the cast. Hence, there is tendency of heat loss by convection and the leakages of liquid through the pores of the metal cast. In some refractory metals, thermal shock depends on porosity of the material. The porous materials (compressed powder metals)

TABLE 1: Particle size (μm) distribution of moulding sand.

		Coarse sand							Fine sand			
Sieve range (μm)	+4750	−4750 +2360	−2360 +1180	−1180 +850	−850 +600	−600 +425	−425 +300	−300 +212	−212 +180	−180 +150	−150 +75	−75
% distribution	0.82	1.55	3.57	8.61	9.01	9.75	13.81	14.03	13.21	11.86	8.05	5.72

TABLE 2: Pouring temperatures and mixing ratios of coarse and fine moulding sand.

Temperature °C	Mixing ratios coarse (−1180 +300 μm) and fine (−300 +75 μm) moulding sand			
	Set 1	Set 2	Set 3	Set 4
700	80 : 20	60 : 40	40 : 60	20 : 80
750	80 : 20	60 : 40	40 : 60	20 : 80
800	80 : 20	60 : 40	40 : 60	20 : 80

have higher resistance to spall than the highly compacted shapes. Therefore, depending on the application, there should be a balance between the porosity and compactness of the material.

Hence, the effects of sand permeability via particle sizes of the moulding sand and variation of pouring temperatures on the properties (strength, hardness, and porosity) of cast aluminium alloy (AA6061) were studied in this report.

2. Materials

The materials used in the experiment include the foundry moulding sands (coarse and fine), 1000 kg of aluminium alloy (AA6061) scrap sourced from the brake master cylinder pistons, and powdered magnesium ferrosilicon inoculant.

3. Method

3.1. Procedure for Sand Cast Specimen. The moulding sand was prepared from different proportions of coarse and fine sand particle sizes. Table 1 shows the sand particles size distribution of moulding sand used. Three sets of moulds were prepared from each of the 80 : 20, 60 : 40, 40 : 60, and 20 : 80 ratios of coarse (+4750 +300 μm) and fine (−300 −75 μm) moulding sand particle size mixtures (Table 2 and Figures 1-2). The sand was properly rammed with adequate vent holes. The moulds were left to dry at room temperature (27°C).

Aluminium scrap was charged into the melting crucible and fired. The molten Al alloy was held at three pouring temperature ranges (700 ± 10°C, 750 ± 10°C, and 800 ± 10°C). 17 g of powdered magnesium ferrosilicon inoculant was added per 1 kg mass of molten metal in the melting pot and before casting [3]. The moulding flasks were preheated before the casting process. To study the effect of variation in the moulding sand permeability and pouring temperatures, twelve specimens were cast at the three pouring temperatures (700 ± 10°C, 750 ± 10°C, and 800 ± 10°C) using the prepared ratios of coarse-fine moulds (80 : 20, 60 : 40, 40 : 60, and 20 : 80). The method used for the determination of the

moulding sand permeability has been described by Jimoh et al. [9].

The aluminium cast was left to solidify and cool to room temperature in the mould. The cast samples were removed from the mould and fettled. It is lightly machined on the lathe to rod of 300 mm long by 30 mm diameter (Figure 3(a)) from which samples were cut for hardness test and tensile strength test (Figure 3(b)). The cross cut section was also examined under microscope. The cast samples from each of the sand mould were designated as TS11–TS14; TS21–TS24; and TS31–TS34 for each set of cast Al alloy poured at 700 ± 10°C, 750 ± 10°C, and 800 ± 10°C, respectively (Figures 3-9).

To determine the porosity of the cast samples, 25 mm × 25 mm × 25 mm cube size test samples were cut from the core of the cylindrical shape cast Al alloy to be tested. The test samples were cleaned from dust and other particles adhering to the surfaces and fired at 110°C in oven to a dry weight (D). The dried specimen is placed in the vacuum desiccator which is then evacuated to a pressure of 2.5 mm Hg. The specimen is immersed in liquid paraffin (boiling above 200°C). The test samples were soaked in the liquid under reduced pressure for 10 hours suspended by a sling thread and were weighed (S) while still suspended in immersion liquid. The test specimen is then lifted up slowly from the immersion liquid by means of the sling thread liquid, and drops appearing on the surface are removed by lightly contacting with a piece of blotting ensuring that it does not make physical contact with the specimen surface itself. The soaked specimen is latter weighed (W) while keeping it suspended in air. The apparent porosity (P) is then calculated as follows:

$$P = \left(\frac{W - D}{W - S} \right) \times 100 \,(\%). \tag{1}$$

3.2. Chemical and Physical Characterisation of Scrap and Cast Samples. The hardness tests of aluminium alloy samples (the scrap and cast Al alloy) were also determined using Brinell Hardness Testing Machine. The test was conducted by pressing a tungsten carbide sphere 10 mm in diameter into the test plate surface for 10 seconds with a load of 1500 kg, and then the diameter of the resulting depression is measured. An average of four BHN tests was carried out over an area of the specimen surface. The BHN is calculated using (2) and average HBN values from the result are presented in Figures 3(a) and 3(b):

$$\text{BHN} = \frac{F}{\left[\pi D/2 \left(D - \sqrt{D^2 - D_i{}^2} \right) \right]}, \tag{2}$$

where BHN is the Brinell hardness number, F is imposed load in kg, D is diameter of the spherical indenter in mm, and D_i is diameter of the resulting indenter impression in mm.

(a)

(b)

FIGURE 1: (a) Photographs showing (A) coarse size (+4750 +300 μm), (B) fine size (−300 −75 μm), and (C) moulding sand mixtures. (b) Photographs showing (A, B) matrix of twelve moulding flasks.

FIGURE 2: Photographs showing (a) machined cast aluminium alloy and (b) cast specimens used for photomicroscopy and SEM analyses.

The grain sizes of the microscopic particles of the purchased piston and cast aluminium alloy samples were determined using XRD. The powder of each sample was produced for XRD study under higher resolution X-ray using X-Ray Minidiffractometer MD-10 model with digital facilities. Each of the peak values in the diffractograms was interpreted by comparing the values with the standard values in the database of compounds under this radiation using the "search and match" technique.

The surfaces of cast samples were examined under high resolution microscopic camera using Samsung ST65-HD5X-14.2 model. Microstructures of the scrap and cast samples were examined under higher resolution metallurgical microscope with digital camera (Accu-Scope microscope model) in the laboratory at ×800 magnification.

The cast samples were further characterised by Atomic Absorption Spectroscopy (AAS-Thermo series 2000 model), X-Ray diffraction (XRD) (Minidiffractometer MD-10 model with digital facilities), and Scanning Electron Microscopy (SEM) and Energy Dispersive X-Ray (EDX) analyses (Jeol JSM-7600F Field Emission). The macrophotographs, micrographs, diffractograms, SEM/EDX spectra line, and AAS data generated were used to interpret the results.

4. Results and Discussions

The chemical compositions of scrap and cast samples are presented in Table 3. The chemical analysis by AAS shows that scrap aluminium alloy contains 98.665% Al as matrix and the following alloying elements: 0.686% Si, 0.403% Mg, 0.001%

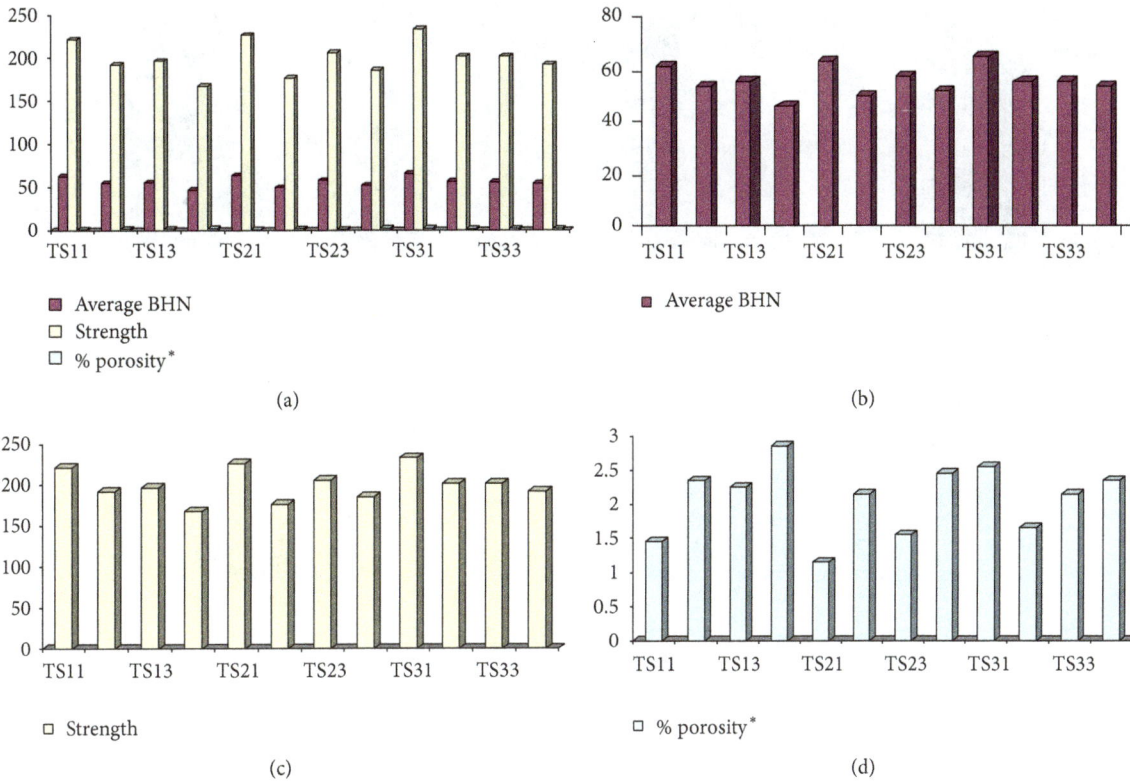

(a)

(b)

(c)

(d)

FIGURE 3: (a) Properties of cast aluminium alloy (TS11–TS34) at varying pouring temperatures (700, 750, and 800°C) and different moulding sand ratios. (b) BHN of cast aluminium alloy (TS11–TS34) at varying pouring temperatures (700, 750, and 800°C) and different moulding sand ratios. (c) Strength of cast aluminium alloy (TS11–TS34) at varying pouring temperatures (700, 750, and 800°C) and different moulding sand ratios. (d) Porosity of cast aluminium alloy (TS11-TS34) at varying pouring temperatures (700, 750, and 800°C) and different moulding sand ratios.

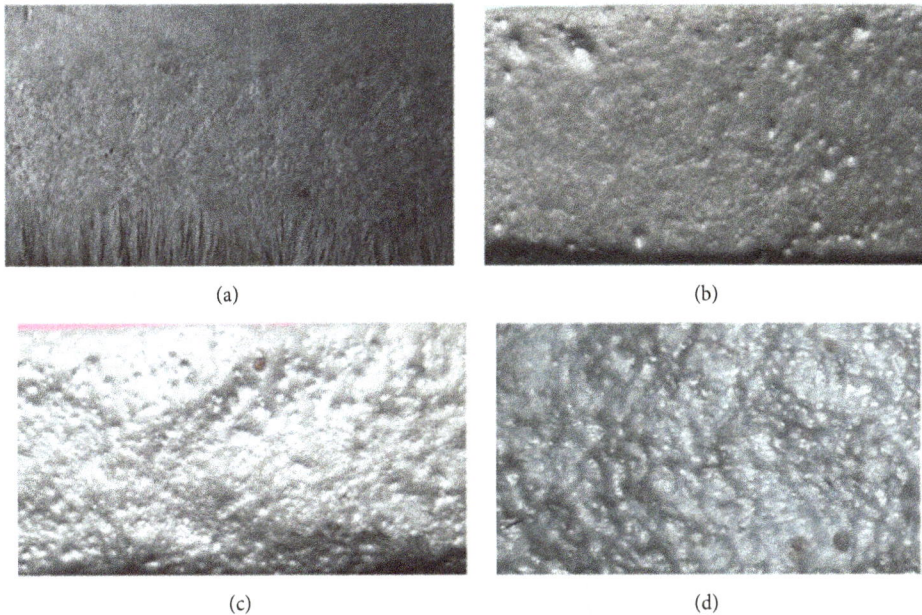

(a)

(b)

(c)

(d)

FIGURE 4: Surfaces of as-cast AA6061 aluminium alloy (TS11–TS14) at 700 ± 10°C pouring temperature at different moulding sand mixing ratios (a) 80 : 20, (b) 60 : 40, (c) 40 : 60, and (d) 20 : 80 (×10 mag.).

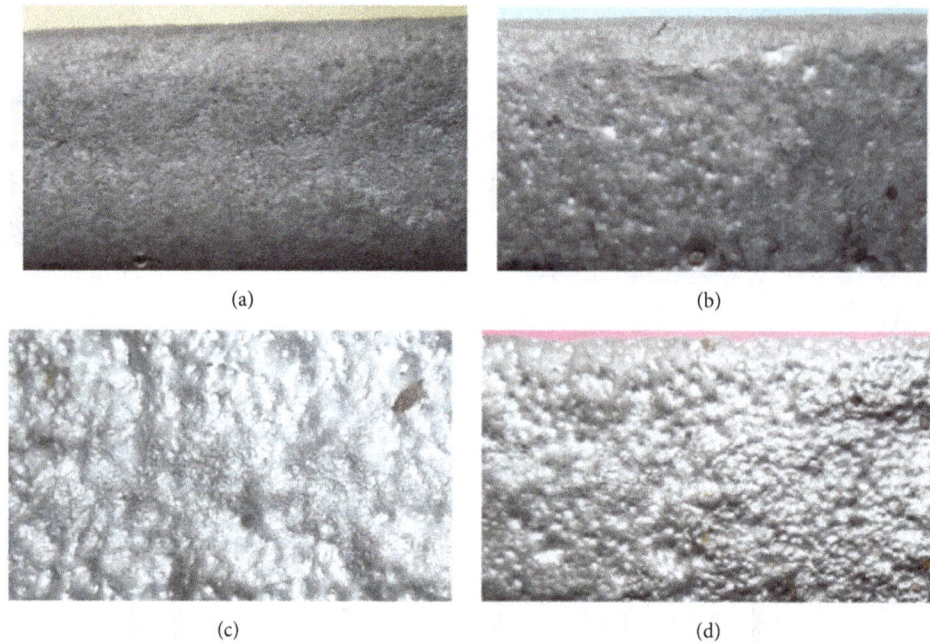

FIGURE 5: Surfaces of as-cast AA6061 aluminium alloy (TS21–TS24) at $750 \pm 10°C$ pouring temperature at different moulding sand mixing ratios (a) 80 : 20, (b) 60 : 40, (c) 40 : 60, and (d) 20 : 80 (×10 mag.).

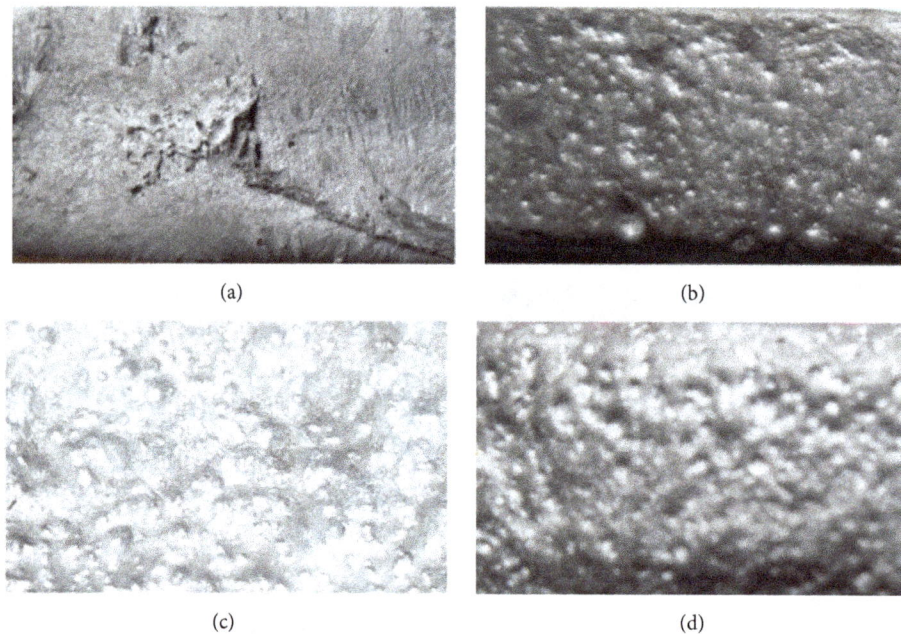

FIGURE 6: Surfaces of as-cast AA6061 aluminium alloy (TS31–TS34) at $800 \pm 10°C$ pouring temperature at different moulding sand mixing ratios (a) 80 : 20, (b) 60 : 40, (c) 40 : 60, and (d) 20 : 80 (×10 mag.).

Cu, 0.001% Zn, 0.001% Ti, 0.001% Mn, 0.001% Cr, and 0.232% Fe.

4.1. Influence of Sand Permeability and Pouring Temperatures on Cast Al Samples. The right casting process starts with the laying hold on the control of the chemistry of the melt. Casting aluminium alloy entails proper handling of the charge materials and the equipment. This ranges from the

type of furnace and fuel, the melting pot, the selection of fluxing additives, and alloying elements.

The choice of the moulding sand permeability (moulding sand particles size mixing ratios) was used as one of the measures to obtain enhanced metallurgical properties (high HBN and eutectic microstructure) of the aluminium cast. The moulding sand was prepared from the combination of high coarse sand (+1180 +300 μm) to low fine (+300 +75 μm) sand

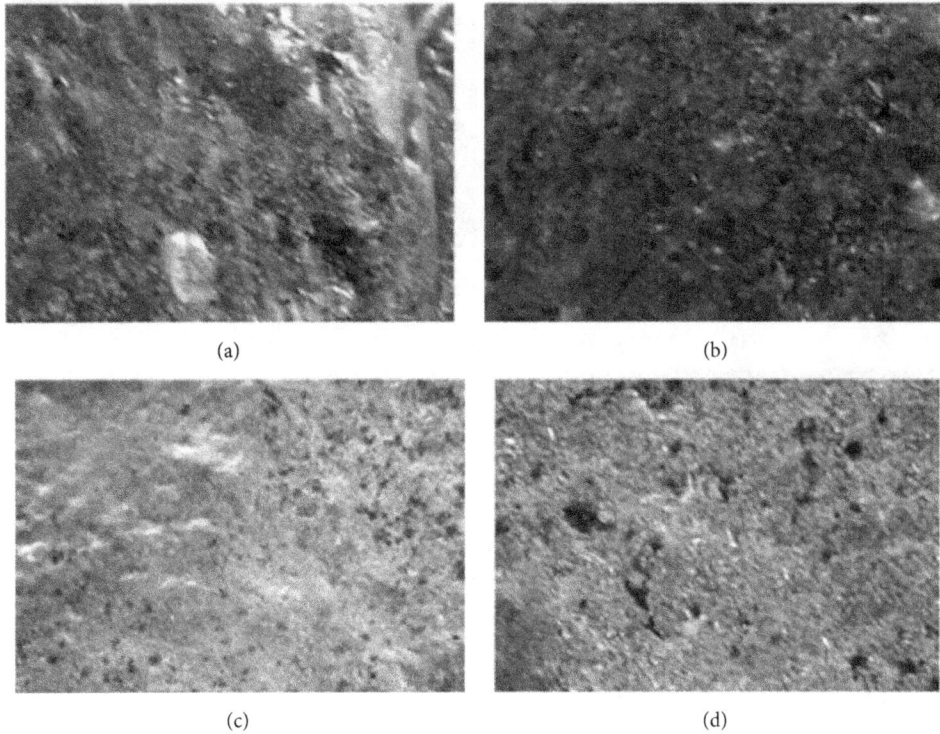

FIGURE 7: Enlarged micrographs showing porous cast aluminium alloy (TS11–TS14) at $700 \pm 10°C$ pouring temperature at different moulding sand mixing ratios (a) 80 : 20, (b) 60 : 40, (c) 40 : 60, and (d) 20 : 80 (×100 mag.).

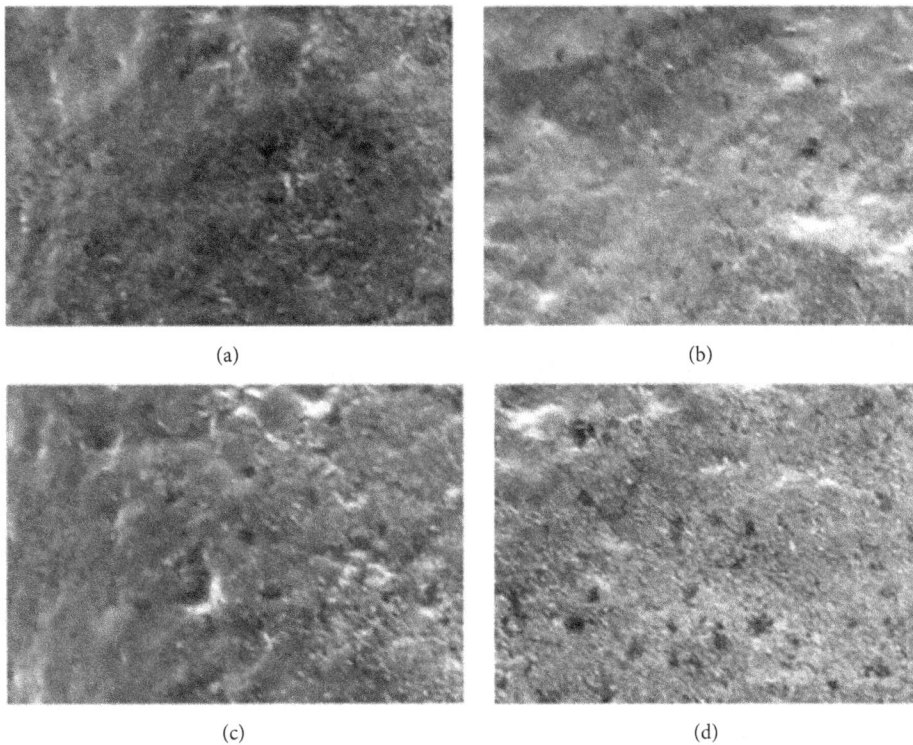

FIGURE 8: Enlarged micrographs showing porous cast aluminium alloy (TS21–TS24) at $750 \pm 10°C$ pouring temperature at different moulding sand mixing ratios (a) 80 : 20, (b) 60 : 40, (c) 40 : 60, and (d) 20 : 80 (×100 mag.).

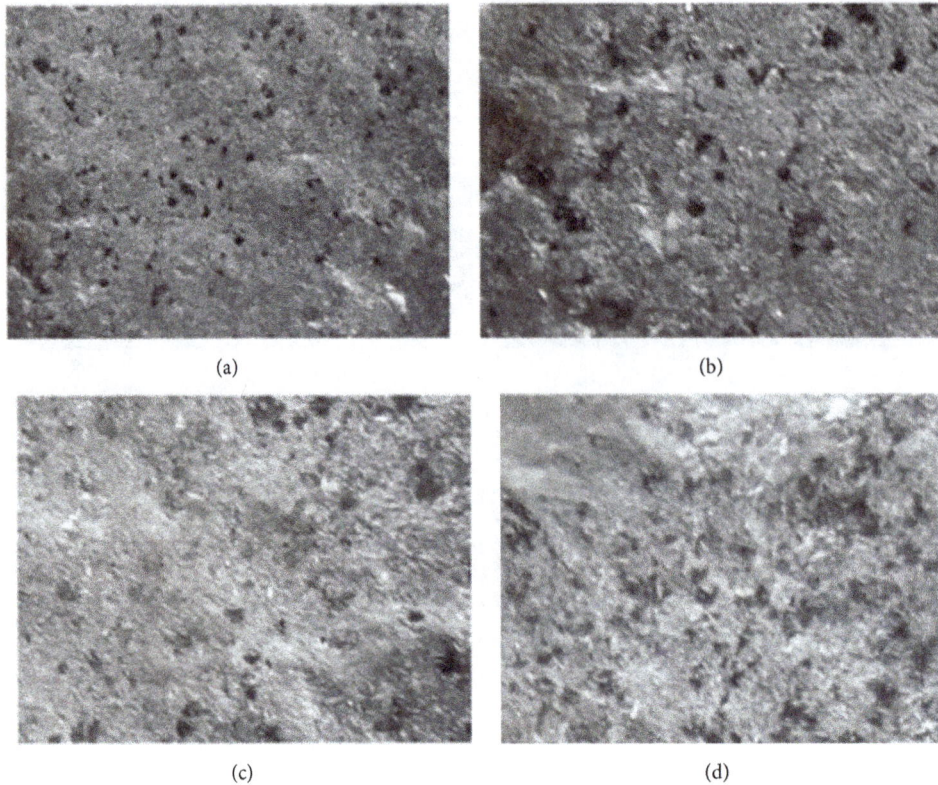

FIGURE 9: Enlarged micrographs showing porous cast aluminium alloy (TS31–TS34) at $800 \pm 10°C$ pouring temperature at different moulding sand mixing ratios (a) $80:20$, (b) $60:40$, (c) $40:60$, and (d) $20:80$ (×100 mag.).

particle sizes to give moderately high permeability [10]. High permeability of moulding sand ($80:20$ coarse-fine ratios) allowed the escape of gases and air bubbles that would have entrapped in the mould thereby increasing the porosity of the cast piston.

The micrographs showing the microstructures of scrap and cast aluminium alloys poured at $700 \pm 10°C$, $750 \pm 10°C$, and $800 \pm 10°C$, respectively, are presented in Figures 4–14. The microstructural examinations revealed the microstructures of the alloys and compared the similarities and differences between the grain sizes and structures of scrap and cast samples (Figures 15-16), as influenced by the pouring temperature and the mixing ratio (permeability) of moulding sand.

Pouring an aluminium alloy at temperature above $690°C$ often gives enhanced metallurgical properties. The microstructure and phases of the aluminium cast were moderated and refined grains were obtained in agreement with Apelian [11]. The structure of cast sample poured at $700°C$ using $80:20$ coarse-fine sand mixing ratio is characterised by fine grains as revealed by the SEM analysis with few pores and fine transitioned Al-Si eutectic (Figure 14) as compared with the scrap material [11].

As the pouring temperature of molten metal is higher, gas content will increase especially for molten aluminium. The choice of pouring at $750 \pm 10°C$ was strategized to moderate the gas content in the molten aluminium alloy and to obtain eutectic structure of cast alloy [10].

FIGURE 10: Microstructure of as-received scrap Al substrate (magnifications ×800).

The cast samples obtained from aluminium alloy held and poured at elevated temperature $750°C$ are more superheated than what was obtainable at $700°C$.

Figure 11(c) shows the macrographs of aluminium cast poured at more elevated temperature ($800°C$) using similar sets of moulding sands as in Figures 11(a) and 11(b). At temperature as high as $800°C$, there is subsequent structural change occurring in the cast aluminium alloy.

The differences in the mechanical properties (hardness, porosity, and strength) are reflection of the effects of variations in moulding sand permeability and pouring/casting temperatures on the products. Porosity may be regarded as

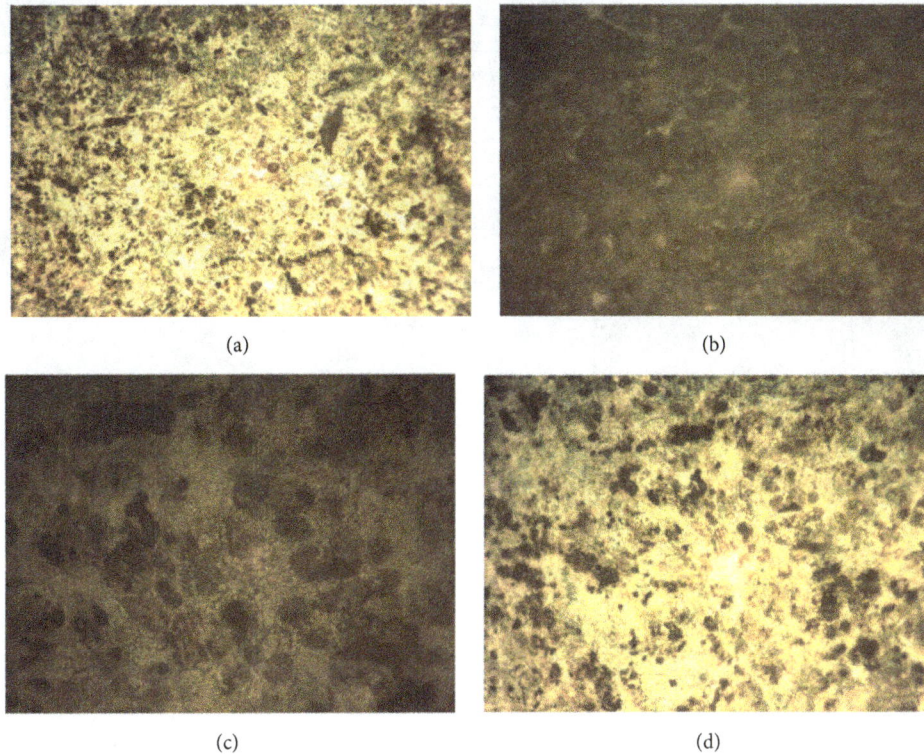

FIGURE 11: Microstructures of cast substrate poured at (a) 700 ± 10°C, (b) 750 ± 10°C, (c) 800 ± 10°C using 80 : 20, and (d) 750°C using 20 : 80 sand mixing ratios with MgFeSi (magnifications ×800).

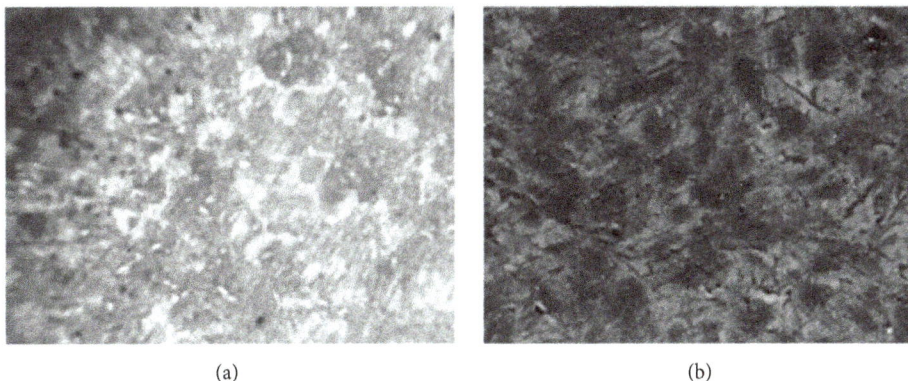

FIGURE 12: Nucleation identified in cast alloy piston (a, b).

problem as in most casting products and in other instance being advantage in shaped porous metals. Most solid metals have higher densities than the liquid and hence the liquid metal flows in the direction of solidifying area in order to avert the formation of voids. In sand moulding, the sand permeability can be affected by the unregulated high amount of mixing water, poor compaction practice, and uneven drying which may result in the mould retaining moisture. It is much possible that when liquid metal runs across the spongy (mushy) zone to feed solidification shrinkage, the molten metal pressure in this mushy zone drops low, below the exterior atmospheric pressure. And as a result microporosity forms in the cast when the local (confined) pressure in the mushy region drops below a critical value [12].

The results of average HBN obtained from the four point hardness tests on the scrap and cast samples are presented in Figures 3(a) and 3(b). The hardness tests show that scrap sample has lower HBN than the cast piston sample. Tensile strengths of samples were determined as 154.78 Mpa for scrap sample and 226.49 Mpa for cast sample.

The addition of MgFeSi initiated nucleation in collaboration with other factors such as alloy composition, cooling rate, temperature gradient in the melt, and casting method that affect the ultimate cast grain size. With the addition of MgFeSi, the AAS characterisation shows that about 97.432% Al, 1.293% Si, 0.598% Mg, 0.202% Cu, 0.001% Zn, 0.051% Ti, 0.051% Mn, 0.041% Cr, and 0.331% Fe were contained in the cast alloy. The increase in the % Fe composition in the

TABLE 3: Chemical analysis of aluminium alloy samples by AAS.

Samples	Matrix	Major elements			Neutral	Microstructure modifier			Impurity
	Al	Si	Mg	Cu	Zn	Ti	Mn	Cr	Fe
Scrap Al	98.665	0.686	0.403	0.001	0.001	0.001	0.001	0.001	0.232
Cast Al	97.432	1.293	0.598	0.202	0.001	0.051	0.051	0.041	0.331

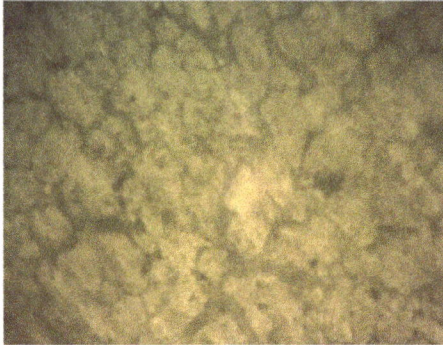

FIGURE 13: Microstructure of cast alloy poured at $750 \pm 10°C$ (without inoculants) using $80 : 20$ coarse-fine sand mixing ratio (magnification ×800).

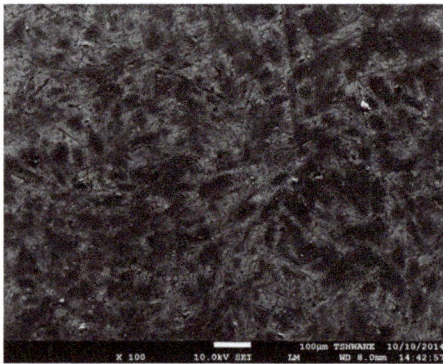

FIGURE 14: SEM image of 100 μm size cast alloy poured at $750 \pm 10°C$ (with inoculant) using $80 : 20$ coarse-fine sand mixing ratio (mag. ×100).

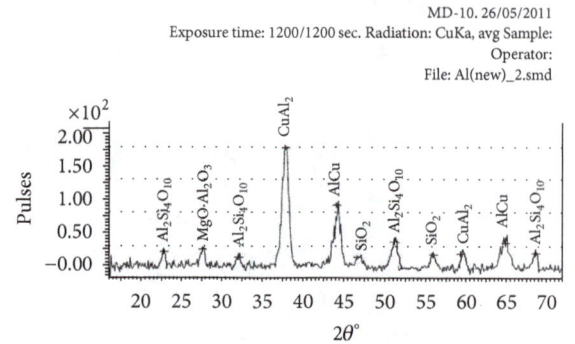

FIGURE 15: The diffractograms of XRD analysis of scrap sample.

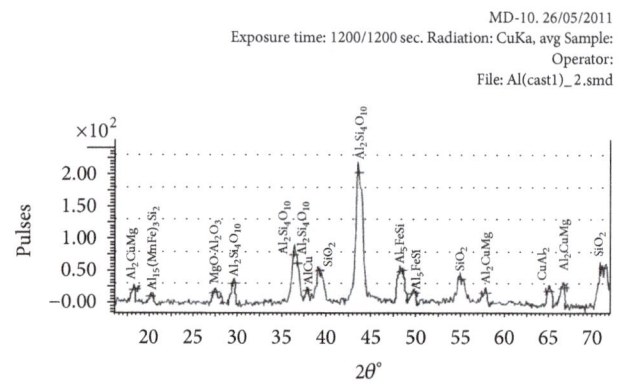

FIGURE 16: The diffractograms of XRD analysis of cast sample (with inoculant).

cast sample has been influenced by both the MgFeSi addition and the Fe pick-up from the melting pot. This is traced to higher solubility of Fe in molten aluminium [13]. Fine grained aluminium alloy was obtained which has better strength by inoculating the melt with MgFeSi powder which forms insoluble compound particles, thus helping to increase the rate of nucleation. In the MgFeSi modified Al-Si alloys, the growth of the Al-Si-Fe phase is cut short resulting in a large number of equiaxed Al grains formation assuming broom-resembling structures different from cast without MgFeSi [14–18] as observed in the SEM image (Figure 14).

The grain refinement of Al and its alloys using inoculants that boost heterogeneous nucleation is an important structure modification method used in many industries. A fine grain size in metal alloy castings guarantees (i) homogeneous mechanical properties, (ii) distribution of second phases and microporosity on a fine scale, (iii) superior machinability because of (ii), (iv) enhanced uniform anodizable face, (v) improved strength, toughness, and fatigue life, and (vi) superior corrosion resistance. The literature has quite a lot of mechanisms proposed for grain refinement and critical reviews exist in reports of Perhpezko [19] and McCartney [20].

In the present study, the initial scrap charge invariably contains significant amounts of iron in addition to MgFeSi inoculant, which takes an imperative function in the nucleation process. High Fe content in the charge also encourages the formation of Al-Si-Fe phase. The SEM/EDX characterisation shows that about 1.49% Si, 96.35% Al, 0.18% Cu, 0.01% Ti, 0.02% Mn, 0.02% Cr, 0.63% Fe, 0.53% Mg, and 0.01% Zn were contained in the cast alloy.

4.1.1. Variation of Pouring Temperatures and Sand Permeability on the Casting. The varying degrees of properties of cast

TABLE 4: SEM/EDX spectra analysis of cast alloy poured at $750 \pm 10°C$ (with inoculant).

Sample	C	O	Si	Al	Cu	Ti	Mn	Cr	Fe	Mg	Zn
Cast Al	0.45	0.31	1.49	96.35	0.18	0.01	0.02	0.02	0.63	0.53	0.01

samples at different pouring temperatures using coarse-fine sand particle sizes mixtures are reported in Figures 3(a)–3(d).

In all cases, at any constant pouring temperature within the range of 700–800°C, the average BHN values (Figure 3(b)) and the strength (Figure 3(c)) of samples measured reduce with the reduction in sand permeability. Meanwhile, in Figure 3(d), there is increase in the trend of the cast porosity as the sand permeability reduces (Figure 3(d)). At increasing pouring temperature, the % porosity reduced with higher degree of sand permeability with respect to sand particle size ratios. The moulds made from more coarse sand have lower % porosity.

From Figures 3(a) and 3(d), it is obvious that the cast samples were characterised by quantity of moderately low % porosity (about 1.15–2.15) and moderately high % porosity (2.15–2.85), with varying sizes of both tiny pores (less than 10 μm) and large pores (above 10 μm) as observed under the microscope.

The casting voids most frequently called porosity are caused by gas formation, solidification shrinkage, or non-metallic compound formation in the molten metal. Blows or blowholes are bulky gas-related voids caused by entrapped mould or core gases in the molten metal. They are large enough and resemble bubbles with smooth internal surfaces and are buoyant and float close to the top of the casting; they can also get trapped on the bottom surface of a core lower in the mould. Moreover, pinholes are caused by gases (atoms) dissolved in molten metal (that connects and become molecules). They remain small (less than 10 μm) but float to a top surface somewhere in the casting [21].

In recent times, there are increasing interests in research on performance of castings with porosity. Most notable research is on the fracture mechanics of microplasticity models of fatigue and failure to incorporate effects of inclusions, microporosity, macroporosity, and microstructure to cast aluminium alloy components [22]. The pores are large enough to be seen even with the naked eyes on some specimen as observed in Figures 7–9. Hence, the simple evacuation approach to apparent porosity measurement was applied and the result was taken as estimate of the % apparent porosity [23]. To really calculate the fraction and size of porosity after solidification is finished, a more complex analysis may be necessary. Reports have it that the quantity and size of the porosity produced in Al-4.5 wt% Cu plate castings containing hydrogen experimentally were calculated by Kubo and Pehlke [24] while Poirier et al. [13, 25] presented such calculations for Al-Cu in directional solidification geometry.

4.2. The Surface Morphology and Porosity of Cast Aluminium Alloy Specimens. The macrographs of the surface appearances of cast aluminium alloy specimens poured at 700 ± 10, 750 ± 10, and 800 ± 10°C using four different sets of moulding sands are shown in Figures 4–6. The porosity of cast aluminium alloy obtained at different pouring temperatures and moulding sand mixing ratios are also presented in Figures 7–9.

4.3. Microstructural Examination of the Scrap and Cast Al Substrates. The microstructures obtained from the scrap and the cast Al substrates using higher resolution metallurgical microscope with digital camera under ×800 magnification are shown in Figures 10–13. The SEM image and EDX analyses of the cast sample with inoculant are presented in Figure 14 and Table 4, respectively. Figure 14 shows the SEM image of 100 μm size cast alloy poured at 750 ± 10°C (with inoculant) using 80 : 20 coarse-fine sand mixing ratio at magnification of ×100.

The images in Figures 4–6 are the photographs of the surfaces of as-cast AA6061 aluminium alloy specimens observed at ×10 magnification under High Resolution Microscopic Camera ST65-HD5X-14.2 model, while for the purpose of better clarification of the pores Figures 7–9 show the enlarged micrographs of same set of surfaces examined at ×100 magnification.

Moreover, the images in Figures 10–13 are the microstructures of 10 mm size section of the cast Al alloy examined at ×800 magnification using a metallurgical microscope, while a more magnificent image of the microstructure of a 100 μm size target was examined at ×100 magnification under the SEM (Figure 14) with the view of clarifying both the microstructure (shapes) and the elemental composition (Table 4) of the cast Al alloy specimen.

The analyses of the mechanical properties (hardness, strength) based on combination of the composition and microstructures revealed the presence of voids and inclusions in the metal cast. The primary inclusions include solids in the melt above the liquidus temperature of the alloy such as (i) the exogenous inclusions (dross, entrapped mould material, slag, and refractories); (ii) salts and fluxes suspended in the melt resulting from a previous melt-treatment processes; and (iii) suspended oxides of the melt (entrapped within by turbulence or on top of the melt). Secondary inclusions include those formed after the solidification of the main metallic phase.

4.4. Characterisation by X-Ray Diffraction Analyses of the Scrap and Cast Samples. In addition to the microphotographic examinations of the materials, the purpose of the XRD analyses in the present study is to make a distinction among the phases and the grain sizes of the microstructure with the view of appreciating and elucidating reasons for their mechanical properties (hardness and strength) and forecasting their wear behaviour.

The XRD method is based on Bragg's diffraction law [26] as follows:

$$n\lambda = 2d \sin\theta, \tag{3}$$

TABLE 5: XRD analysis of scrap Al sample.

S/N	Peak	Diffraction angle 2θ	Grain size (Å)	Crystal structures	Phases
1	0.20	23.0124	0.39	Triclinic	$Al_2Si_4O_{10}$
2	0.25	27.6803	0.05	Tetragonal	$MgO \cdot Al_2O_3$
3	0.15	32.2546	0.43	Triclinic	$Al_2Si_4O_{10}$
4	1.80	37.8674	0.14	Tetragonal	$CuAl_2$
5	0.95	46.1875	0.24	Monoclinic	$AlCu$
6	0.10	46.7107	0.60	Monoclinic	SiO_2
7	0.35	51.2504	0.43	Triclinic	$Al_2Si_4O_{10}$
8	0.15	56.0038	0.60	Monoclinic	SiO_2
9	0.20	59.7511	0.47	Cubic	$CuAl_2$
10	0.35	64.7793	0.10	Monoclinic	$AlCu$
11	0.20	68.2625	0.31	Triclinic	$Al_2Si_4O_{10}$

TABLE 6: XRD analysis of cast Al sample (with MgFeSi inoculant).

S/N	Peak	Diffraction angle 2θ	Grain size Å	Crystal structures	Phases
1	0.20	18.4317	0.18	Cubic	Al_2CuMg
2	0.10	20.1235	0.15	Cubic	$Al_{15}(MnFe)_3Si_2$
3	0.15	27.3567	0.15	Tetragonal	$MgO \cdot Al_2O_3$
4	0.03	29.5332	0.05	Triclinic	$Al_2Si_4O_{10}$
5	0.70	36.3582	0.04	Triclinic	$Al_2Si_4O_{10}$
6	2.00	36.8058	0.38	Triclinic	$Al_2Si_4O_{10}$
7	0.15	38.0476	0.10	Monoclinic	$AlCu;$
8	0.60	39.2411	0.60	Monoclinic	SiO_2
9	0.45	43.6492	0.34	Triclinic	$Al_2Si_4O_{10}$
10	0.50	47.1035	0.29	Cubic	Al_5FeSi
11	0.15	49.0033	0.25	Cubic	Al_5FeSi
12	0.35	55.0069	0.14	Tetragonal	SiO_4
13	0.15	56.3417	0.21	Tetragonal	Al_2CuMg
14	0.15	65.2541	0.15	Tetragonal	$CuAl_2$
15	0.15	66.7506	0.13	Cubic	Al_2CuMg
16	0.55	71.1864	0.11	Hexagonal	SiO_2

where n is the order of X-ray reflection, d is inter granular space, and λ is X-ray wavelength.

In addition to this, the average grain sizes of phases in the Al alloy samples presented in Tables 5 and 6 are determined by using Scherrer's equation [27] which is given by

$$\tau = \frac{0.9\lambda}{(\beta \cos\theta)}, \qquad (4)$$

where 0.9 is the shape factor, λ is the X-ray wavelength, β is the line amplification at half of the maximum intensity (in radians), θ is Bragg's angle, and τ is the mean size of the ordered (crystalline) domains. 2θ is diffraction angle.

The grain size D is also related to the diffraction angle by

$$D = \frac{0.9\lambda}{\Delta(2\theta)\cos\theta}, \qquad (5)$$

where D is the grain size and λ is the wavelength; 2θ is the diffraction angle. The parameters such as peak values, the diffraction angles, grain sizes, and the crystal structures are analysed and illustrated in Tables 5 and 6 and diffractograms (Figures 15-16). The phases of compounds at diffraction angles (2θ) and peak values which are found to be present in the samples are shown in Tables 5 and 6 and diffractograms (Figures 15-16). At a constant wavelength, the grain sizes (D) of different phase compounds are calculated from (3). Tables 5 and 6 present the XRD analyses for the scrap and cast aluminium alloy samples. By comparing the results in Tables 5 and 6, it is clear that fine grains are present more than coarse grains in the cast sample than in the as-received scrap sample. Fine grains are usually characterised by high BHN values and tensile strength properties as obtained in Figures 3(a)–3(d). Hence, there is no doubt that such cast material will possess better wear resistance property than the as-received scrap Al alloy material as compared with the previous findings [14, 16, 28]. Relatively, sets of 0.05; 0.10–0.14; 0.24; 0.31–0.39; 0.43–0.47; 0.60 Å and 0.04–0.05; 0.10–0.18; 0.21–0.29; 0.34–0.38; and 0.60 Å grain particle sizes were obtained, respectively, for the as-received scrap and the cast samples. Higher fraction was obtained from the 0.04–0.29 Å size than the 0.34–0.60 Å sizes for cast sample.

The overall effects of variation in the pouring temperatures in combination with the moulding sand mixing ratios on the surface morphology, porosity, hardness, and strength of cast samples are illustrated Figures 3–9.

Higher casting temperature causes gassing resulting from the boiling of molten Al alloy and escape of some volatile oxides which increased the metal cast porosity hence producing porous cast with corresponding reduction in strength and hardness measured.

The best set of results were obtained from the cast specimens at $750 \pm 10°C$ using $80:20$ ratio of coarse and fine sand particle sizes mixture as presented in Figure 3. Under this condition, there are sufficient pores in the mould which allow the timely escape of heat, reducing the possibility of gases being entrapped in the metal cast. This explains the reasons for the wear behaviour of the cast Al alloy samples previously reported by the authors [29].

5. Conclusions

The effects of moulding sand particles mixing ratio (with respect to sand permeability) and the pouring temperatures on the hardness, porosity, and microstructures of cast aluminium pistons used in hydraulic brake master cylinder have been studied.

From the combinations of the sand mixtures made from the coarse ($+4750 +300\,\mu m$) and fine ($-300 -75\,\mu m$) moulding sand particle size, $80:20$ ratio gave the best result in terms of the surface morphology. The best of the metallurgical properties such as the hardness, porosity, and microstructure were also obtained from the combination of $80:20$ coarse-fine sand ratio and $750 \pm 10°C$ pouring temperature. An $8:25$ ratio of coarse-grain to fine grain eutectic aluminium alloy was obtained based on the SEM examination results. Higher BHN and strength values were also obtained by inoculating the melt with MgFeSi which forms insoluble compound particles. The SEM image of $100\,\mu m$ size cast alloy poured at $750 \pm 10°C$ shows a large number of fine Al grains formation assuming broom-resembling structures as examined in the SEM image.

Conflict of Interests

The authors declare that there is no conflict of interests regarding the publication of this paper.

Acknowledgments

The authors acknowledge the staff and management of the Premier Wings Engineering Services, Ado Ekiti, for providing the workshop services for the production and preparation of materials used for the study. Electrochemical & Materials Characterization Research Laboratory of the Tshwane University of Technology, Pretoria, South Africa, is also appreciated by the authors for the SEM analysis.

References

[1] O. O. Ajibola and B. O. Jimoh, "Aluminium recycling industries in Nigeria: entrepreneurship challenges and opportunities," in *Proceedings of the 7th Engineering Forum*, vol. 2, pp. 238–247, Ado Ekiti, Nigeria, November 2011.

[2] O. O. Ajibola, B. O. Adewuyi, and D. T. Oloruntoba, "Design and performance evaluation of wear test jig for aluminium alloy substrate in hydraulic fluid," in *Proceedings of the 8th Engineering Forum*, vol. 1, pp. 85–96, Ado Ekiti, Nigeria, October 2012.

[3] J. O. Alasoluyi, J. A. Omotoyinbo, J. O. Borode, S. O. O. Olusunle, and O. O. Adewoye, "Influence of secondary introduction of carbon and ferrosilicon on the microstructure of rotary furnace produced ductile iron," *The International Journal of Science & Technology*, vol. 2, no. 2, pp. 211–217, 2013.

[4] A. M. Hassan, A. Alrashdan, M. T. Hayajneh, and A. T. Mayyas, "Wear behavior of Al–Mg–Cu–based composites containing SiC particles," *Tribology International*, vol. 42, no. 8, pp. 1230–1238, 2009.

[5] J. E. Gruzleski and B. M. Closset, *The Treatment of Liquid Aluminium-Silicon Alloys*, AFS, Des Plaines, Ill, USA, 1st edition, 1990.

[6] P. R. Gibson, A. J. Clegg, and A. A. Das, "Wear of cast Al-Si alloys containing graphite," *Wear*, vol. 95, no. 2, pp. 193–198, 1984.

[7] P. Shanmughasundaram and R. Subramanian, "Wear behaviour of eutectic Al-Si alloy-graphite composites fabricated by combined modified two-stage stir casting and squeeze casting methods," *Advances in Materials Science and Engineering*, vol. 2013, Article ID 216536, 8 pages, 2013.

[8] S. Mahipal, B. Manjinder, S. Rohit, and A. Hitesh, "Behaviour of aluminium alloy casting with the variation of pouring temperature and permeability of sand," *International Journal of Scientific and Engineering Research*, vol. 4, no. 6, pp. 1497–1502, 2013.

[9] B. O. Jimoh, O. O. Ajibola, and S. G. Borisade, "Suitability of selected sand mine in Akure, Nigeria for use as foundry sands in aluminium alloy casting," *The International Journal of Engineering & Technology*, vol. 5, no. 8, pp. 485–495, 2015.

[10] R. J. Fruehan, "Gases in metals," in *Casting*, vol. 15 of *ASM Handbook*, ASM International, Geauga County, Ohio, USA, 9th edition, 1992.

[11] D. Apelian, *Aluminium Cast Alloy: Enabling Tools for Improved Performance*, North American Die Casting Association, Wheeling, Ill, USA, 2009.

[12] R. W. Cahn and P. Haasen, *Physical Metallurgy*, vol. 1, Elsevier, Amsterdam, The Netherlands, 4th edition, 1996.

[13] T. H. Ludwig, P. L. Schaffer, and L. Arnberg, "Influence of some trace elements on solidification path and microstructure of Al-Si foundry alloys," *Metallurgical and Materials Transactions A*, vol. 44, no. 8, pp. 3783–3796, 2013.

[14] O. O. Ajibola, D. T. Oloruntoba, and B. O. Adewuyi, "Metallurgical study of cast aluminium alloy used in hydraulic master brake calliper," *International Journal of Innovation and Scientific Research*, vol. 8, no. 2, pp. 324–333, 2014.

[15] S. Das, S. V. Prasad, and T. R. Ramachandran, "Microstructure and wear of cast (Al-Si alloy)-graphite composites," *Wear*, vol. 133, no. 1, pp. 173–187, 1989.

[16] O. O. Ajibola, B. O. Adewuyi, and D. T. Oloruntoba, "Wear behaviour of sand cast eutectic Al-Si alloy in hydraulic brake fluid," *International Journal of Innovation and Applied Studies*, vol. 6, no. 3, pp. 420–430, 2014.

[17] O. O. Ajibola, D. T. Oloruntoba, and B. O. Adewuyi, "Effects of hard surface grinding and activation on electroless-nickel plating on cast aluminium alloy substrates," *Journal of Coatings*, vol. 2014, Article ID 841619, 10 pages, 2014.

[18] O. O. Ajibola, D. T. Oloruntoba, and B. O. Adewuyi, "Design and performance evaluation of wear test jig for aluminium alloy substrate in hydraulic fluid," in *Proceedings of the Inaugural African Corrosion Congress and Exhibition (AfriCORR '14)*, Pretoria, South Africa, July 2014.

[19] J. H. Perhpezko, "Casting," in *ASM Metals Handbook*, p. 101, ASM International, Geauga County, Ohio, USA, 1988.

[20] D. G. McCartney, "Grain refining of aluminium and its alloys using inoculants," *International Materials Reviews*, vol. 34, no. 1, pp. 247–260, 1989.

[21] American Foundry Society (AFS), "Understanding porosity," in *About AFs and Metal Casting*, American Foundry Society (AFS), Schaumburg, Ill, USA, 2014, http://www.afsinc.org/.

[22] R. A. Hardin and C. Beckermann, "Effect of porosity on mechanical properties of 8630 cast steel," in *Proceedings of the 58th SFSA Technical and Operating Conference*, Paper no. 4, Steel Founders' Society of America, Chicago, Ill, USA, 2004.

[23] Determination of apparent porosity, http://www.che.iitb.ac.in/online/files/MS-207.pdf.

[24] K. Kubo and R. D. Pehlke, "Mathematical modeling of porosity formulation in solidification," *Metallurgical Transaction B*, vol. 16, pp. 359–366, 1985.

[25] D. R. Poirier, "Permeability for flow of interdendritic liquid in columnar-dendritic alloys," *Metallurgical Transaction B*, vol. 18, pp. 245–255, 1987.

[26] https://en.wikipedia.org/wiki/Bragg%27s_law, 2015.

[27] https://en.wikipedia.org/wiki/Scherrer_equation, 2015.

[28] A. B. Gurcan and T. N. Baker, "Wear behaviour of AA6061 aluminium alloy and its composites," *Wear*, vol. 188, no. 1-2, pp. 185–191, 1995.

[29] O. O. Ajibola, D. T. Oloruntoba, and B. O. Adewuyi, "Design and performance evaluation of wear test jig for aluminium alloy substrate in hydraulic fluid," *African Corrosion Journal*, vol. 1, no. 1, pp. 40–45, 2015.

6

Microstructure, Strength, and Fracture Topography Relations in AISI 316L Stainless Steel, as Seen through a Fractal Approach and the Hall-Petch Law

Oswaldo Antonio Hilders,[1] Naddord Zambrano,[2] and Ramón Caballero[3]

[1]*Department of Physical Metallurgy, School of Metallurgical Engineering and Materials Science, Central University of Venezuela, Apartado 47514, Los Chaguaramos, Caracas 1041-A, Distrito Capital, Venezuela*
[2]*Foundation for Professional Development, The Venezuelan College of Engineering, Caracas 1050, Venezuela*
[3]*Failure Analysis Laboratory, School of Metallurgical Engineering and Materials Science, Central University of Venezuela, Apartado 47514, Los Chaguaramos, Caracas 1041-A, Distrito Capital, Venezuela*

Correspondence should be addressed to Oswaldo Antonio Hilders; ohilders@hotmail.com

Academic Editor: Carlos Garcia-Mateo

The influence of the fracture surface fractal dimension D_F and the fractal dimension of grain microstructure D_M on the strength of AISI 316L type austenitic stainless steel through the Hall-Petch relation has been studied. The change in complexity experimented by the net of grains, as measured by D_M, is translated into the respective fracture surface irregularity through D_F, in such a way that the higher the grain size (lower D_M values) the lower the fracture surface roughness (lower values of D_F) and the shallower the dimples on the fractured surfaces. The material was heat-treated at 904, 1010, 1095, and 1194°C, in order to develop equiaxed grain microstructures and then fractured by tension at room temperature. The fracture surfaces were analyzed with a scanning electron microscope, D_F was determined using the slit-island method, and the values of D_M were taken from the literature. The relation between grain size, D_M, mechanical properties, and D_F, developed for AISI 316L steel, could be generalized and therefore applied to most of the common micrograined metal alloys currently used in many key engineering areas.

1. Introduction

Many steels and conventional metallic alloys in general still fill an important place in engineering technology. Although nanocrystalline materials show promise for applications in several fields [1–4], their use is generally restricted for large-scale applications [3]. On the other hand, many important engineering applications of materials involve the use of conventional metallic alloys in polycrystalline form. For these alloys, the individual grains generally ranged between 10 and 300 μm. Conventional metallic alloys are widely used in several engineering areas in which they will be difficult to replace in the near future. The knowledge related with microscopic grained metallic alloys is constantly updated. Some examples of these alloys can be seen in the recent literature [5–13].

In metallic polycrystalline alloys, the relation between the fracture surface features and the underlying microstructure is very well known [14–16]. As the mechanical properties depend on the microstructure, it is clear that the topography of the fractured surfaces is also related to the mechanical properties. On the other hand, in view of the usefulness of the fractal geometry to study the relation between fracture surface tortuosity and mechanical properties [17–27] and that of the Hall-Petch relationship to relate microstructure and mechanical properties [28–32], it is understandable that the microstructure-fracture topography-mechanical properties relationship can be studied by combining both approaches. So far only a few bridges have been built between these two approaches (see, e.g., [28, 33]). The aim of this work is to establish a quantitative correlation between the fracture surface fractal dimension D_F (a measure of tortuosity of a fracture surface) and the fractal dimension of grain microstructure D_M (a measure of complexity of the internal net of grains) in AISI 316L type steel, as both can be related through the Hall-Petch law. The link between D_M and D_F can provide an understanding of the role of microstructure on the mechanics of crack propagation. This link arises because microstructure has a major influence on the topography of the fracture surface. The correlation between D_F and D_M could be a useful tool to analyze the connection among microstructure, design, fabrication, and performance, in both conventional [28–32] and nanocrystalline metallic alloys [1, 2, 4, 34, 35].

2. Materials and Methods

The material used in this work was austenitic type AISI 316L stainless steel fabricated into hot-rolled bar with diameter of 25.4 mm provided by a commercial supplier. The chemical composition of the steel is 16.9Cr, 12.0Ni, 2.52Mo, 1.5Mn, 0.35Si, 0.025C, 0.035N, 0.030P, and 0.030S (wt.%). Four slices and eight tensile samples of 25.4 mm gage length were taken from the as-received bar and heat-treated at four different temperatures: 904, 1010, 1095, and 1194°C, in order to develop equiaxed grain microstructures (one slice and two tensile samples for each temperature). The temperatures were selected according to the work of Colás [36]. After the heat treatments, the slices and tensile samples were water-quenched at room temperature. Then, the slices were ground and polished by standard metallographic methods, while the microstructure was revealed by electrolytic etching.

An automatic image analyzer was used in order to perform grain size measurements according to the mean linear intercept method. At least ten different fields of view were analyzed for each metallographic sample. Before performing the measurements of grain size, the grain boundaries were extracted and enhanced by means of image processing techniques [37]. Briefly, well-defined grain boundaries were obtained transforming our 256 gray level images to two gray values: black and white (thresholding). Then, a specialized operation that prevents the separation of grain boundaries while eroding away pixels is performed (skeletonization). Finally, an image processing was done to eliminate impurities, particles, and so forth in the grain interiors (hole filling).

The tensile samples were deformed at room temperature at a nominal strain rate of 3.5×10^{-4}/s in an Instron tensile machine until fracture. A 10 mm section from both the cup and the cone portions of fractured tension samples was removed and the fracture surfaces were analyzed with a scanning electron microscope (SEM), which was operated at 20 Kv. The fractographic features were studied in the central region of the cup portion of broken samples using several micrographs for each case.

The values of the fracture surface fractal dimension D_F have been determined using the so-called slit-island method (SIM) [17, 19, 24, 38–41] in the central region of the respective cone portions. Each cone was cold molding using epoxy resin, which was pouring over the sample (which was previously attached to a cylindrical support of convenient size). Each

TABLE 1: Average grain size and D_F data.

T (°C)	904	1010	1095	1194
d (μm)	21.4 ± 3.1	27.2 ± 3.3	67.9 ± 4.9	148.8 ± 7.8
D_F	1.281	1.260	1.225	1.142

sample was positioned face up, allowing the epoxy to cover all the fracture surface. Grinding and polishing operations were performed parallel to the mean plane of fracture, developing a number of successive layers in which part of the fracture surface becomes visible ("islands"). As the layers increased in number, the islands do, and growth and coalescence of islands take place. For a particular jth layer with n islands, Pi and Ai represent the perimeter and the area of the ith island, respectively. Taking into account all the islands in this layer, the total perimeter and the total area are ΣPi and ΣAi, respectively. For all the layers, a full logarithmic scale diagram of ΣPi versus ΣAi leads to obtaining a straight curve, from which D_F = 2 × slope. Figure 1 shows an example of a sequence of 4 nonconsecutive partial layers (out of 26), to calculate the value of D_F = 1.142.

On the other hand, the values of the microstructural fractal dimension were taken from the work of Colás [36]. In order to estimate the values of D_M, Colás employed the box-counting method [42]. In this method, a square grid containing boxes of a given side length h is superimposed on the grain boundary pattern. Then, the number of boxes $N(h)$ containing boundary contours is counted. This process is repeated to find $N(h)$ for smaller values of h. Asymptotically, in the limit of small h,

$$N(h) = N_o h^{-D_M}, \tag{1}$$

where N_o is a constant. For a fractal pattern, the slope of the straight curve $\log N(h)$ versus $\log h$ is the microstructural fractal dimension D_M whose values are between 1.0 and 2.0.

3. Results and Discussion

3.1. The Relation between Yield Stress and D_F

3.1.1. *Grain Size-D_F Relationship.* Figure 2 shows optical micrographs of the microstructures of AISI 316L steel heat-treated at four different temperatures. Figure 3 shows the enhanced microstructures of AISI 316L steel after the image processing.

The data of fracture surface fractal dimension D_F and average grain size d are listed in Table 1. On the other hand, Figure 4 shows the fracture surface fractal dimension D_F plotted against the average grain size d. As can be seen, there is a negative linear correlation between D_F and d, that is, higher fracture surface fractal dimension for lower average grain size. The corresponding equation is

$$D_F = \lambda - \beta d, \tag{2}$$

where λ = 1.30 and β = 0.001/μm. Equation (2) represents the connection between the microstructure (grain size) and the irregularity of the fracture surface (measured by D_F). The

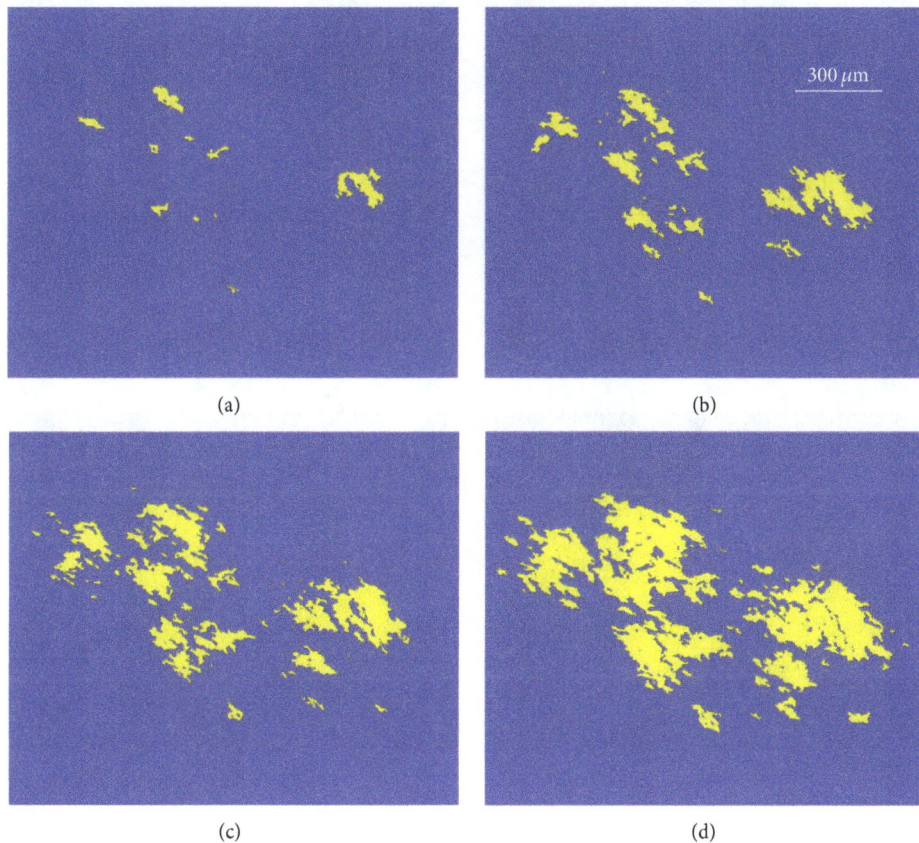

FIGURE 1: Optical micrographs of metallic islands of AISI 316L stainless steel, developed according to the SIM. (a) Layer number 1, (b) layer number 7, (c) layer number 14, and (d) layer number 24.

results predicted in Figure 4 are consistent with the general observation that grain size reduction is a means to increase the toughness of a metallic alloy [43, 44], since, as many studies have been strongly supported, the higher the fracture surface fractal dimension is, the higher the toughness is [19, 21, 39, 45–47].

3.1.2. Fracture Surface Characteristics. Figure 5 shows several fractographs of the fractured tensile samples, which correspond to the four heat treatment temperatures. The microvoid coalescence mechanism of separation was observed for all experimental conditions. Although the values of D_F increase as the grain size decreases according to Figure 4, the corresponding values of dimple size (as seen on the mean plane of fracture in Figure 5) were somewhat the same, which implies, in principle, that the dimple size (as measured by the surface dimple diameter) is not related with the grain size. In view of this fact, some factor must exist for the decrease in D_F as the grain size increases.

As D_F is a measure of the irregularity of the fracture surface, it is suggested that the dimples become shallow as the grain size increases (lower values of D_F) which was confirmed by in situ extensive analysis (SEM). This can be checked in Figure 5, at least for the extreme values of grain size developed in the present work: Figure 5(a): lower grain size ("deep dimples") and Figure 5(d): higher grain size ("shallow dimples").

The last view is supported by the fact that for smaller grain size the plastic deformation spreads out to the microstructure more easily than for larger grain size (smaller grain material stores more energy than larger grain material), creating a rougher fracture surface ("deep dimples") with a higher fractal dimension D_F. The rougher the fracture surface, the higher the stored energy and the tougher the material. Obviously, for this case the internal area of the "deep dimples" is also higher. Note that for a totally brittle material (which is not the case in any of the studied conditions) the absorbed energy is zero, and the fracture surface is flat.

Currently, a relationship between grain size and the deep of dimples in ductile fracture can be established indirectly, through the toughness. Note that, for a tougher material tested in tension, the plasticity is higher and the dimples are more enlarged in the axial direction ("deep dimples"). Figure 6 can illustrate these concepts.

On the other hand, the reason for dimples to remain about the same diameter is that dimple size is controlled by the size and population of particles (precipitates and/or inclusions) in the interior of grains [48, 49]. After the nucleation of dimples begins from particles, their size increases until the coalescence with other dimples, which inhibits an additional growth. Provided the density and size of particles were the same for all experimental conditions, the corresponding average dimple size becomes roughly the same.

FIGURE 2: Optical micrographs of microstructures of AISI 316L austenitic stainless steel heat-treated at (a) 904°C, (b) 1010°C, (c) 1095°C, and (d) 1194°C.

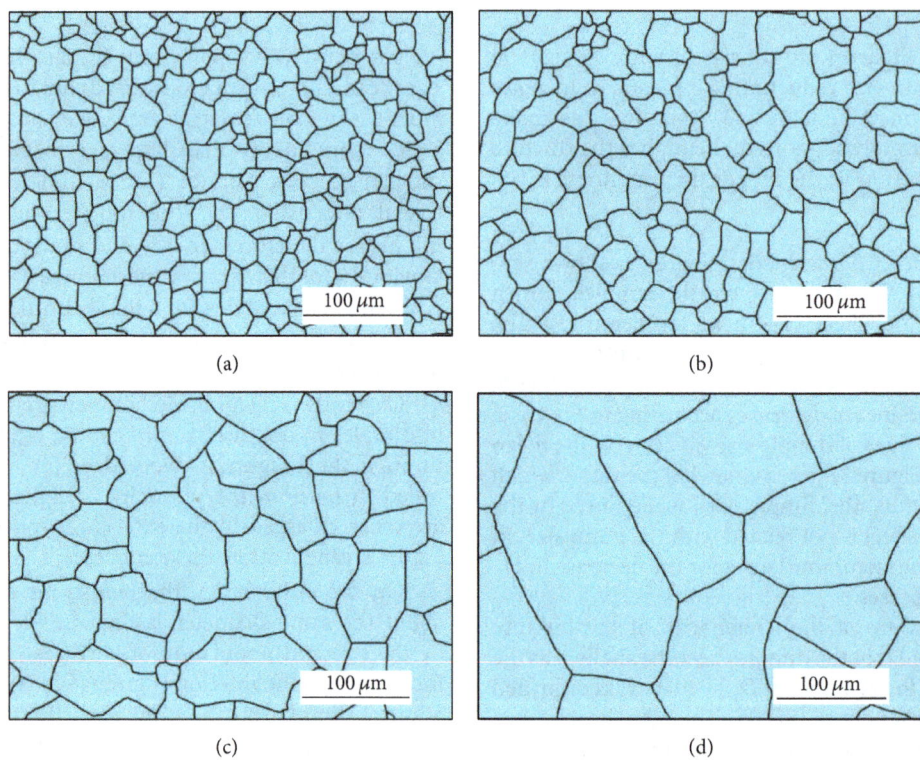

FIGURE 3: The processed images of grain microstructures of AISI 316L austenitic stainless steel corresponding to the micrographs of Figure 2.

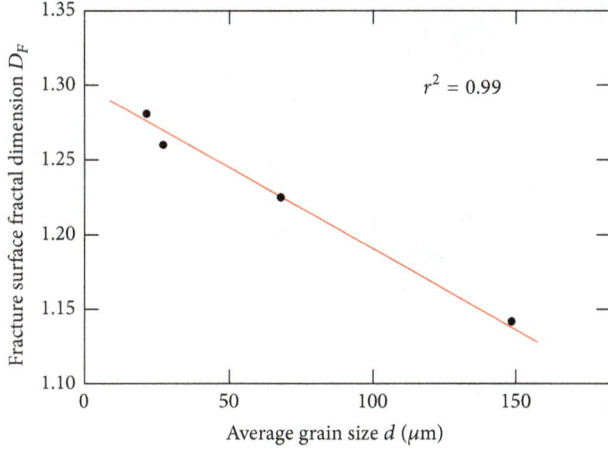

FIGURE 4: Variation of the fracture surface fractal dimension with the average grain size for AISI 316L austenitic stainless steel.

TABLE 2: Average grain size and D_M data [36].

T (°C)	904	1010	1095	1194
d (μm)	19.6 ± 3.6	29.6 ± 4.8	59.3 ± 5.2	158.4 ± 8.2
D_M	1.504	1.470	1.423	1.365

The value of $\sigma_y = 207.46$ MPa represents the yield stress for AISI 316L steel broken in tension, whose fracture surface is totally flat ($D_F = 1$). Theoretically, for this case the value of d should be 300 μm, (see (2)). For the curve σ_y versus D_F in Figure 7, the limit conditions are the same, although the approach to $\sigma_y = \infty$ as $D_F \to 1.30$ is of a nonlinear nature. The general Hall-Petch type relationship between σ_y and D_F (see (4)) can be potentially useful to relate the fracture surface fractal dimension D_F with mechanical properties in many commercial alloys.

The relative absence of particles inside the dimples in Figure 5 could be related with one or more of several factors: some particles remain attached to the matting fracture surface, some particles were lost during the fracture event, or simply some voids (few of them) nucleate homogeneously. Note that the relation between grain size and D_F is easier to explain for the case of intergranular fracture (which was not obtained in any case in the present work). For intergranular separation, as the path of the fracture surface follows the contour of grains, a lower grain size material will have a higher area of grains and correspondingly a higher area of the fracture surface. In this case, the value of D_F will be higher too.

3.2. The Relation between Yield Stress and D_M

3.2.1. Grain Size-D_M Relationship. According to Colás [36], the relationship between the average grain size d and its microstructural fractal dimension D_M for AISI 316L can be described by means of the following equation:

$$D_M = \alpha d^{-\eta}, \tag{5}$$

where $\alpha = 1.716\ \mu$m$^\eta$ and $\eta = 0.045$. This equation predicts an increase in D_M as the grain size decreases. The data of microstructural fractal dimension and the range of the investigated average grain size upon which (5) was developed are listed in Table 2 [36].

3.2.2. Hall-Petch Type Relation for D_M. According to (5), the average grain size is $d = (\alpha/D_M)^{1/\eta}$. This microstructural fractal dimension dependence for d is substituted into (3), to predict the yield stress as a function of D_M according to

$$\sigma_y = \sigma_o + k'' D_M^{1/(2\eta)}, \tag{6}$$

where $k'' = k/\alpha^{1/(2\eta)}$ is a new constant. From the values of k, α, and η, $k'' = 1.91$ MPa. It can be seen that the yield stress increases as the microstructural fractal dimension increases, which in turn means a decrease in the average grain size. For the present case, and based on (6), once again two Hall-Petch type relations can be plotted (Figure 8): a linear relation σ_y versus $(D_M)^{1/(2\eta)}$, ($\eta = 0.045$), and σ_y versus D_M. Three subscales for the variable $(D_M)^{1/(2\eta)}$ have been used in Figure 8 in order to preserve a natural arithmetic scale, which facilitates a good visualization of the fractal dimension values. This is performed according to the "level" of D_M. The first zone is defined for $1.0 \le (D_M)^{1/(2\eta)} \le 10$, ($1.0 \le D_M \le 1.23$), so from (6) two values of σ_y can be defined, $\sigma_y = 164.91$ MPa, $((D_M)^{1/(2\eta)} = 1, D_M = 1)$ and $\sigma_y = 182.1$ MPa, $((D_M)^{1/(2\eta)} = 10, D_M = 1.23)$; thus, the curve for this zone can be traced. It is suggested that the first zone could be identified with low complex microstructures. The second zone ranged between $(D_M)^{1/(2\eta)} = 10$ and some value around 100. The last limit can

3.1.3. Hall-Petch Type Relation for D_F.

The relation between the fracture surface fractal dimension and mechanical properties has been established through the Hall-Petch equation:

$$\sigma_y = \sigma_o + kd^{-1/2}, \tag{3}$$

where σ_y is the yield stress, σ_o is the friction stress which opposes dislocation motion, k is a constant related with the difficulty in spreading yielding from grain to grain, and d is the average grain size. From (2) the average grain size is $d = (\lambda - D_F)/\beta$, and then (3) is therefore rearranged to predict the yield stress as

$$\sigma_y = \sigma_o + k' \left(\lambda - D_F\right)^{-1/2}, \tag{4}$$

where $k' = k\beta^{1/2}$ is a constant. For AISI 316L $\sigma_o = 163$ MPa and $k = 0.77$ MPa m$^{1/2}$ [50], k' becomes 24.35 MPa. Then, smaller grain size corresponds to higher fracture surface fractal dimension and so to higher yield stress. Based on (4), two Hall-Petch type relations have been represented in Figure 7: a linear relation of σ_y versus $(\lambda - D_F)^{-1/2}$, ($\lambda = 1.30$), and σ_y versus D_F. In the first case, the theoretical values corresponding to $(\lambda - D_F)^{-1/2}$ ranged between 1.83 ($D_F = 1$) and ∞ ($D_F = 1.30$), being the values of σ_y, 207.46 MPa, and ∞, respectively.

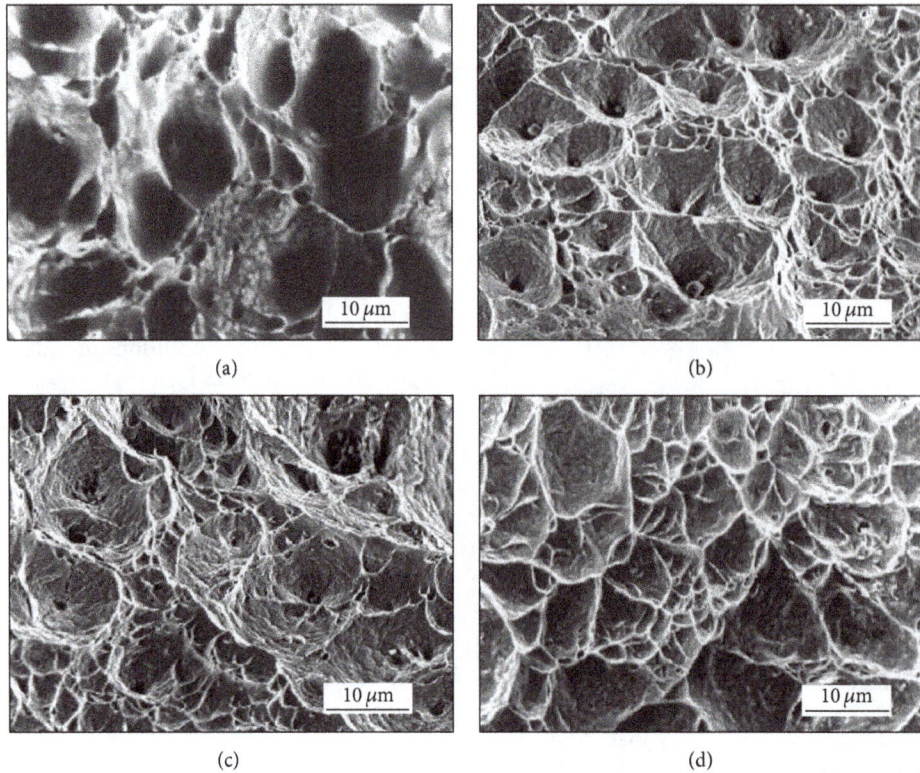

FIGURE 5: Scanning electron fractographs of fractured tensile samples heat-treated at four different temperatures: (a) 904°C: $d = 21.4\,\mu$m, $D_F = 1.281$; (b) 1010°C: $d = 27.2\,\mu$m, $D_F = 1.260$; (c) 1095°C: $d = 67.9\,\mu$m, $D_F = 1.225$; (d) 1194°C: $d = 148.8\,\mu$m, $D_F = 1.142$.

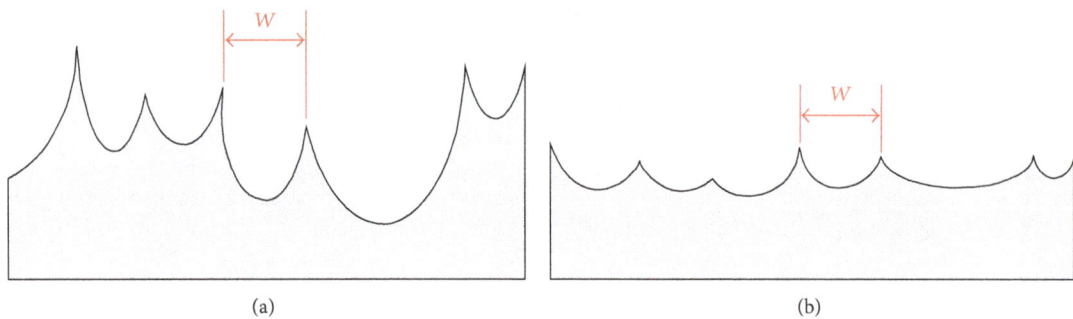

FIGURE 6: Ductile tension fracture surface profiles, showing the same average dimple size W (axial direction is $\updownarrow$). (a) "Deep dimples" for smaller grain size, in a tougher material with a higher D_F. (b) "Shallow dimples" for a higher grain size, in a material with smaller toughness and lower D_F.

move more or less freely in a narrow range, since it represents an uncertainty of the value for $(D_M)^{1/(2\eta)}$ (and therefore, the value of D_M) from which the related microstructure starts to be very complex in the third zone. For $(D_M)^{1/(2\eta)} = 100$, $D_M = 1.51$, so for the second zone, $1.23 \leq D_M \leq 1.51$. The second zone can represent microstructures with an average complexity. The curve for this zone can be traced using (6) for any two values of the corresponding scale. The curve for the third zone can be defined using one more time (6) and any two values of the third scale, for example, $(D_M)^{1/(2\eta)} = 2,212$ $(D_M = 2)$, which corresponds to $\sigma_y = 4,387.84$ MPa, and $(D_M)^{1/(2\eta)} = 2,000$ $(D_M = 1.98)$, for $\sigma_y = 3,983$ MPa.

The value of $\sigma_y = 164.91$ MPa represents the yield stress for AISI 316L steel broken in tension, for a grain size of $\approx 162,740.33\,\mu$m ≈ 16 cm (see (5)). We can write, as a first approximation (see (6)), that for $D_M = 1$, $\sigma_y \approx \sigma_o = 163$ MPa. Correspondingly, we could consider the material as an individual grain of 16 cm (an infinite system as compared to our real grains), which is consistent with the notion of σ_o as a friction stress below which dislocations will not move in the material in the absence of grain boundaries. For $D_M = 2$, $d = 0.03\,\mu$m (see (5)) and $\sigma_y = 4,387.84$ MPa. From a theoretical point of view (see (6)), this microstructure can be related to such a high value of σ_y. Truly, a loss of strengthening for

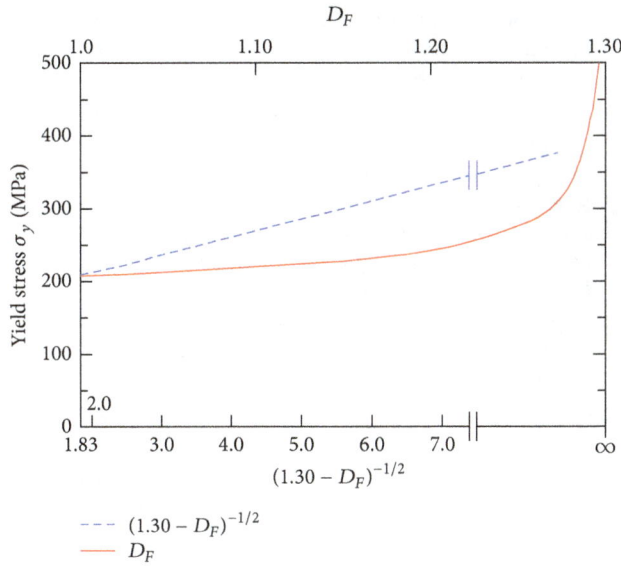

FIGURE 7: Hall-Petch type relations between yield stress and fracture surface fractal dimension for AISI 316L austenitic stainless steel.

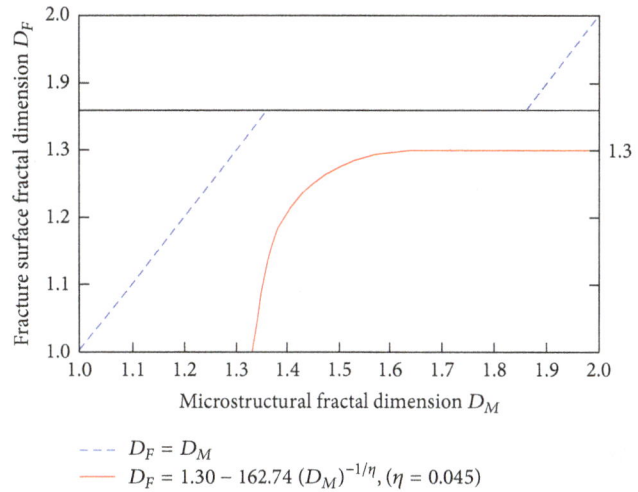

FIGURE 8: Hall-Petch type relations between yield stress and microstructural fractal dimension for AISI 316L austenitic stainless steel.

$d = 0.03\,\mu$m (30 nm) which falls into the so-called "inverse Hall-Petch dependence zone" [51, 52] should occur. For the curve σ_y versus D_M in Figure 8, two arithmetic subscales have been introduced which encompass the full theoretical range of D_M ($1 \le D_M \le 2$). The natural link between the fractal dimension of grain boundaries and mechanical properties has been proven to be very important in metallic materials engineering [53–55]. On the other hand, the Hall-Petch type relationship between σ_y and D_M (see (6)) facilitates the comprehension of this link.

3.3. *Relation between D_F and D_M*. The relation between the fracture surface fractal dimension D_F and the microstructural fractal dimension D_M can be found by equating (2) and (5), which leads to

$$D_F = \lambda - \beta\,(\alpha)^{1/\eta}\,(D_M)^{-1/\eta} \qquad (7)$$

which in turn, taking into account the values of the constants α, β, η and λ, gives

$$D_F = 1.30 - 162.74\,(D_M)^{-22.22}, \qquad (8)$$

where $22.22 \approx 1/0.045$ and 162.74 is a constant without dimensions. The relation between D_F and D_M is represented in Figure 9 and compared with $D_F = D_M$. As can be seen, the values of D_F are smaller than the values of D_M, being the limit values: $D_M = 1.328$ for $D_F = 1$ and $D_M = 2$ for $D_F = 1.30$. The experimental values for D_M and D_F ranged between the intervals $1.365 \le D_M \le 1.504$ and $1.142 \le D_F \le 1.281$ as have been quoted in Tables 1 and 2, respectively.

The very nature of the relationship between D_M and D_F possesses great difficulties in analysis. Nevertheless, the present results confirm that an increase in D_M or D_F involves an increase in the yield stress as the grain size becomes small. Although both the box counting method [42], which

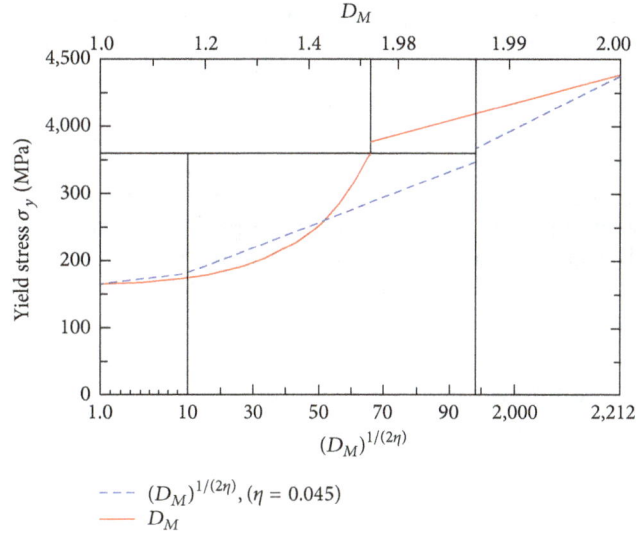

FIGURE 9: Microstructural fractal dimension D_M versus fracture surface fractal dimension D_F as compared with $D_F = D_M$.

has been used by Colás [36] to determine D_M, and the slit-island method [17, 19, 24, 38, 39, 41], used in the present work to determine D_F, are based on 2D metallographic image obtained by grinding the specimen surface flat, the kind of microstructures in which they were applied is essentially different. Note that the different methods to determine D_F are, in theory, equivalent, but the SIM method was selected because it is more suitable for the analysis of a rough surface. In addition, a great part of the D_F data in the literature is based on this method, which facilitates comparison. In spite of the above, the changes in D_F and D_M for the corresponding range of grain sizes were the same: $\Delta D_F = (1.281) - (1.142) = 0.139$ and $\Delta D_M = (1.504) - (1.365) = 0.139$ (Tables 1 and 2). Although these results can be regarded as fortuitous, they suggest, in principle, that the increase in complexity experienced by the net of grains between 1194°C and 904°C (increasing the value of D_M) was completely translated into the respective fracture surface. No previous results for the correlation between D_M and D_F exist for AISI 316L stainless steel, which makes a comparison difficult to achieve.

Estimating the influence of the microstructural fractal dimension on the fracture surface fractal dimension can be very important, theoretically and in practical applications. The connection between fractal characteristics of materials and mechanical properties can be easily established through equations such as (4) and (6).

4. Conclusions

From the present study, it can be seen that the fracture surface fractal dimension D_F and the fractal dimension of the grain microstructure D_M in AISI 316L austenitic stainless steel can be related to the strength of the material through the Hall-Petch law, which provides a well-established and sound platform to study and analyze this relation. The present results indicate the strong interplay between micrograins (microstructure), yield stress (mechanical property), and fracture topography (fracture behavior) for the studied material. The increase in complexity of the microstructure as the grain size decreases is measured by D_M and translated into the fracture surfaces, which become more irregular as indicated by the values of D_F. The relation between grain size, D_M, mechanical properties, and D_F, developed for AISI 316L steel, should be generalized and applied to most of the commercial metallic alloys of technological importance.

Conflict of Interests

The authors declare that there is no conflict of interests regarding the publication of this paper.

Acknowledgments

The authors would like to thank the staff of the Electron Microscopy Center of the Faculty of Science of the Central University of Venezuela for their assistance, the Ferrum C.A. for providing test material, and the Scientific and Humanistic Development Council of the Central University of Venezuela for financial support.

References

[1] G. Cao and Y. Wang, *Nanostructures and Nanomaterials: Synthesis, Properties and Applications*, World Scientific Publishing, Singapore, 2011.

[2] H. Conrad and J. Narayan, "Mechanisms governing the plastic deformation of nanocrystalline materials, including grain-size softening," in *Mechanical Properties of Nanocrystalline Materials*, J. C. M. Li, Ed., Pan Stanford Publishing, Danvers, Mass, USA, 2011.

[3] H. K. D. H. Bhadeshia, "The first bulk nanostructured metal," *Science and Technology of Advanced Materials*, vol. 14, no. 1, Article ID 014202, 2013.

[4] T. J. Rupert, "Solid solution strengthening and softening due to collective nanocrystalline deformation physics," *Scripta Materialia*, vol. 81, pp. 44–47, 2014.

[5] S. Huang, Y.-X. Zhao, and C.-F. He, "Shear fracture of advanced high strength steels," *Journal of Iron and Steel Research International*, vol. 21, no. 10, pp. 938–944, 2014.

[6] F. Barlat, G. Vincze, J. J. Grácio, M.-G. Lee, E. F. Rauch, and C. N. Tomé, "Enhancements of homogenous anisotropic hardening model and application to mild and dual-phase steels," *International Journal of Plasticity*, vol. 58, pp. 201–218, 2014.

[7] I. B. Okipnyi, P. O. Maruschak, V. I. Zakiev, and V. S. Mocharskyi, "Fracture mechanism analysis of the heat-resistant steel 15Kh2MFA(II) after laser shok-wave processing," *Journal of Failure Analysis and Prevention*, vol. 14, no. 5, pp. 668–674, 2014.

[8] J. C. Stinville, J. Cormier, C. Templier, and P. Villechaise, "Modeling of the lattice rotations induced by plasma nitriding of 316L polycrystalline stainless steel," *Acta Materialia*, vol. 83, pp. 10–16, 2015.

[9] H. Mirzadeh, "Constitutive behaviors of magnesium and Mg-Zn-Zr alloy during hot deformation," *Materials Chemistry and Physics*, vol. 152, pp. 123–126, 2015.

[10] M. Jobba, R. K. Mishra, and M. Niewczas, "Flow stress and work-hardening behaviour of Al-Mg binary alloys," *International Journal of Plasticity*, vol. 65, pp. 43–60, 2015.

[11] T. Dursun and C. Soutis, "Recent developments in advanced aircraft aluminium alloys," *Materials and Design*, vol. 56, pp. 862–871, 2014.

[12] M. Harooni, J. Ma, B. Carlson, and R. Kovacevic, "Two-pass laser welding of AZ31B magnesium alloy," *Journal of Materials Processing Technology*, vol. 216, pp. 114–122, 2014.

[13] B. Chen, S. Cao, H. Xu, Y. Jin, S. Li, and B. Xiao, "Effect of processing parameters on microstructure and mechanical properties of 90W-6Ni-4Mn heavy alloy," *International Journal of Refractory Metals and Hard Materials*, vol. 48, pp. 293–300, 2015.

[14] A. Srivastava, L. Ponson, S. Osovski, E. Bouchaud, V. Tvergaard, and A. Needleman, "Effect of inclusion density on ductile fracture toughness and roughness," *Journal of the Mechanics and Physics of Solids*, vol. 63, pp. 62–79, 2014.

[15] R. Bidulský, J. Bidulská, and M. A. Grande, "Correlation between microstructure/fracture surfaces and material properties," *Acta Physica Polonica A*, vol. 122, no. 3, pp. 548–552, 2012.

[16] M. May, *A new approach to discovering the fundamental mechanisms of hydrogen failure [Ph.D. thesis]*, University of Illinois Urbana-Champaign, Champaign, Ill, USA, 2013.

[17] B. B. Mandelbrot, D. E. Passoja, and A. J. Paullay, "Fractal character of fracture surfaces of metals," *Nature*, vol. 308, no. 5961, pp. 721–722, 1984.

[18] R. H. Dauskardt, F. Haubensak, and R. O. Ritchie, "On the interpretation of the fractal character of fracture surfaces," *Acta Metallurgica et Materialia*, vol. 38, no. 2, pp. 143–159, 1990.

[19] V. Y. Milman, N. A. Stelmashenko, and R. Blumenfeld, "Fracture surfaces: a critical review of fractal studies and a novel morphological analysis of scanning tunneling microscopy measurements," *Progress in Materials Science*, vol. 38, no. 1, pp. 425–474, 1994.

[20] O. A. Hilders, L. Sáenz, M. Ramos, and N. D. Peña, "Effect of 475°C embrittlement on fractal behavior and tensile properties

of a duplex stainless steel," *Journal of Materials Engineering and Performance*, vol. 8, no. 1, pp. 87–90, 1999.

[21] O. A. Hilders, M. Ramos, N. D. Peña, and L. Sàenz, "Fractal geometry of fracture surfaces of a duplex stainless steel," *Journal of Materials Science*, vol. 41, no. 17, pp. 5739–5742, 2006.

[22] M. Ostoja-Starzewski and J. Li, "Fractal materials, beams, and fracture mechanics," *Zeitschrift für Angewandte Mathematik und Physik*, vol. 60, no. 6, pp. 1194–1205, 2009.

[23] W. Tang and Y. Wang, "Fractal characterization of impact fracture surface of steel," *Applied Surface Science*, vol. 258, no. 10, pp. 4777–4781, 2012.

[24] L. R. Carney and J. J. Mecholsky, "Relationship between fracture toughness and fracture surface fractal dimension in AISI 4340 steel," *Materials Sciences and Applications*, vol. 4, no. 4, pp. 258–267, 2013.

[25] A. Vinogradov, I. S. Yasnikov, and Y. Estrin, "Stochastic dislocation kinetics and fractal structures in deforming metals probed by acoustic emission and surface topography measurements," *Journal of Applied Physics*, vol. 115, no. 23, Article ID 233506, 2014.

[26] O. A. Hilders and N. Zambrano, "The effect of aging on impact toughness and fracture surface fractal dimension in SAF 2507 super duplex stainless steel," *Journal of Microscopy and Ultrastructure*, vol. 2, no. 4, pp. 236–244, 2014.

[27] G. G. Savenkov, B. K. Barakhtin, and K. A. Rudometkin, "Multifractal analysis of structural modifications in a cumulative jet," *Technical Physics*, vol. 60, no. 1, pp. 96–101, 2015.

[28] N. Hirota, F. Yin, T. Azuma, and T. Inoue, "Yield stress of duplex stainless steel specimens estimated using a compound Hall-Petch equation," *Science and Technology of Advanced Materials*, vol. 11, no. 2, pp. 1–11, 2010.

[29] N. Stanford, U. Carlson, and M. R. Barnett, "Deformation twinning and the Hall-Petch relation in commercial purity Ti," *Metallurgical and Materials Transactions A*, vol. 39, no. 4, pp. 934–944, 2008.

[30] H. Izadi, R. Sandstrom, and A. P. Gerlich, "Grain growth behavior and Hall–Petch strengthening in friction stir processed Al 5059," *Metallurgical and Materials Transactions A: Physical Metallurgy and Materials Science*, vol. 45, no. 12, pp. 5635–5644, 2014.

[31] S. Thangaraju, M. Heilmaier, B. S. Murty, and S. S. Vadlamani, "On the estimation of true Hall-Petch constants and their role on the superposition law exponent in Al alloys," *Advanced Engineering Materials*, vol. 14, no. 10, pp. 892–897, 2012.

[32] W. Yuan, S. K. Panigrahi, J.-Q. Su, and R. S. Mishra, "Influence of grain size and texture on Hall–Petch relationship for a magnesium alloy," *Scripta Materialia*, vol. 65, no. 11, pp. 994–997, 2011.

[33] S. Hui, Z. Yugui, and Y. Zhenqi, "Fractal analysis of microstructures and properties in ferrite-martensite steels," *Scripta Metallurgica et Materiala*, vol. 25, no. 3, pp. 651–654, 1991.

[34] R. W. Armstrong, "Engineering science aspects of the Hall–Petch relation," *Acta Mechanica*, vol. 225, no. 4-5, pp. 1013–1028, 2014.

[35] M. A. Tschopp, H. A. Murdoch, L. J. Kecskes, and K. A. Darling, "'Bulk' nanocrystalline metals: review of the current state of the art and future opportunities for copper and copper alloys," *Journal of the Minerals, Metals & Materials Society*, vol. 66, no. 6, pp. 1000–1019, 2014.

[36] R. Colás, "On the variation of grain size and fractal dimension in an austenitic stainless steel," *Materials Characterization*, vol. 46, no. 5, pp. 353–358, 2001.

[37] M. S. Nixon and A. S. Aguado, *Feature Extraction & Image Processing for Computer Vision*, Academic Press, London, UK, 2012.

[38] L. V. Meisel, "Perimeter-area analysis, the slit-island method and the fractal characterization of metallic fracture surfaces," *Journal of Physics D: Applied Physics*, vol. 24, no. 6, pp. 942–952, 1991.

[39] O. A. Hilders and D. Pilo, "On the development of a relation between fractal dimension and impact toughness," *Materials Characterization*, vol. 38, no. 3, pp. 121–127, 1997.

[40] O. A. Hilders, M. Ramos, N. D. Peña, L. Sáenz, and R. A. Caballero, "Plasticity-fractal-behavior trends for different aluminum alloys tested in tension," in *Aluminium Alloys Their Physical and Mechanical Properties*, J. F. Nie, A. J. Morton, and B. C. Muddle, Eds., The Institute of Materials Engineering Australasia, Brisbane, Australia, 2004.

[41] P. Mazurek and D. O. Mazurek, "From the slit-island method to the ising model: analysis of irregular grayscale objects," *International Journal of Applied Mathematics and Computer Science*, vol. 24, no. 1, pp. 49–63, 2014.

[42] J. Feder, *Fractals*, Plenum Press, New York, NY, USA, 1988.

[43] C. Wang, M. Wang, J. Shi, W. Hui, and H. Dong, "Effect of microstructure refinement on the strength and toughness of low alloy martensitic steel," *Journal of Materials Science & Technology*, vol. 23, no. 5, pp. 659–664, 2007.

[44] Z. Fan, "The grain size dependence of ductile fracture toughness of polycrystalline metals and alloys," *Materials Science and Engineering A*, vol. 191, no. 1-2, pp. 173–183, 1995.

[45] J. M. Li, L. Lü, M. O. Lai, and B. Ralph, *Image-Based Fractal Description of Microstructures*, Kluwer Academic Publishers, Norwell, Mass, USA, 2003.

[46] Z. Q. Mu and C. W. Lung, "Studies on the fractal dimension and fracture toughness of steel," *Journal of Physics D: Applied Physics*, vol. 21, no. 5, pp. 848–850, 1988.

[47] D. Shi, J. Jiang, E. Tian, and C. Lung, "Perimeter-area relation and fractal dimension of fracture surfaces," *Journal of Materials Science & Technology*, vol. 13, no. 5, pp. 416–420, 1997.

[48] D. Ma, D. Chen, S. Wu, H. Wang, C. Cai, and G. Deng, "Dynamic experimental verification of void coalescence criteria," *Materials Science and Engineering A*, vol. 533, pp. 96–106, 2012.

[49] W. M. Garrison Jr. and A. L. Wojcieszynski, "A discussion of the spacing of inclusions in the volume and of the spacing of inclusion nucleated voids on fracture surfaces of steels," *Materials Science and Engineering A*, vol. 505, no. 1-2, pp. 52–61, 2009.

[50] C. S. Kusko, J. N. Dupont, and A. R. Marder, "The influence of microstructure on fatigue crack propagation behavior of stainless steel welds," *Welding Journal*, vol. 83, no. 1, pp. 6–14, 2004.

[51] M. A. Meyers, A. Mishra, and D. J. Benson, "Mechanical properties of nanocrystalline materials," *Progress in Materials Science*, vol. 51, no. 4, pp. 427–556, 2006.

[52] K. A. Padmanabhan, S. Sripathi, H. Hahn, and H. Gleiter, "Inverse Hall–Petch effect in quasi- and nanocrystalline materials," *Materials Letters*, vol. 133, pp. 151–154, 2014.

[53] E. Hornbogen, "Fractals in microstructure of metals," *International Materials Reviews*, vol. 34, no. 6, pp. 277–296, 1989.

[54] R. J. Mitchell, H. Y. Li, and Z. W. Huang, "On the formation of serrated grain boundaries and fan type structures in an advanced polycrystalline nickel-base superalloy," *Journal of Materials Processing Technology*, vol. 209, no. 2, pp. 1011–1017, 2009.

[55] K. J. Kim, H. U. Hong, and S. W. Nam, "Investigation on the formation of serrated grain boundaries with grain boundary characteristics in an AISI 316 stainless steel," *Journal of Nuclear Materials*, vol. 393, no. 2, pp. 249–253, 2009.

7

Experimental Investigation of the Corrosion Behavior of Friction Stir Welded AZ61A Magnesium Alloy Welds under Salt Spray Corrosion Test and Galvanic Corrosion Test Using Response Surface Methodology

A. Dhanapal,[1] **S. Rajendra Boopathy,**[2] **V. Balasubramanian,**[3]
K. Chidambaram,[1] **and A. R. Thoheer Zaman**[1]

[1] Department of Mechanical Engineering, Sri Ramanujar Engineering College, Vandalur, Chennai, Tamil Nadu 600 048, India
[2] Department of Mechanical Engineering, College of Engineering, Anna University, Chennai 600 025, India
[3] Center for Materials Joining & Research (CEMAJOR), Department of Manufacturing Engineering, Annamalai University, Annamalai Nagar, Chidambaram 608 002, India

Correspondence should be addressed to A. Dhanapal; sridhanapal2010@gmail.com

Academic Editor: Chi Tat Kwok

Extruded Mg alloy plates of 6 mm thick of AZ61A grade were butt welded using advanced welding process and friction stir welding (FSW) processes. The specimens were exposed to salt spray conditions and immersion conditions to characterize their corrosion rates on the effect of pH value, chloride ion concentration, and corrosion time. In addition, an attempt was made to develop an empirical relationship to predict the corrosion rate of FSW welds in salt spray corrosion test and galvanic corrosion test using design of experiments. The corrosion morphology and the pit morphology were analyzed by optical microscopy, and the corrosion products were examined using scanning electron microscope and X-ray diffraction analysis. From this research work, it is found that, in both corrosion tests, the corrosion rate decreases with the increase in pH value, the decrease in chloride ion concentration, and a higher corrosion time. The results show the usage of the magnesium alloy for best environments and suitable applications from the aforementioned conditions. Also, it is found that AZ61A magnesium alloy welds possess low-corrosion rate and higher-corrosion resistance in the galvanic corrosion test than in the salt spray corrosion test.

1. Introduction

Magnesium alloys have received extensive recognition due to their excellent physical properties, including light weight, high strength/weight ratio, high thermal conductivity, and good electromagnetic shielding characteristics; thus, become promising candidates to replace steel and aluminum alloys in many structural and mechanical applications due to their attractive mechanical and metallurgical properties [1, 2]. The joining of magnesium components made from this alloy is, however, still limited. Unfortunately, conventional fusion welding of magnesium alloys often produces porosity and hot cracks in the welded joint. This deteriorates both the mechanical properties and corrosion resistance [3, 4]. Hence,

it will be of extreme benefit if a solid state joining process, that is, one which avoids bulk melting of the base materials, hot cracking, and porosity, can be developed and carried out for the joining of magnesium alloys.

FSW is a solid state welding process without emission of ration or dangerous fumes, and it avoids the formation of solidification defects like hot cracking and porosity. Moreover, it significantly improved the weld properties and had been extensively applied in the joining of magnesium alloys [5]. The application of Mg alloy in the structural members is still limited due to its conventional fusion welding resulting in many solidifications related problems such as hot cracking, porosity, alloy segregation, and partial melting zone. To overcome the previously said problems, FSW process had been used which is a solid state autogenous process, and, hence, there are no melting and solidification defects.

However, the corrosion resistance of the Mg-based alloys is generally inadequate due to the low-standard electrochemical potential −2.37 V compared to the (SHE) standard hydrogen electrode [6], and this limits the range of applications for Mg and its alloys. Therefore, the study of corrosion behavior of magnesium alloys in active media, especially those containing aggressive ions, is crucial to the understanding the corrosion mechanisms and, hence, to improving the corrosion resistance under various service conditions. Salt spray testing is the main technique for corrosion studies, which was employed in this research in an effort to expose the AZ61 Mg alloy to an environment similar to that experienced by automotive engine blocks [7]. It is well known that Mg alloys are susceptible to corrosion such as pitting and stress cracking corrosion (SCC). Major studies show that the SCC susceptibility of Mg alloys is increased in solutions containing chloride [8, 9]. The galvanic corrosion of magnesium using a (GCA) galvanic corrosion assembly which systematically investigates the influence of cathode materials the distance between anode and cathode, also for the anode/cathode area ratio. This study identified important effects such as the "alkalization," "passivation," poisoning, and shortcuts effect as well as the effectiveness of an insulating spacer in reduced galvanic corrosion [10]. It is well known that Mg alloys are susceptible to corrosion such as pitting and stress cracking corrosion (SCC). More studies show that the SCC susceptibility of Mg alloys is increased in solutions containing chloride [11, 12]. The welding process inevitably causes changes in the original microstructure of the alloy due to welding thermal cycles. These microstructural changes can affect the localized corrosion behavior of the alloy [13]. Thus, the present study contributed towards the galvanic effect of the friction stir welded AZ61A magnesium alloy weld in contact with AZ61A Mg base metal couple. Galvanic corrosion, which is originally defined as the enhanced corrosion between two or more electrically connected metals. It is one of the most common forms of corrosion considering the real world engineering structures [14].

This research focused on comparing salt spray testing with galvanic corrosion testing, which are the two main techniques for the corrosion studies in an effort to expose the magnesium alloy and its welds to environments similar to those environments experienced for automotive and structural applications. Moreover, galvanic corrosion has never been investigated using identical couple electrodes; so in this present investigation, a new method is enhanced to predict the galvanic corrosion of FSW AZ61A Magnesium alloy. From the literature review, it was understood that most of the published information on corrosion behavior of Mg alloys was focused on general corrosion and pitting corrosion of unwelded base alloys. Very few investigations have been conducted so far on corrosion behavior of FSW joints of Mg alloys. The aim of this research is to investigate the occurrence of salt spray corrosion in FSW welds and galvanic corrosion in weld zone with parent alloy of AZ61A Mg alloy. Hence, the present investigation was carried out to study the effect of pH value, chloride ion concentration, and corrosion time

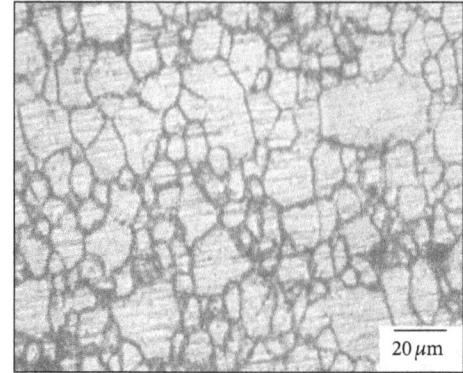

FIGURE 1: Optical micrograph of AZ61A base metal.

TABLE 1: (a) Chemical composition (wt%) of AZ61A Mg alloy and (b) mechanical properties of AZ61A Mg alloy.

(a)

Al	Zn	Mn	Mg
5.45	1.26	0.17	Balance

(b)

Yield strength (MPa)	Ultimate tensile strength (MPa)	Elongation (%)	Vickers hardness at 0.05 kg load (Hv)
177	272	8.40	57

on corrosion rate of AZ61A magnesium alloy welds and the galvanic couple.

2. Experimental Procedure

2.1. Test Materials. The material used in this study was an AZ61A magnesium alloy in the form of an extruded condition and supplied in plates of 6 mm thickness. The chemical composition and mechanical properties of the base metal are presented in Tables 1(a) and 1(b). The optical micrograph of the base metal is shown in Figure 1.

2.2. Fabricating the Joints and Preparing the Specimens. The plate was cut to a required size (300 mm × 150 mm) by power hacksaw followed by milling. A square-butt joint configuration was prepared to fabricate the joints. The initial joint configuration was obtained by securing the plates in position using mechanical clamps. The direction of welding was normal to the extruded direction. Single pass welding procedure was followed to fabricate the joints. A nonconsumable tool made of high carbon steel was used to fabricate joints. An indigenously designed and developed computer numerical controlled friction stir welding machine (22 kW; 4000 RPM; 60 kN) was used to fabricate joints. The FSW parameters were optimized by conducting trial runs, and the welding conditions which produced defect-free joints were taken as optimized welding conditions. The optimized welding conditions used to fabricate the joints in this investigation are presented in Table 2. From the base metal and welded joints,

TABLE 2: Optimized welding conditions and process parameters used to fabricate the joints.

Rotational speed (rpm)	Welding speed (mm/min)	Axial force (kN)	Tool shoulder diameter (mm)	Pin diameter (mm)	Pin length (mm)	Pin profile
1000	75	3	18	6	5	Left hand thread of 1 mm pitch

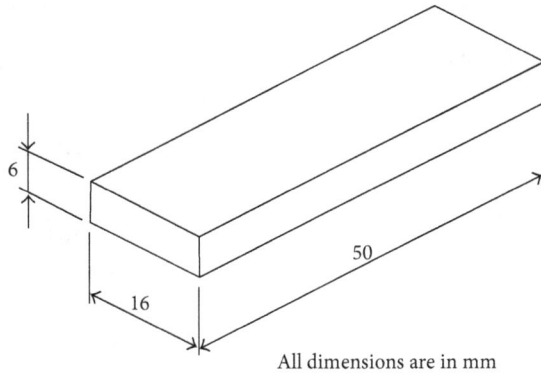

FIGURE 2: Dimensions of corrosion test specimen.

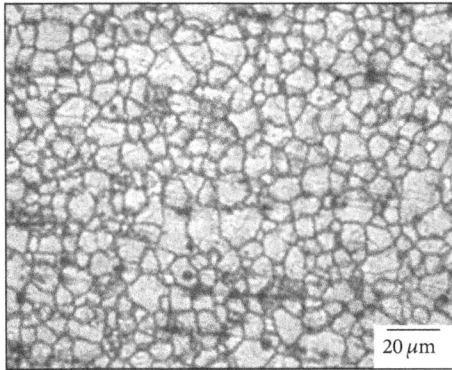

FIGURE 3: Microstructure of friction stir welded stir zone before the corrosion test.

the corrosion test specimens were sliced to the dimensions of 50 mm × 16 mm × 6 mm shown in Figure 2. The specimens were ground with $500^\#$, $800^\#$, $1200^\#$, $1500^\#$ grit SiC paper. Finally, it was cleaned with acetone and washed in distilled water then dried by warm flowing air. The optical micrograph of the stir zone of the FSW joint of AZ61A magnesium alloy is shown in Figure 3.

2.3. Finding the Limits of Corrosion Test Parameters.

From the literature [15, 16], the predominant factors that have a greater influence on corrosion behavior of AZ61A magnesium alloy are identified. They are (i) pH value of the solution, (ii) exposure time, and (iii) chloride ion concentration. Large numbers of trial experiments were conducted to identify the feasible testing conditions using friction stir welded AZ61A magnesium alloy joints under galvanic test conditions. The following inferences are obtained.

(1) If the pH value of the solution was less than 3 therefore, the change in chloride ion concentration did not considerably affect the corrosion.

(2) If the pH value was between 3 and 11 therefore, there were inhibition of the corrosion process and stabilization of the protective layer.

(3) If the pH value was greater than 11 therefore, blocking of further corrosion by the active centers of the partially protective layer.

(4) If the chloride ion concentration was less than 0.2 M therefore, the visible corrosion did not occur in the experimental period.

(5) If the chloride ion concentration was between 0.2 M and 1 M therefore, there was a reasonable fluctuation in the corrosion rate.

(6) If the chloride ion concentration was greater than 1 M therefore, the rise in corrosion rate slightly decreased a little.

(7) If the exposure time was less than an hour therefore, the surface would be completely covered with thick and rough corrosion products.

(8) If the exposure time was between 1 and 9 hours therefore, the tracks of the corrosion could be predicted.

(9) If the exposure time was greater than 9 hours therefore, the tracks of corrosion film were difficult to identify.

2.4. Developing the Experimental Design Matrix.

Owing to a wide range of factors, the use of three factors and central composite rotatable design matrix was chosen to minimize the number of experiments. The assay conditions for the reaction parameters were taken at zero level (center point) and one level (+1) and (1). The design was extended up to a $\pm\alpha$ (axial point) of 1.68. The center values for variables were carried out at least six times for the estimation of error and single runs for each of the other combinations; twenty runs were done in a totally random order. The design would consist of the eight corner points of the 2^3 cube, the six star points, and m center points. The star points would have $a = 8^\wedge(1/4) = 1.682$. Design matrix consisting of 20 sets of coded conditions (comprising a full replication three factorial of 8 points, six corner points, and six center points) was chosen in this investigation. Table 3 represents the ranges of factors considered, and Table 4 shows the 20 sets of coded and actual values used to conduct the experiments. For the convenience of recording and processing experimental data, the upper and lower levels of the factors were coded here as +1.682 and −1.682, respectively. The coded values of any intermediate values could be calculated using the following relationship:

$$X_i = \frac{1.682\left[2X - (X_{max} - X_{min})\right]}{(X_{max} - X_{min})}, \tag{1}$$

TABLE 3: Important factors and their levels.

S. no.	Factor	Unit	Notation	Levels				
				−1.682	−1	0	+1	+1.682
1	pH value		P	3	4.62	7	9.38	11
2	Corrosion time	Hours (h)	T	1	2.62	5	7.38	9
3	Cl^- concentration	Mole (M)	C	0.2	0.36	0.6	0.84	1

TABLE 4: Design matrix and experimental results.

EX. no.	Coded values			Actual values			Corrosion rate (salt spray tests) (mm/yr)	Corrosion rate (galvanic corrosion tests) (mm/yr)
	pH (P)	Time (T)	Conc. (C)	pH (P)	Time (T) (hour)	Conc. (C) (Mole)		
1	−1	−1	−1	4.62	2.62	0.36	14.62 (0.11)	0.0397 (0.16)
2	+1	−1	−1	9.38	2.62	0.36	10.23 (0.56)	0.0254 (0.13)
3	−1	+1	−1	4.62	7.38	0.36	11.89 (0.21)	0.0340 (0.20)
4	+1	+1	−1	9.38	7.38	0.36	8.99 (0.18)	0.0111 (0.18)
5	−1	−1	+1	4.62	2.62	0.84	15.82 (0.16)	0.0456 (0.14)
6	+1	−1	+1	9.38	2.62	0.84	11.31 (0.1)	0.0425 (0.1)
7	−1	+1	+1	4.62	7.38	0.84	12.92 (0.26)	0.0519 (0.52)
8	+1	+1	+1	9.38	7.38	0.84	10.75 (0.05)	0.0376 (0.05)
9	−1.682	0	0	3	5	0.60	17.96 (0.26)	0.0612 (0.08)
10	+1.682	0	0	11	5	0.60	6.88 (0.41)	0.0267 (0.11)
11	0	−1.682	0	7	1	0.60	11.23 (0.56)	0.0598 (0.07)
12	0	+1.682	0	7	9	0.60	8.51 (0.42)	0.0282 (0.11)
13	0	0	−1.682	7	5	0.20	6.66 (0.23)	0.0184 (0.05)
14	0	0	+1.682	7	5	1	15.28 (0.4)	0.0534 (0.08)
15	0	0	0	7	5	0.60	9.89 (0.36)	0.0432 (0.03)
16	0	0	0	7	5	0.60	8.56 (0.02)	0.0365 (0.02)
17	0	0	0	7	5	0.60	9.79 (0.03)	0.0370 (0.21)
18	0	0	0	7	5	0.60	9.97 (0.05)	0.0377 (0.24)
19	0	0	0	7	5	0.60	8.93 (0.04)	0.0379 (0.38)
20	0	0	0	7	5	0.60	9.62 (0.09)	0.0381 (0.12)

The values presented in brackets are the standard deviation.

where X_i is the required code values of a variable X, and X is any values of the variable from X_{min} to X_{max}; X_{min} is the lower level of the variable, and X_{max} is the upper level of the variable.

2.5. Salt Spray Corrosion Test (Recording the Response). Solutions of NaCl with concentrations of 0.2 M, 0.361 M, 0.6 M, 0.838 M, and 1 M were prepared. The pH values of the solutions were maintained as pH 3, pH 4.619, pH 7, pH 9.38, and pH 11 with concentrated HCl and NaOH, respectively. The pH value was measured using a digital pH meter. The test method consists of exposing the specimens in a salt spray chamber as per ASTM B 117 standards and evaluating the corrosion tested specimen with the method as per ASTM G1-03. Basically, the salt spray test procedure involves the spraying of a salt solution onto the samples being tested. This was done inside a temperature controlled chamber. The glass racks were contained in the salt fog chamber (3″ high, 3″ deep, and 5″ wide). The samples under test were inserted into the chamber, following which the salt-containing solution was sprayed as a very fine fog mist over the samples. NaCl in tapped water was pumped from a reservoir to spray nozzles. The solution was mixed with humidified compressed air at the nozzle, and this compressed air and atomized the NaCl solution into a fog at the nozzle. Heaters were maintained at 35°C cabinet temperature. Within the chamber, the samples were rotated frequently so that all samples were exposed uniformly to the salt spray mist. Since the spray was continuous, the samples were continuously wet and therefore, uniformly subjected to corrosion. The corrosion rate of the friction stir welded AZ61A alloy specimen was estimated by weight loss measurement. The original weight (w_o) of the specimen was recorded, and then the specimen was sprayed with the solution of NaCl for different spraying times of 1, 2.62, 5, 7.38, and 9 hours. The corrosion products were removed by immersing the specimens for one minute in a solution prepared by using 50 gm chromium trioxide (CrO_3),

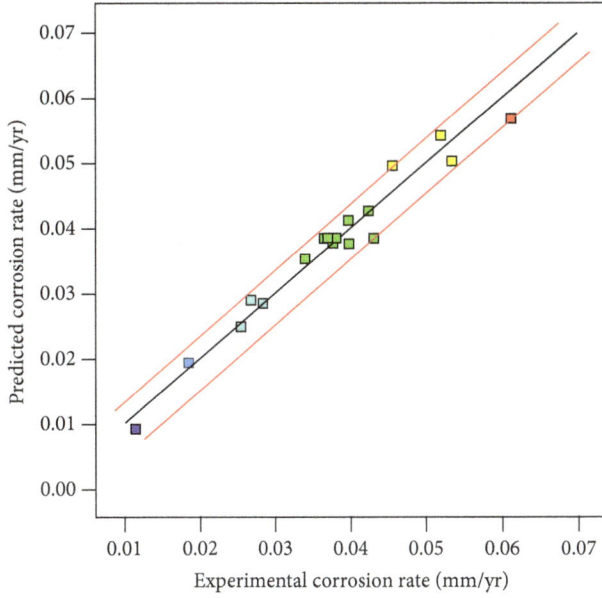

FIGURE 4: Correlation graph for response (salt spray corrosion test).

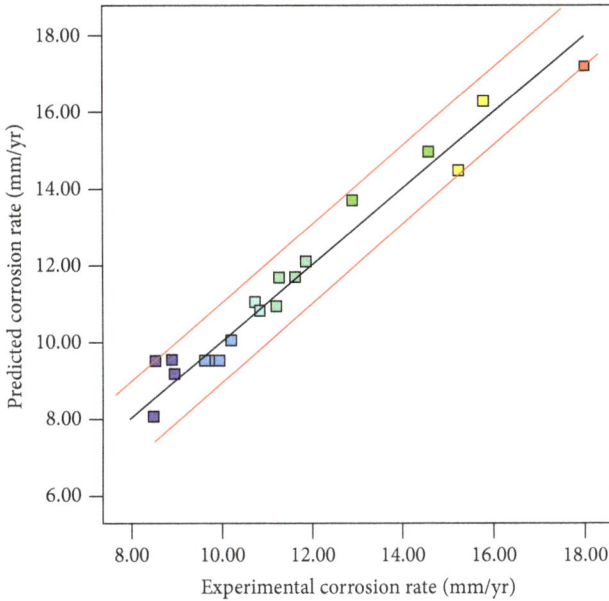

FIGURE 5: Correlation graph for response (galvanic corrosion test).

2.5 gm silver nitrate ($AgNO_3$), and 5 gm barium nitrate ($Ba(NO_3)_2$) in 250 mL distilled water. Finally, the specimens were washed with distilled water, dried, and weighed again to obtain the final weight (w_1). The weight loss (w) can be measured using the following relation:

$$w = (w_o - w_1), \qquad (2)$$

where w is weight loss in grams, w_o is original weight before test in grams, and w_1 is final weight after test in grams.

The corrosion rate of FSW joints of AZ61A was calculated by using the following equation as per the ASTM standards B117:

$$\text{corrosion rate (mm/year)} = \frac{8.76 \times 10^4 \times w}{A \times D \times T}, \qquad (3)$$

where w is weight loss in grams, A is surface area of the specimen in cm^2, D is density of the material (1.72 gm/cm^3), and T is spraying time in hours.

2.6. Galvanic Corrosion Test (Recording the Response). The test method consisted of exposing the specimens in a specially designed apparatus as per ASTM G 82-98 standards and evaluating the corrosion tested specimen with the method as per ASTM G 102-89. The galvanic samples were prepared in the following way. A saturated calomel electrode and graphite electrode were used as the reference and auxiliary electrode, respectively. The working electrodes were the friction stir welded AZ61A magnesium alloy welds coupled with AZ61A magnesium alloy base metal. An electrical contact was made between the galvanic couple, where a Teflon insulation of the same thickness was inserted between the electrodes to avoid the direct contact between the electrodes. The galvanic couple was immersed in NaCl solution with different pH and chloride ion concentration for different immersion times of 1, 2.62, 5, 7.38, and 9 hours. When the mixed potential theory was applied to the individual reactions, the uncoupled corrosion rates were i_{corr} (A) for AZ61A magnesium alloy base metal and i_{corr} (B) for AZ61A friction stir welds. When equal areas of AZ61A Mg base metal and AZ61A friction stir welds were coupled, the resultant mixed potential of the system E_{corr} (AB) was at the intersection where the total oxidation rate equals the total reduction rate. The rate of oxidation of the individual coupled metals was such that the base metal corroded at a reduced rate i_{corr} (A), and AZ61A friction stir welds corroded at an increased rate I_{corr} (B). Hence, the AZ61A friction stir weld acts as an anode, and AZ61A base alloy acts as a cathode. Half-cell reactions were carried out constituting a single cell. Thus, the current i_{corr} (AB) was the galvanic current which can be measured by a zero resistance ammeter (ZRA). Free corrosion potential of both metals was found individually and from the potential difference; FSW AZ61A magnesium alloy was considered to be an anode, because of its more negative potential than AZ61 base alloy, where the latter was the cathode. Corrosion current values may be obtained from the ZRA measurements. The corrosion rate can be calculated using Faraday's Law in terms of penetration rates as follows:

$$\text{corrosion rate (mm/yr)} = \frac{K \times I_{corr} \times EW}{\rho}, \qquad (4)$$

where K is corrosion constant (K is 0.00327 if corrosion rate in (mm/yr)), I_{corr} is current density in mA/cm^2, EW is equivalent weight of the alloy, and ρ is density of the FSW AZ61A alloy (1.72 gm/cm^3).

2.7. Metallography. Microstructural analysis of the corroded specimens was carried out using a light optical microscope

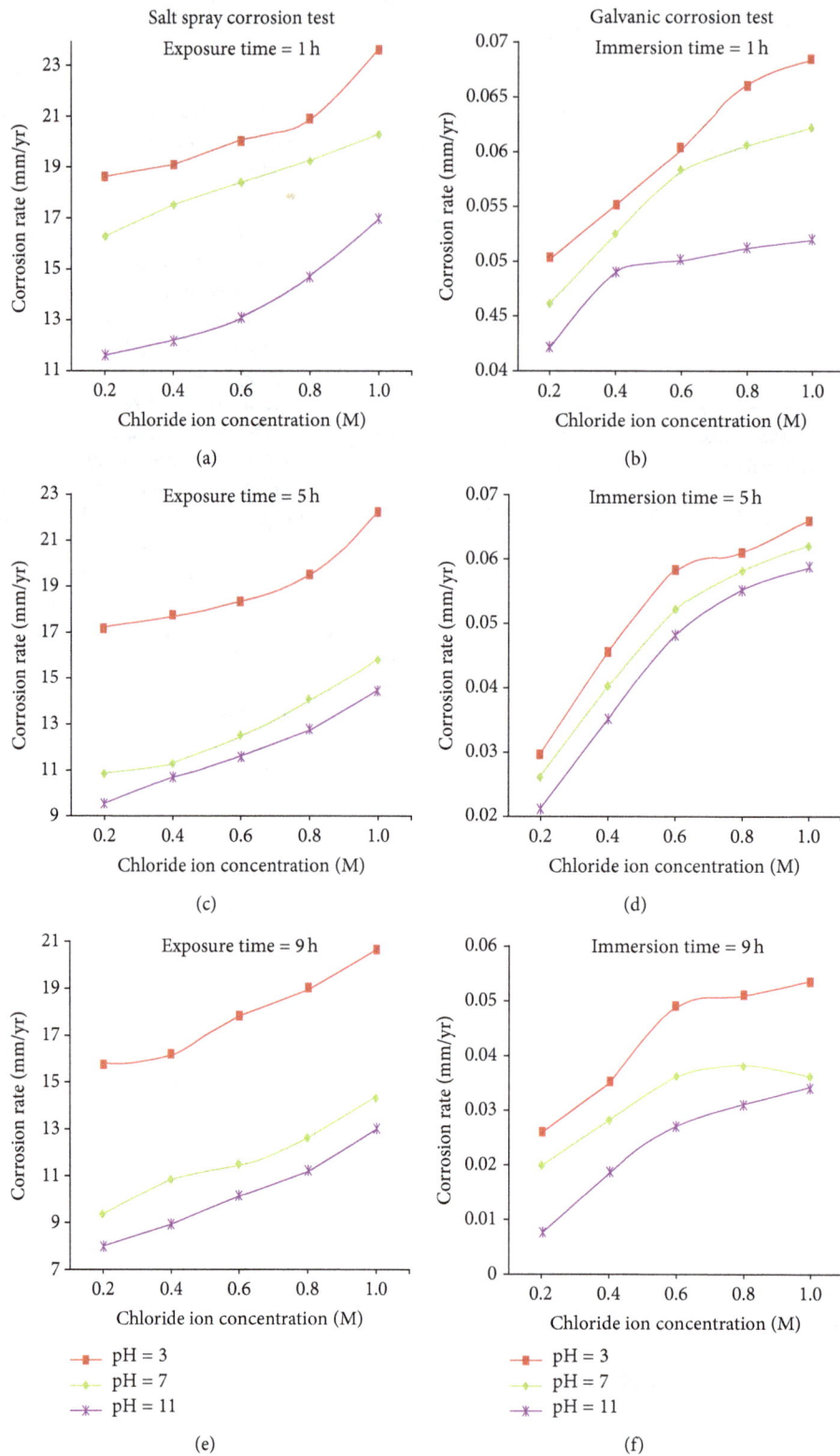

FIGURE 6: Effect of pH value on corrosion rate.

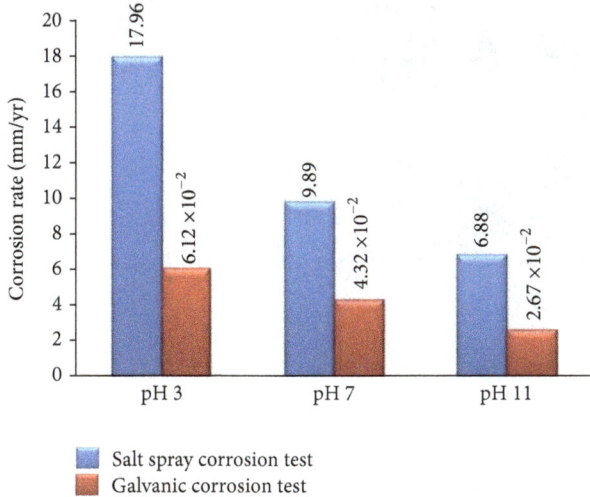

FIGURE 7: Comparative estimation of corrosion rate with respect to pH value.

(Union Optics, Japan; model: Versamet-3) incorporated with an image analyzing software (Clemex-vision). The exposed specimen surface was prepared for the microexamination in the "as polished" conditions. The corrosion test specimens were polished in disc-polishing machine with minor polishing, and the surface was observed at 200x magnification. The corrosion products were analyzed by SEM-EDX and XRD analysis.

3. Developing an Empirical Relationship

The response surface methodology (RSM) approach was adopted in this study because of its following advantages: (1) the ability to evaluate the effects of interactions between tested parameters and (2) the benefit of limiting the number of actual experiments to be carried out, in comparison to a classical approach for the same number of estimated parameters [17–20]. In the present investigation, to correlate the potentiodynamic polarization test parameters and the corrosion rate of AZ61A welds, a second-order quadratic model was developed. The response (corrosion rate of AZ61A welds) is a function of pH values (P), exposure time (T), and chloride ion concentration (C), and it could be expressed as

$$\text{corrosion rate} = f(P, T, C). \tag{5}$$

In order to study the combined effects of these parameters, experiments were conducted at different combinations using statistically designed experiments. The empirical relationship must include the main and interaction effects of all factors and hence the selected polynomials are expressed as follows:

$$Y = b_o + \sum b_i x_i + \sum b_{ii} x_i^2 + \sum b_{ij} x_i x_j. \tag{6}$$

For three factors, the selected polynomial could be expressed as

$$\text{corrosion rate} = \left\{ b_0 + b_1(P) + b_2(T) + b_3(C) \right.$$

$$+ b_{11}(P^2) + b_{22}(T^2)$$

$$+ b_{33}(C^2) + b_{12}(PT)$$

$$+ b_{13}(PC) + b_{23}(TC) \bigg\}, \tag{7}$$

where b_0 is the average of responses (corrosion rate), and $b_1, b_2, b_3 \ldots b_{11}, b_{12}, b_{13} \ldots b_{22}, b_{23}, b_{33}$ are the coefficients that depend on their respective main and interaction factors, which are calculated using the expression given as follows:

$$B_i = \frac{\sum (X_i, Y_i)}{n}, \tag{8}$$

where "I" varies from 1 to n, in which X_i is the corresponding coded value of a factor, Y_i is the corresponding response output value (corrosion rate) obtained from the experiment, and "n" is the total number of combinations considered. All the coefficients were obtained applying central composite rotatable design matrix including the Design Expert statistical software package. After determining the significant coefficients (at 95% confidence level), the final relationship was developed including only these coefficients. The final empirical relationship obtained by the above procedure to estimate the corrosion rate of friction stir welds of AZ61A magnesium alloy is given as follows.

Salt spray corrosion test,

$$\text{corrosion rate} = \left\{ 9.48 - 1.89(P) - 0.88(T) + 0.82(C) \right.$$

$$\left. + 1.60(P^2) + 1.27(C^2) \right\} \text{mm/yr}; \tag{9}$$

Galvanic corrosion test,

$$\text{corrosion rate} = \left\{ 0.056 - 7.33 \times 10^{-3}(P) \right.$$

$$+ 2.75 \times 10^{-3}(T) + 0.016(C)$$

$$- 4.25 \times 10^{-3}(PT) + 4.26 \times 10^{-3}(PC)$$

$$+ 4.6 \times 10^{-3}(TC)$$

$$\left. - 3.74 \times 10^{-4}(T^2) \right\} \text{mm/yr}. \tag{10}$$

3.1. Checking the Adequacy of the Model Salt Spray Testing. The Analysis of Variance (ANOVA) technique was used to find the significant main and interaction factors. The results of the second-order response surface model fitting in the form of Analysis of Variance (ANOVA) are given in Table 5. The determination coefficient (r^2) indicated the goodness of fit for the model. The model F value of 31.30 implies the model is significant. There is only a 0.01% chance that a "model F value" this large could occur due to noise. Values of "Prob > F" less than 0.0500 indicate that model terms are significant. In this case P, T, C, P^2, and C^2 are significant model terms. Values greater than 0.1000 indicate that the model terms are not significant. If there are many insignificant model terms (not counting those required to support hierarchy), model reduction may improve the model. The "lack of fit F value" of 1.69

Salt spray corrosion test

Galvanic corrosion test

pH 3

(a)

pH 3

(b)

pH 7

(c)

pH 7

(d)

pH 11

(e)

pH 11

(f)

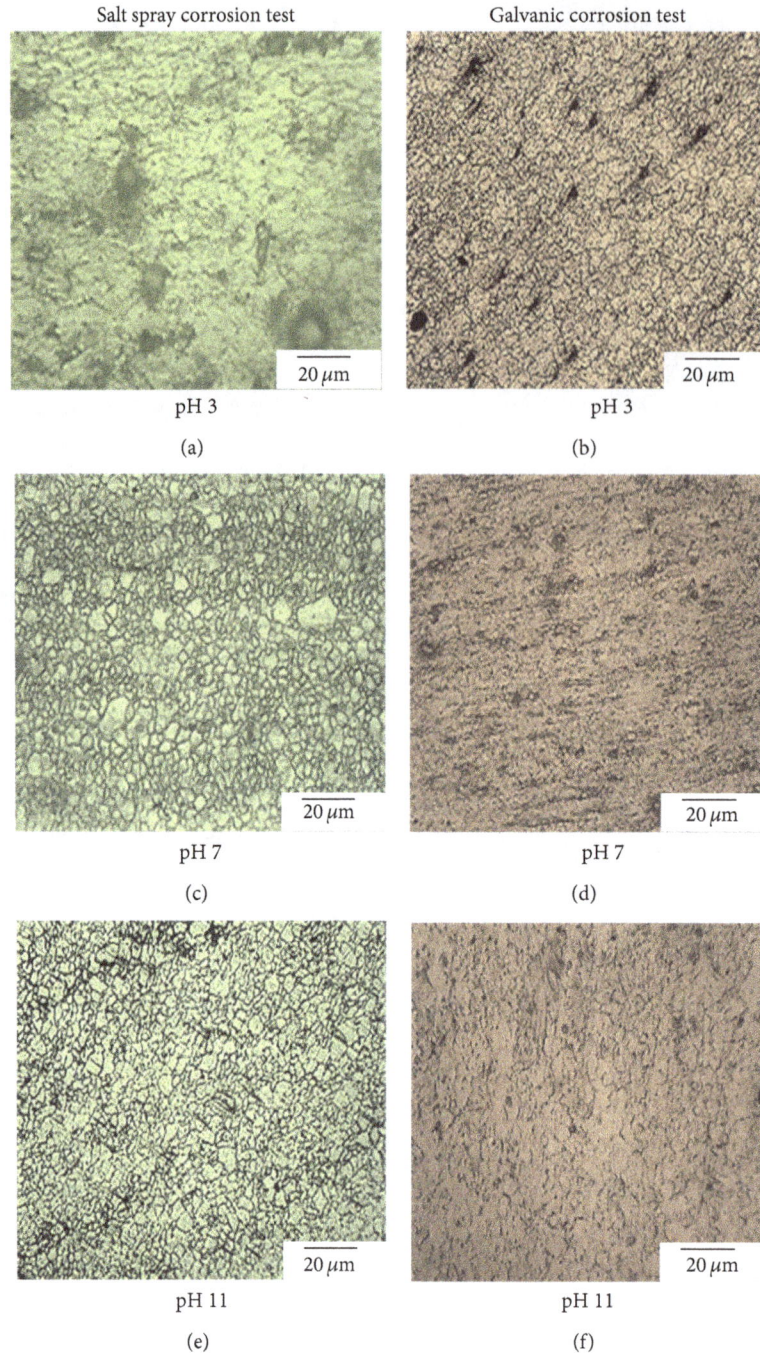

FIGURE 8: Effect of pH on pit morphology.

implies that the lack of fit is not significant relative to the pure error. There is a 28.93% chance that a "lack of fit F value" this large could occur due to noise. Nonsignificant lack of fit is good. The "Pred R-Squared" of 0.8176 is in reasonable agreement with the "Adj R-Squared" of 0.9349. "Adeq Precision" measures the signal to noise ratio. P ratio greater than 4 is desirable. Our ratio of 19.440 indicates an adequate signal. All of this indicated an excellent suitability of the regression model. Each of the observed values compared with the experimental values are shown in Figure 4.

3.2. Checking the Adequacy of the Model Galvanic Corrosion Testing. The Analysis of Variance (ANOVA) technique was used to find the significant main and interaction factors. The results of second-order response surface model fitting in the form of Analysis of Variance (ANOVA) are given in Table 6. The determination coefficient (r^2) indicated the goodness of fit for the model. The model F value of 27.66 implies the model is significant. There is only a 0.01% chance that a "model F value" this large could occur due to noise. Values of "Prob > F" less than 0.0500 indicate that model terms are significant.

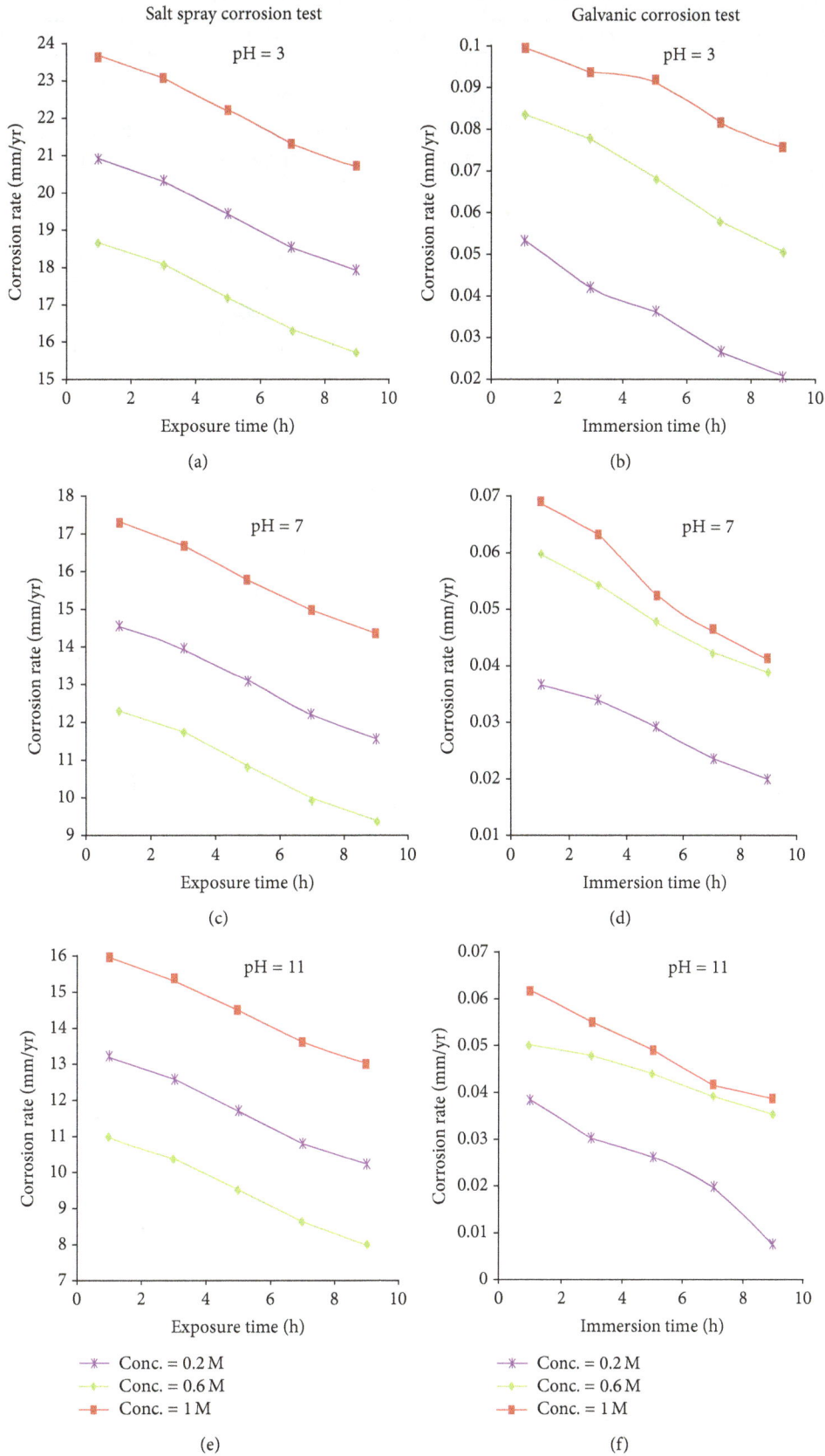

FIGURE 9: Effect of chloride ion concentration on corrosion rate.

TABLE 5: ANOVA test results for salt spray corrosion test.

Source	Sum of squares	df	Mean square	F value	P value	Prob > F
Model	126.60	9	14.07	31.30	<0.0001	**Significant**
P	49.03	1	49.03	109.11	<0.0001	
T	10.55	1	10.55	23.48	0.0007	
C	9.12	1	9.12	20.29	0.0011	
PT	1.83	1	1.83	4.08	0.0710	
PC	0.047	1	0.047	0.10	0.7543	
TC	0.033	1	0.033	0.072	0.7934	
P^2	37.07	1	37.07	82.48	<0.0001	
T^2	$3.4E-004$	1	$3.4E-004$	$7.6E-004$	0.9785	
C^2	23.17	1	23.17	51.55	<0.0001	
Residual	4.49	10	0.45			
Lack of fit	2.82	5	0.53	1.69	0.2893	**Not significant**
Pure error	1.67	5	0.33			
Cor total	131.09	19				

TABLE 6: ANOVA test results for galvanic corrosion test.

Source	Sum of squares	df	Mean square	F value	P value	Prob > F
Model	$2.472E-003$	9	$2.472E-003$	27.66	<0.0001	**Significant**
P	$9.229E-004$	1	$9.229E-004$	92.93	<0.0001	
T	$1.057E-004$	1	$1.057E-004$	10.64	0.0085	
C	$1.163E-004$	1	$1.163E-004$	117.12	<0.0001	
PT	$4.642E-005$	1	$4.642E-004$	4.67	0.0559	
PC	$4.700E-005$	1	$4.700E-005$	4.73	0.0547	
TC	$5.634E-005$	1	$5.634E-005$	5.67	0.0385	
P^2	$2.804E-005$	1	$2.804E-005$	2.82	0.1238	
T^2	$6.470E-005$	1	$6.470E-005$	6.51	0.0287	
C^2	$3.072E-005$	1	$3.072E-005$	3.09	0.1091	
Residual	$9.931E-005$	10	$9.931E-005$			
Lack of fit	$6.999E-005$	5	$1.400E-005$	2.39	0.1808	**Not significant**
Pure error	$2.932E-005$	5	$5.864E-005$			
Cor total	$2.572E-005$	19				

In this case P, T, C, TC, and T^2 are significant model terms. Values greater than 0.1000 indicate that the model terms were not significant. If there are many insignificant model terms (not counting those required to support hierarchy), model reduction may improve the model.

The "lack of fit F value" of 2.39 implies that the lack of fit was not significant relative to the pure error. There was an 18.08% chance that a "lack of fit F value" this large could occur due to noise. Nonsignificant lack of fit is good. The "Pred R-Squared" of 0.7763 is in reasonable agreement with the "Adj R-Squared" of 0.9266. "Adeq Precision" measures the signal to noise ratio. P ratio greater than 4 is desirable. Our ratio of 21.393 indicates an adequate signal. Each of the observed values compared with the experimental values are shown in Figure 5, and it had a good agreement between the observed values and the experimental values.

4. Results and Discussion

4.1. Effect of pH on Corrosion Rate. Figure 6 shows the graph representing the effect of pH on corrosion rate during salt spray testing and galvanic corrosion testing. For both corrosion tests, the graph shows clearly that the corrosion rate decreased with the increase in pH value. At every chloride ion concentration and immersion time, the FS welds usually exhibited a decrease in corrosion rate with increase in pH. In neutral pH, the corrosion rate remained approximately constant, and comparatively low corrosion rate was observed in alkaline solution. It was seen that the influence of pH was more at higher concentration as compared to lower concentration in neutral and alkaline solutions.

On comparing the corrosion rate of both, the corrosion tested specimen was represented as bar diagram in Figure 7.

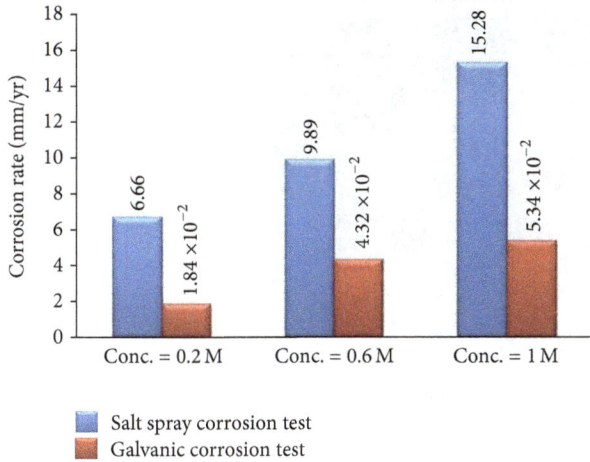

FIGURE 10: Comparative estimation of corrosion rate with respect to chloride ion concentration.

It was found that the corrosion rates obtained from the salt spray testing were much higher than the rates obtained from the galvanic corrosion tests. This was due to spraying effect where recycling of the solution could not be taken into account, while in galvanic corrosion testing; there is a substantial increase in the pH of the solution during immersion testing causing alkalization or basification of the solution with the increase in reactivity and time. Thus, the corrosion rate was much higher in salt spray testing than in the galvanic corrosion testing. So, the couple were galvanically a good couple, and can be suitable for good applications [20–22].

Figure 8 shows the effect of pH on pit morphology of the corroded specimen exposed in 0.6 M concentration of NaCl for 5 hours with different pH values of pH 3, pH 7, and pH 11 for both salt spray testing and galvanic corrosion testing. During salt spray testing, the density of the pit formed in exposing lower pH (acidic) solution is quite high, compared with the neutral and alkaline solution. It was observed that the matrix shows the pitting marks and the pitting corrosion that has taken place at the friction stir welded microstructure. The particles are Mn-Al compound and fragmented $Mg_{17}Al_{12}$. The numbers of pits were more in the joints when it is sprayed with the solution of low pH. Hence, the corrosion rate increases with the decrease in pH value. Since the increase of grain and grain boundary in the joints, the grain boundary acts cathodic to grain causing a microgalvanic effect. The presence of microgalvanic effect between the α phase and the β phase that formed was due to the presence of aluminum. During galvanic corrosion tests, the grain boundaries of the anodic specimen got attacked, and its gravity varies with the parameters used in the experiment. Corrosion tends to be concentrated in the area adjacent to the grain boundary until eventually the grain may be undercut and fall out [23].

4.2. Effect of Chloride Ion Concentration on Corrosion Rate. Figure 9 shows the graph representing the effect of chloride ion concentration on corrosion rate during salt spray testing and galvanic corrosion testing. However, it was observed that, with the increase in chloride ion concentration, the rising rate of corrosion rate decreased. The increase in corrosion rate with increasing chloride ion concentration may be attributed to the participation of chloride ions in the dissolution reaction for both corrosion tests [24]. Figure 10 represents the comparison chart for the corrosion rate obtained from both corrosion tests. This is consistent with the detailing of the protective layer. With the increase of chloride ion concentration, the protective layer $Mg(OH)_2$ changed into soluble $MgCl_2$ layer in salt spray corrosion and $Mg(OH)Cl_2$ in galvanic corrosion. The corrosion rate was quite higher in salt spray corrosion test than in the galvanic corrosion test. It states that the $MgCl_2$ was highly soluble compared to $Mg(OH)Cl_2$.

Figure 11 shows the effect of chloride ion concentration on pit morphology of the corroded specimen exposed in pH 7 for 5 hours with different chloride ion concentration of 0.2 M, 0.6 M, and 1 M for both salt spray testing and galvanic corrosion testing. During salt spray testing, it showed that the alloy exhibited a rise in corrosion rate with the increase in Cl^- concentration and thus, the change of Cl^- concentration affected the corrosion rate much more in higher concentration solutions than that in lower concentration solutions. When more Cl^- in NaCl solution promoted the corrosion, the corrosive intermediate (Cl^-) would be rapidly transferred through the outer layer and reach the substrate of the alloy surface. Hence, the corrosion rate increased [20]. But in galvanic corrosion tests, the anodic specimen exhibited a rise in corrosion rate with increase in Cl^- concentration and thus, the change of Cl^- concentration affected the corrosion rate much more in higher concentration solutions than that in lower concentration solutions. Chloride ions were aggressive for magnesium. The adsorption of chloride ions to oxide covered magnesium surface transformed $Mg(OH)_2$ to easily soluble $MgCl_2$. It was considered that the corrosion becomes severe owing to the penetration of the hydroxide film by Cl^- ion and thereby caused the formation of metal hydroxyl chloride complex which governed the following reaction:

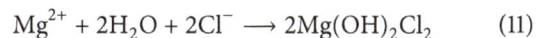

$$Mg^{2+} + 2H_2O + 2Cl^- \longrightarrow 2Mg(OH)_2Cl_2 \qquad (11)$$

4.3. Effect of Corrosion Time on Corrosion Rate. Figure 12 shows the graph representing the effect of corrosion time on corrosion rate during salt spray testing and galvanic corrosion testing. During salt spray testing, the graph shows clearly that the corrosion rate was decreased with the increase in exposure time. It resulted in an increase in hydrogen evolution with the increasing exposure time, which tends to increase the concentration of OH^- ions strengthening the surface from causing further corrosion. Thus, the rate of corrosion decreases with the increase in corrosion time. During galvanic corrosion testing, the corrosion rate decreases with the increase in immersion time. The increase in immersion time enhanced the tendency to form the corrosion products, which accumulated over the surface of the samples. These corrosion products in turn depressed the corrosion rate due to the passivation in the medium immersion [25]. It resulted in an increase in hydrogen evolution with the increasing

Salt spray corrosion test   Galvanic corrosion test

Chloride ion concentration = 0.2 M   Chloride ion concentration = 0.2 M

(a)   (b)

Chloride ion concentration = 0.6 M   Chloride ion concentration = 0.6 M

(c)   (d)

Chloride ion concentration = 1 M   Chloride ion concentration = 1 M

(e)   (f)

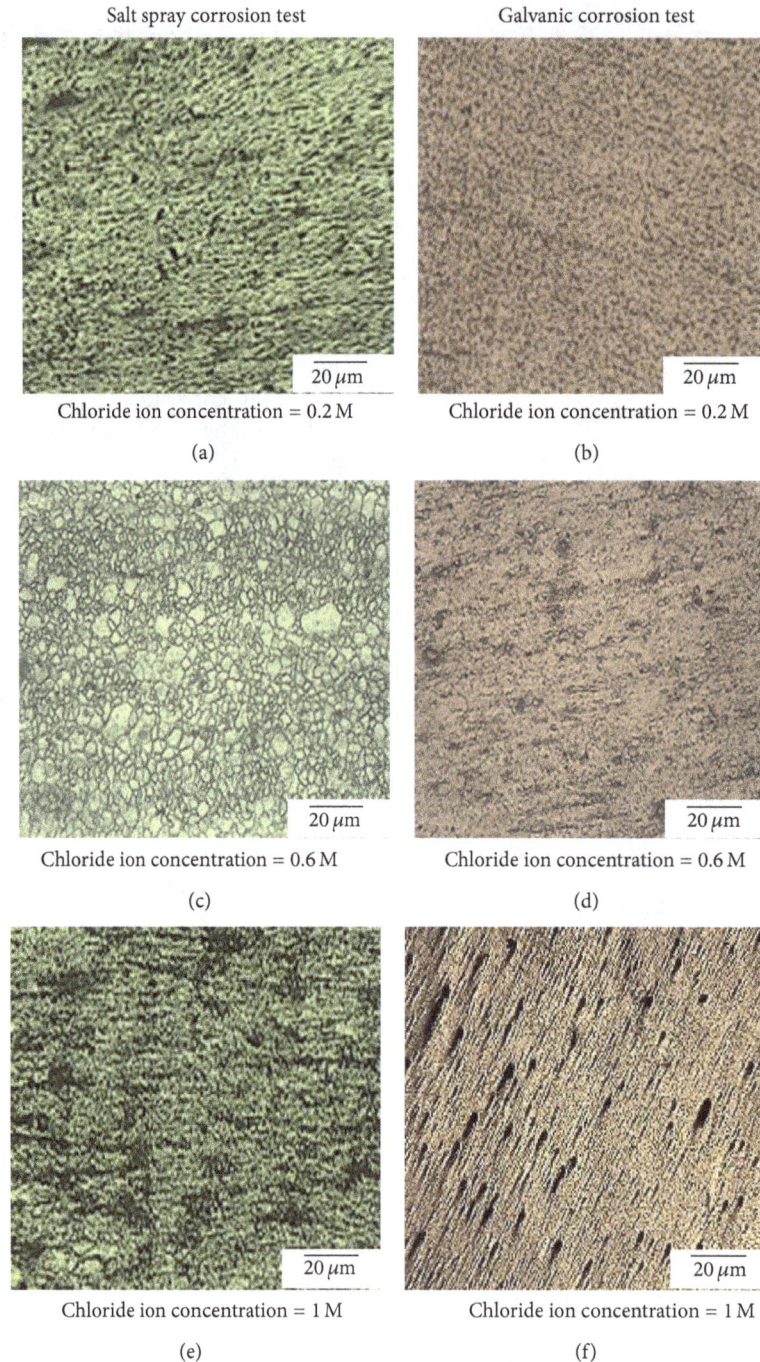

FIGURE 11: Effect of chloride ion concentration on pit morphology.

immersion time, which tends to increase the concentration of OH$^-$ ions strengthening the surface from causing further corrosion. The strength of the electrolyte reduces from acidity to alkalinity with the increase of time.

This is attributed to corrosion occurs over an increasing fraction on the surface leaving the white flakes, which is the insoluble corrosion product [24]. The insoluble corrosion products on the surface of the alloy could slow down the corrosion rate:

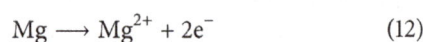

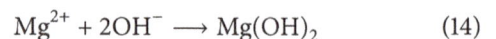

$$Mg \longrightarrow Mg^{2+} + 2e^- \qquad (12)$$

$$2H_2O + 2e^- \longrightarrow 2OH^- + H_2 \qquad (13)$$

$$Mg^{2+} + 2OH^- \longrightarrow Mg(OH)_2 \qquad (14)$$

Figure 13 shows the comparison of the corrosion rate obtained during the salt spray and galvanic corrosion test. With the increase of corrosion time the corrosion rate decreases for both specimens. It proved that the protective layer made a predominant role to strike against corrosion with the increment of time. The corrosion rate seems higher in salt spray corrosion test due to the spraying effect, while

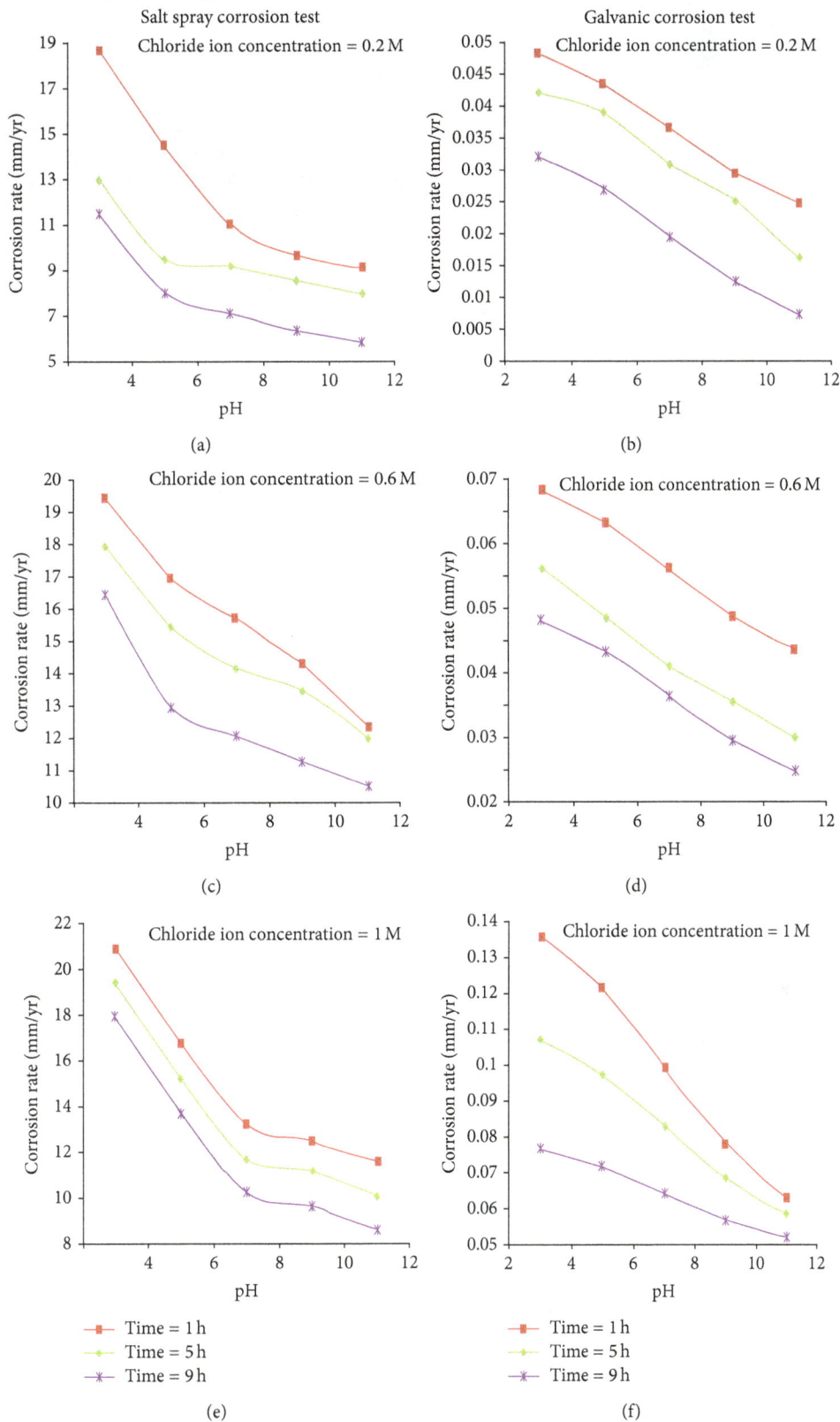

FIGURE 12: Effect of corrosion time on corrosion rate.

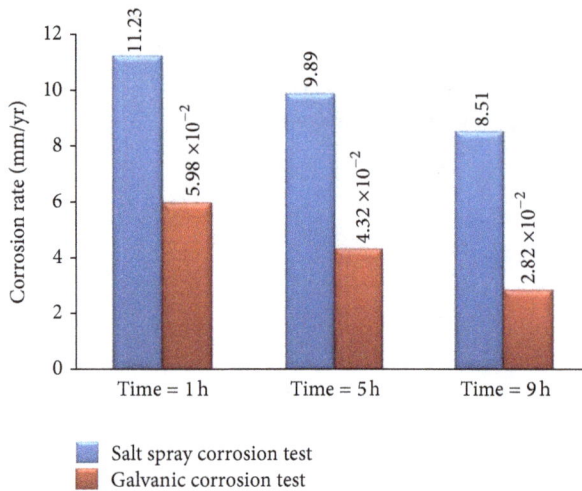

FIGURE 13: Comparative estimation of corrosion rate with respect to corrosion time.

in immersed condition, the protective layer formed during galvanic corrosion was enhanced by the alkalization of the solution.

Figure 14 shows the effect of corrosion time on pit morphology of the corroded specimen exposed in pH 7 with chloride ion concentration of 0.6 M NaCl exposed, 1 h, 5 h, and 9 h for both salt spray testing and galvanic corrosion testing. The mode of microstructural features was comparatively the same during corrosion testing for both tests as its corrosion time is taken into account. The FS welded specimens possess refined grain, and quite a lot of β particles were distributed continually along the grain boundary.

In this case, β phase particles cannot be easily destroyed and, with the increase of corrosion time, the quantity of β phases in the exposed surface would increase and finally play the role of corrosion barrier [26]. Although there are some grains of α phase still being corroded, most of the remaining α phase grains are protected under the β phase barrier, so the corrosion rate decreased with the increase in corrosion time. Thus, the corrosion morphology of the alloy was predominantly controlled by the β phase distribution [27].

4.4. SEM and XRD Analysis. Figure 15 shows the surface texture of the specimens that underwent salt spray corrosion and galvanic corrosion tests which were observed under SEM as corrosion time was a factor. Figure 15(a) shows the specimens exposed to 1 hour comprised of more localized attack. With corrosion time as a factor, the less spraying time tends to attack more locally on the surface, and later, it penetrates to the substrates, causing higher corrosion behavior and corrosion rate. It was redeems to spalling of corrosion products. Thus, serious pitting occurred in the surface of the weldment with less exposure time for both corrosion-tested specimens. The quantity of corrosion products formed was quite comparably larger in the galvanic corrosion test caused due to immersion.

Figure 15(b) shows the specimens exposed to 9 hours composed of more corrosion products. With the increase

in corrosion time, the hydroxide layer formed is the dominant factor to avoid further corrosion. This is attributed to corrosion occurring over an increasing fraction of the surface, which is the insoluble corrosion product, $Mg(OH)_2$. Thus, the corrosion rate decreases with the increase of corrosion time. It was observed that the corrosion products were thick and adherent in the specimen that underwent galvanic corrosion test while the specimen that underwent salt spray test exhibited lamellar corrosion products, which seems higher attack and less protective. This showed higher corrosion rate in the salt spray test than the galvanic corrosion test.

Figure 16 shows the XRD analysis to predict the composition of corrosion products and phase in the specimen subjected to salt spray and galvanic corrosion tests. Figure 16(a) shows that the specimen underwent the salt spray corrosion test; the characteristic peaks originate from the metallic Mg substrate and the β phase. The detected number of peaks relates to the β phase ($Mg_{17}Al_{12}$) which is higher in its intensities. It symbolizes the hydroxide layer which finds it hard to form during the salt spray corrosion test due to its spraying effects. However, $Mg(OH)_2$ and MgO phases are detected in Figure 16(b), where the specimen underwent the galvanic corrosion test; $Mg(OH)_2$ is the dominant product in the corrosion zone of the anodic specimen that underwent the galvanic corrosion test. $Mg(OH)_2$ (brucite) has a hexagonal crystal structure and easily undergoes basal cleavage causing cracking and curling in the film, which could play a major role in reducing the corrosion behavior and the corrosion rate. It signifies during immersion that the alkalization effect tends to strengthen the formation of the hydroxide layer.

5. Conclusions

(1) A mathematical model has been developed here to predict the corrosion rate using salt spray tests and galvanic corrosion with a 95% confidence level. The corrosion rates obtained were quite different in both tests; it found that, the corrosion rate was much higher in salt spray test than in the galvanic corrosion test.

(2) In this investigation, it was proved from both tests, for every pH value; that the FS weld metal exhibited a rise in corrosion rate with decrease in pH value. In the neutral pH, the corrosion rate remained approximately constant in neutral solutions, and a comparatively low corrosion rate was observed in alkaline solutions.

(3) Chloride ions were aggressive on magnesium alloy. The adsorption of chloride ions to oxide covered the magnesium surface and transformed $Mg(OH)_2$ to easily soluble $MgCl_2$ in salt spray corrosion testing while in galvanic corrosion testing, it readily formed $Mg(OH)_2Cl_2$ due to the immersion criterion.

(4) It resulted in an increase in hydrogen evolution with the increasing corrosion time, which tended to increase the concentration of OH^- ions; thereby, an increasing fraction of the surface was observed, which is the insoluble corrosion products. The insoluble

Salt spray corrosion test

Galvanic corrosion test

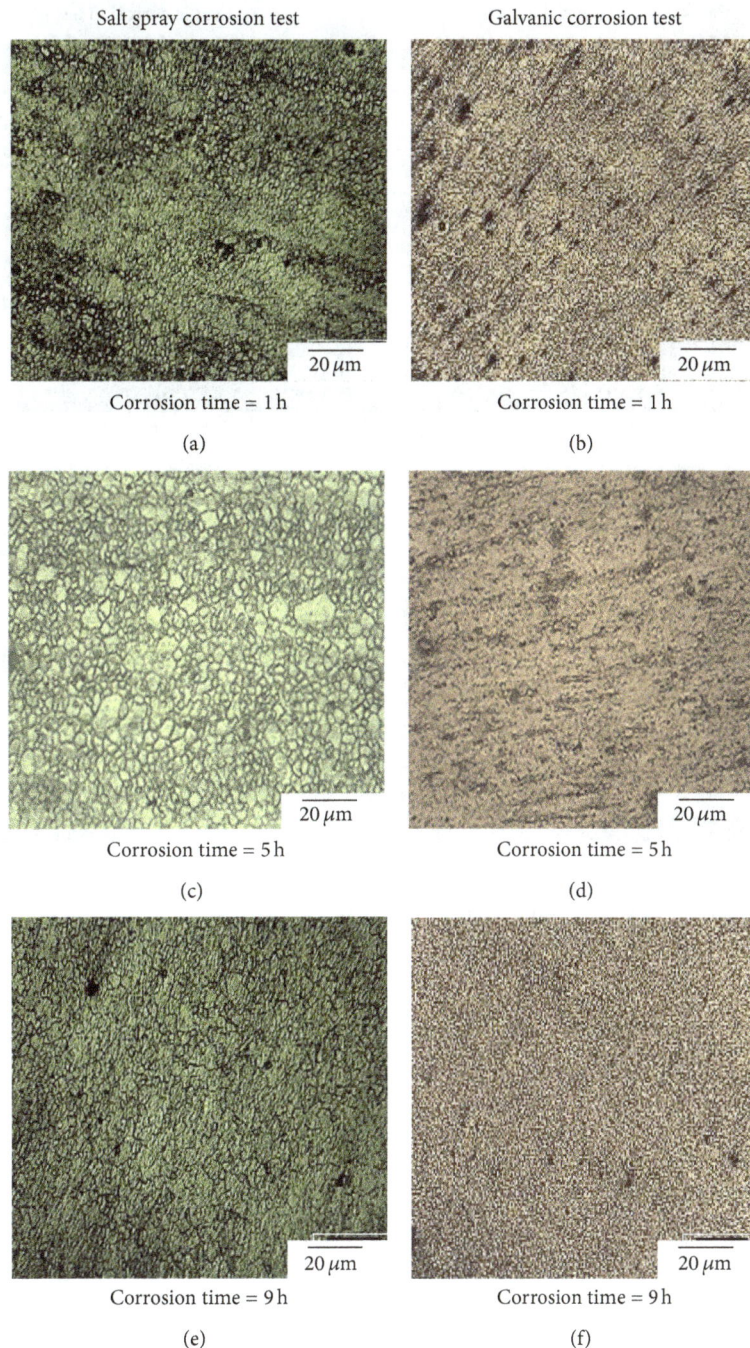

Corrosion time = 1 h

(a)

Corrosion time = 1 h

(b)

Corrosion time = 5 h

(c)

Corrosion time = 5 h

(d)

Corrosion time = 9 h

(e)

Corrosion time = 9 h

(f)

FIGURE 14: Effect of corrosion time on pit morphology.

corrosion products on the surface of the alloy could slow down the corrosion rate.

(5) In this investigation, it was found that the corrosion rate obtained from the salt spray tests was much higher than the rates obtained from the galvanic corrosion tests. It was due to spraying effect where stagnation of the solution could not be taken into account while in galvanic corrosion testing, there is a substantial increase in the pH of the solution during corrosion reactions due to the migration of ions in the electrolyte under immersion. The corrosion rate of galvanic couple ranges from 0.03 to 0.06 mm/yr, which is quite negligible, and shows excellent property of corrosion resistant as per corrosion handbooks and guides. So, the couple were galvanically a good couple, and they can be suitable for good applications.

Conflict of Interests

The authors declare that there is no conflict of interests.

Salt spray corrosion test

Galvanic corrosion test

(a) Corrosion time = 1 h

(b) Corrosion time = 9 h

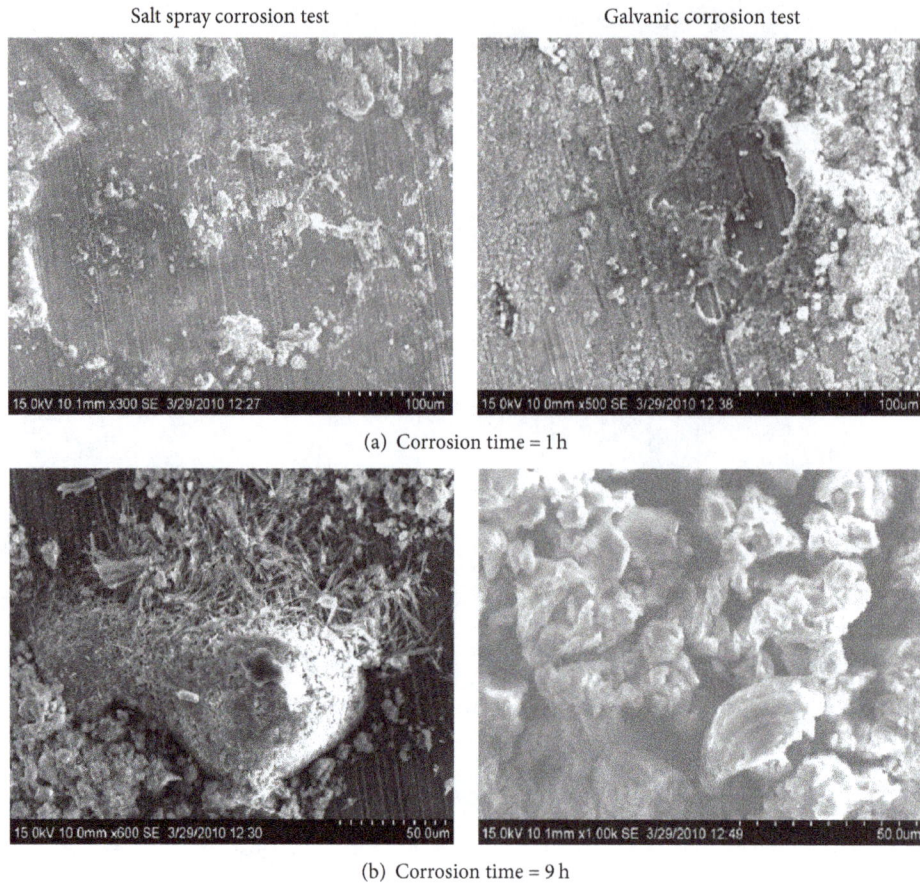

FIGURE 15: Scanning electron micrograph of corrosion test specimens underwent salt spray corrosion test and galvanic corrosion test.

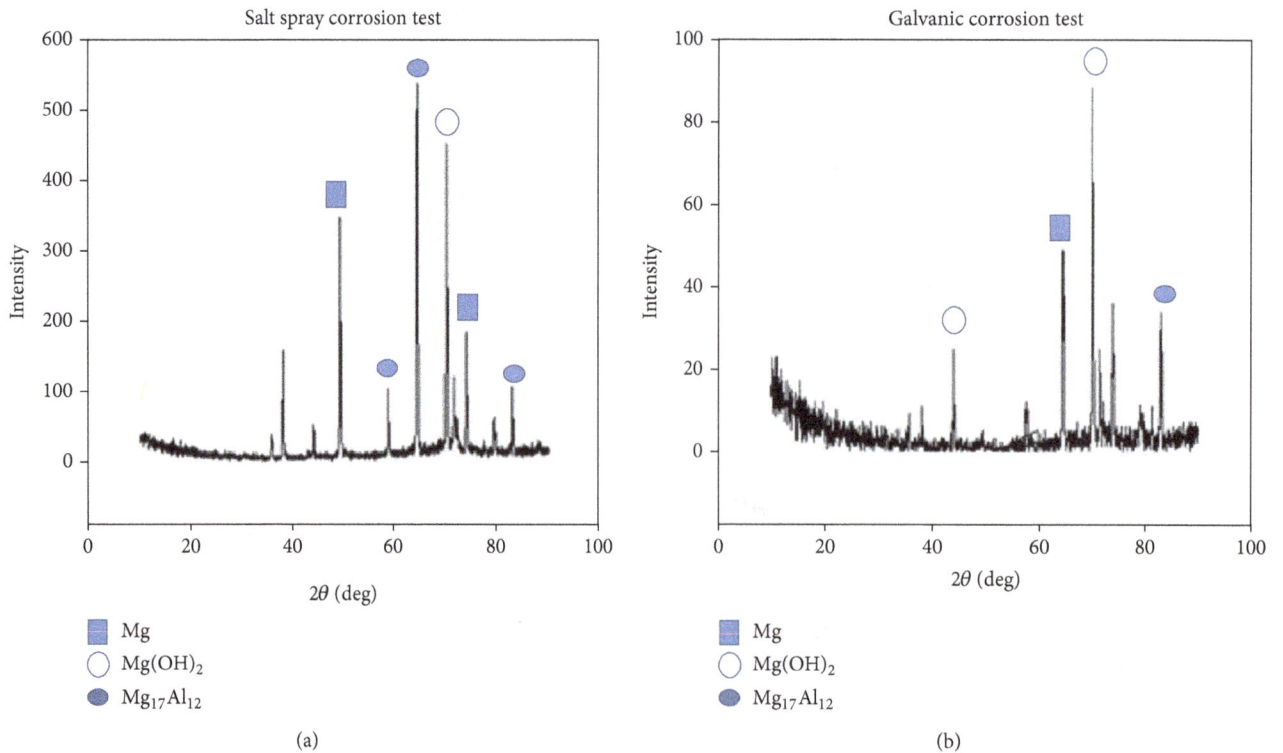

(a)

(b)

FIGURE 16: XRD pattern of corrosion test specimens underwent salt spray corrosion test and galvanic corrosion test.

Acknowledgment

The authors would like to thank the Centre for Materials Joining & Research (CEMAJOR), Department of Manufacturing Engineering, Annamalai University, Annamalai Nagar, India for extending the facilities of Materials Joining Laboratory and Corrosion Testing Laboratory to carry out this investigation.

References

[1] B. L. Mordike and T. Ebert, "Magnesium Properties, applications, potential," *Materials Science and Engineering A*, vol. 302, no. 1, pp. 37–45, 2001.

[2] R. C. Zeng, W. Dietzel, R. Zettler, J. Chen, and K. U. Kainer, "Microstructure evolution and tensile properties of friction-stir-welded AM50 magnesium alloy," *Transactions of Nonferrous Metals Society of China*, vol. 18, no. 1, pp. s76–s80, 2008.

[3] R. C. Zeng, J. Zhang, W. J. Huang et al., "Review of studies on corrosion of magnesium alloys," *Transactions of Nonferrous Metals Society of China*, vol. 16, supplement 2, pp. s763–s771, 2006.

[4] T. Nagasawa, M. Otsuka, T. Yokota, and T. Ueki, "Structure and mechanical properties of friction stir weld Joints of magnesium alloy AZ31," in *Magnesium Technology 2000*, H. I. Kaplan, J. Hryn, and B. Clow, Eds., pp. 383–387, TMS, Warrendale, Pa, USA, 2000.

[5] W. Xu, J. Liu, and H. Zhu, "Pitting corrosion of friction stir welded aluminum alloy thick plate in alkaline chloride solution," *Electrochimica Acta*, vol. 55, no. 8, pp. 2918–2923, 2010.

[6] M. Zhao, S. Wu, J. R. Luo, Y. Fukuda, and H. Nakae, "A chromium-free conversion coating of magnesium alloy by a phosphate-permanganate solution," *Surface and Coatings Technology*, vol. 200, no. 18-19, pp. 5407–5412, 2006.

[7] B. A. Shaw, "Corrosion resistance of magnesium alloys," in *ASM Handbook, vol. 13A: Corrosion*, L. J. Korb, Ed., p. 692, ASM International Handbook Committee, Metals Park, Ohio, USA, 9th edition, 2003.

[8] D. L. Hawke, J. E. Hillis, M. pekguleryuz, and I. Nkatusugawa, "Corrosion behavior," in *Magnesium and Magnesium Alloys*, M. M. Avedesian and H. Baker, Eds., pp. 194–1210, ASM International, Materials Park, Ohio, USA, 1999.

[9] G. Song and A. Atrens, "Recent insights into the mechanism of magnesium corrosion and research suggestions," *Advanced Engineering Materials*, vol. 9, no. 3, pp. 177–183, 2007.

[10] G. Song, B. Johanesson, S. Hagupoda, and D. StJohn, "Galvanic corrosion of magnesium alloy AZ91D in contact with an aluminium alloy, steel and zinc," *Corrosion Science*, vol. 46, no. 4, pp. 955–977, 2004.

[11] M. Jönsson, D. Persson, and D. Thierry, "Corrosion product formation during NaCl induced atmospheric corrosion of magnesium alloy AZ91D," *Corrosion Science*, vol. 49, no. 3, pp. 1540–1558, 2007.

[12] R. G. Song, C. Blawert, W. Dietzel, and A. Atrens, "A study on stress corrosion cracking and hydrogen embrittlement of AZ31 magnesium alloy," *Materials Science and Engineering A*, vol. 399, no. 1-2, pp. 308–317, 2005.

[13] M. B. Kannan, W. Dietzel, C. Blawert, S. Riekehr, and M. Koçak, "Stress corrosion cracking behavior of Nd:YAG laser butt welded AZ31 Mg sheet," *Materials Science and Engineering A*, vol. 444, no. 1-2, pp. 220–226, 2007.

[14] R. Baboian, "Electrochemical techniques for corrosion engineering," in *Corrosion '76*, p. 114, NACE, 1976.

[15] H. Altun and S. Sen, "Studies on the influence of chloride ion concentration and pH on the corrosion and electrochemical behaviour of AZ63 magnesium alloy," *Materials and Design*, vol. 25, no. 7, pp. 637–643, 2004.

[16] Y. Song, D. Shan, R. Chen, and E. H. Han, "Effect of second phases on the corrosion behaviour of wrought Mg-Zn-Y-Zr alloy," *Corrosion Science*, vol. 52, no. 5, pp. 1830–1837, 2010.

[17] K. H. Goh, T. T. Lim, and P. C. Chui, "Evaluation of the effect of dosage, pH and contact time on high-dose phosphate inhibition for copper corrosion control using response surface methodology (RSM)," *Corrosion Science*, vol. 50, no. 4, pp. 918–927, 2008.

[18] N. Aslan, "Application of response surface methodology and central composite rotatable design for modeling and optimization of a multi-gravity separator for chromite concentration," *Powder Technology*, vol. 185, no. 1, pp. 80–86, 2008.

[19] J. S. Cowpe, J. S. Astin, R. D. Pilkington, and A. E. Hill, "Application of response surface methodology to laser-induced breakdown spectroscopy: Influences of hardware configuration," *Spectrochimica Acta B*, vol. 62, no. 12, pp. 1335–1342, 2007.

[20] A. Dhanapal, S. R. Boopathy, and V. Balasubramanian, "Developing an empirical relationship to predict the corrosion rate of friction stir welded AZ61A magnesium alloy under salt fog environment," *Materials and Design*, vol. 32, no. 10, pp. 5066–5072, 2011.

[21] B. D. Craig and D. B. Anderson, *Handbook of Corrosion Data*, ASM International, 1995.

[22] H. H. Uhlig, *The Corrosion Handbook*, John Wiley, 1948.

[23] N. Hara, Y. Kobayashi, D. Kagaya, and N. Akao, "Formation and breakdown of surface films on magnesium and its alloys in aqueous solutions," *Corrosion Science*, vol. 49, no. 1, pp. 166–175, 2007.

[24] G. Song, A. Atrens, and M. Dargusch, "Influence of microstructure on the corrosion of diecast AZ91D," *Corrosion Science*, vol. 41, no. 2, pp. 249–273, 1998.

[25] Z. M. Zhang, H. Y. Xu, and B. C. Li, "Corrosion properties of plastically deformed AZ80 magnesium alloy," *Transactions of Nonferrous Metals Society of China*, vol. 20, no. 2, pp. s697–s702, 2010.

[26] Y. Song, D. Shan, R. Chen, and E. H. Han, "Effect of second phases on the corrosion behaviour of wrought Mg-Zn-Y-Zr alloy," *Corrosion Science*, vol. 52, no. 5, pp. 1830–1837, 2010.

[27] *Corrosion Tests and Standards: Application and Interpretation*, ASTM international, 2005.

8

Band Gap Engineering of $Cd_{1-x}Be_xSe$ Alloys

**Djillali Bensaid,[1] Mohammed Ameri,[1] Nadia Benseddik,[2]
Ali Mir,[2] Nour Eddine Bouzouira,[1] and Fethi Benzoudji[2]**

[1] *Laboratory of Physical Chemistry of Advanced Materials, University of Djillali Liabes, BP 89, 22000 Sidi Bel Abbes, Algeria*
[2] *Physics Department, Science Faculty, University of Sidi Bel Abbes, 22000 Sidi Bel Abbes, Algeria*

Correspondence should be addressed to Djillali Bensaid; djizer@yahoo.fr

Academic Editor: Velimir Radmilovic

The structural and electronic properties of the ternary $Cd_{1-x}Be_xSe$ alloys have been calculated using the full-potential linear muffin-tin-orbital (FP-LMTO) method based on density functional theory within local density approximation (LDA). The calculated equilibrium lattice constants and bulk moduli are compared with previous results. The concentration dependence of the electronic band structure and the direct and indirect band gaps are investigated. Moreover, the refractive index and the optical dielectric constant for $Cd_{1-x}Be_xSe$ are studied. The thermodynamic stability of the alloys of interest is investigated by means of the miscibility. This is the first quantitative theoretical prediction to investigate the effective masses, optical and thermodynamic properties for $Cd_{1-x}Be_xSe$ alloy, and still awaits experimental.

1. Introduction

In recent years, the wide-gap II–VI compounds are widely investigated because of the attractive applications in fabricating blue-green and blue optoelectronic devices, such as light-emitting diodes and laser diodes [1–4]. The applications include the use of II–VI compound based materials as light sources, in full colour displays, and for increasing the information density in optical recording [5, 6]. The beryllium containing II–VI compounds had been found to possess an enhanced ability to significantly reduce the defect propagation due to a greater prevalence of strong covalent bonding and lattice hardening in the materials [7, 8]. The strong covalent bonding in beryllium-based II–VI compounds achieves a considerable lattice hardening, which avoids multiplication of defects during the operation of II–VI semiconductor laser devices [9, 10] $Be_xCd_{1-x}Se$ alloys have attracted great attention because they are promising for the fabrication of full-colour visible optical devices due to a large difference in the energy gaps E_g of the binary constituents (CdSe, E_g = 1.74 eV; BeSe, E_g = 5.5 eV) [11].

In the present theoretical work, band gap of zinc-blende CdSe is varied systematically by alloying with Be. In order to investigate the optoelectronic nature of these alloys, their structural, electronic, and optical properties are calculated. All calculations are based on density functional full-potential linear muffin-tin orbital (FP-LMTO) method with perdew-wang local density approximation (LDA).

2. Method of Calculations

The calculations reported here were carried out using the *ab initio* full-potential linear muffin-tin orbital (FP-LMTO) method [12–15] as implemented in the Lmtart code [16]. The exchange and correlation potential was calculated using the local density approximation (LDA) [17]. The FP-LMTO is an improved method compared to previous LMTO techniques, and it treats muffin-tin spheres and interstitial regions on the same footing, leading to improvements in the precision of the eigenvalues. At the same time, the FP-LMTO method, in which the space is divided into an interstitial region (IR) and nonoverlapping muffin-tin spheres (MTS) surrounding the atomic sites, uses a more complete basis than its predecessors. In the IR regions, the basis set consists of plane waves. Inside the MT spheres, the basis set is described by radial solutions of the one particle Schrödinger equation (at fixed energy) and their energy derivatives multiplied by spherical harmonics.

TABLE 1: Lattice constants a and bulk modulus B of $Cd_{1-x}Be_xSe$ compared with experimental results, Vegard's law, and other theoretical calculations.

x	Lattice constant a(Å)				Bulk modulus B (GPa) [B']		
	This work	Exp.	Vegard's law	Other calc.	This work	Exp.	Other calc.
0	6.06	6.052[a]		6.025[b]	55.05 [4.17]	55[c]	54[d]
0.25	5.90		5.82		57.72 [3.93]		
0.5	5.71		5.58		63.19 [4.15]		
0.75	5.46		5.34		66.6 [3.85]		
1	5.10	5.14[e]		5.13[f]–5.04[g]	81.25 [3.77]	92[e]	77 [3.55][f]–80 [3.11][g]

[a]Ref [25], [b]Ref [26], [c]Ref [27], [d]Ref [28], [e]Ref [29, 30], [f]Ref [31], [g]Ref [32].

The charge density and the potential are represented inside the MTS by spherical harmonics up to $l_{max} = 6$. The integrals over the Brillouin zone are performed up to 35 special k-points for binary compounds and 27 special k points for the alloys in the irreducible Brillouin zone (IBZ) using Blochl's modified tetrahedron method [18]. The self-consistent calculations are considered to be converged when the total energy of the system is stable within 10^{-6} Ry. In order to avoid the overlap of atomic spheres, the MTS radius for each atomic position is taken to be different for each composition. We point out that the use of the full-potential calculation ensures that the calculation is not completely independent of the choice of sphere radii.

Structural properties of $Cd_{1-x}Be_xSe$ are calculated using Murnaghan's equation of state [19] as follows:

$$E(V) = E_0 + \frac{B_0 V}{B_0'} \left(\frac{(V_0/V)^{B_0'}}{B_0' - 1} + 1 \right) - \frac{B_0 V_0}{B_0' - 1}, \quad (1)$$

where E_0 is the total energy of the supercell, V_0 is the unit volume, B_0 is the bulk modulus at zero pressure, and B_0' is the derivative of bulk modulus with pressure.

Optical properties of $Cd_{1-x}Be_xSe$ are calculated using a fine k mesh of 1500 points for the present calculation. The dielectric function of a crystal depends on the electronic band structure and its investigation by optical spectroscopy which is a powerful tool in the determination of the overall optical behavior of a compound. It can be divided into two parts, real and imaginary as follows:

$$\varepsilon(\omega) = \varepsilon_1(\omega) + i\varepsilon_2(\omega). \quad (2)$$

The imaginary part of the complex dielectric function, $\varepsilon_2(\omega)$, in cubic symmetry compounds can be calculated by the following relation [20, 21]:

$$\varepsilon_2(\omega) = \frac{8}{2\pi\omega^2} \sum_{nn'} \int |pnn'(k)|^2 \frac{dSk}{\nabla\omega nn'(k)}, \quad (3)$$

while $\varepsilon_1(\omega)$ is used to calculate the real part of the complex dielectric function as follows:

$$\varepsilon_1(\omega) = 1 + \frac{2}{\pi} p \int_0^\infty \frac{\omega' \varepsilon_2(\omega')}{\omega'^2 - \omega'^2} d\omega'. \quad (4)$$

Refractive index is calculated in terms of real and imaginary parts of dielectric function by the following relation

$$n(\omega) = \frac{1}{\sqrt{2}} \left[\left\{ \varepsilon_1(\omega)^2 + \varepsilon_2(\omega)^2 \right\}^{1/2} + \varepsilon_1(\omega) \right]^{1/2}. \quad (5)$$

3. Result and Discussion

In order to study the structural properties of $Cd_{1-x}Be_xSe$ ($0 < x < 1$), alloys are modeled at various compositions of Be with a step of 0.25. Structure optimization of each compound is performed by minimizing the total energy with respect to the unit cell volume and c/a ratio using Murnaghan's equation of state [19]. The crystal structure of CdSe and BeSe zinc-blende with space group $F\bar{4}3m$ (no. 216). Structural parameters such as lattice constant, a (Å), are calculated from the stable volume and are presented in Table 1. It is clear from the table that our calculated results for the binary compounds are in good agreement with the available experimental and calculated data.

Figures 1 and 2 show the variation of the calculated equilibrium lattice constant and bulk modulus as a function of concentrations x for the $Cd_{1-x}Be_xSe$ alloy. The obtained results for the composition dependence of the calculated equilibrium lattice parameter almost follow Vegard's law [22]. In going from CdSe to BeSe, when the Be-content increases, the values of the lattice parameters of the $Cd_{1-x}Be_xSe$ alloy decrease. This is due to the fact that the size of the Be atom is smaller than the Cd atom. On the opposite side, one can see from Figure 2 that the value of the bulk modulus increases as the Be concentration increases.

The calculated band structure and partial density of states for $Cd_{1-x}Be_xSe$ ($0 < x < 1$) are presented in Figure 3. It is clear from the figure that $Cd_{1-x}Be_xSe$ ($0 < x < 1$) is a direct band gap material. The substitution of Be does not affect the direct band gap nature of the compound but increases the gap, which is clear from Figure 2(d). The direct band gap varies from 0.39 to 4.45 eV and the indirect band gap also increases from 1.89 to 3.71 eV with the increase in Be concentration between 0.25 and 0.75. It is obvious from the data presented in Table 2 that our calculated values for the band gaps of CdSe and BeSe are closer to the experimental results than the other calculated ones. The reason for our better results is the use of effective Perdew and Wang potential in the LDA scheme [17] and high k-points (1500). DFT always underestimates the band gaps; the origin of band structures is

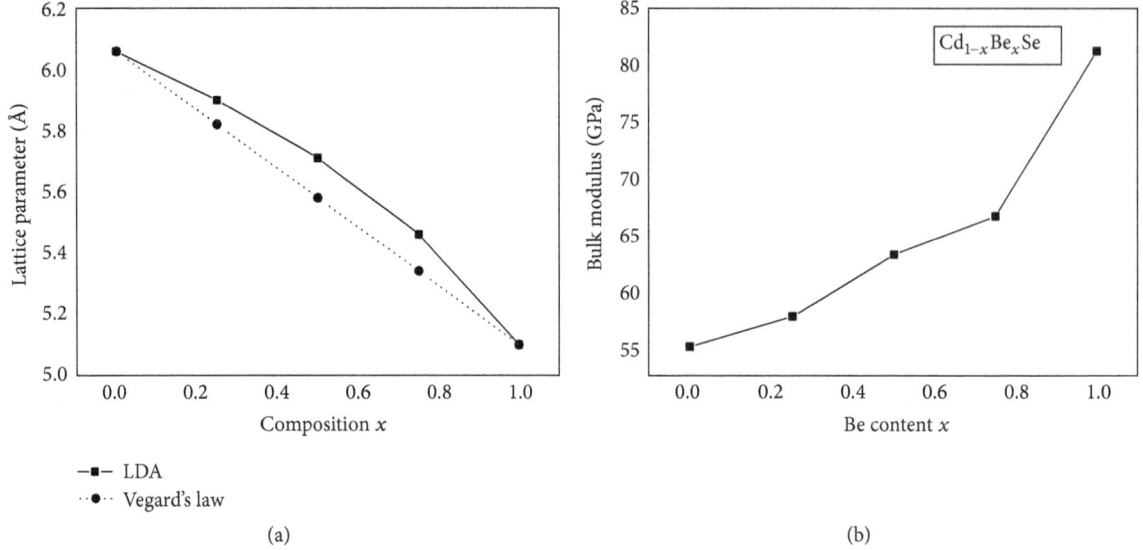

FIGURE 1: Variation in lattice constants a and bulk modulus B of $Cd_{1-x}Be_xSe$ as a function of composition x.

TABLE 2: Fundamental direct and indirect band gaps of $Cd_{1-x}Be_xSe$ compared with the experimental and other calculations.

x	Direct energy band gap, $E_g^{\Gamma-\Gamma}$			Indirect energy band gap, $E_g^{\Gamma-X}$			Band gap bowing
	This work	Exp.	Other calc.	This work	Exp.	Other calc.	
0	0.39	1.75[a]	0.34[b], 0.26[c]	2.31	5.4[d]	5.4[e], 3.82[f], 3.79[g]	
0.25	0.84	2.12[h], 2.56[h] [x = 0.2]		1.89			3.41
		2.65[i] [0.24], 2.45[j] [x = 0.35]					
0.5	1.44	2.65[j] [x = 0.46]		2.74			2.44
0.75	2.47			3.71			6.32
1	4.45		4.72[k]–4.04[l]	2.47	4–4.5[m]	2.39[k], 2.31[l]	

[a]Ref [33], [b]Ref [26], [c]Ref [27], [d]Ref [34], [e]Ref [35], [f]Ref [36], [g]Ref [37], [h]Ref [38], [i]Ref [39], [j]Ref [40], [k]Ref [41], [l]Ref [42], [m]Ref [43].

presented in Figure 3. In general, the band structures of these three compositions x (0.25, 0.5, and 0.75) are very similar. The uppermost valence band is mainly formed by Se 4p states.

The valence band maximum appears to be almost degenerate at the Γ and X k points for three concentrations (0.25, 0.5, and 0.75), the energy at X being (1.04, 1.29, and 1.24) eV lower than that at Γ, respectively. On the other hand, the conduction band minimum occurs at Γ, so there is a direct gap of 0.84, 1.44, and 2.47 eV at Γ and an indirect Γ–X gap of 1.89, 2.74, and 3.71 eV for three concentrations (0.25, 0.5, and 0.75), respectively. To the best of our knowledge, there are no theoretical or experimental data on the energy band gaps for $x = 0.25$, 0.5, and 0.75 available in the literature to make a meaningful comparison. It is clear from the results that the conduction band is mainly composed of Be-2p state for all ternary alloys. Figures 3(a) and 3(b) show that the lower part of the valence band is composed of Cd-4d and the upper part is mainly dominated by Se-4p state. This is due to the nature of bonding. At $x = 0.25$, 0.5, and 0.75 partial covalent bond is stronger so the charge is shared by Se-4p and Be-2s states.

The calculated band gap versus concentration was fitted by a polynomial equation. The results are shown in Figure 2(d) and are summarised as follows:

$$Cd_{1-x}Be_xSe \longrightarrow \begin{cases} E_{\Gamma-\Gamma} = 0.466 - 0.088x + 3.99x^2 \\ E_{\Gamma-X} = 1.978 + 2.593x - 1.737x^2. \end{cases} \quad (6)$$

The variation in the band gap of $Cd_{1-x}Be_xSe$ provides promising results of the use of the compound in optoelectronic devices working in visible to ultraviolet regions. Depending on the need and requirement of a particular application, any desired band gap between 0.39 and 4.45 eV can be achieved.

We have calculated the frequency dependent imaginary dielectric function and real dielectric function. The effects of using k points in the BZ have already been discussed in the earlier work by Khan et al. [23]. The knowledge of both the real and the imaginary parts of the dielectric function allows the calculation of important optical functions. In this work, we also present and analyse the refractive index $n(\omega)$ given by (4).

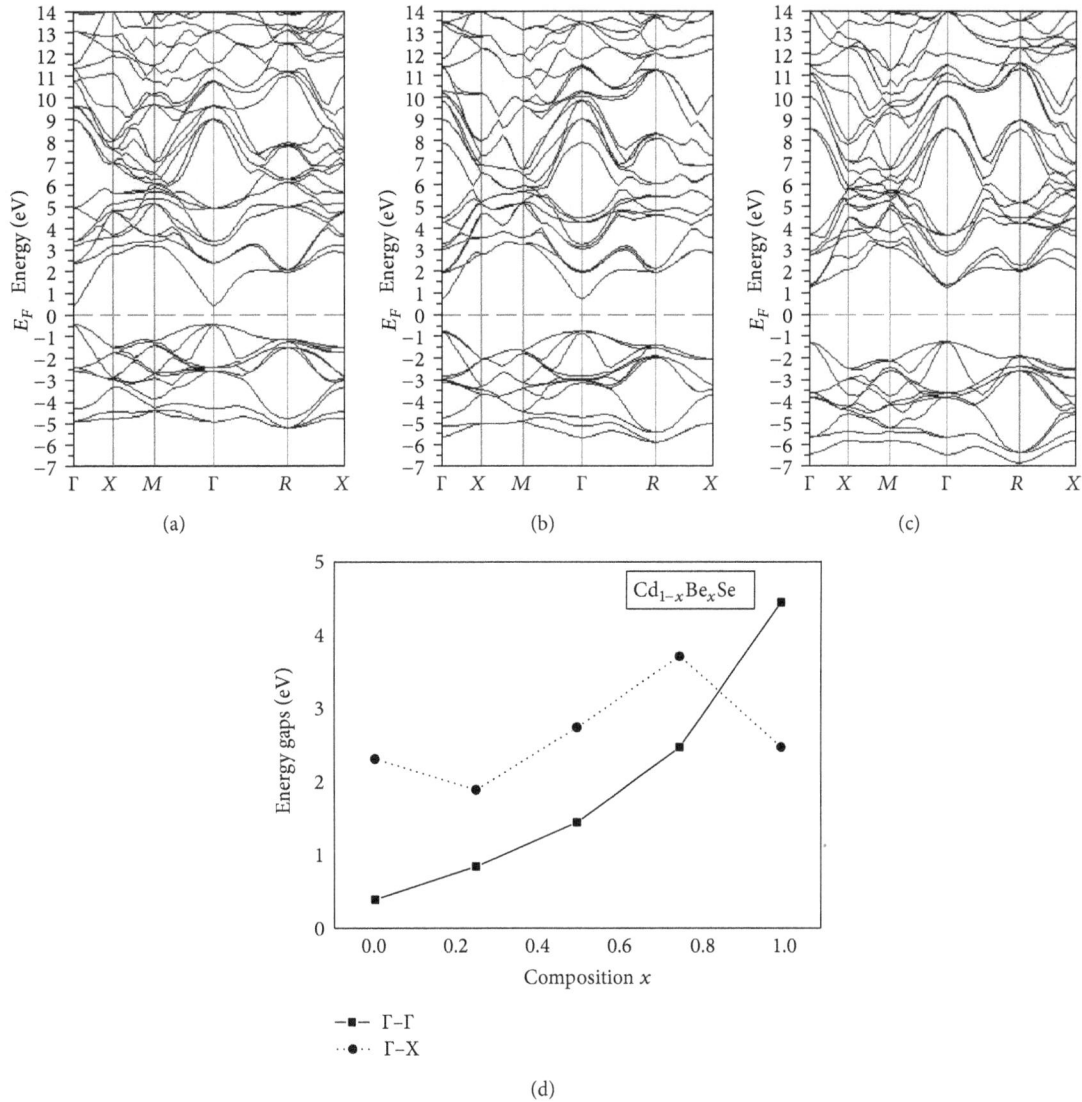

FIGURE 2: Calculated band structure of (a) $Cd_{0.75}Be_{0.25}Se$, (b) $Cd_{0.5}Be0_{0.5}Se$, (c) $Cd_{0.25}Be_{0.75}Se$, and (d) band gap as a function of x.

The calculated imaginary part of the dielectric function for $Cd_{1−x}Be_xSe$ (x = 0.25, 0.5 and 0.75) in the energy range 0–12 eV is shown in Figure 3. It is clear from the figure that for x = 0.25, 0.50, 0.75, and 1.0 the critical points in the imaginary part of the dielectric function occur at about 0.85, 1.45, 2.48, and 4.45 eV, respectively. These points are closely related to the direct band gaps $E_G^{Γ-Γ}$; 0.84, 1.44, 2.47, and 4.45 eV of $Cd_{1−x}Be_xSe$ for the corresponding values of x = 0. 25, 0.50, 0.75 and 1.

The calculated real parts of the complex dielectric function $\varepsilon_1(\omega)$ for $Cd_{1−x}Be_xSe$ are presented in Figure 3. It is clear from the figure that the static dielectric constant, $\varepsilon_1(\omega)$, is strongly dependent on the band gap of the compound. The calculated values of $\varepsilon_1(\omega)$ for $Cd_{1−x}Be_xSe$ at x = 0, 0.25, 0.50, 0.75 and 1.0 are 4.35, 3.94, 3.74, 3.57, and 3.42 for corresponding direct band gaps 0.39, 0.84, 1.44, 2.47, and 4.45 eV, respectively. These data explain that the smaller energy gap yields larger $\varepsilon_1(0)$ value. This inverse relation of

$\varepsilon_1(\omega)$ with the band gap can be explained by the Penn model [24] as follows:

$$\varepsilon_1(0) \approx 1 + \left(\frac{\hbar\omega_p}{E_g}\right)^2. \qquad (7)$$

The calculated values of the optical dielectric constant $\varepsilon(\omega)$ and refractive index $n(\omega)$ are listed in Table 3; comparison with the available data has been made where possible. As compared with other calculations, it seems that the values of $n(\omega)$ obtained from FP-LMTO method for the end-point compounds (i.e., CdSe and BeSe) are in good agreement with the theoretical results, together with the refractive index $n(\omega) = \sqrt{\varepsilon}$ at zero pressure. Note that ε is obtained from the zero-frequency limit of $\varepsilon_1(\omega)$, and it corresponds to the electronic part of the static dielectric constant of the material, a parameter of fundamental importance in many aspects of materials properties. It is clear from Figure 4 that

(a)

(b)

(c)

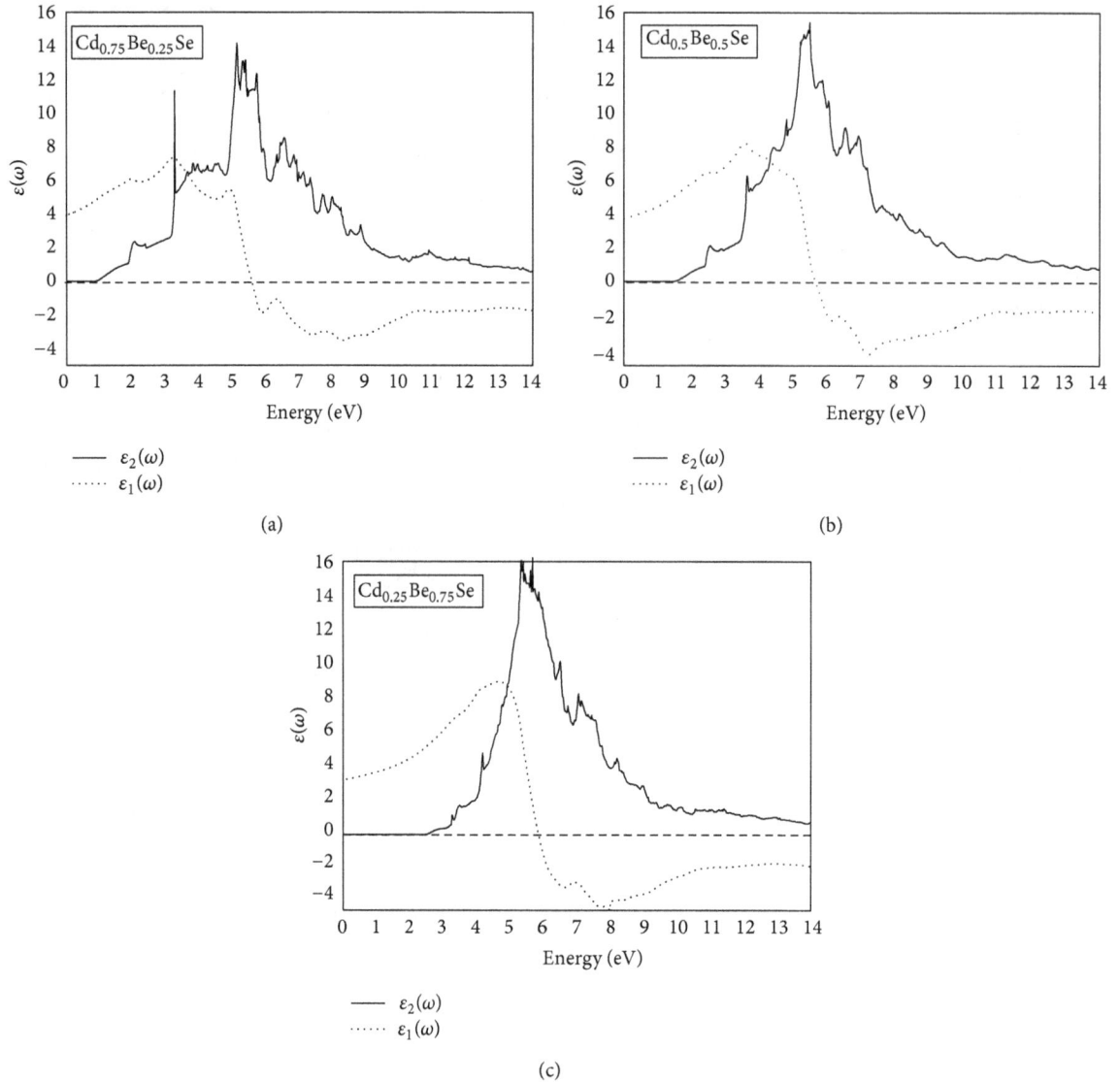

FIGURE 3: Frequency dependent imaginary part and real part of dielectric functions of $Cd_{1-x}Be_xSe$.

TABLE 3: Refractive index, optical dielectric constant of $Cd_{1-x}Be_xSe$ alloys for different compositions x.

x	Refractive index n			Optical dielectric constant ε		
	This work	Exp.	Other calc.	This work	Exp.	Other calc.
0	2.08	2.64[a]	2.47[b]	4.36	5.2[c]	5.05[d], 4.89[e]
0.25	1.98			3.947		
0.5	1.93			3.74		
0.75	1.89			3.57		
1	1.85			3.43	6.1[f]	6.09[g]

[a]Ref [44], [b]Ref [45], [c]Ref [46], [d]Ref [47], [e]Ref [48], [f]Ref [48], [g]Ref [49].

the refractive index of the material decreases with the increase in the Be concentration.

Figure 4 shows the variation of the computed static optical dielectric constant and static refractive index versus composition for $Cd_{1-x}Be_xSe$ alloys. The computed static optical dielectric constant and static refractive index versus composition were fitted by polynomial equation. The results are summarized as follows:

$$Cd_{1-x}Be_xSe \longrightarrow \begin{cases} \varepsilon(0) = 4.3400 - 1.5610x + 0.6662x^2 \\ n(0) = 2.0745 - 0.3685x + 0.1485x^2. \end{cases}$$

(8)

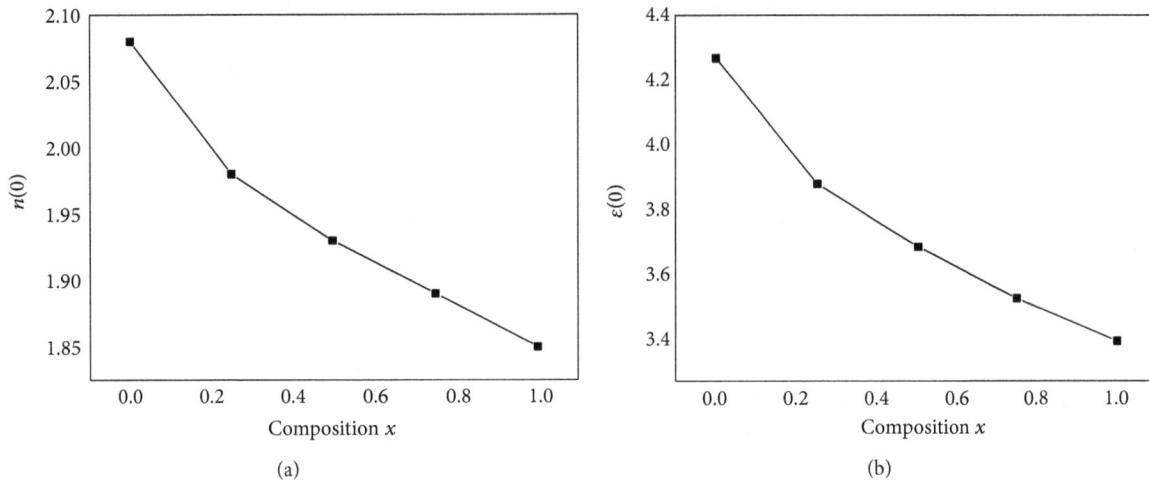

FIGURE 4: Computed static optical dielectric constant and static refractive index as function of composition for $Cd_{1-x}Be_xSe$.

4. Conclusions

Density functional calculations are carried out for the first time to investigate structural and optoelectronic properties $Cd_{1-x}Be_xSe$. Structure as well as bonding nature of the material significantly varies with Be concentration. The lattice constant of the crystal decreases linearly with x. The calculated band structure predicts that the alloys have direct band gaps, which increase with the increase in x. On the basis of wide range of fundamental direct band gaps (0.39–4.45 eV) and indirect band gaps (1.89–3.71 eV) between 0.25 and 0.75, it can be concluded that the material can be used in optoelectronic devices working in the IR, visible, and UV regions of the spectrum.

Conflict of Interests

The authors declare that there is no conflict of interests regarding the publication of this paper.

References

[1] H. Luo and J. K. Furdyna, "The II-VI semiconductor blue-green laser: challenges and solution," *Semiconductor Science and Technology*, vol. 10, no. 8, p. 1041, 1995.

[2] M. C. Tamargo, W. Lin, S. P. Guo, Y. Guo, Y. Luo, and Y. C. Chen, "Full-color light-emitting diodes from ZnCdMgSe/ZnCdSe quantum well structures grown on InP substrates," *Journal of Crystal Growth*, vol. 214-215, pp. 1058–1063, 2000.

[3] M. A. Haase, J. Qiu, J. M. DePuydt, and H. Cheng, "Blue-green laser diodes," *Applied Physics Letters*, vol. 59, no. 11, pp. 1272–1274, 1991.

[4] E. Kato, H. Noguchi, M. Nagai, H. Okuyama, S. Kijima, and A. Ishibashi, "Significant progress in II-VI blue-green laser diode lifetime," *Electronics Letters*, vol. 34, no. 3, pp. 282–284, 1998.

[5] F. Vigué, E. Tournié, and J. P. Faurie, "Zn(Mg)BeSe-based p-i-n photodiodes operating in the blue-violet and near-ultraviolet spectral range," *Applied Physics Letters*, vol. 76, p. 242, 2000.

[6] V. Bousquet, E. Tournié, M. Laugt, P. Vennegues, and J. P. Faurie, "Structural and optical properties of lattice-matched ZnBeSe layers grown by molecular-beam epitaxy onto GaAs substrates," *Applied Physics Letters*, vol. 70, p. 3564, 1997.

[7] A. Waag, F. Fischer, K. Schüll et al., "Laser diodes based on beryllium-chalcogenides," *Applied Physics Letters*, vol. 70, no. 3, pp. 280–282, 1997.

[8] J. Y. Zhang, D. Z. Shen, X. W. Fan, B. J. Yang, and Z. H. Zhang, "ZnBeSe epitaxy layers grown by photo-assisted metalorganic chemical vapor deposition," *Journal of Crystal Growth*, vol. 214-215, pp. 100–103, 2000.

[9] A. Muñoz, P. Rodríguez-Hernández, and A. Mujica, "Ground-state properties and high-pressure phase of beryllium chalcogenides BeSe, BeTe, and BeS," *Physical Review B*, vol. 54, p. 11861, 1996.

[10] A. Waag, F. Fischer, H. J. Lugauer et al., "Molecular-beam epitaxy of beryllium-chalcogenide-based thin films and quantum-well structures," *Journal of Applied Physics*, vol. 80, no. 2, pp. 792–796, 1996.

[11] O. Maksimov, S. P. Guo, and M. C. Tamargo, "Be-Chalcogenide Alloys for Improved R-G-B LEDs: $Be_xZn_yCd_{1-x-y}Se$ on InP," *Physica Status Solidi B*, vol. 229, no. 2, pp. 1005–1009, 2002.

[12] S. Y. Savrasov and D. Y. Savrasov, "Full-potential linear-muffin-tin-orbital method for calculating total energies and forces," *Physical Review B*, vol. 46, no. 19, pp. 12181–12195, 1992.

[13] S. Y. Savrasov, "Linear-response theory and lattice dynamics: a muffin-tin-orbital approach," *Physical Review B*, vol. 54, p. 16470, 1996.

[14] P. Hohenberg and W. Kohn, "Inhomogeneous electron gas," *Physical Review B*, vol. 136, no. 3, pp. B864–B871, 1964.

[15] W. Kohn and L. J. Sham, "Self-consistent equations including exchange and correlation effects," *Physical Review A*, vol. 140, no. 4, pp. A1133–A1138, 1965.

[16] S. Y. Savrasov, "Program LMTART for electronic structure calculations," *Zeitschrift fur Kristallographie*, vol. 220, no. 5-6, pp. 555–557, 2005.

[17] J. P. Perdew and Y. Wang, "Pair-distribution function and its coupling-constant average for the spin-polarized electron gas," *Physical Review B*, vol. 46, no. 20, pp. 12947–12954, 1992.

[18] P. Blochl, O. Jepsen, and O. K. Andersen, "Improved tetrahedron method for Brillouin-zone integrations," *Physical Review B*, vol. 49, p. 16223, 1994.

[19] F. D. Murnaghan, "The compressibility of media under extreme pressures," *Proceedings of the National Academy of Sciences of the United States*, vol. 30, no. 9, pp. 244–247, 1944.

[20] F. Wooten, *Optical Properties of Solids*, Academic Press, New York, NY, USA, 1972.

[21] M. Fox, *Optical Properties of Solids*, Oxford University Press, 2001.

[22] L. Vegard, "Die konstitution der mischkristalle und die raum-füllung der atome," *Zeitschrift für Physik*, vol. 5, no. 1, pp. 17–26, 1921.

[23] M. A. Khan, A. Kashyap, A. K. Solanki, T. Nautiyal, and S. Auluck, "Interband optical properties of Ni3Al," *Physical Review B*, vol. 48, no. 23, pp. 16974–16978, 1993.

[24] D. Penn, "Wave-number-dependent dielectric function of semiconductors," *Physical Review*, vol. 128, p. 2093, 1962.

[25] O. Zakharov, A. Rubio, X. Blase, M. L. Cohen, and S. G. Louie, "Quasiparticle band structures of six II-VI compounds: ZnS, ZnSe, ZnTe, CdS, CdSe, and CdTe," *Physical Review B*, vol. 50, no. 15, pp. 10780–10787, 1994.

[26] J. Heyd, J. E. Peralta, G. E. Scuseria, and R. L. Martin, "Energy band gaps and lattice parameters evaluated with the Heyd-Scuseria-Ernzerhof screened hybrid functional," *The Journal of Chemical Physics*, vol. 123, p. 174101, 2005.

[27] P. Y. Yu and M. Cardona, *Fundamentals of Semiconductors*, Springer, Berlin, Germany, 2001.

[28] Y. H. Chang, C. H. Park, K. Sato, and H. Katayama-Yoshida, "First-principles study of the effect of the superexchange interaction in (Ga, Mn)V ($V =$ N, P, As and Sb)," *Journal of the Korean Physical Society*, vol. 49, no. 1, pp. 203–208, 2006.

[29] H. Luo, K. Ghandehari, R. G. Greene, A. L. Ruoff, S. S. Trail, and F. J. DiSalvo, "Phase transformation ofBeSe and BeTe to the NiAs structure at high pressure," *Physical Review B*, vol. 52, p. 7058, 1995.

[30] C. Narayana, V. J. Nesamony, and A. L. Ruoff, "Phase transformation of BeS and equation-of-state studies to 96 GPa," *Physical Review B*, vol. 56, p. 14338, 1997.

[31] S. Laref and A. Laref, "Thermal properties of BeX ($X =$ S, Se and Te) compounds from *ab initio* quasi-harmonic method," *Computational Materials Science*, vol. 51, no. 1, pp. 135–140, 2012.

[32] M. Gonzalez-Diaz, P. Rodriguez-Hernandez, and A. Munoz, "Elastic constants and electronic structure of beryllium chalcogenides BeS, BeSe, and BeTefrom first-principles calculations," *Physical Review B*, vol. 55, p. 14043, 1997.

[33] N. Samarth, H. Luo, J. K. Furdyna et al., "Growth of cubic (zinc blende) CdSe by molecular beam epitaxy," *Applied Physics Letters*, vol. 54, no. 26, pp. 2680–2682, 1989.

[34] I. M. Tsidilkovski, *Band Structure of Semiconductors*, Elsevier Science & Technology Books, Amsterdam, The Netherlands, 1982.

[35] M. Cardona, "Fundamental reflectivity spectrum of semiconductors with zinc-blende structure," *Journal of Applied Physics*, vol. 32, no. 10, pp. 2151–2155, 1961.

[36] J. C. Salcedo-Reyes, "Electronic band structure of the ordered $Zn_{0.5}Cd_{0.5}Se$ alloy calculated by the semi-empirical tight-binding method considering second-nearest neighbor," *Universitas Scientiarum*, vol. 13, no. 2, pp. 198–207.

[37] O. Zakharov, A. Rubio, X. Blase, M. L. Cohen, and S. G. Louie, "Quasiparticle band structures of six II-VI compounds: ZnS, ZnSe, ZnTe, CdS, CdSe, and CdTe," *Physical Review B*, vol. 50, no. 15, pp. 10780–10787, 1994.

[38] P. J. Huang, Y. S. Huang, F. Firszt et al., "Photoluminescence and contactless electroreflectance characterization of $Be_xCd_{1-x}Se$ alloys," *Journal of Physics: Condensed Matter*, vol. 19, no. 2, 026208 pages, 2007.

[39] S. V. Ivanov et al., in *Abstracts of the 9th InternationalConference on ll-VI Compounds*, p. 209, Kyoto, Japan, 1999, (to be published in Journal of Crystal Growth, 2000).

[40] S. V. Ivanov, O. V. Nekrutkina, V. A. Kaygorodov et al., "Optical and structural properties of BeCdSe/ZnSe QW heterostructures grown by MBE," in *Proceedings of the 8th Nanostructures: Physics and Technology International Symposium*, St.Petersburg, Russia, June 2000.

[41] M. Gonzalez-Diaz, P. Rodriguez-Hernandez, and A. Munoz, "Elastic constants and electronic structure of beryllium chalcogenides BeS, BeSe, and BeTefrom first-principles calculations," *Physical Review B*, vol. 55, pp. 14043–14046, 1997.

[42] A. Fleszar and W. Hanke, "Electronic excitations in beryllium chalcogenides from the ab initio GW approach," *Physical Review B*, vol. 62, no. 4, pp. 2466–2474, 2000.

[43] W. M. Yim, J. B. Dismakes, E. J. Stofko, and R. J. Paff, "Synthesis and some properties of BeTe, BeSe and BeS," *Journal of Physics and Chemistry of Solids*, vol. 33, no. 2, pp. 501–505, 1972.

[44] N. A. Hamizi and M. R. Johan, "Optical properties of CdSe quantum dots via Non-TOP based route," *International Journal of Electrochemical Science*, vol. 7, pp. 8458–8467, 2012.

[45] S. Ouendadji, S. Ghemid, H. Meradji, and F. E. H. Hassan, "Density functional study of $CdS_{1-x}Se_x$ and $CdS_{1-x}Te_x$ alloys," *Computational Materials Science*, vol. 48, no. 1, pp. 206–211, 2010.

[46] T. M. Bieniewski and S. J. Czyzak, "Refractive indexes of single hexagonal ZnS and CdS crystals," *The Journal of the Optical Society of America*, vol. 53, no. 4, pp. 496–497, 1963.

[47] F. Kootstra, P. L. De Boeij, and J. G. Snijders, "Application of time-dependent density-functional theory to the dielectric function of various nonmetallic crystals," *Physical Review B*, vol. 62, p. 7071, 2000.

[48] G. P. Srivastava, H. M. Tutuncu, and N. Gunhan, "First-principles studies of structural, electronic, and dynamical properties of Be chalcogenides," *Physical Review B*, vol. 70, Article ID 085206, 2004.

[49] V. Wagner, J. J. Liang, R. Kruse et al., "Lattice dynamics and bond polarity of be-chalcogenides a new class of II-VI materials," *Physica Status Solidi B*, vol. 215, no. 1, pp. 87–91, 1999.

9

Herbal Plant Synthesis of Antibacterial Silver Nanoparticles by *Solanum trilobatum* and Its Characterization

M. Vanaja,[1] **K. Paulkumar,**[1] **G. Gnanajobitha,**[1] **S. Rajeshkumar,**[2]
C. Malarkodi,[1] **and G. Annadurai**[1]

[1] *Environmental Nanotechnology Division, Sri Paramakalyani Centre for Environmental Sciences,
Manonmaniam Sundaranar University, Alwarkurichi, Tamil Nadu 627412, India*
[2] *PG and Research Department of Biochemistry, Adhiparasakthi College of Arts and Science, Kalavai,
Tamil Nadu 632506, India*

Correspondence should be addressed to G. Annadurai; annananoteam@gmail.com

Academic Editor: Hao Wang

Green synthesis method of nanomaterials is rapidly growing in the nanotechnology field; it replaces the use of toxic chemicals and time consumption. In this present investigation we report the green synthesis of silver nanoparticles (AgNPs) by using the leaf extract of medicinally valuable plant *Solanum trilobatum*. The influence of physical and chemical parameters on the silver nanoparticle fabrication such as incubation time, silver nitrate concentration, pH, and temperature is also studied in this present context. The green synthesized silver nanoparticles were characterized by UV-vis spectroscopy, X-ray diffraction (XRD), scanning electron microscope (SEM), energy dispersive X-ray (EDX), and transmission electron microscope (TEM). The SEM and TEM confirm the synthesis of spherical shape of nanocrystalline particles with the size range of 2–10 nm. FTIR reveals that the carboxyl and amine groups may be involved in the reduction of silver ions to silver nanoparticles. Antibacterial activity of synthesized silver nanoparticles was done by agar well diffusion method against different pathogenic bacteria. The green synthesized silver nanoparticles can be used in the field of medicine, due to their high antibacterial activity.

1. Background

Nanotechnology deals with the synthesis of nanoparticles with controlled size, shape, and dispersity of materials at the nanometer scale length [1, 2]. Nanoparticles possess high surface area to volume ratio. Nanoparticles such as silver, gold, cadmium sulfide, zinc sulfide, and zinc oxide play important role in various fields [3–6]. Recently fabrication of silver nanoparticles has drawn considerable attention due to their physical and chemical properties and application in biomedicine, antiangiogenic activity against bovine retinal endothelial cells, anticancer activity against lung carcinoma cells [7], controlling HIV infection [8], detection of bacterial pathogens [9], and good catalytic activity [10]. Silver nanoparticles are having good history in the field of antimicrobial properties. The silver nanoparticles are vigorously involved in the antimicrobial activity against a lot of disease causing food borne and water borne pathogenic bacteria and fungus [6]. Some pathogenic microbes killed by silver nanoparticles are *Bacillus subtilis*, *Klebsiella planticola*, *Bacillus* sp., *Pseudomonas* sp. [11, 12], *S. aureus*, *Vibrio cholerae*, *Proteus vulgaris* and *P. aeruginosa* [13], *Shigella dysenteriae* type I, *Staphylococcus aureus*, *Citrobacter* sp., *Escherichia coli*, *Candida albicans*, and *Fusarium oxysporum* [14].

Synthesis of silver nanoparticles has been proved by various biological and green materials such as bacteria both gram positive and gram negative like *Klebsiella pneumonia* and *Bacillus subtilis* [15, 16], *Cladosporium cladosporioides* [17], marine algae *Padina tetrastromatica* and *Turbinaria conoides* [18, 19], the green waste peels of banana fruits [20], carbohydrate molecules like polysaccharide and disaccharides starch, sucrose, and maltose, and monosaccharides like glucose and fructose [21–23]. In the green materials mediated nanoparticles synthesis plant sources have major role in the past ten years. Many plants are used for synthesizing nanoparticles including *Cinnamomum camphora* [24], *Azadirachta* indica [25], Nelumbo nucifera [26], Garcinia mangostana [27], pomegranate, and grape fruit extracts [28, 29].

Plants have a lot of phytochemicals in their parts; they are applied in various fields. The biochemicals may play an important role in the nanoparticles synthesis [27]. Some biochemical compounds involved in synthesis of silver nanoparticles are proteins/enzymes and secondary metabolites such as terpenoids [30], some water-soluble polyhydroxy components such as alkaloids, flavonoids, and polysaccharose [31], metabolites (like organic acids and quinones) or metabolic fluxes and other oxidoreductively labile metabolites like ascorbates or catechol/photocatacheuic acid [32], verbascoside, isoverbascoside, luteolin, and chrysoeriol-7-O-diglucuronide [33], quercetin, and other phenolic compounds [34].

Solanum trilobatum Linn is an important medicinal plant of the family Solanaceae. The leaves contain rich amount of calcium, iron, phosphorus, carbohydrates, fat, crude fiber, and minerals [35]. This herb is used to treat common cold, cough, and asthma. This plant was used in the Siddha medicine system and treatment of respiratory diseases, chronic febrile infections, tuberculosis, and cardiac and liver diseases [36]. This plant also possesses the antibiotic, antimitotic, antibacterial [37, 38], and anticancer activities [39]. Sobatum, β-solamarine, solaine, solasodine, glycoalkaloid and diosgenin, and tomatidine are the constituents isolated from this plant [40]. Its pivotal action is cardiac, tonic, and carminative [41]. This plant has strong immunostimulatory effect due to the presence of alkaloids and carbohydrates. Due to these antioxidant and antibiotic properties this was used to synthesize silver nanoparticles.

In this present investigation the medically important plant is used for the synthesis of medically valued silver nanoparticles. The morphological, crystalline, and biochemical characters of green synthesized silver nanoparticles were analyzed by UV-vis spectrophotometer, scanning electron microscope, X-ray diffraction assay, transmission electron microscope, and Fourier transform infrared spectroscopy. Finally the medical property of the silver nanoparticle was characterized using antibacterial assay against Klebsiella planticola, Klebsiella pneumonia, Bacillus subtilis, E. coli, Serratia sp.,and Streptococcus sp.

2. Materials and Methods

Silver nitrate, Luria Bertani agar, and nutrient broth were purchased from Himedia, Mumbai. Leaves of Solanum trilobatum were collected from Sri Paramakalyani Centre for Environmental Sciences, MS University, Alwarkurichi.

2.1. Preparation of Leaf Extract. About 10 g of fresh leaves of S. trilobatum was thoroughly washed 2-3 times with distilled water for surface cleaning, and surface sterilized with 0.1% HgCl$_2$ for 1 min to reduce microbial contamination [42]. The sterile leaves were cut into fine pieces and boiled with 100 mL of double distilled water for 15 min at 60°C and filtered through Whatman number 1 filter paper and stored at 4°C in refrigerator for 2 weeks.

2.2. Synthesis of Silver Nanoparticles. In the typical synthesis of silver nanoparticles, 10 mL of leaf extract was treated with 90 mL of 1 mM silver nitrate solution and kept in room temperature. Subsequently the synthesis of silver nanoparticles was initially identified by brown colour formation and further monitored by measuring UV-vis spectra of the reaction mixture.

To study the effect of parameters such as reaction time, silver nitrate concentration, pH, and temperature on the nanoparticles synthesis the reaction was carried out by the following experiments. Silver nitrate and leaf extract reaction mixture was kept at room temperature and formation of nanoparticles was recorded at different functional times. Influences of silver nitrate concentration (1 to 5 mM, pH: 5.5, temperature: 35°C), pH (3.5, 4.5, 5.5, 7.5, and 9.5, silver nitrate: 1 mM, temperature: 35°C), and temperature (20°C, 35°C, 45°C, and 70°C, silver nitrate: 1 mM, pH: 5.5) were performed to find their effects on nanoparticles synthesis.

2.3. Characterization of Synthesized Silver Nanoparticles. Synthesis of silver nanoparticles was initially characterized by position of SPR band by measuring double beam UV-vis spectroscopy at different wavelengths from 360 to 700 nm. Crystal structure was characterized by XRD at 2θ ranges from 10 to 90° (Philips PW 1830); shape and size were analysed by using SEM (Philip model CM 200) and TEM (JEOL3010). Elemental composition was performed by EDAX (Philips XL-30). FTIR spectrum of silver nanoparticles was obtained on a SHIMADZU instrument with the sample as KBR pellet in thewave number region of 500–4,000 cm^{-1}.

2.4. Antibacterial Activity of Synthesized Silver Nanoparticles. The antibacterial activity of synthesized silver nanoparticles was performed by agar well diffusion method against pathogenic bacteria, Klebsiella planticola, Klebsiella pneumonia, Bacillus subtilis, E. coli, Serratia sp., and Streptococcus sp. Fresh overnight culture of each strain was swabbed uniformly onto the individuals' plates containing sterile Luria Bertani agar and 5 wells were made with the diameter of 6 mm. Then 25 μL of purified silver nanoparticles, leaf extract, and silver nitrate solution were poured into each well and commercial antibiotic discs are placed as control and incubate for 24 h at 37°C. After incubation the different levels of zonation formed around the well and it was measured. This experiment was repeated for three times.

3. Results and Discussion

3.1. Visual Observation. Silver nanoparticles formation was primarily identified by colour change visually. S. trilobatum leaf extract was treated with silver nitrate aqueous solution showed a colour change from yellow to brown within 2 min (Figure 1). The colour change was clear indication for the formation of silver nanoparticles [43]. This brown colour of silver nanoparticles arises due to the surface plasmon vibrations in the aqueous solution [44].

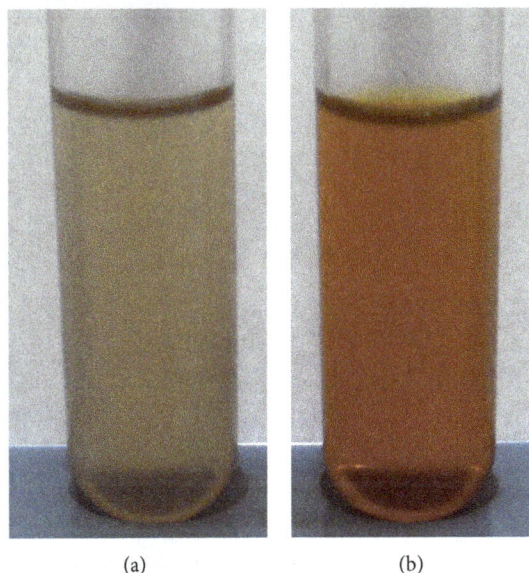

FIGURE 1: Formation of silver nanoparticles was identified by colour change (a) and leaf extract of *S. trilobatum* (b) after the reduction of silver nitrate.

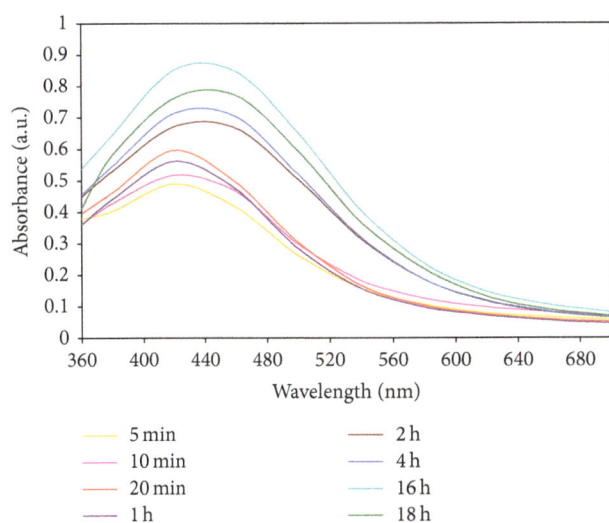

FIGURE 2: Effect of reaction time on silver nanoparticles synthesis by leaves extract of *S. trilobatum*.

3.2. UV-Vis Spectrophotometer

3.2.1. Effect of Reaction Time. Figure 2 shows the time dependant synthesis of silver nanoparticles. The UV-vis spectroscopy method can be used to track the size evolution of silver nanoparticles based on localized surface plasmon resonance band exhibited at different wavelengths. The optical properties of silver nanoparticles are related to excitation of plasmon resonance or interband transmission particularly on the size effect. Figure 2 shows the UV-vis spectra obtained from solution at different reaction times. The spectra show peaks at 420 nm at the time of 20 min. With the increase in reaction time, UV-vis spectra show sharp narrow peak

in 20 min which indicates the formation of disaggregated nanoparticles. This single and strong band indicates that the particles are isotropic in shape and uniform size [44]. After 20 min the peak shifts to 440 nm. By increasing the reaction time the synthesis of nanoparticles also increased by the leaf extract of *S. trilobatum*. Maximum production of nanoparticles was confirmed by maximum absorption which occurs in the UV-vis spectra. The narrow peak and increasing absorbance were observed from 5 min to 4 h without any shift of plasmon resonance band. The absorbance was increased at the incubation time of 16 h and broad band was formed at 460 nm. The reaction is completed at 4 h and is visually identified by appearance of precipitation in the bottom of the flask.

3.2.2. Effect of Silver Nitrate Concentration. In Figure 3 UV-vis spectra show the SPR band at 440 nm in the 1 mM concentration of silver nitrate. The silver nanoparticles were formed at 1 mM silver nitrate solution without aggregation and also it shows monodispersed nanoparticles formation. But the 2–4 mM concentration shows the band at 460 nm with aggregation; 5 mM concentration of silver nitrate solution treated with leaf extract shows the band at 500 nm with broad peak which indicates that the particles are polydispersed. The SPR spectra of silver nanoparticles broadened with the increase of the initial $AgNO_3$ concentration. Similarly reported by Huang et al. [45], synthesize the silver nanoparticles at different initial concentrations using leaf extract of *Cacumen platycladi*.

3.2.3. Effect of pH. Figure 4 shows the effect of pH on the synthesis of silver nanoparticles. The pH of leaf extract is found to be 6.8; the pH of extract was altered to 5.8, 7.8, and 8.8 to attain the maximum synthesis of silver nanoparticles. The lower pH suppresses the nanoparticles formation due

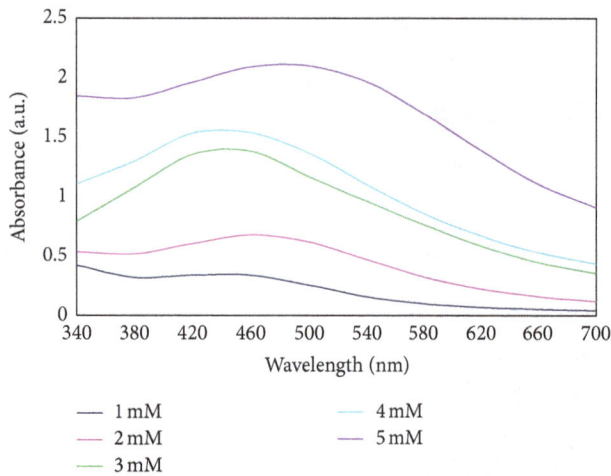

FIGURE 3: Effect of silver nitrate concentration on silver nanoparticles formation.

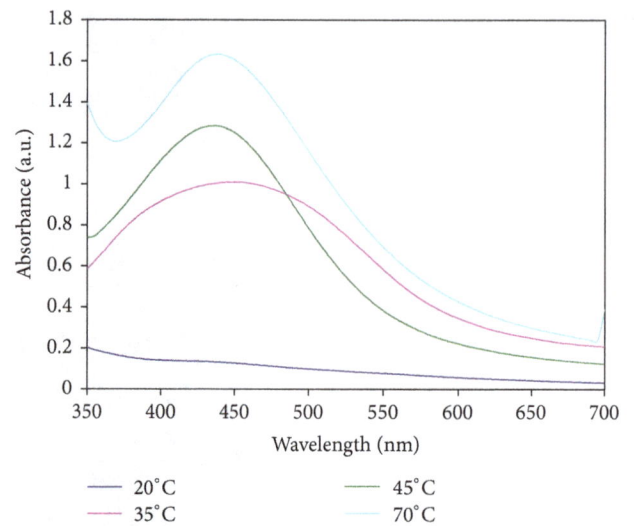

FIGURE 5: Effect of temperature on nanoparticles synthesis.

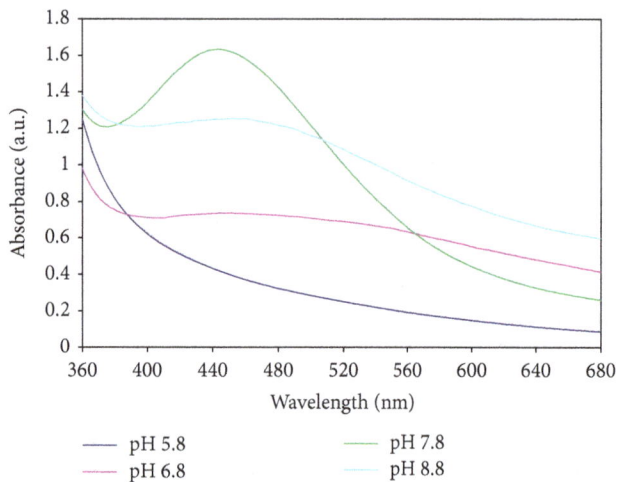

FIGURE 4: Effect of pH on silver nanoparticles synthesis.

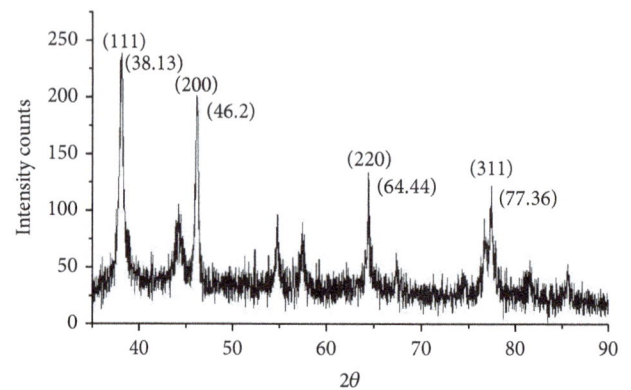

FIGURE 6: XRD analysis of *S. trilobatum* leaf extract mediated synthesized AgNPs.

to low availability of functional groups of leaf extract. At high pH of 8.8 the absorption band was obtained at 480 nm with broad peak. But the pH of 7.8 shows the narrow peak positioned at 440 nm with maximum production of silver nanoparticles.

3.2.4. Effect of Temperature. UV-vis spectra are obtained from solution at different reaction temperatures (Figure 5). In the low temperature of 20°C there is no SPR band for silver nanoparticles. The spectra show peaks at 440 nm (70°C), 450 nm (45°C), and 460 nm (35°C). These peaks are characteristic Plasmon band for silver nanoparticles. With the increase in reaction temperature UV-vis spectra show sharp narrow peak at lower wavelength regions 440 nm. Maximum production of silver nanoparticles was obtained at high temperature. The SPR peaks of the AgNPs significantly underwent blue shift, suggesting that the same particle shape strongly influences the SPR band in the aqueous solution at higher temperature [46].

3.3. X-Ray Diffraction Analysis. Structural and crystalline nature of the silver nanoparticles has been performed using XRD analysis. Figure 6 shows that the biosynthesized silver nanostructure by using *S. trilobatum* leaf extract was demonstrated and confirmed by the four characteristic peaks observed in the XRD image at 2θ values ranging from 30 to 90. The four intense peaks are 38.13°, 46.2°, 64.44°, and 77.36° corresponding to the planes of (111), (200), (220), and (311), respectively. These lattice planes were observed which may be indexed based on the face-centered crystal structure of silver (JCPDS file number 04-0783). The XRD pattern thus clearly showed that the Ag-NPs are crystalline in nature. Similar report was obtained using cell filtrate of *Streptomyces* sp. ERI-3 synthesized extracellularly [40].

3.4. Scanning Electron Microscopy. The SEM image (Figure 7) showing the high density Ag-NPs synthesized by using the leaf extract of *S. trilobatum* further confirmed the development of silver nanostructures. Obtained nanoparticle showed that Ag-NPs are spherical shaped and monodispersed and

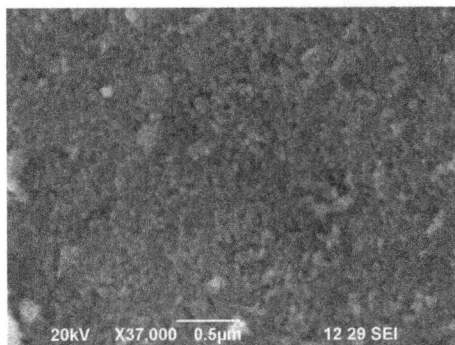

FIGURE 7: Morphology of silver nanoparticles showed by SEM.

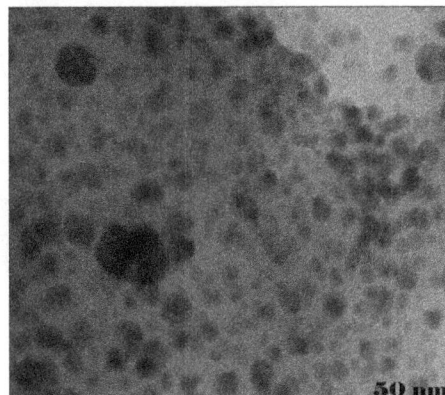

FIGURE 9: TEM analysis of synthesized silver nanoparticles shows well dispersed spherical shape nature.

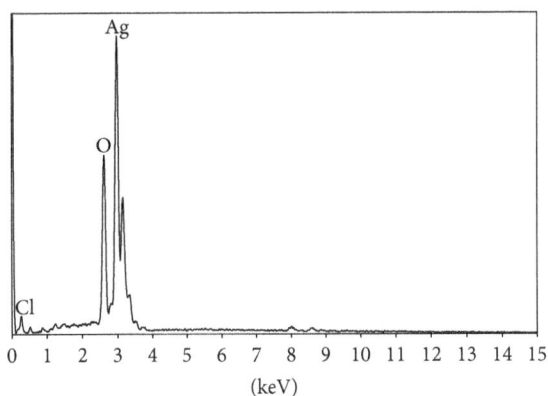

FIGURE 8: Elemental analysis of silver nanoparticles using leaf extract of S. trilobatum.

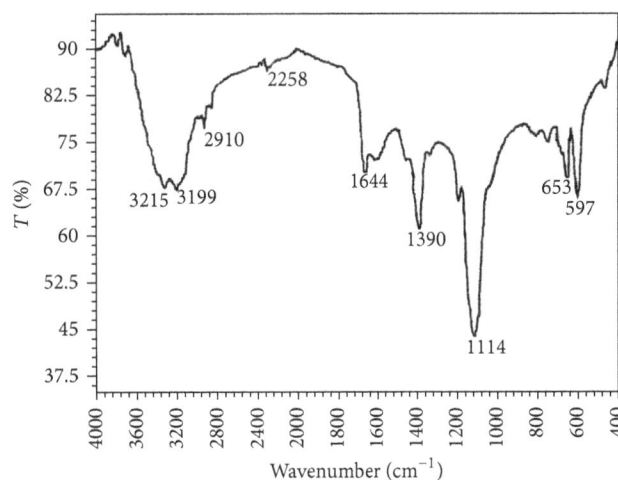

FIGURE 10: FT-IR spectrum of silver nanoparticles synthesized by S. trilobatum.

well distributed with aggregation in the size range about 50–70 nm (scale bar 500 nm). Similarly monodispersed silver nanoparticle was reported by using the extract of Coccinia grandis leaf extract [43].

3.5. Energy Dispersive X-Ray (EDX) Analysis.

Analysis through energy dispersive X-ray (EDX) spectrometers confirmed the presence of elemental silver signal of the silver nanoparticles. The vertical axis displays the number of X-ray counts whilst the horizontal axis displays energy in keV. The EDX spectrum (Figure 8) observed a strong signal from the silver atoms in the nanoparticles at 3 keV and weak signal from "Cl" and "O." These weak signals are from the plant organic constituents. This analysis revealed that the nanostructures formed were solely of silver.

3.6. Transmission Electron Microscopy.

Morphological and size characters of S. trilobatum mediated synthesized silver nanoparticles were characterized by TEM (Figure 9). TEM image revealed that the nanoparticles were evenly distributed without agglomeration. Synthesized silver nanoparticles were mostly spherical and their dimensional ranges were from 2 to 10 nm. Thus TEM characterization studies confirm that the synthesized silver nanoparticles were in nanometer size range.

3.7. Fourier Transform Infrared Spectroscopy.

Figure 10 shows that the FTIR image of S. trilobatum leaf mediated synthesized silver nanoparticles indicates presence of biomolecules involved in the reduction process. The minor peak found at 3215 cm^{-1} represents the –NH stretch of primary and secondary amines or amides. The same peak may be due to the –OH stretch of alcohols and phenols; the peak at 3199 cm^{-1} is due to the –OH stretch of carboxylic acids; the smaller peak at 2920 cm^{-1} shows –CH stretch of alkanes and –OH stretch of carboxylic acids, 2258 cm^{-1} corresponding to –CN stretch of nitriles; 1644 cm^{-1} indicates –CC– stretch of alkenes and –NH bend of primary amines, 1390 cm^{-1} for CH$_2$ and CH$_3$ deformation, CH$_3$ deformation, and CH$_2$ rocking; 1114 cm^{-1} shows –CN stretch of aliphatic amines; 653 and 597 cm^{-1} are corresponding to C–Cl and C–Br stretch of alkyl halides and –CH bend of alkynes. FT-IR reveals that carboxyl and amine groups may be involved in the reduction and stabilizing mechanism.

TABLE 1: Zone of inhibition against pathogenic bacteria using green synthesized silver nanoparticles.

Antibacterial agent	Zone inhibition (mm in diameter)					
	B. subtilis	*Streptococcus* sp.	*Serratia* sp.	*E. coli*	*K. planticola*	*K. pneumonia*
Leaf extract	6.4 ± 0.23	6.77 ± 0.14	7.00 ± 0.57	6.84 ± 0.12	6.63 ± 0.33	6.71 ± 0.18
Silver nitrate solution	9.0 ± 0.57	12.73 ± 0.14	11.40 ± 0.21	10.43 ± 0.18	10.3 ± 1.2	12.97 ± 0.12
Silver nanoparticles	11.0 ± 0.30	13.13 ± 0.13	12.33 ± 1.20	14.9 ± 0.07	14.1 ± 0.17	13.9 ± 0.09
Commercial antibiotic disc	10.8 ± 0.16	12.57 ± 0.23	11.33 ± 0.66	10.00 ± 0.57	10.3 ± 0.33	11.17 ± 0.60

3.8. Antibacterial Activity of Silver Nanoparticles. Antibacterial activity of synthesized silver nanoparticles was performed against *Streptococcus* sp., *Serratia* sp., *Bacillus subtilis*, *K. pneumonia*, *K. planticola*, and *E. coli* by well diffusion method. The antibacterial activity of synthesized silver nanoparticles was compared with plant extract, silver nitrate, and commercial antibiotic disc. The zone of inhibition was measured and denoted in millimeter (mm) in diameter. The zone of inhibition in diameter was tabulated by performing triplicate experiments (Table 1). Among the four antibacterial agents, silver nanoparticles highly inhibit the growth of pathogenic bacteria. Highest inhibition was noted against *E. coli*, *Streptococcus* sp., and *K. pneumonia*.

4. Conclusion

In this study we successfully demonstrated that *S. trilobatum* leaf extract has the ability to synthesize the nanoparticles. Nanoparticles synthesis initiated within 5 min and gradually increased up to 16 h. At pH 7.8 and temperature 70°C maximum yield of nanoparticles was observed. The size of the silver nanoparticles ranges from 50 to 70 nm, predominantly spherical shapes with crystalline nature. Elemental analysis by EDX shows that strong peak at 3 keV confirms the presence of silver nanoparticles. Carboxyl and amine groups from plant extract may be involved in the bioreduction process of silver ions to nanoparticles and stabilizing mechanism confirmed by FT-IR analysis. The silver nanoparticles show high antibacterial activity when assayed by agar well diffusion method. This green synthesized nanoparticle could be used in the medical field against human diseases due to their high efficiency as antibacterial agent.

Conflict of Interests

The authors declare that there is no conflict of interests regarding the publication of this paper.

Acknowledgment

The authors gratefully acknowledge the DST-FIST sponsored programme, DST, New Delhi, India, for funding the research development (reference no. S/FST/ESI-101/2010) to carry out this work.

References

[1] K. Kalishwaralal, V. Deepak, S. Ramkumarpandian, H. Nellaiah, and G. Sangiliyandi, "Extracellular biosynthesis of silver nanoparticles by the culture supernatant of *Bacillus licheniformis*," *Materials Letters*, vol. 62, no. 29, pp. 4411–4413, 2008.

[2] K. Paulkumar, G. Gnanajobitha, M. Vanaja et al., "*Piper nigrum* leaf and stem assisted green synthesis of silver nanoparticles and evaluation of its antibacterial activity against agricultural plant pathogens," *The Scientific World Journal*, vol. 2014, Article ID 829894, 9 pages, 2014.

[3] C. Malarkodi, S. Rajeshkumar, K. Paulkumar, M. Vanaja, G. Gnanajobitha, and G. Annadurai, "Biosynthesis and antimicrobial activity of semiconductor nanoparticles against oral pathogens," *Bioinorganic Chemistry and Applications*, vol. 2014, Article ID 347167, 11 pages, 2014.

[4] M. Vanaja, K. Paulkumar, M. Baburaja et al., "Degradation of methylene blue using biologically synthesized silver nanoparticles," *Bioinorganic Chemistry and Applications*, vol. 2014, Article ID 742346, 8 pages, 2014.

[5] S. Rajeshkumar, C. Malarkodi, K. Paulkumar, M. Vanaja, G. Gnanajobitha, and G. Annadurai, "Algae mediated green fabrication of silver nanoparticles and examination of its antifungal activity against clinical pathogens," *International Journal of Metals*, vol. 2014, Article ID 692643, 8 pages, 2014.

[6] K. Paulkumar, S. Rajeshkumar, G. Gnanajobitha et al., "Biosynthesis of silver chloride nanoparticles using *Bacillus subtilis* MTCC 3053 and assessment of its antifungal activity," *ISRN Nanomaterials*, vol. 2013, Article ID 317963, 8 pages, 2013.

[7] M. Valodkar, P. S. Nagar, R. N. Jadeja, M. C. Thounaojam, R. V. Devkar, and S. Thakore, "Euphorbiaceae latex induced green synthesis of non-cytotoxic metallic nanoparticle solutions: a rational approach to antimicrobial applications," *Colloids and Surfaces A: Physicochemical and Engineering Aspects*, vol. 384, no. 1–3, pp. 337–344, 2011.

[8] J. L. Elechiguerra, J. L. Burt, J. R. Morones et al., "Interaction of silver nanoparticles with HIV-1," *Journal of Nanobiotechnology*, vol. 3, article 6, 2005.

[9] X. Zhao, L. R. Hilliard, S. J. Mechery et al., "A rapid bioassay for single bacterial cell quantitation using bioconjugated nanoparticles," *Proceedings of the National Academy of Sciences of the United States of America*, vol. 101, no. 42, pp. 15027–15032, 2004.

[10] A. Nagy and G. Mestl, "High temperature partial oxidation reactions over silver catalysts," *Applied Catalysis A: General*, vol. 188, no. 1-2, pp. 337–353, 1999.

[11] M. Vanaja, G. Gnanajobitha, K. Paulkumar, S. Rajeshkumar, C. Malarkodi, and G. Annadurai, "Phytosynthesis of silver nanoparticles by *Cissus quadrangularis*: influence of physicochemical parameters," *Journal of Nanostructure in Chemistry*, vol. 3, no. 17, pp. 1–8, 2013.

[12] M. Vanaja and G. Annadurai, "Coleus aromaticus leaf extract mediated synthesis of silver nanoparticles and its bactericidal activity," *Applied Nanoscience*, vol. 3, pp. 217–223, 2013.

[13] N. Prabhu, D. T. Raj, K. Yamuna Gowri, S. Ayisha Siddiqua, and D. Joseph Puspha Innocent, "Synthesis of silver phyto nanoparticles and their antibacterial efficacy," *Digest Journal of Nanomaterials and Biostructures*, vol. 5, no. 1, pp. 185–189, 2010.

[14] J. Musarrat, S. Dwivedi, B. R. Singh, A. A. Al-Khedhairy, A. Azam, and A. Naqvi, "Production of antimicrobial silver nanoparticles in water extracts of the fungus *Amylomyces rouxii* strain KSU-09," *Bioresource Technology*, vol. 101, no. 22, pp. 8772–8776, 2010.

[15] A. R. Shahverdi, S. Minaeian, H. R. Shahverdi, H. Jamalifar, and A.-A. Nohi, "Rapid synthesis of silver nanoparticles using culture supernatants of *Enterobacteria*: a novel biological approach," *Process Biochemistry*, vol. 42, no. 5, pp. 919–923, 2007.

[16] K. Paulkumar, S. Rajeshkumar, G. Gnanajobitha, M. Vanaja, C. Malarkodi, and G. Annadurai, "Eco-friendly synthesis of silver chloride nanoparticles using *Klebsiella planticola* (MTCC 2277)," *International Journal of Green Chemistry and Bioprocess*, vol. 3, no. 1, pp. 12–16, 2013.

[17] D. S. Balaji, S. Basavaraja, R. Deshpande, D. B. Mahesh, B. K. Prabhakar, and A. Venkataraman, "Extracellular biosynthesis of functionalized silver nanoparticles by strains of *Cladosporium cladosporioides* fungus," *Colloids and Surfaces B: Biointerfaces*, vol. 68, no. 1, pp. 88–92, 2009.

[18] S. Rajeshkumar, C. Kannan, and G. Annadurai, "Green synthesis of silver nanoparticles using marine brown Algae turbinaria conoides and its antibacterial activity," *International Journal of Pharma and Bio Sciences*, vol. 3, no. 4, pp. 502–510, 2012.

[19] S. Rajeshkumar, C. Kannan, and G. Annadurai, "Synthesis and characterization of antimicrobial silver nanoparticles using marine brown seaweed Padina tetrastromatica," *Drug Invention Today*, vol. 4, no. 10, pp. 511–513, 2012.

[20] A. Bankar, B. Joshi, A. R. Kumar, and S. Zinjarde, "Banana peel extract mediated novel route for the synthesis of silver nanoparticles," *Colloids and Surfaces A: Physicochemical and Engineering Aspects*, vol. 368, no. 1–3, pp. 58–63, 2010.

[21] D. Manno, E. Filippo, M. Di Giulio, and A. Serra, "Synthesis and characterization of starch-stabilized Ag nanostructures for sensors applications," *Journal of Non-Crystalline Solids*, vol. 354, no. 52–54, pp. 5515–5520, 2008.

[22] E. Filippo, A. Serra, A. Buccolieri, and D. Manno, "Green synthesis of silver nanoparticles with sucrose and maltose: morphological and structural characterization," *Journal of Non-Crystalline Solids*, vol. 356, no. 6–8, pp. 344–350, 2010.

[23] S. Panigrahi, S. Kundu, S. K. Ghosh, S. Nath, and T. Pal, "Sugar assisted evolution of mono- and bimetallic nanoparticles," *Colloids and Surfaces A: Physicochemical and Engineering Aspects*, vol. 264, no. 1–3, pp. 133–138, 2005.

[24] J. Huang, Q. Li, D. Sun et al., "Biosynthesis of silver and gold nanoparticles by novel sundried *Cinnamomum camphora* leaf," *Nanotechnology*, vol. 18, no. 10, Article ID 105104, 2007.

[25] S. S. Shankar, A. Rai, A. Ahmad, and M. Sastry, "Rapid synthesis of Au, Ag, and bimetallic Au core-Ag shell nanoparticles using Neem (*Azadirachta indica*) leaf broth," *Journal of Colloid and Interface Science*, vol. 275, no. 2, pp. 496–502, 2004.

[26] P. Karthiga, R. Soranam, and G. Annadurai, "Alpha-mangostin, the major compound from *Garcinia mangostana* Linn. Responsible for synthesis of Ag Nanoparticles: its characterization and Evaluation studies," *Research Journal of Nanoscience and Nanotechnology*, vol. 2, no. 2, pp. 46–57, 2012.

[27] T. Santhoshkumar, A. A. Rahuman, G. Rajakumar et al., "Synthesis of silver nanoparticles using *Nelumbo nucifera* leaf extract and its larvicidal activity against malaria and filariasis vectors," *Parasitology Research*, vol. 108, no. 3, pp. 693–702, 2011.

[28] G. Gnanajobitha, S. Rajeshkumar, G. Annadurai, and C. Kannan, "Preparation and characterization of fruit-mediated silver nanoparticles using pomegranate extract and assessment of its antimicrobial activities," *Journal of Environmental Nanotechnology*, vol. 2, no. 1, pp. 4–10, 2013.

[29] G. Gnanajobitha, K. Paulkumar, M. Vanaja et al., "Fruit-mediated synthesis of silver nanoparticles using *Vitis vinifera* and evaluation of their antimicrobial efficacy," *Journal of Nanostructure in Chemistry*, vol. 3, no. 67, pp. 1–6, 2013.

[30] S. S. Shankar, A. Ahmad, and M. Sastry, "Geranium leaf assisted biosynthesis of silver nanoparticles," *Biotechnology Progress*, vol. 19, no. 6, pp. 1627–1631, 2003.

[31] L. Lin, W. Wang, J. Huang et al., "Nature factory of silver nanowires: plant-mediated synthesis using broth of *Cassia fistula* leaf," *Chemical Engineering Journal*, vol. 162, no. 2, pp. 852–858, 2010.

[32] A. K. Jha, K. Prasad, and A. R. Kulkarni, "Plant system: nature's nanofactory," *Colloids and Surfaces B: Biointerfaces*, vol. 73, no. 2, pp. 219–223, 2009.

[33] D. Cruz, P. L. Falé, A. Mourato, P. D. Vaz, M. Luisa Serralheiro, and A. R. L. Lino, "Preparation and physicochemical characterization of Ag nanoparticles biosynthesized by *Lippia citriodora* (Lemon Verbena)," *Colloids and Surfaces B: Biointerfaces*, vol. 81, no. 1, pp. 67–73, 2010.

[34] A. I. Lukman, B. Gong, C. E. Marjo, U. Roessner, and A. T. Harris, "Facile synthesis, stabilization, and anti-bacterial performance of discrete Ag nanoparticles using *Medicago sativa* seed exudates," *Journal of Colloid and Interface Science*, vol. 353, no. 2, pp. 433–444, 2011.

[35] M. Jawhar, G. AmalanRabert, and M. Jeyaseelan, "Rapid proliferation of multiple shoots in *Solanum trilobatum* L," *Plant Tissue Culture*, vol. 14, no. 2, pp. 107–112, 2004.

[36] M. Shahjahan, K. E. Sabitha, M. Jainu, and C. S. S. Devi, "Effect of *Solanum trilobatum* against carbon tetrachloride induced hepatic damage in albino rats," *Indian Journal of Medical Research*, vol. 120, no. 3, pp. 194–198, 2004.

[37] S. J. P. Jacob and S. Shenbagaraman, "Evaluation of antioxidant and antimicrobial activities of the selected green leafy vegetables," *International Journal of PharmTech Research*, vol. 3, no. 1, pp. 148–152, 2011.

[38] A. Doss, H. M. Mubarack, and R. Dhanabalan, "Antibacterial activity of tannins from the leaves of *Solanum trilobatum* Linn," *Indian Journal of Science and Technology*, vol. 2, no. 2, pp. 41–43, 2009.

[39] P. V. Mohanan and K. S. Devi, "Cytotoxic potential of the preparations from *Solanum trilobatum* and the effect of sobatum on tumour reduction in mice," *Cancer Letters*, vol. 110, no. 1-2, pp. 71–76, 1996.

[40] J. Subramani, P. C. Josekutty, A. R. Mehta, and P. N. Bhatt, "Solasodine levels in *Solanum sisymbriifolium* Lam," *Indian Journal of Experimental Biology*, vol. 27, no. 2, p. 189, 1989.

[41] A. Ahmad, P. Mukherjee, S. Senapati et al., "Extracellular biosynthesis of silver nanoparticles using the fungus *Fusarium oxysporum*," *Colloids and Surfaces B: Biointerfaces*, vol. 28, no. 4, pp. 313–318, 2003.

[42] V. Singh, A. Tyagi, P. K. Chauhan, P. Kumari, and S. Kaushal, "Identification and prevention of bacterial contimination on

explant used in plant tissue culture labs," *International Journal of Pharmacy and Pharmaceutical Sciences*, vol. 3, no. 4, pp. 160–163, 2011.

[43] R. Arunachalam, S. Dhanasingh, B. Kalimuthu, M. Uthirappan, C. Rose, and A. B. Mandal, "Phytosynthesis of silver nanoparticles using *Coccinia grandis* leaf extract and its application in the photocatalytic degradation," *Colloids and Surfaces B: Biointerfaces*, vol. 94, pp. 226–230, 2012.

[44] B. Mahitha, B. D. P. Raju, G. R. Dillip et al., "Biosynthesis, characterization and antimicrobial studies of Ag NPs extract from *Bacopa monniera* whole plant," *Digest Journal of Nanomaterials and Biostructures*, vol. 6, pp. 135–142, 2011.

[45] J. Huang, G. Zhan, B. Zheng et al., "Biogenic silver nanoparticles by *Cacumen Platycladi* extract: synthesis, formation mechanism, and antibacterial activity," *Industrial and Engineering Chemistry Research*, vol. 50, no. 15, pp. 9095–9106, 2011.

[46] M.-C. Daniel and D. Astruc, "Gold nanoparticles: assembly, supramolecular chemistry, quantum-size-related properties, and applications toward biology, catalysis, and nanotechnology," *Chemical Reviews*, vol. 104, no. 1, pp. 293–346, 2004.

10

Corrosion Behavior of Carbon Steel in Synthetically Produced Oil Field Seawater

Subir Paul, Anjan Pattanayak, and Sujit K. Guchhait

Department of Metallurgical and Material Engineering, Jadavpur University, Kolkata 700032, India

Correspondence should be addressed to Subir Paul; spmet4@gmail.com

Academic Editor: Chi Tat Kwok

The life of offshore steel structure in the oil production units is decided by the huge corrosive degradation due to SO_4^{2-}, S^{2-}, and Cl^-, which normally present in the oil field seawater. Variation in pH and temperature further adds to the rate of degradation on steel. Corrosion behavior of mild steel is investigated through polarization, EIS, XRD, and optical and SEM microscopy. The effect of all 3 species is huge material degradation with FeS_x and $FeCl_3$ and their complex as corrosion products. EIS data match the model of Randle circuit with Warburg resistance. Addition of more corrosion species decreases impedance and increases capacitance values of the Randle circuit at the interface. The attack is found to be at the grain boundary as well as grain body with very prominent sulphide corrosion crack.

1. Introduction

The severe corrosion of the submersed structures in the oil field at the production site and crude oil transportation is unpredictable and is a major component of the total corrosion loss in oil and gas industries. The corrosion species in the aqueous oil field seawater are CO_3^{2-}, S^{2-}, Cl^-, SO_4^{2-}, and O (Table 1) [1] which are also influenced by the variation of pH and temperature. CO_3^{2-} and S^{2-} are formed from CO_2 and H_2S of the oil in the aqueous environment. And Cl^-, SO_4^{2-}, and O are present in the seawater. Besides these parameters, there are fluid dynamics of sea water and suspended solids and sands, influencing the erosion corrosion of the marine structures. Crude oil and natural gas can carry various high-impurity products which are inherently corrosive. In the case of oil and gas wells and pipelines, such highly corrosive media are carbon dioxide (CO_2), hydrogen sulfide (H_2S), and free water [2].

The effect of any individual parameter on corrosion rate has been studied extensively [3–6]. But the conjoint effect of the above mentioned parameters and interfering effects and interactions are complex and are not very well understood. The salts and sulfide compounds dissolved in crude oil can provoke the formation of a corrosive aqueous solution whose chemical composition involves the presence

of both hydrochloric acid (HCI) and hydrogen sulfide (H_2S) [3, 4]. Corrosion mitigation in the oil field industry has traditionally been performed by combining methods for measuring the corrosion rates such as corrosion coupons and regular pipeline inspections with prevention strategies [6]. But that required years to get empirical results and could not be applied to other geographical locations of different sea water chemistry. All the factors make the corrosion mechanisms in the oil fields very complex with high degree of interaction among the species. Several previous studies have been performed related to the corrosion process of iron and steel in H_2S solutions [4, 7–13]. These works studied the influence of H_2S on the corrosion phenomena at ambient temperature. In H_2S-containing solutions, the corrosion process of metal may be accompanied by the formation of a sulfide film on the metal surface and leads to more complicated corrosion behavior. Previous researches [14–16] have shown that H_2S had a remarkable acceleration effect on both the anodic iron dissolution and the cathodic evolution in most cases but H_2S may exhibit an inhibitive effect on the corrosion of iron or steel weld. Recently, the influence of H_2S concentration on the corrosion behavior of carbon steel at 90°C has been investigated [15]. Physical modeling of ships and offshore structures in ocean water by Melchers et al. [17–19] and Shehadeh and Hassan [20] adds to

TABLE 1: Ions present in typical oil field seawater.

Species typically found in oil field brines		Element		Concentration mg/L
CO_2	Dissolved carbon dioxide	Barium	Ba^{2+}	31
H_2CO_3	Carbonic acid	Boron	B	6
HCO_3^-	Bicarbonate ion	Calcium	Ca^{2+}	284
CO_3^{2-}	Carbonate ion	Iron	Fe^{3+}	55.85
H^+	Hydrogen ion	Magnesium	Mg^{2+}	24.31
OH^-	Hydroxide ion	Phosphorous	P^{3-}	1
Fe^{2+}	Iron ion	Potassium	K	50
Cl^-	Chloride ion	Sodium	Na	4770
Na^+	Sodium ion	Strontium	Sr^{2+}	83
K^+	Potassium	Chloride	Cl^-	7480
Ca^{2+}	Calcium ion	Bromide	Br^-	20
Mg^{2+}	Magnesium ion	Sulphate	SO_4^{2-}	21
Ba^{2+}	Barium ion	Nitrate	NO_3^-	0.50
Sr^{2+}	Strontium ion	Hydroxyl	OH^-	0
CH_3COOH (HAc)	Acetic acid	Carbonate	CO_3^{2-}	0
CH_3COO^- (Ac^-)	Acetate ion	Bicarbonate	HCO_3^-	500
HSO_4^-	Bisulphate ion	Dissolved CO_2	CO_2	92.4
SO_4^{2-}	Sulphate ion	Specific gravity		1.014
		pH		6.58
		Resistivity		0.4405 Ohm
		Total dissolved solids 1		3.453 mg

better understanding of the present investigation. However, little research has been done on the corrosion behavior of carbon steel in the presence of both H_2S and NaCl at ambient and elevated temperature.

Corrosion mechanisms in oil field systems are complex and are showing high degrees of interaction between corrosion species, products, and oil field metallurgies. The interactions of sulfate and chloride are of interest in this work, since presence of sulfate ions, in oilfield produced water, strongly influence corrosion mechanisms. While there are many research works on the effects of CO_2 and H_2S on corrosion of carbon steel, those of conjoint effects of S^{2-}, Cl^-, and SO_4^{2-} are much less. The present investigation aims to study the conjoint effects of S^{2-}, Cl^-, and SO_4^{2-} along with variation of pH and temperature on carbon steel. The corrosive species included are sulfate, chloride, hydrogen sulfide, temperature, and pH. Sulphate and chloride were added as Na_2SO_4 and NaCl. Hydrogen sulfide was introduced to the corrosion cell with the following reaction:

$$Na_2S + H_2SO_4 = H_2S + FeSO_4 \qquad (1)$$

The effect on corrosion of these species was examined through polarization experimentation using a three-electrode glass corrosion cell and potentiostat. Electrochemical AC impedance spectroscopy studies were also carried out for better understanding of electrochemical effects of corrosive species on electrical phenomenon occurring at metal-solution interface. The corroded and uncorroded substrates were characterized by XRD. The morphology of the corroded surface was investigated by optical microscopy and SEM.

2. Experimental Methods

2.1. Polarization Studies. Electrochemical measurements were conducted using Gamry Potentiostat instrument coupled with Echem analyst software, controlled by a personal computer, in a conventional three-electrode cell systems. The working electrode was carbon steel, the counter electrode was graphite, and a saturated calomel electrode (SCE) acted as the reference electrode. Experiments were performed in different concentrations of Cl^-, SO_4^{2-}, and S^{2-} solutions, at preselected pH and temperature, to determine the corrosion potential E_{corr} and corrosion current i_{corr}. The potential was scanned between −1.5 V and 1 V at a scan rate of 1 mV/s.

2.2. Electrochemical Impedance Spectroscopy (EIS). The experimental arrangement was the same as that of polarization studies. The electrochemical cell was connected to an impedance analyzer (EIS300 controlled by Echem analyst software) for electrochemical impedance spectroscopy. The electrochemical impedance spectra were obtained at frequencies between 300 kHz and 0.01 Hz. The amplitude of the sinusoidal wave was 10 mV. The following results and information are obtained from the EIS experiments: Polarization resistance (R_p), electrolyte resistance (R_u), double layer capacitance (C_{dl}), capacitive load or constant phase element, CPE(Y), and α which is defined from the capacitive impedance equation $Z = 1/C(jw)^{-\alpha}$.

Capacitors in EIS experiments often do not behave ideally. Instead, they act like a constant phase element (CPE). The exponent $\alpha = 1$ for pure capacitance. For a constant phase element, the exponent α is less than one. The "double layer capacitor" on real cells often behaves like a CPE instead of like a pure capacitor.

2.3. X-Ray Diffraction (XRD) Analysis. The X-ray diffraction technique is used to define the crystalline structure and the crystalline phases. This test was done using a Rigaku Ultima III X-Ray Diffractometer for recording the diffraction traces of the samples with monochromatized Cu K_α radiation, at room temperature; the scan region (2θ) ranged from 10° to 100° at a scan rate of 5° min^{-1}.

2.4. Scanning Electron Microscope (SEM) Morphology. The electron micrographs were studied by SEM with accelerating voltage 30 kV, magnification up to 300,000x, and resolution of 3.5 nm. The images of the corroded samples were photographed at low and high magnification.

3. Results and Discussions

The effects of Cl$^-$, SO$_4^{2-}$, S^{2-}, pH, and temperature on degradation behavior of carbon steel were studied by potentiostatic polarization to determine corrosion current and corrosion potential. The various electrical properties at the metal-solution interface were determined by electrochemical impedance spectroscopy (EIS). The presence of different elements on corroded surface was detected by XRD. The morphology of the degraded surfaces was characterized by optical microscopy and SEM. Before going into the experimental findings of the effects of different interfering ions, it is worthwhile to discuss the basic electrochemical reactions of aqueous corrosion of steel in the presence of those ions.

The main electrochemical anodic and cathodic reactions for the corrosion of carbon steel in aqueous oil fields environments in presences of the ions are as follows.

Half Cell Reactions (E versus SCE). Consider

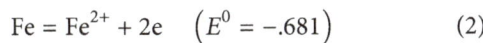

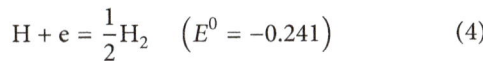

$$Fe = Fe^{2+} + 2e \quad \left(E^0 = -.681\right) \tag{2}$$

$$O_2 + 2H_2O + 4e^- \longrightarrow 4OH^- \quad \left(E^0 = 0.579\right) \tag{3}$$

$$H + e = \frac{1}{2}H_2 \quad \left(E^0 = -0.241\right) \tag{4}$$

In presence of SO$_4^{2-}$, present reactions are

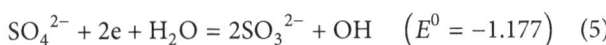

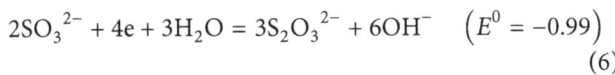

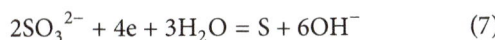

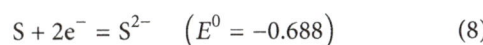

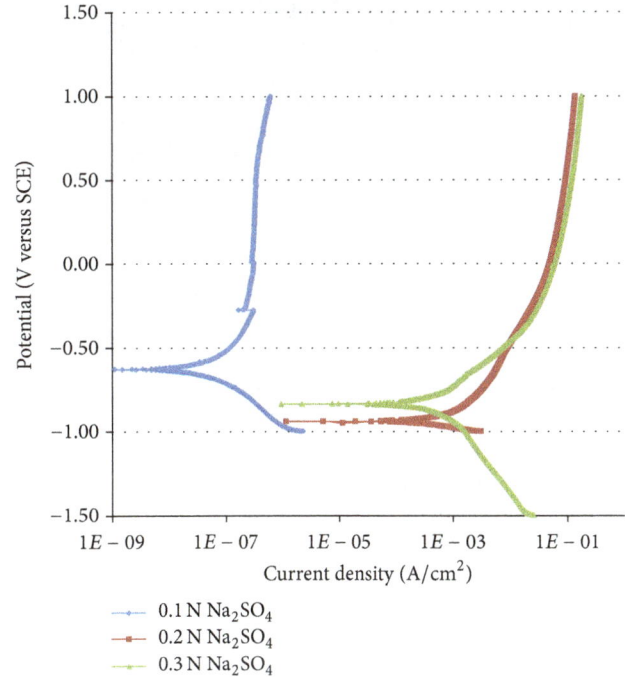

$$SO_4^{2-} + 2e + H_2O = 2SO_3^{2-} + OH \quad \left(E^0 = -1.177\right) \tag{5}$$

$$2SO_3^{2-} + 4e + 3H_2O = 3S_2O_3^{2-} + 6OH^- \quad \left(E^0 = -0.99\right) \tag{6}$$

$$2SO_3^{2-} + 4e + 3H_2O = S + 6OH^- \tag{7}$$

$$S + 2e^- = S^{2-} \quad \left(E^0 = -0.688\right) \tag{8}$$

FIGURE 1: Potentiodynamic polarization curves of low carbon steel in different concentration of Na$_2$SO$_4$ solution at pH 6 and temperature 25°C.

In presence of Na$_2$S or H$_2$S (Na$_2$S was added in state of H$_2$S to understand the effect of s^{2-}),

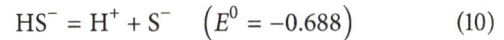

$$H_2S = H^+ + HS^- \tag{9}$$

$$HS^- = H^+ + S^- \quad \left(E^0 = -0.688\right) \tag{10}$$

In the presence of Cl$^-$, that is (NaCl/HCl),

$$Fe^{2+} + 2Cl^- = FeCl_2 \tag{11}$$

$$FeCl_2 + 2H_2O = Fe(OH)_2 + 2H^+ + 2Cl^- \tag{12}$$

$$FeCl_2 + Cl^- = FeCl_3 + e \tag{13}$$

$$Fe + 2H^+ = Fe^{2+} \tag{14}$$

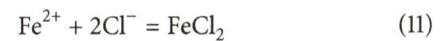

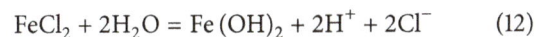

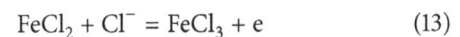

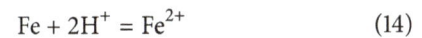

The above equations would help in better understanding of the effects of the corrosion species found in the experimental results.

3.1. Polarization Studies

3.1.1. Effect of SO$_4^{2-}$. Figure 1 shows the potential dynamic polarization curve with increasing SO$_4^{2-}$ concentration at pH 6. The pH of sea water normally varies from 7.5 to 8.2. but in the oil fields due to the presence of few acidic substances, namely, carbonic acid, H$_2$S, and other organic acids, pH may shift from near neutral towards the acidic side between 6 and 4. It is seen here that the corrosion rate increases with increase in concentration of SO$_4^{2-}$. It is seen from (5)–(8) that

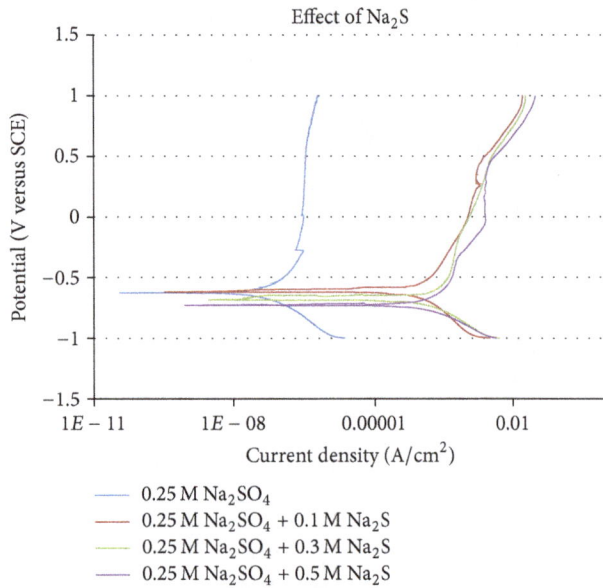

FIGURE 2: Effect of S^{2-} on potentiodynamic polarization curves of carbon steel in 0.25 M Na_2SO_4 solution at pH 6 and temperature 25°C.

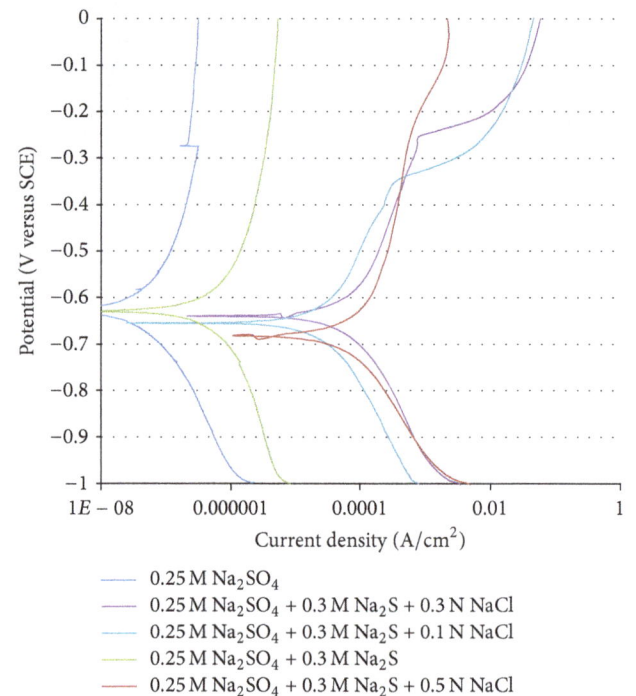

FIGURE 3: Effect of Cl^- on potentiodynamic polarization curves of carbon steel in 0.25 M Na_2SO_4 solution at pH 8 and temperature 25°C.

the cathodic reduction of SO_4^{2-} ions produces thiosulfate and sulphide ions, both of which are aggressive corrosion species and hence degrade the steel surface in sea water in the acidic pH range below the neutral medium. The corrosion product in this case should be iron sulphide or thiosulfate.

3.1.2. Effect of SO_4^{2-} + S^{2-}. Addition of S^{2-} to the solution containing SO_4^{2-} further enhances the corrosion rate as can be seen from Figure 2. And the pH has a strong effect on it. It can be seen from (11) and (12) above that the conjoint effect of SO_4^{2-} and S^- is the production of increasing amount of S^-, as well as corrosive sulphur compound.

3.1.3. Effect of SO_4^{2-} + Cl^-. Corrosion rate also increases with addition of Cl^- to the solution containing SO_4^{2-} (Figure 3). The rate increases with increase in Cl^- concentration. It is seen from (11)–(14) above that the Cl^- ions attack Fe/Fe^{2+} with the formation of corrosion products $FeCl_2$ and $FeCl_3$. $FeCl_2$ is unstable and may hydrolyse or further react with Cl^- ions to form $Fe(OH)_2$ or $FeCl_3$, respectively.

3.1.4. Effect of SO_4^{2-} + S^{2-} + Cl^-. The conjugate of all 3 ions which are normally present in the oil field sea water is the degradation of carbon steel structure at the highest level. It is seen from Figure 4 that i_{corr} values have shifted to the right and E_{corr} towards the active potential with increasing the concentration of both S^{2-} and Cl^-.

3.1.5. Effect of pH. There is not any much significant effect of polarization curves with change in pH except at pH 11 under alkaline condition (Figure 5), when the corrosion rate is very low. The steel is in the passive region at this pH. The pH of

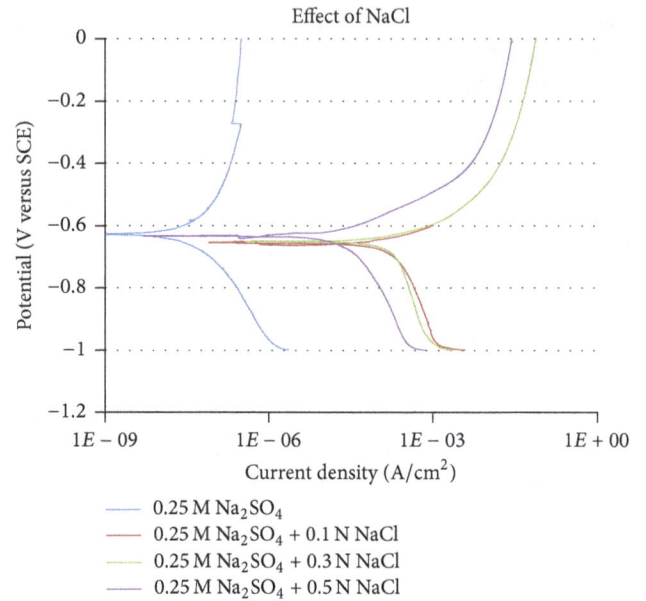

FIGURE 4: Effect of Cl^- + S^{2-} on potentiodynamic polarization curves of carbon steel in 0.25 M Na_2SO_4 solution at pH 8 and temperature 25°C.

the oil field water is in the range of 4–6 when the corrosion rate is high.

3.1.6. Effect of Temperature. Temperature aggravates the material degradation (Figure 6) by increasing diffusion and mass transfer coefficient of the aggressive ions corroding

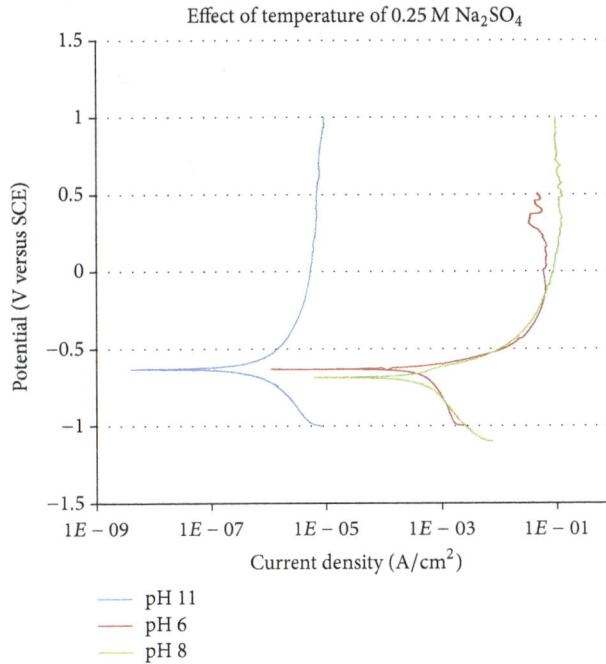

FIGURE 5: Potentiodynamic polarization curves of carbon steel in an aqueous solution of 0.25 M Na$_2$SO$_4$ and 0.3 M Na$_2$S at temperature 25°C, at different pH.

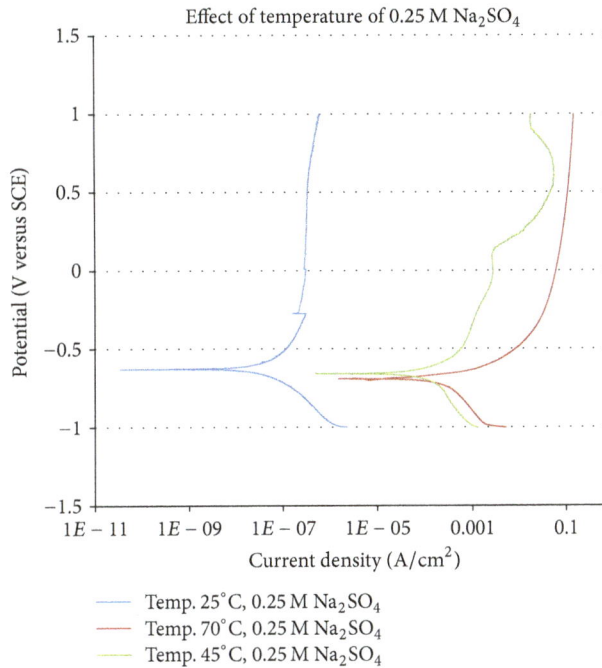

FIGURE 6: Potentiodynamic polarization curves of low carbon steel in 0.25 M Na$_2$SO$_4$ solution at (pH = 8) at different temperature.

the metallic surface. Temperature also increases the I_L, the limiting current density of the concentration polarization, and hence shifts the polarization curves to the right.

3.2. *Electrochemical Impedance Spectroscopy (EIS)*. The phenomenon at the interface of the solid metal and aqueous electrolyte is a complex process consisting of a line of positively

and negatively charged ions, capacitance due to double layer, corroded product or film formation on surfaces, polarization resistance (R_p), pore resistance (R_{po}), and various types of impedance due to diffusion of ions, movement of charge in or away from metal surface, and adsorption of cation and anion. The whole phenomenon can be represented by an equivalent AC electrical circuit. The phenomenon can be interpreted

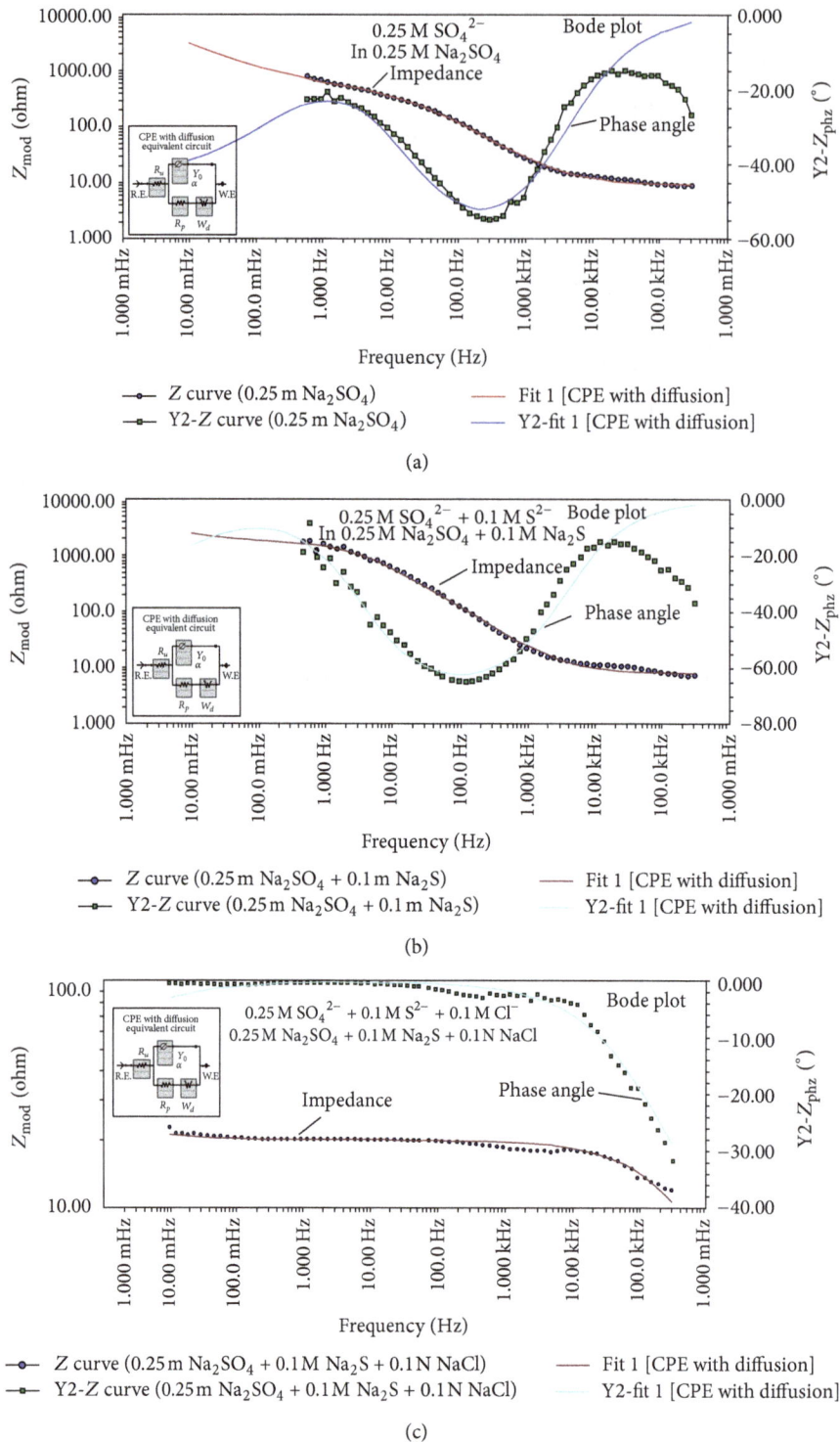

FIGURE 7: Bode plots of carbon steel in SO_4^{2-}, $SO_4^{2-} + S^{2-}$, and $SO_4^{2-} + S^{2-} + Cl^-$ solution at pH 8 and temperature 25°C.

from the Bode plots, which are depicted and discussed in the following section for various corrosive species.

Figure 7 displays the Bode plots of carbon steel in solutions of SO_4^{2-}, $SO_4^{2-} + S^{2-}$, and $SO_4^{2-} + S^{2-} + Cl^-$. It is to be noticed that, in all the solutions, EIS data match the model of Randle circuit with a Warburg resistance, W_d

(given in the inset of each figure), that prevails at the metal-solution interface. It is seen that the impedance decreases with addition of different aggressive ions compared to those with base solution of only SO_4^{2-}. This decrease in impedance leads to more current flow across the interface and hence increase in corrosion rate. There is a phase angle shift with

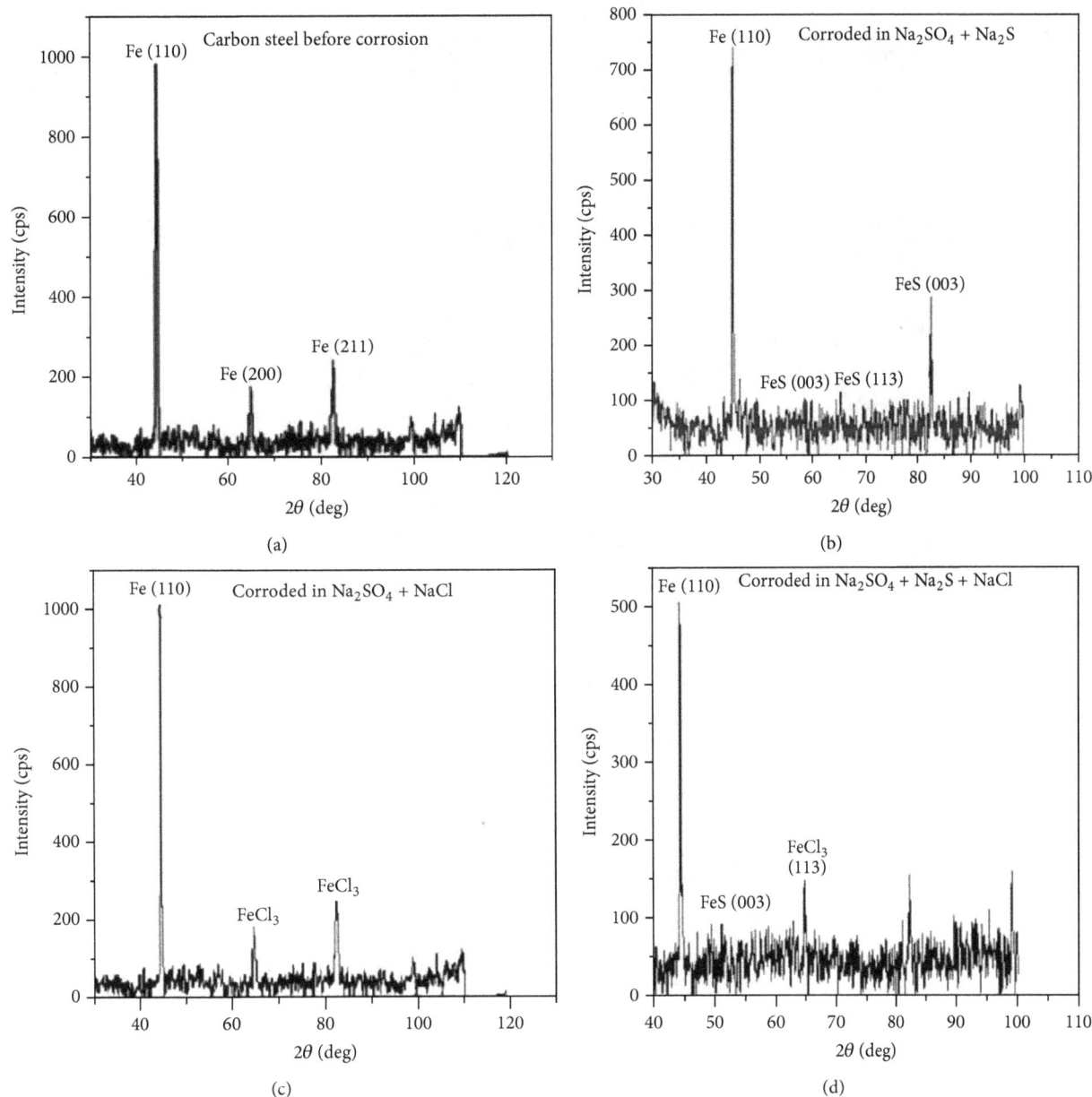

FIGURE 8: X-ray diffraction showing peaks of different corroded phases.

frequency. The minimum phase angle reaches much less than 90 degrees, indicating the capacitance in the circuit is not a pure capacitance but constant phase element, CPE(Y), and is given by the capacitive impedance equation $Z = 1/C(jw)^{-\alpha}$, where α is fraction varying from 0 to 1; the value less than one indicates that it does not behave ideally as pure capacitance. At high frequency, the value of impedance Z_{mod} is roughly equal to R_S, while at low frequency the value is ($R_S + R_P$). Both can be determined from the blots. Table 1 depicted the computed values of the EIS parameters. It is seen that polarization resistance decreases with addition of more types of ions which support the polarization results of corrosion rates increase as found in Figures 2, 3, and 4. The increase in corrosion rate is supported by the EIS data, increase in Y_0 which behaves like capacitance, and decrease in

polarization resistance (Table 2). The values of α indicate that the capacitance behaves like constant phase element rather than pure capacitance. The Warburg resistance W_d which signifies resistance to the flow of ions from the solution to corroded metal surface also decreases with addition of more types of ions. All the facts confirm the enhanced corrosion rates of carbon steel in the oil field sea water with presence of $SO_4^{2-} + S^{2-} + Cl^-$.

3.3. X-Ray Diffraction.
The presence of corrosion products of FeS, $FeCl_2$, and $FeCl_3$ is clearly indicated by the XRD peak intensities in Figure 8. This supports the corrosion enhancement by SO_4^{2-}, S^{2-}, and Cl^- ions as found in polarization and EIS studies.

TABLE 2: Computed EIS parameters.

Sample	Corrosive ions	R_s (ohm)	R_p (ohm)	$Y_0 \mu F$	α	W_d (ohm)
Carbon steel	SO_4^{2-}	9.443	8500.0	75.0	0.720	10.430
Carbon steel	$SO_4^{2-} + S^{2-}$	8.268	3700	572.2	0.769	4.187
Carbon steel	$SO_4^{2-} + S^{2-} + Cl^-$	13.31	95	905.6	0.602	2.432

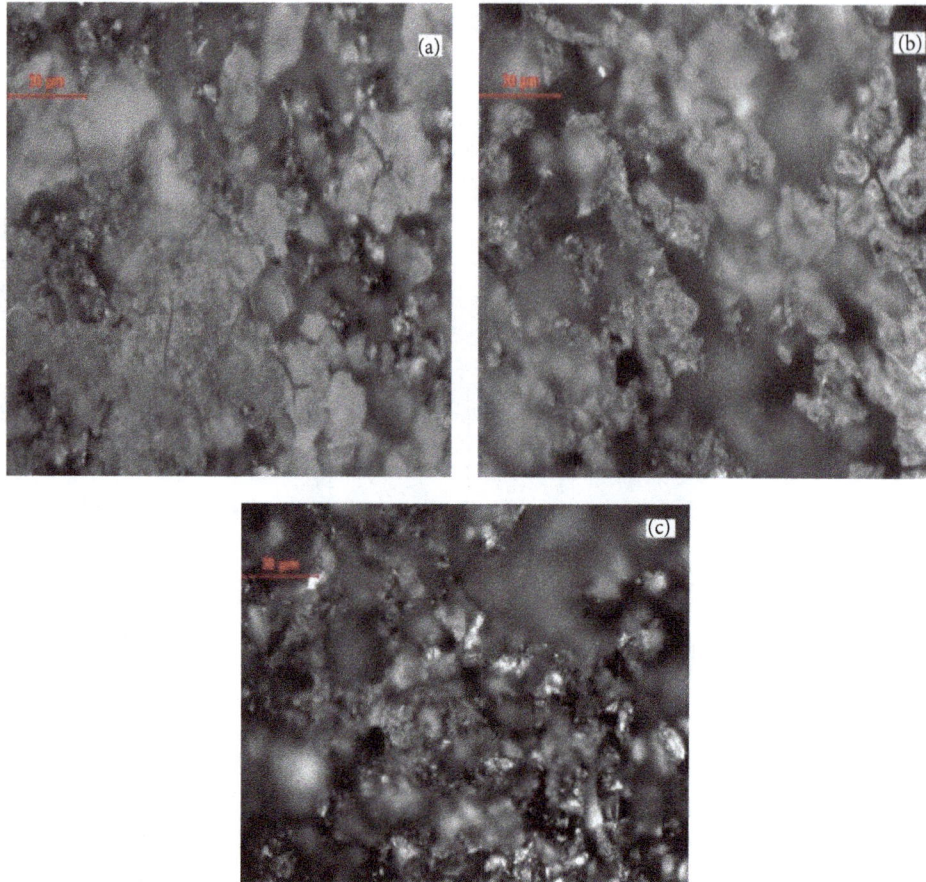

FIGURE 9: Optical microscopy images of corroded steel. (a) Na_2SO_4, (b) $Na_2SO_4 + Na_2S$, and (c) $Na_2SO_4 + Na_2S + NaCl$.

3.4. *Optical and SEM Microscopy Images.* The microstructures by optical microscopy clearly show (Figure 9) the corroded structure of steel with products of corrosion over it. The form of corrosion seems to be uniform, not localized. SEM images (Figure 10) distinctly reveal how the degree of degradation increases with presence of ions in the solutions from SO_4^{2-} to $SO_4^{2-} + S^{2-}$ to $SO_4^{2-} + S^{2-} + Cl^-$. It is interesting to observe from the morphology of SEM image at higher magnification (Figure 11) that the corrosion has taken place at grain boundary as well as grain body but the attack at grain boundary is very prominent with sulphide (S) corrosion crack. The corrosion products of sulphide compounds as well as element sulphur are clearly revealed. The structure of SEM images shows almost catastrophic failure with the presence of minor and major cracks when the seawater is enriched with the presence of all three corrosive ions, SO_4^{2-}, S^{2-}, and Cl^-. The morphology of SEM image analysis along with

XRD is in complete agreement with the corrosion data of the polarization experiments and EIS.

From (5)–(8), it is seen that SO_4^{2-} is cathodically reduced to give rise to S^{2-}. It is observed from the polarization curve (Figure 1) that the increase of SO_4^{2-} concentration depolarizes the cathodic curves much more compared to anodic ones, shifting E_{corr} towards the negative potential. The release of S^{2-} from the reaction produces corrosion products: iron sulfides (FeS_x) on carbon steel surface. They are nonstoichiometric iron sulfide films mainly composed of mackinawite and pyrrhotite Berner [21–26]. The black color of mackinawite phase is also seen as corrosion products in optical microstructures (Figures 9(a) and 9(b)).

The deterioration of metal due to contact with S^{2-} (H_2S) and moisture is sour corrosion. The forms of sour corrosion are uniform (Figure 9(b)), pitting, and stepwise cracking (Figures 10(c) and 11(b)).

FIGURE 10: SEM images of corroded steel (a) before corrosion, (b) in Na_2SO_4, (c) in $Na_2SO_4 + Na_2S$, and (d) in $Na_2SO_4 + Na_2S + NaCl$.

The general equation of sour corrosion can be expressed as follows [21]:

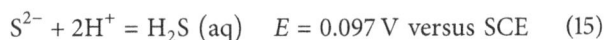

$$S^{2-} + 2H^+ = H_2S \ (aq) \quad E = 0.097 \text{ V versus SCE} \quad (15)$$

Hydrogen sulphide dissociates to produce proton and bisulphide as described [22] by the following equations:

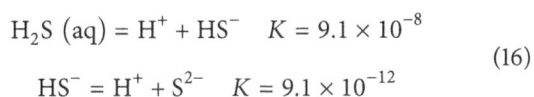

$$H_2S \ (aq) = H^+ + HS^- \quad K = 9.1 \times 10^{-8}$$
$$HS^- = H^+ + S^{2-} \quad K = 9.1 \times 10^{-12}$$
$$(16)$$

This reaction scheme shows that the presence of hydrogen sulphide can contribute to the concentration of sulfide at the surface by dissociation rather than charge exchange with the surface. The sulfide concentration at the surface is, therefore, dependent on the concentration of aqueous hydrogen sulfide in the electrolyte as well as the reduction processes of sulfate.

H_2S exhibits the different role in anodic process of carbon steel depending on the pH value in the solutions. The local supersaturation of FeS_x could be formed on the carbon steel surface via the following reaction, with the nucleation and growth of one or more of the iron sulfide, mackinawite:

$$H_2S + Fe + H_2O \longrightarrow FeS_x + 2H + H_2O \quad (17)$$

The black corrosion products (FeS_x) formed on the steel surface in the H_2S-containing solutions could be observed (Figure 9(b)). The addition of chloride to the corrosion system did not exhibit any localized corrosion; rather it drastically increased the i_{corr} values and shifted E_{corr} towards the negative potential (Figures 3 and 4), depolarizing the anodic reactions. The increase in i_{corr} with Cl addition seems to have modified the anodic substrate area (Figures 9(c) and 10(d)) and has enhanced the exchange current density for the iron oxidation process.

4. Conclusion

All three corrosive ions SO_4^{2-}, S^{2-}, and Cl^- have a strong effect on increasing the corrosion rate of carbon steel.

Cathodic reduction of SO_4^{2-} generates elemental S or S^{2-} ions, in addition to S^{2-} from H_2S in oil. The species

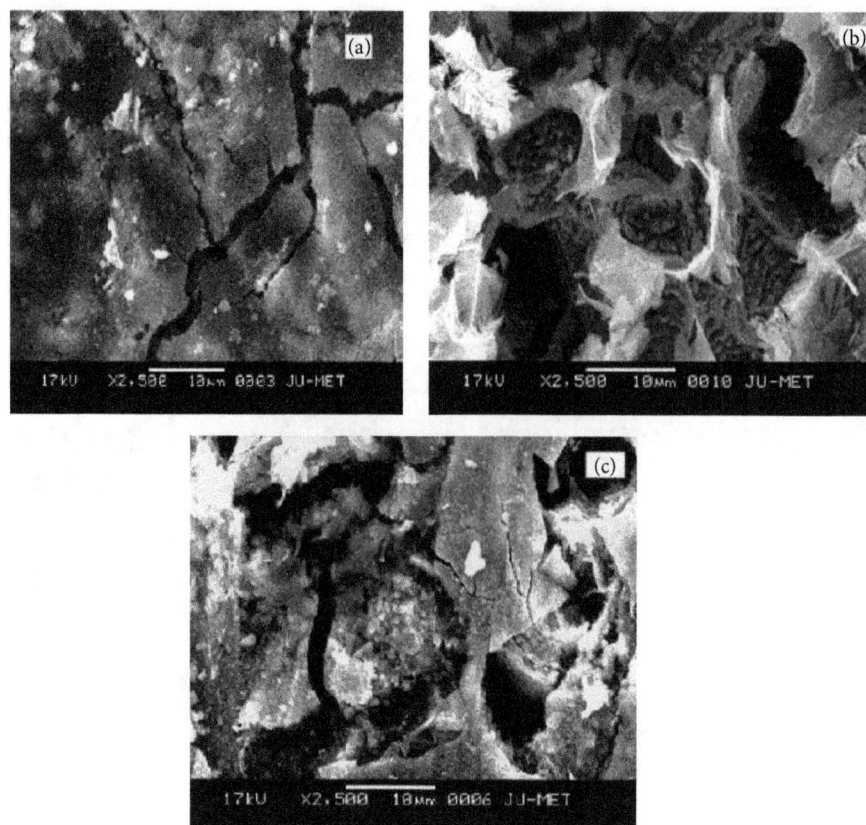

FIGURE 11: SEM images of corroded steel (a) in Na_2SO_4, (b) in $Na_2SO_4 + Na_2S$, and (c) in $Na_2SO_4 + Na_2S + NaCl$.

cause major corrosion with formation of FeS_x as corrosion products. The attack is found at grain boundary with sulphide cracking as well as some grain body degradation. The effect of addition of Cl^- is the increase of i_{corr} values by hundreds of times possibly due to enhancement of exchange current density of the anodic reactions.

Conflict of Interests

The authors declare that there is no conflict of interests regarding the publication of this paper.

Acknowledgments

The authors would like to acknowledge "COE, TEQUIP" in Jadavpur University for the support of this work.

References

[1] S. Nešić, "Key issues related to modelling of internal corrosion of oil and gas pipelines—a review," *Corrosion Science*, vol. 49, no. 12, pp. 4308–4338, 2007.

[2] L. T. Popoola, A. S. Grema, G. K. Latinwo, B. Gutti, and A. S. Balogun, "Corrosion problems during oil and gas production and its mitigation," *International Journal of Industrial Chemistry*, vol. 4, aticle 35, 2013.

[3] B. W. A. Sherar, P. G. Keech, and D. W. Shoesmith, "The effect of sulfide on the aerobic corrosion of carbon steel in near-neutral pH saline solutions," *Corrosion Science*, vol. 66, pp. 256–262, 2013.

[4] B. W. A. Sherar, I. M. Power, P. G. Keech, S. Mitlin, G. Southam, and D. W. Shoesmith, "Characterizing the effect of carbon steel exposure in sulfide containing solutions to microbially induced corrosion," *Corrosion Science*, vol. 53, no. 3, pp. 955–960, 2011.

[5] F. M. Song, "A comprehensive model for predicting CO_2 corrosion rate in oil and gas production and transportation systems," *Electrochimica Acta*, vol. 55, no. 3, pp. 689–700, 2010.

[6] M. Hairil Mohd and J. K. Paik, "Investigation of the corrosion progress characteristics of offshore subsea oil well tubes," *Corrosion Science*, vol. 67, pp. 130–141, 2013.

[7] J. Tang, Y. Shao, T. Zhang, G. Meng, and F. Wang, "Corrosion behaviour of carbon steel in different concentrations of HCl solutions containing H_2S at 90 °C," *Corrosion Science*, vol. 53, no. 5, pp. 1715–1723, 2011.

[8] Z. A. D. Foroulis, "Role of solution pH on wet H_2S cracking in hydrocarbon production," *Corrosion Prevention and Control*, vol. 40, no. 4, pp. 84–89, 1993.

[9] M. A. Veloz and I. González, "Electrochemical study of carbon steel corrosion in buffered acetic acid solutions with chlorides and H_2S," *Electrochimica Acta*, vol. 48, no. 2, pp. 135–144, 2002.

[10] S. Arzola and J. Genescá, "The effect of H_2S concentration on the corrosion behavior of API 5L X-70 steel," *Journal of Solid State Electrochemistry*, vol. 9, no. 4, pp. 197–200, 2005.

[11] H. Y. Ma, X. L. Cheng, S. H. Chen et al., "Theoretical interpretation on impedance spectra for anodic iron dissolution in acidic solutions containing hydrogen sulfide," *Corrosion*, vol. 54, no. 8, pp. 634–640, 1998.

[12] H. Ma, X. Cheng, S. Chen, C. Wang, J. Zhang, and H. Yang, "An ac impedance study of the anodic dissolution of iron in sulfuric acid solutions containing hydrogen sulfide," *Journal of Electroanalytical Chemistry*, vol. 451, no. 1-2, pp. 11–17, 1998.

[13] H. Ma, X. Cheng, G. Li et al., "The influence of hydrogen sulfide on corrosion of iron under different conditions," *Corrosion Science*, vol. 42, no. 10, pp. 1669–1683, 2000.

[14] H.-H. Huang, W.-T. Tsai, and J.-T. Lee, "Electrochemical behavior of the simulated heat-affected zone of A516 carbon steel in H$_2$S solution," *Electrochimica Acta*, vol. 41, no. 7-8, pp. 1191–1199, 1996.

[15] J. Tang, Y. Shao, J. Guo, T. Zhang, G. Meng, and F. Wang, "The effect of H$_2$S concentration on the corrosion behavior of carbon steel at 90 ∘C," *Corrosion Science*, vol. 52, no. 6, pp. 2050–2058, 2010.

[16] P. Smith, S. Roy, D. Swailes, S. Maxwell, D. Page, and J. Lawson, "A model for the corrosion of steel subjected to synthetic produced water containing sulfate, chloride and hydrogen sulfide," *Chemical Engineering Science*, vol. 66, no. 23, pp. 5775–5790, 2011.

[17] R. E. Melchers, "Development of new applied models for steel corrosion in marine applications including shipping," *Ships and Offshore Structures*, vol. 3, no. 2, pp. 135–144, 2008.

[18] M. H. Mohd, D. K. Kim, D. W. Kim, and J. K. Paik, "A time-variant corrosion wastage model for subsea gas pipelines," *Ships and Offshore Structures*, vol. 9, no. 2, pp. 161–176, 2014.

[19] R. E. Melchers and J. K. Paik, "Effect of flexure on rusting of ship's steel plating," *Ships and Offshore Structures*, vol. 5, no. 1, pp. 25–31, 2010.

[20] M. Shehadeh and I. Hassan, "Study of sacrificial cathodic protection on marine structures in sea and fresh water in relation to flow conditions," *Ships and Offshore Structures*, vol. 8, no. 1, pp. 102–110, 2013.

[21] R. A. Berner, "Thermodynamic stability of sedimentary iron sulfides," *The American Journal of Science*, vol. 265, pp. 773–785, 1967.

[22] R. H. Hausler, L. A. Goeller, R. P. Zimmerman, and R. H. Rosenwald, "Contribution to the "filming amine" theory: an interpretation of experimental results," *Corrosion*, vol. 28, no. 1, pp. 7–16, 1972.

[23] P. Taylor, "The stereochemistry of iron sulfides—a structural ration for the crysta llization of some metastable phases from aqueous solution," *American Mineralogist*, vol. 65, pp. 1026–1030, 1980.

[24] R. A. Berner, "Iron sulfides formed from aqueous solution at low temperatures and atmospheric pressure," *The Journal of Geology*, vol. 72, pp. 293–306, 1964.

[25] D. T. Rickard, "The chemistry of iron sulphide formation at low temperatures," *Stockholm Contributions in Geology*, vol. 20, pp. 67–95, 1969.

[26] A. J. Bard, R. Parsons, and J. Jordan, *Standard Potentials in Aqueous Solution*, CRC Press, 1st edition, 1985.

11

Effect of Electrode Types on the Solidification Cracking Susceptibility of Austenitic Stainless Steel Weld Metal

J. U. Anaele, O. O. Onyemaobi, C. S. Nwobodo, and C. C. Ugwuegbu

Department of Materials and Metallurgical Engineering, Federal University of Technology, PMB 1526, Owerri, Nigeria

Correspondence should be addressed to J. U. Anaele; krimsysage@yahoo.com

Academic Editor: Peiqing La

The effect of electrode types on the solidification cracking susceptibility of austenitic stainless steel weld metal was studied. Manual metal arc welding method was used to produce the joints with the tungsten inert gas welding serving as the control. Metallographic and chemical analyses of the fusion zones of the joints were conducted. Results indicate that weldments produced from E 308-16 (rutile coated), E 308-16(lime-titania coated) electrodes, and TIG welded joints fall within the range of $1.5 \leq Cr_{eq.}/Ni_{eq.} \leq 1.9$ and solidified with a duplex mode and were found to be resistant to solidification cracking. The E 308-16 weld metal had the greatest resistance to solidification cracking. Joints produced from E 310-16 had $Cr_{eq.}/Ni_{eq.}$ ratio < 1.5 and solidified with austenite mode. It was found to be susceptible to solidification cracking. E 312-16 produced joints having $Cr_{eq.}/Ni_{eq.}$ ratio > 1.9 and solidified with ferrite mode. It had a low resistance to solidification cracking.

1. Introduction

Stainless steel is a common name for steel alloys that consist of 10.5 weight percent or more of chromium (Cr) and more than 50 weight percent of iron (Fe). Stainless steels may be classified by their crystalline structure into three main types: austenitic, ferritic, and martensitic stainless steel. Austenitic stainless steel (ASS) contains a maximum of 0.15 percent carbon, a minimum of 16 percent chromium, and sufficient nickel and/or manganese to retain an austenitic structure at all temperatures from the cryogenic temperature to the melting point of the alloy.

Austenitic stainless steels have become the most widely used stainless steels and correspond to about 70 percent of all the stainless steel produced worldwide, as a result of their mechanical and metallurgical properties and their good weldability [1]. The excellent properties of ASS which include high tensile strength, good impact resistance, excellent ductility, corrosion, and wear resistances have found various applications in domestic as well as in many engineering industries, some of which are cooking utensils, food processing equipment, equipment for chemical industry, truck trailers, kitchen sinks, exterior architecture, pressure boilers and vessels, fossil-fired power plant, fuel gas desulphurization equipment, evaporator tubing, super heater and reheating tubing, steam headers, and pipes, among others [2].

In recent times, advancement has been made in such joining process as adhesives, mechanical fasteners, brazing, and soldering. However, welding remains the most important metal joining process, even as arc welding is the most widely used fusion-welding process. In the fabrication of austenitic stainless steel components, welding is one of the most employed methods [3, 4]. Despite the good weldability property exhibited by ASS, hot cracking has been the major metallurgical problem encountered during welding of austenitic stainless steel components. It is caused by the formation of low melting eutectics at the grain boundaries during welding, which cause failure under the action of shrinkage stresses associated with solidification. Solidification cracking is a type of hot cracking which depends on mechanical restraint and metallurgical susceptibility [5]. It consists of fractures at the interdendritic and/or intergranular weld metal boundaries in the solidification process, during which the liquid phase of the mushy melt becomes rich in impurities, mainly sulphur (S), and phosphorus (P). This phenomenon reduces the mechanical strength at the grain and dendritic boundaries,

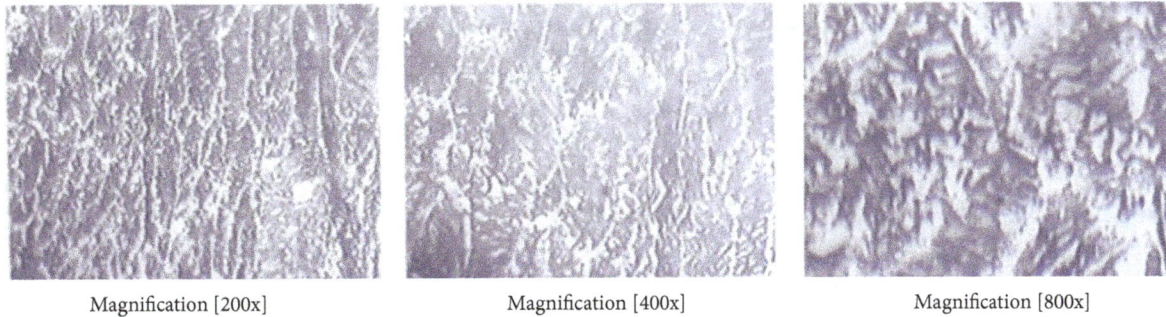

| Magnification [200x] | Magnification [400x] | Magnification [800x] |

FIGURE 1: Fusion zone micrograph of E 312-16/10 welded joint.

rendering them susceptible to cracking and failure eventually [6]. One of such failures is the corrosion cracking of a grade 304 stainless steel pipe improperly seam welded and meant for the conveyance of glucose solution in Illinois USA [7].

In view of the problem of solidification cracking in ASS weldment, many works have been carried out in order to explain the phenomenon of solidification cracking and ways of preventing it. As early as 1941, Scherer et al. found that crack resistance in ASS weld metal may be improved by adjusting the composition to 5–35 percent ferrite in the completed weld. Hull [8] confirmed this by stating that when ferrite content in the completed weld increases beyond 35 weight percent, the weld metal would become susceptible to solidification cracking, but the mechanism by which crack resistance is achieved by the effect of retained ferrite in the weld metal is still not completely understood.

Good attempts, however, have been made to explain the effect. Borland and Younger [9] suggested that the higher solubility for impurity elements in delta ferrite leads to less inter-dendritic segregation and reduces cracking tendency. Thier et al. [10] found that the volume contraction associated with ferrite-austenite transformation reduces tensile stresses close to the crack tip, which decreases cracking tendency. Apart from the effect of retained delta ferrite in the control of solidification cracking in ASS weldment, Baldev et al. [5] and Borland [11] suggested that solidification cracking in ASS weld metal could be minimized by the various practices which reduce mechanical restraint in the completed weld metal. As can be seen in some of the research works cited above, solidification cracking in austenitic stainless steel weldment is partly a function of the weld metal composition. A well designed product, for example, can fail by cracking if the weld rod selected results in the weld zone having lower alloy content than that of the parent metal. Therefore, there is need to determine how the electrode type affects the solidification cracking susceptibility of ASS weldments. The main aim of this work is, therefore, to investigate the effect of electrode types on the microstructural susceptibility of the austenitic stainless steel weldment to solidification cracking.

2. Material and Methods

The base metal of the test specimens used for this study was type 304H austenitic stainless steel and the nominal chemical composition of the material is shown in Table 1.

Two methods of welding were adopted, namely, Shielded Metal Arc Welding (SMAW) and Tungsten Inert Gas (TIG)

TABLE 1: Chemical composition (wt.%) of the austenitic stainless steel material.

Element	wt.%
C	0.0570
Cr	18.5500
Ni	8.7200
Si	0.4400
Mn	1.7200
S	0.0075
P	0.0230
Mo	1.7200
Al	0.0057
Cu	0.2010
Co	0.1110
Nb	0.0270
V	0.0750
B	0.0022
Sn	0.0086
As	0.0870
Ca	0.0004
Fe	69.8000

Welding. The welding operations were conducted under constant condition as shown in Table 2.

The variable parameter in this study was the welding electrodes while the welded joints produced from TIG autogenous welding served as the control or reference for comparism. The chemical composition of the electrodes, according to American Welding Society (AWS) electrode classification, is shown in Table 3. The joints produced were subjected to metallographic test and chemical analysis.

3. Results and Discussion

3.1. Metallographic Analysis of the Weldments. The results obtained from the metallographic test conducted on the Fusion Zone of the each weldment were analyzed.

3.1.1. Analysis of E 312-16/10 Micrograph. The micrograph of E 312-16/10 fusion zone test specimen shown in Figure 1 revealed a primary ferrite (dark) matrix containing secondary austenite (white) and carbide precipitation at the grain boundaries. The ferrite dendrites being the first to solidify had a lathy morphology and partly transforms into austenite after solidification by diffusion controlled mechanism.

TABLE 2: Welding parameters held constant in the welding operation.

Welding parameters	Tungsten inert gas welding	Manual metal arc welding
Welding current	110 A	110 A
Welding speed*	60 mm/min.	60 mm/min.
Voltage	40 V	40 V
Polarity	DC electrode negative	DC electrode negative
Heat source	Arc	Arc
Weld pool shield	Argon gas	Electrode flux
Filler rod	304 H stainless steel wire	Welding electrode
Argon gas pressure	10 bars	—

*Since the welding process was carried out manually, the welding speed is approximate and represents the average values.

TABLE 3: Chemical composition of the electrodes.

Elements (wt.%)	E 308-16 (rutile coated) Electrodes (10 & 12)	E 308-16 (lime-titania) Electrode	E 310-16 Electrode	E 312-16 Electrode
C	0.08	0.08	0.08–0.12	0.15
Mn	0.7–2.0	0.5–2.5	1.0–2.5	0.7–2.0
Si	0.3–0.85	0.9	0.3–0.7	0.3–0.9
Cr	18–21	18–21	25–28	28–32
Ni	9–11	9–11	20–22	8–10.5
S	0.03	—	0.03	0.03
P	0.03	—	0.03	0.03
Mo	0.5	—	0.5	0.75
Cu	0.75	—	0.75	0.75

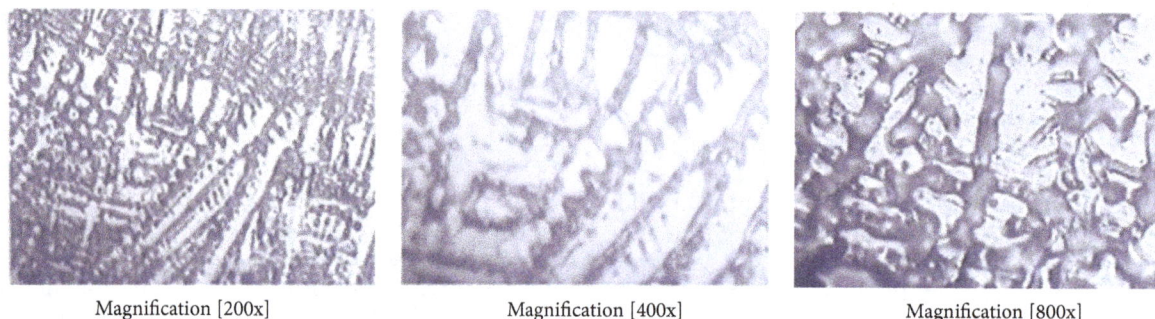

Magnification [200x]   Magnification [400x]   Magnification [800x]

FIGURE 2: Fusion zone micrograph of E 310-16/10 welded joint.

3.1.2. Analysis of E 310-16/10 Micrograph. E 310-16/10 fusion zone micrograph is shown in Figure 2. As can be seen from the micrograph, the primary austenite (white) formed directly from the liquid as a primary dendritic phase as well as a secondary phase around ferrite. The interdendritic ferrite (dark) had a vermicular morphology engulfed in the austenite matrix, with carbide precipitation along the grain boundaries.

3.1.3. Analysis of E 308-16/12 (Lime-Titania) Micrograph. Fu et al. [12] noted that ferrite-austenite (FA) solidification duplex mode is characterized by the formation of primary ferrite plus three phase (ferrite, austenite, and liquid) reactions at the terminal solidification stage. The fusion zone micrograph of E 308-16/12 (lime-titania) joint shown in Figure 3 revealed a plenty of fine colonies of lathy ferrite (dark) embedded in

austenite (white) matrix. The result was a duplex microstructure consisting of thin lathy ferrite and austenite.

3.1.4. Analysis of TIG Micrograph. The fusion zone micrograph of TIG joint specimen shown in Figure 4 revealed primary equiaxed dendritic and lathy ferrite (dark) enclosed in austenite (white) matrix, with the precipitation of carbides along grain boundaries.

3.1.5. Analysis of E 308-16/12 (Rutile) Micrograph. The E 308-16/12 (rutile) weld joint micrograph is shown in Figure 5. The figure revealed a duplex structure of ferrite and austenite. The primary ferrite (dark) dendrites having a combination of lathy and vermicular ferrite morphology contained in austenite (white) matrix and a precipitation of carbide along

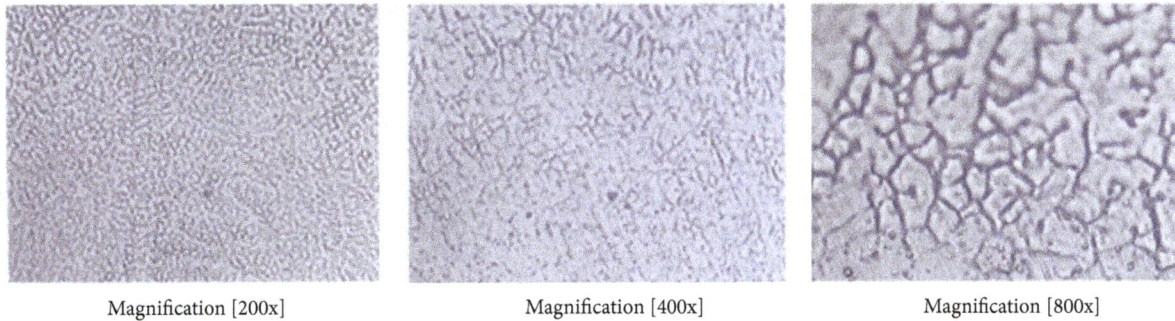

| Magnification [200x] | Magnification [400x] | Magnification [800x] |

FIGURE 3: Fusion zone micrograph of E 308-16/12 (lime-titania) welded joint.

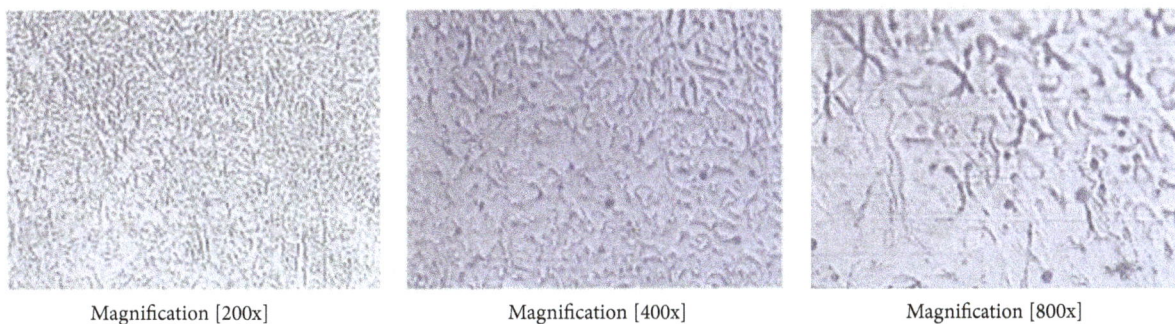

| Magnification [200x] | Magnification [400x] | Magnification [800x] |

FIGURE 4: Fusion zone micrograph of TIG welded joint.

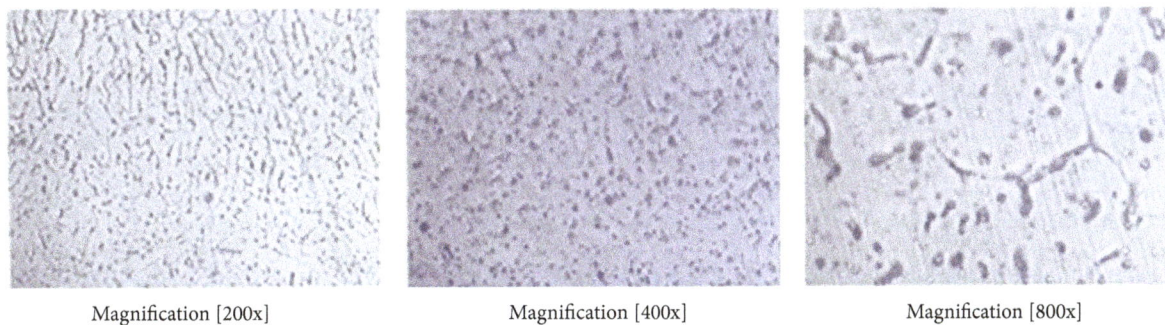

| Magnification [200x] | Magnification [400x] | Magnification [800x] |

FIGURE 5: Fusion zone micrograph of E 308-16/12 (rutile) welded joint.

grain boundaries. The amount of retained ferrite in E 308-16/12 (rutile) weld metal was found to be less than TIG and E 308-16/12 (lime-titania) weld metals, respectively, but more than that observed in E 308-16/10 (rutile) weld metal.

3.1.6. Analysis of E 308-16/10 (Rutile) Micrograph. The fusion zone micrograph of the E 308-16/10 (rutile) joint shown in Figure 6 revealed a duplex structure consisting of ferrite (dark) and austenite (white). The primary ferrite had a thin lathy morphology engulfed by austenite which grew epitaxially and fills the interdendritic region of the primary ferrite. Carbide precipitation was also observed along the grain boundaries.

3.2. Effect of Electrode Types on the Solidification Mode of Austenitic Stainless Steel Weld Metal. The results displayed in Table 4 showed that the type of electrode selected in the welding of 304H stainless steel component affects the solidification microstructure of the weld metal. It was found that the TIG autogenous weld (Figure 4) had nearly the same solidification microstructure (FA) with the parent material (Figure 7), a result which was highly anticipated since there was no filler dilution in the completed weld as the tungsten electrode was nonconsumable. Weldments produced from E 308-16/12 (rutile), E 308-16/10 (rutile), and E 308-16/12 (lime-titania) electrodes had a duplex structure of ferrite-austenite (FA) with more or less amount of retained ferrite. Filler rod or electrode dilution is believed to be responsible for the evolved microstructure and solidification mode as shown in the results of the fusion zones micrographs of the E 308-16/12 (lime-titania), E 308-16/12 (rutile), and E 308-16/10 (rutile) joints displayed in Figures 3, 5, and 6, respectively. Joints produced from E 310-16/10 electrode (Figure 2) had austenite solidification mode whereas the joints made from

TABLE 4: Values of $Cr_{eq.}$, $Ni_{eq.}$, $Cr_{eq.}/Ni_{eq.}$ ratio, and (P + S) wt.% of the tested weld joints.

Weld joints	$Cr_{eq.}$	$Ni_{eq.}$	$Cr_{eq.}/Ni_{eq.}$	(P + S) wt.%	Ferrite number (FN)	Solidification mode
E 308-16/12 (rutile) weld joint	17.3019	9.7480	1.7750	0.0171	6	Ferrite-austenite (FA)
E 308-16/10 (rutile) weld joint	17.2042	9.8678	1.7435	0.0094	5-6	Ferrite-austenite (FA)
E 308-16/12 (lime-titania) joint	19.8223	11.3915	1.7401	0.0086	10	Ferrite-austenite (FA)
E 310-16/10 weld joint	24.0273	22.7530	1.0560	0.0150	0-1	Austenite (A)
E 312-16/10 weld joint	26.8069	12.9698	2.0669	0.0140	50–55	Ferrite (F)
TIG weld joint	18.6387	10.9375	1.7041	0.0068		

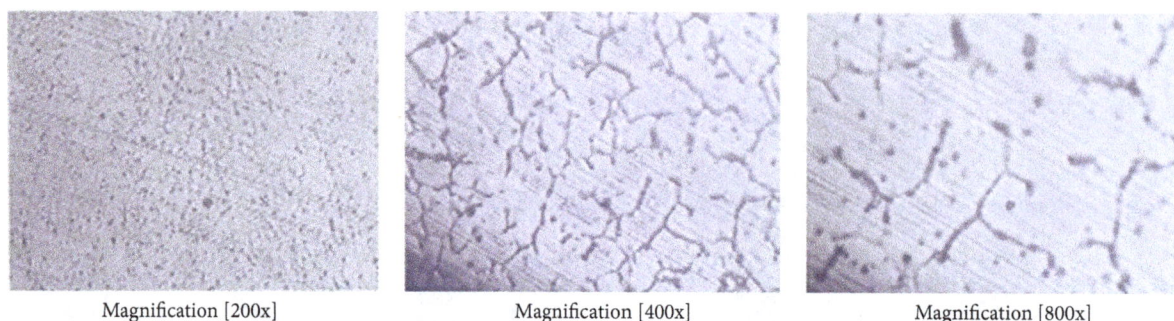

| Magnification [200x] | Magnification [400x] | Magnification [800x] |

FIGURE 6: Fusion zone micrograph of E 308-16/10 (rutile) welded joint.

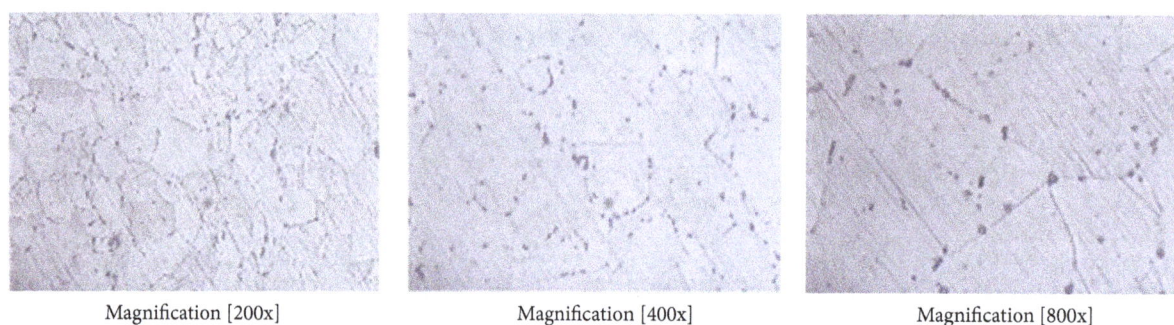

| Magnification [200x] | Magnification [400x] | Magnification [800x] |

FIGURE 7: Micrograph of the unwelded parent material.

E 312-16/10 (Figure 1) solidified with a primary ferrite solidification mode. The compromise reached between the parent material composition and filler rod or electrode dilution was found to be the major factor which determined the weld metal final microstructure and solidification mode. The findings of this research were found to be in line with the results of many researchers [5, 6, 12–15].

3.3. Effect of Electrode Types on Weld Metal Composition and Cracking Propensity.

The results of chemical analysis (presented in Table 5) carried out on the weldments showed that electrode types have effect on the weld metal composition. The TIG autogenous weldment had nearly the same composition and Chromium-Nickel equivalence as that of the unwelded parent metal. However, remarkable difference in weld metal constitution was observed in the joints produced from the various electrodes relative to the Chromium-Nickel equivalence of the parent material. The Chromium-Nickel equivalence results, calculated for each weld joints and presented in Table 4, were obtained by using the the

1992 Welding Research Council model equation culled from Kotecki and Siewert [16]:

$$Cr_{eq.} = Cr + Mo + 0.7Nb$$
$$Ni_{eq.} = Ni + 35C + 20N + 0.25Cu, \tag{1}$$

where, $Cr_{eq.}$ = Chromium equivalent and $Ni_{eq.}$ = Nickel equivalent.

The results showed that solidification cracking of the weld joints was sensitive to $Cr_{eq.}/Ni_{eq.}$ ratio and solidification mode of the welds. E 308-16/12 (lime-titania), TIG, E 308-16/12 (rutile), and E 308-16/10 (rutile) welds with primary ferrite-austenite solidification modes and 1.5 < $Cr_{eq.}/Ni_{eq.}$ < 1.9 are immune to solidification cracking in the order of decreasing resistance to solidification cracking, respectively. E 312-16/10 weld with ferrite solidification mode and $Cr_{eq.}/Ni_{eq.}$ > 1.9 has low susceptibility to solidification cracking, whilst E 310-16/10 weld with primary austenite solidification mode, $Cr_{eq.}/Ni_{eq.}$ < 1.5, and (P + S) wt.% = 0.015 may be susceptible to solidification cracking. Filler rod

TABLE 5: Chemical composition of the weld joints.

Elements wt.%	E 308-16/12 (rutile) weld	E 308-16/10 (rutile) weld	E 308-16/12 (lime-titania) weld	E 310-16/10 weld	E 312-16/10 weld	TIG weld
Carbon, C	0.0710	0.0730	0.0740	0.1350	0.0920	0.0600
Silicon, S	0.6900	0.6300	0.4900	0.5100	0.9100	0.3760
Manganese, Mn	1.2200	1.1800	1.3900	1.8700	1.3200	1.6200
Phosphorus, P	0.0061	0.0010	0.0010	0.0010	0.0010	0.0022
Sulphur, S	0.0110	0.0084	0.0076	0.0140	0.0130	0.0046
Chromium, Cr	17.1500	17.0200	19.5700	23.8500	26.7100	18.4500
Nickel, Ni	7.1900	7.2100	8.7400	17.9900	9.7300	8.7900
Molybdenum, Mo	0.1330	0.1660	0.2320	0.1570	0.0640	0.1740
Aluminium, Al	0.0055	0.0049	0.0056	0.0170	0.0072	0.0055
Copper, Cu	0.2920	0.4110	0.2460	0.1520	0.0790	0.1900
Cobalt, Co	0.0830	0.0960	0.1120	0.0770	0.0640	0.1170
Titanium, Ti	0.0240	0.0190	0.0110	0.4250	0.0160	0.0010
Niobium, Nb	0.0270	0.0260	0.0290	0.0290	0.0470	0.0210
Vanadium, V	0.0680	0.0700	0.1050	0.1080	0.1190	0.0770
Tungsten, W	0.0100	0.0100	0.0100	0.0100	0.0100	0.0560
Lead, Pb	0.0034	0.0030	0.0030	0.0030	0.0030	0.0038
Boron, B	0.0021	0.0018	0.0016	0.0017	0.0012	0.0022
Tin, Sn	0.0100	0.0098	0.0090	0.0075	0.0110	0.0110
Arsenic, As	0.0860	0.0810	0.0930	0.0960	0.1160	0.0950
Bismuth, Bi	0.0015	0.0015	0.0015	0.0015	0.0015	0.0015
Calcium, Ca	0.0002	0.0002	0.0003	0.0024	0.0003	0.0003
Iron, Fe	72.9000	73.0000	68.9000	54.6000	60.7000	69.9000

or electrode dilution is one of the factors which determined the final weld metal composition and solidification mode. Since solidification cracking is sensitive to weld metal composition and solidification mode, it therefore follows that the type of electrode used during welding of ASS materials determines the weld metal solidification cracking propensity.

These results were compared with the cracking susceptibility of 300 series stainless steel based on Cr-Ni equivalence according to Hammar and Svensson [17] and found to be consistent and also in line with the findings of Arantes and Trevisan [6], Baldev et al. [5], Korinko and Malene [18], and Brooks and Thompson [19] who affirmed that the propensity for solidification cracking in austenitic stainless steel is sensitive to the $Cr_{eq.}/Ni_{eq.}$ ratio, $(P + S)$ wt.%, and ferrite number of the weld metal and maintained that weld metal with solidified FA mode in the range of $1.5 < Cr_{eq.}/Ni_{eq.} < 1.9$ is immune to solidification cracking, while those in the region of $Cr_{eq.}/Ni_{eq.} > 1.9$ and $Cr_{eq.}/Ni_{eq.} < 1.5$ have low resistance and susceptible to solidification cracking, respectively. Generally, it was found that $(P + S)$ wt.% values were less than 0.02 in the final composition of the respective welded joints, which is below the critical level suggested by Arantes and Trevisan [6], necessary to induce cracking.

The results also suggest that the type of electrode coating has effect on the weld metal properties. The electrodes used for the welding were designated "-16" which denotes rutile coating for stainless steel electrodes. Rutile coatings are titania-type based electrodes containing little proportion of other additives. However, E 308-16/12 (lime-titania) electrodes were coated with titanium calcium and contain lime which makes it distinct from the rutile category. The presence of lime (which is a slag former) in E 308-16/12 (lime-titania) electrode was relevant in slowing down the cooling rate of both the weld pool and the just solidified weld metal of the resultant weldment. This suggests the reason for the slight difference observed in properties (such as ductility and strain hardening exponent) of weldments produced from the E 308-16 (lime titania) with respect to E 308-16 (rutile) welding electrodes. Consequently, the weldment produced from E 308-16/12 (lime-titania) electrode has a higher ductility of about 36% (in terms of percentage elongation) compared to 26% and 18% obtained from weldments produced from E 308-16/10 (rutile) and E 308-16/12 (rutile) electrodes, respectively. This result is confirmed in the micrograph of E 308-16/12 (lime-titania) weldment which suggests that the ferrite dendrite had more time for growth in the region where delta ferrite is most stable (due to slower cooling rate offered by lime in the electrode coating) compared to the micrographs of E 308-16/10 (rutile) and E 308-16/12 (rutile) weldments.

4. Conclusion

The cracking of austenitic stainless steel (ASS) material during welding was successfully reviewed while investigating the microstructural propensity of an ASS component to solidification cracking. It was found that fabricated ASS components produced from E 308-16/12 (rutile), E 308-16/10 (rutile), E 308-16/12 (lime-titania) electrodes, and TIG joints (all having FA duplex mode of solidification and ratio of Cr to Ni equivalence in the range $1.5 < Cr_{eq.}/Ni_{eq.} < 1.9$) are resistant to solidification cracking. The E 308-16/12 (lime-titania) electrode (having ferrite number (FN) of 10) was observed to impact the highest resistance to cracking, followed by TIG joints (with FN of 8), E 308-16/12 (rutile) with FN = 6, and E 308-16/10 (rutile) with FN = 5.5 in that order. ASS components fabricated from E 312-16/10 electrode (produced a $Cr_{eq.}/Ni_{eq.}$ ratio of about 2.01 which is greater than 1.9 and ferrite number of about 53% in the completed weld which is beyond 35% stipulated by Hull [8] and necessary to cause a shift from immunity zone to crack susceptible zone). They solidified with a ferrite mode and were found to show little resistance to solidification cracking. The E 310-16/10 electrode solidified with austenite mode and was found to be somewhat liable to solidification cracking since it produced a $Cr_{eq.}/Ni_{eq.}$ ratio of about 1.01, which is less than 1.5 the value suggested by [6, 19–21] to prevent solidification cracking in the completed weld metal.

Conflict of Interests

The authors declare that there is no conflict of interests regarding the publication of this paper.

References

[1] Y. Cui, C. D. Lundin, and V. Hariharan, "Mechanical behavior of austenitic stainless steel weld metals with microfissures," *Journal of Materials Processing Technology*, vol. 171, no. 1, pp. 150–155, 2006.

[2] A. Galal, N. F. Atta, and M. H. S. Al-Hassan, "Effect of some thiophene derivatives on the electrochemical behaviour of AISI 316 austenitic stainless steel in acidic solutions containing chloride ions," *Materials Chemistry and Physics*, vol. 89, no. 1, pp. 38–48, 2005.

[3] A. S. Afolabi, "Effect of electric arc welding parameters on corrosion behavior of austenitic stainless steel in chloride medium," *AU Journal of Technology*, vol. 11, no. 3, pp. 171–176, 2008.

[4] F. A. Ovat, L. O. Asuquo, and A. J. Anyandi, "Microstructural effects of electrodes types on the mechanical behavior of welded steel joints," *Research Journal in Engineering and Applied Sciences*, vol. 1, no. 3, pp. 171–176, 2012.

[5] R. Baldev, V. Shankar, and A. K. Bhaduri, *Welding Technology for Engineers*, Narosa Publishing House, New Delhi, India, 2006.

[6] F. M. L. Arantes and R. E. Trevisan, "Experimental and theoritical evaluation of solidification cracking in weld metal," *Journal of Achievements in Materials and Manufacturing Engineering*, vol. 20, no. 1-2, 2007.

[7] G. K. James, *Chronology of Corrosion Disaster*, vol. 5, American Society for Metals, New York, NY, USA, 2000.

[8] F. C. Hull, "The effect of δ-ferrite on the hot cracking of stainless steel," *Welding Journal*, vol. 46, pp. 399–409, 1967.

[9] J. C. Borland and R. N. Younger, "Some aspects of cracking in welded Cr–Ni austenitic steels," *British Welding Journal*, vol. 7, pp. 22–59, 1960.

[10] H. Thier, R. Killing, and U. Killing, "Solidification modes of weldmentsIn corrosion resistant steels—how to make them visible," *Metal Construction*, vol. 19, no. 3, pp. 127–130, 1987.

[11] J. C. Borland, "Generalized theory of supersolidus cracking in welds and castings," *British Welding Journal*, vol. 7, pp. 508–512, 1960.

[12] J. W. Fu, Y. S. Yang, and J. J. Guo, "Formation of a blocky ferrite in Fe–Cr–Ni alloy during directional solidification," *Journal of Crystal Growth*, vol. 311, no. 14, pp. 3661–3666, 2009.

[13] A. Di Schino, M. G. Mecozzi, M. Barteri, and J. M. Kenny, "Solidification mode and residual ferrite in low-Ni austenitic stainless steels," *Journal of Materials Science*, vol. 35, no. 2, pp. 375–380, 2000.

[14] G. L. Leone and H. W. Kerr, *The Ferrite to Austenite Transformation in Stainless Steels*, Welding Research Supplement, 1982.

[15] T. Udomphol, "Solidification and phase transformations in welding," Lecture Material, Suranaree University of Technology, India, 2007.

[16] D. J. Kotecki and T. A. Siewert, "WRC-1992 constitution diagram for stainless steel weld metals: a modification of the WRC-1988 diagram," *Welding Journal, Research Supplement*, vol. 71, no. 5, pp. 171–177, 1992.

[17] O. Hammar and U. Svensson, "Influence of steel composition on segregationAnd microstructure during solidification of austenitic stainless steel," in *Solidification and Casting of Metals*, pp. 401–410, The Metal Society, London, UK, 1979.

[18] P. S. Korinko and S. H. Malene, "Considerations for the weldability of types 304L and 316L stainless steel," *Practical Failure Analysis*, vol. 1, no. 4, pp. 61–68, 2001.

[19] J. A. Brooks and A. W. Thompson, "Microstructural development and solidification cracking susceptibility of austenitic stainless steel welds," *International Materials Reviews*, vol. 36, no. 1, pp. 16–44, 1991.

[20] V. Shankar, T. P. S. Gill, S. L. Mannan, and S. Sundaresan, "Solidification cracking in austenitic stainless steel welds," *Sadhana*, vol. 34, no. 3-4, pp. 359–382, 2003.

[21] R. Scherer, G. Riedrich, and H. Hougardy, "Welding rod," US Patent 2240672, 1941.

12

Static Response of Functionally Graded Material Plate under Transverse Load for Varying Aspect Ratio

Manish Bhandari[1] and Kamlesh Purohit[2]

[1] Jodhpur Institute of Technology, Jodhpur, Rajasthan 342003, India
[2] Jai Narain Vyas University, Jodhpur, Rajasthan 342005, India

Correspondence should be addressed to Manish Bhandari; manish.bhandari@jietjodhpur.com

Academic Editor: Yanqing Lai

Functionally gradient materials (FGM) are one of the most widely used materials in various applications because of their adaptability to different situations by changing the material constituents as per the requirement. Nowadays it is very easy to tailor the properties to serve specific purposes in functionally gradient material. Most structural components used in the field of engineering can be classified as beams, plates, or shells for analysis purposes. In the present study the power law, sigmoid law and exponential distribution, is considered for the volume fraction distributions of the functionally graded plates. The work includes parametric studies performed by varying volume fraction distributions and aspect ratio. The FGM plate is subjected to transverse UDL (uniformly distributed load) and point load and the response is analysed.

1. Introduction

The material property of the FGM can be tailored to accomplish the specific demands in various engineering utilizations to achieve the advantage of the properties of individual material. This is possible due to the material composition of the FGM which changes sequentially in a preferred direction with a predefined function. The thermomechanical deformation of FGM structures has attracted the attention of many researchers in the past few years in various engineering applications which include design of aerospace structures, heat engine components, and nuclear power plants. A huge amount of literature has been published about the thermomechanical analysis of functionally gradient material plate using finite element techniques. A number of approaches have been employed to study the static bending problems of FGM plates. The assessment of thermomechanical deformation behaviour of functionally graded plate structures considerably depends on the plate model kinematics. Praveen and Reddy reported that the response of the plates with material properties between those of the ceramic and metal is not necessarily in between to the responses of the ceramic and metal plates [1].

Reddy reported theoretical formulations and finite element analysis of the thermomechanical, transient response of functionally graded cylinders and plates with nonlinearity [2]. Cheng and Batra developed a solution in closed-form for the functionally graded elliptic plate rigidly clamped at the edges. It was found that the in-plane displacements and transverse shear stresses in a functionally graded plate do not agree with those assumed in classical and shear deformation plate theories [3]. Reddy formulated Navier's solutions in conjunction with finite element models of rectangular plates based on the third-order shear deformation plate theory for functionally graded plates [4]. Sankar and Tzeng solved the thermoelastic equilibrium equations for a functionally graded beam in closed-form to obtain the axial stress distribution [5]. Qian et al. analyzed static deformations, free, and forced vibrations of a thick rectangular functionally graded elastic plate by using a higher order shear and normal deformable plate theory and a meshless local Petrov-Galerkin (MLPG) method [6]. Ferreira et al. presented the use of the collocation method with the radial basis functions to analyze several plate and beam problems with a third-order shear deformation plate theory (TSDT) [7]. Tahani et al. developed analytical method to analyze analytically the displacements and stresses in a functionally graded composite beam subjected to transverse load and the results obtained from this method were compared with the finite element solution done by ANSYS [8]. Chi and Chung evaluated the numerical solutions directly from theoretical formulations and calculated the results using MARC program. Besides, they compared the results of P-FGM, S-FGM, and E-FGM [9, 10]. Wang and Qin developed a meshless algorithm to simulate the static thermal stress distribution in two-dimensional (2D) functionally graded materials (FGMs). The analog equation method (AEM) was used to obtain the equivalent homogeneous system to the original nonhomogeneous equation [11]. Shabana and Noda used the homogenization method (HM) based on the finite element method (FEM) to determine the macroscopic effective properties which lead to the same thermomechanical behaviour as one of the materials with the periodic microstructure [12]. Mahdavian derived equilibrium and stability equations of a FGM rectangular plate under uniform in-plane compression [13]. Zenkour and Mashat determined the thermal buckling response of functionally graded plates using sinusoidal shear deformation plate theory (SPT) [14]. Alieldin et al. proposed three transformation procedures of a laminated composite plate to an equivalent single-layer FG plate. The first approach is a curve fitting approach which is used to obtain an equivalent function of the FG material property, the second approach is the effective material property approach, and the third approach is the volume fraction approach in which the FG material property varies through the plate thickness with the power law [15]. Na and Kim reported stress analysis of functionally graded plates using finite element method. Numerical results were compared for three types of materials. The 18-node solid element was selected for more accurate modeling of material properties in the thickness direction [16]. Vanam et al. analyzed the static analysis of an isotropic rectangular plate with various boundary conditions and various types of load applications. Numerical analysis (finite element analysis, FEA) has been carried out by developing programming in mathematical software MATLAB and they compared results with those obtained by finite element analysis software ANSYS [17]. Raki et al. derived equilibrium and stability equations of a rectangular plate made of functionally graded material (FGM) under thermal loads based on the higher order shear deformation plate theory [18]. Talha and Singh reported formulations based on higher order shear deformation theory with a considerable amendment in the transverse displacement using finite element method (FEM) [19]. Srinivas and Shiva Prasad focused on analysis of FGM flat plates under mechanical loading in order to understand the effect variation of material properties on structural response using ANSYS software [20]. Srinivas and Shiva Prasad focused on analysis of FGM flat plates under thermal loading in order to understand the effect variation of material properties on structural response. Results are compared to published results in order to show the accuracy of modelling FGMs using ANSYS software [21]. Alshorbagy et al. worked for the exact neutral plane position and evaluated FSDT model on a plate and presented the effect of heat source intensity

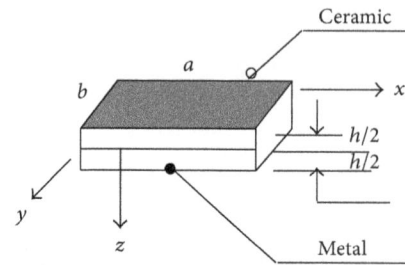

FIGURE 1: FGM plate.

for thermomechanical loading. They reported that FGMs provide a highly stable response for the thermal loading compared to that of the isotropic materials [22].

In the present study, the power law, sigmoid and exponential distribution, is considered for the volume fraction distributions of the functionally graded plates. The work includes parametric studies performed by varying volume fraction distributions and aspect ratio. The FGM plate is subjected to transverse UDL (uniformly distributed load) and point load and the response is analysed. The finite element software ANSYS APDL-13 is used for the modelling and analysis purpose.

2. Material Gradient of FGM Plates

The effective material properties like Young's modulus, Poisson's ratio, coefficient of thermal expansion, and thermal conductivity on the upper and lower surfaces are different but are predefined. However, Young's modulus and Poisson's ratio of the plates vary continuously only in the thickness direction (z-axis); that is, $= E(z)$, $\nu = \nu(z)$.

The FGM plate of thickness "h" is modelled usually with one side of the material as ceramic and the other side as metal (Figure 1). The "z" is varying from "$h/2$" at top face, "0" at the middle of the thickness, to "$-h/2$" at bottom face. However, Young's moduli in the thickness direction of the FGM plates vary with power law functions (P-FGM), exponential functions (E-FGM), or sigmoid functions (S-FGM). A mixture of the two materials composes the through-the-thickness characteristics.

2.1. Power Law Function (P-FGM). The material properties of a P-FGM can be determined by the rule of mixture:

$$P(z) = (P_t - P_b) V_f + P_b, \tag{1}$$

where P denotes a generic material property like modulus, P_t and P_b denote the property of the top and bottom faces of the plate, and V_f is volume fraction.

Material properties are dependent on the volume fraction (V_f) of P-FGM which obeys power law as depicted in

$$V_f = \left(\frac{z}{h} + \frac{1}{2}\right)^n, \tag{2}$$

where n is a parameter that dictates the material variation profile through the thickness known as the volume fraction

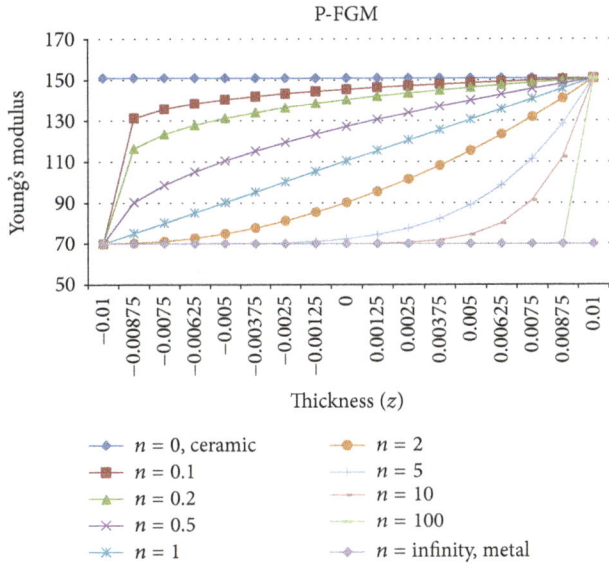

FIGURE 2: Variation of Young's modulus in a P-FGM with "n."

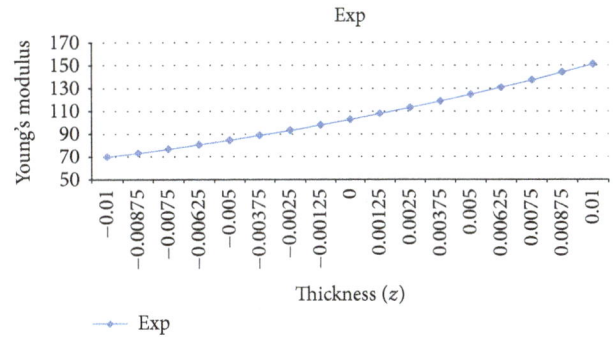

FIGURE 3: Variation of Young's modulus in a S-FGM with "n."

exponent. At the bottom face, $(z/h) = -1/2$; hence $V_f = 0$ and $P(z) = P_b$, and at top face, $(z/h) = 1/2$; hence $V_f = 1$ and $P(z) = P_t$.

At $n = 0$ the plate is a fully ceramic plate while at $n = \infty$ the plate is fully metal. The variation of Young's modulus in the thickness direction of the P-FGM plate is depicted in Figure 2, which shows that Young's modulus changes rapidly near the lowest surface for $n > 1$ and increases quickly near the top surface for $n < 1$.

2.2. Sigmoid Law. In the case of adding an FGM of a single power law function to the multilayered composite, stress concentrations appear on one of the interfaces where the material is continuous but changes rapidly. Therefore, Chung and Chi (2001) defined the volume fraction using two power law functions to ensure smooth distribution of stresses among all the interfaces. The two power law functions are defined by

$$g_1(z) = 1 - \frac{1}{2}\left(\frac{h/2 - z}{h/2}\right)^p, \quad \text{for } 0 \leq z \leq \frac{h}{2},$$

$$g_2(z) = \frac{1}{2}\left(\frac{h/2 + z}{h/2}\right)^p, \quad \text{for } -\frac{h}{2} \leq z \leq 0. \tag{3}$$

By using the rule of mixture, Young's modulus of the S-FGM can be calculated by

$$E(z) = g_1(z) E_1 + \left[1 - g_1(z)\right] E_2, \quad \text{for } 0 \leq z \leq \frac{h}{2},$$

$$E(z) = g_2(z) E_1 + \left[1 - g_2(z)\right] E_2, \quad \text{for } -\frac{h}{2} \leq z \leq 0. \tag{4}$$

The variation of Young's modulus in the thickness direction of the S-FGM plate is depicted in Figure 3, which shows that Young's modulus changes are gradual because of using two power law functions together as described above.

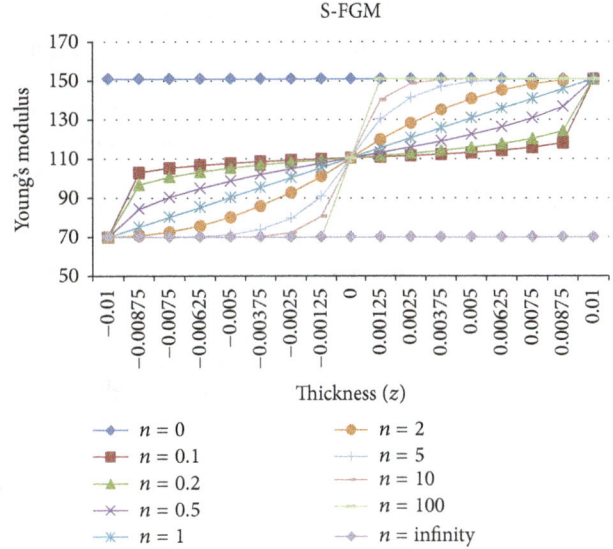

FIGURE 4: Variation of Young's modulus in a E-FGM plate.

2.3. Exponential Law. Many researchers used the exponential function to describe the material properties of FGMs as follows:

$$E(z) = E_2 e^{(1/h)\ln(E_1/E_2)(z+h/2)}. \tag{5}$$

The material distribution in the E-FGM plates is plotted in Figure 4.

3. Finite Element Modelling Technique

The material properties of the FGM change throughout the thickness; the numerical model is to be broken up into various "layers" in order to capture the change in properties. These "layers" capture a finite portion of the thickness and are treated like isotropic materials. Material properties are calculated from the bottom surface using the various volume fraction distribution laws. The "layers" and their associated properties are then layered together to establish the through-the-thickness variation of material properties. Although the layered structure does not reflect the gradual change in material properties, a sufficient number of "layers" can reasonably approximate the material gradation. In present analysis FGM plate has been modelled with 16 layers. In this paper, the finite element analysis has been carried out using minimum total potential energy formulation and modelling of FGM plate is carried out using ANSYS software. ANSYS offers a number of elements to choose from for the modelling of gradient materials. The FGM characteristics under mechanical loads have been studied on a flat plate which was modelled in 3D.

4. Mechanical Analysis

The mechanical analysis is conducted for FGM made of combination of metal and ceramic. The metal and ceramic chosen are aluminium and zirconia, respectively. Young's modulus for aluminium is 70 GPa and that for zirconia is 151 GPa. Poisson's ratio for both of the materials was chosen to be 0.3 for simplicity. The FGM plate is simply supported at all of its edges (SSSS). The thickness of the plate (h) is taken as 0.02 m and one of the side lengths (b) is taken as 1 m. The ratio of the plate side lengths is termed as aspect ratio (a/b). The mechanical analysis was performed by applying uniformly distributed load (UDL) and also for the point load with varying aspect ratio (a/b). The value of UDL and point load chosen was equal to 1×10^6 N/m^2. The gravity quite load is less as compared to the external load and is therefore neglected. The analysis is performed for E-FGM and for various values of the volume fraction exponent (n) in P-FGM and S-FGM. The results are presented in terms of nondimensional parameters, that is, nondimensional deflection ($\overline{u_z}$), nondimensional tensile stress ($\overline{\sigma_x}$), and nondimensional shear stress ($\overline{\sigma_{xy}}$).

The various nondimensional parameters used are nondimensional deflection

$$\overline{u_z} = \frac{\left(100 E_m h^3 u_z\right)}{\left(1 - v^2\right) a^4 p_o} \qquad (6)$$

and nondimensional stress

$$\overline{\sigma_x} = \frac{\sigma h^2}{p_o a^2}, \qquad (7)$$

where "u_z" is maximum deflection, "σ" is maximum stress, "a" and "b" are side lengths of plate, "E_m" is Young's modulus of aluminium, and p_o is applied load (N/m^2).

4.1. Variation of Aspect Ratio (a/b) with Uniformly Distributed Load. This section discusses the results of the analysis performed on FGM plate with varying aspect ratio subject to constant UDL. The results are presented in terms of nondimensional parameters.

4.1.1. Nondimensional Deflection ($\overline{u_z}$). Figures 5, 6, and 7 show the variation of nondimensional deflection (u_z) with aspect ratio (a/b) for simply supported plates under UDL for P-FGM, S-FGM, and E-FGM, respectively. In case of P-FGM and S-FGM the comparison of various values of volume fraction exponent (n) has been presented. In case of E-FGM a single graph is obtained.

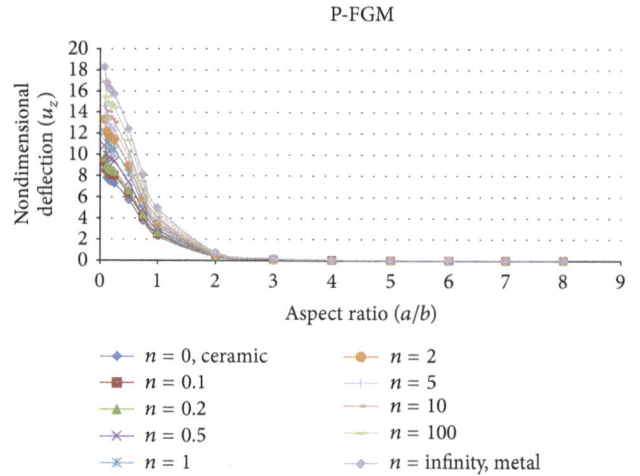

FIGURE 5: Nondimensional deflection (u_z) versus aspect ratio (a/b) for simply supported plates under UDL (P-FGM).

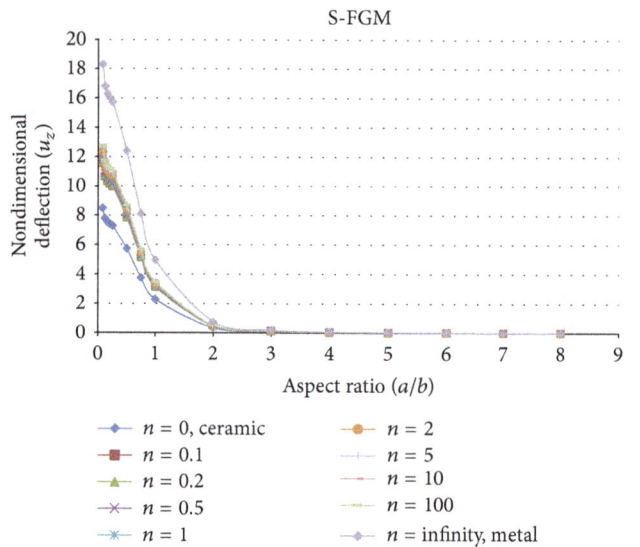

FIGURE 6: Nondimensional deflection (u_z) versus aspect ratio (a/b) for simply supported plates under UDL (S-FGM).

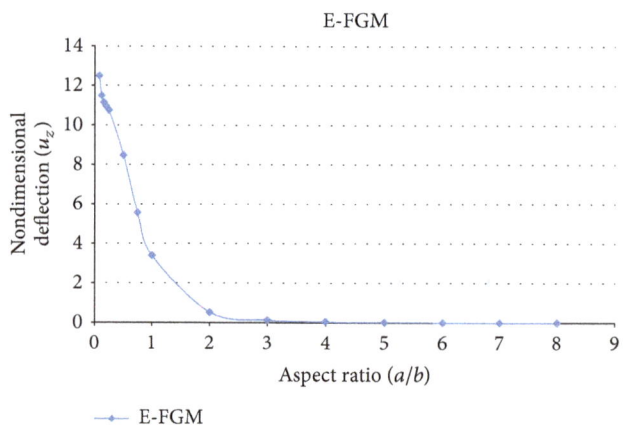

FIGURE 7: Nondimensional deflection (u_z) versus aspect ratio (a/b) for simply supported plates under UDL (E-FGM).

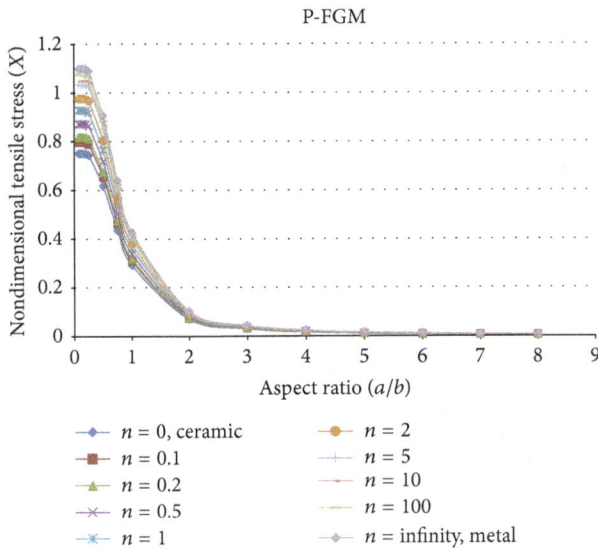

FIGURE 8: Nondimensional tensile stress (σ_x) versus aspect ratio (a/b) for simply supported plates under UDL (P-FGM).

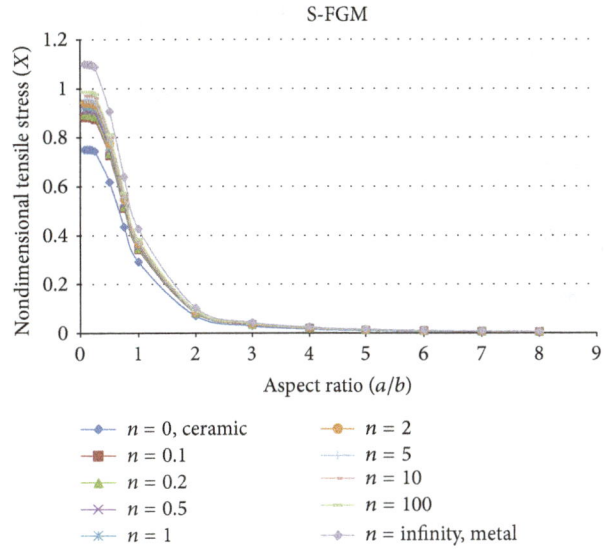

FIGURE 9: Nondimensional tensile stress (σ_x) versus aspect ratio (a/b) for simply supported plates under UDL (S-FGM).

It can be observed that

(a) the nondimensional deflection reduces as the aspect ratio increases up to 3 and it becomes constant as the aspect ratio is increased beyond the value 3. The nondimensional deflection reduces steeply up to aspect ratio 3;

(b) the nondimensional deflection is maximum for the case of pure metal ($n = \infty$) and is minimum for the case of pure ceramic ($n = 0$). As the "n" increases the nondimensional deflection increases. This is due to the fact that the bending stiffness is the maximum for ceramic plate, while it is minimum for metallic plate, and degrades continuously as n increases. The nondimensional deflection for S-FGM remains closer for various values of "n" as compared to that of the P-FGM.

4.1.2. Nondimensional Tensile Stress ($\overline{\sigma_x}$).

Figures 8, 9, and 10 show the variation of nondimensional tensile stress ($\overline{\sigma_x}$) with aspect ratio (a/b) for simply supported plates under UDL for P-FGM, S-FGM, and E-FGM, respectively. In case of P-FGM and S-FGM the comparison of various values of volume fraction exponent (n) has been presented. In case of E-FGM a single graph is obtained.

It can be observed that

(a) the nondimensional tensile stress does not have significant change for the aspect ratio up to 0.25. The nondimensional tensile stress reduces steeply between aspect ratios 0.25 and 2. The nondimensional tensile stress reduces as the aspect ratio increases and it becomes constant as the aspect ratio is increased beyond the value 6;

(b) the nondimensional tensile stress is maximum for the case of pure metal ($n = \infty$) and minimum for the case

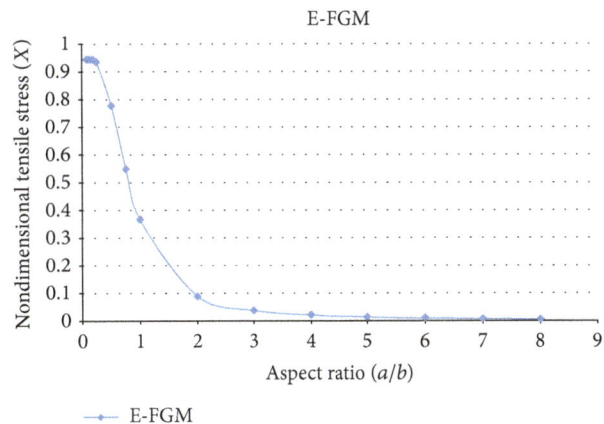

FIGURE 10: Nondimensional tensile stress (σ_x) versus aspect ratio (a/b) for simply supported plates under UDL (E-FGM).

of pure ceramic ($n = 0$). The nondimensional tensile stress for S-FGM remains closer for various values of "n" as compared to that of the P-FGM.

4.1.3. Nondimensional Shear Stress ($\overline{\sigma_{xy}}$).

Figures 11, 12, and 13 show the variation of nondimensional shear stress ($\overline{\sigma_{xy}}$) with aspect ratio (a/b) for simply supported plates under UDL for P-FGM, S-FGM, and E-FGM, respectively. In case of P-FGM and S-FGM the comparison of various values of volume fraction exponent (n) has been presented. In case of E-FGM a single graph is obtained.

It can be observed that the nondimensional shear stress (σ_{xy}) increases as the aspect ratio is increased, it reaches maximum value at aspect ratio 1, and it reduces as the aspect ratio increases beyond 1. The nondimensional shear stress (σ_{xy}) has a steep decline between aspect ratios 1 and 2 and a gradual decline after aspect ratio 2. The nondimensional

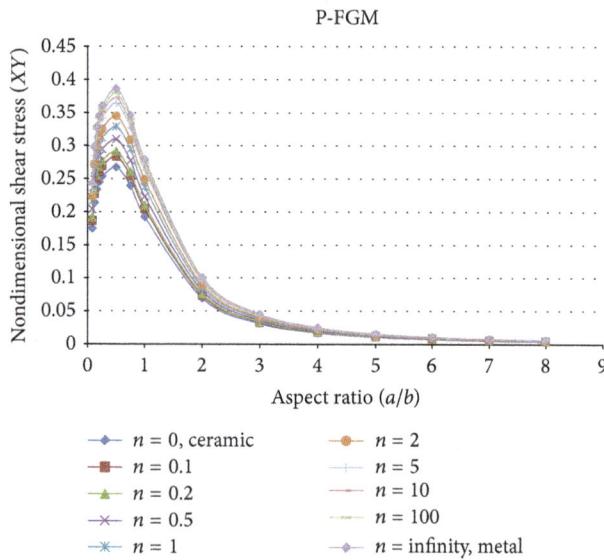

FIGURE 11: Nondimensional shear stress (σ_{xy}) versus aspect ratio (a/b) for simply supported plates under UDL (P-FGM).

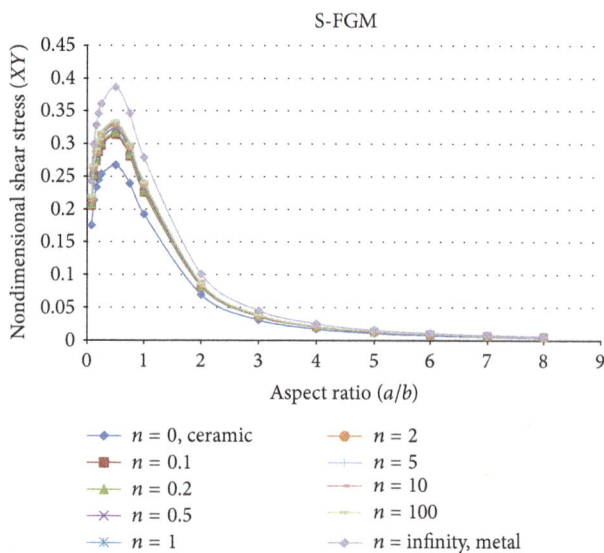

FIGURE 13: Nondimensional shear stress (σ_{xy}) versus aspect ratio (a/b) for simply supported plates under UDL (E-FGM).

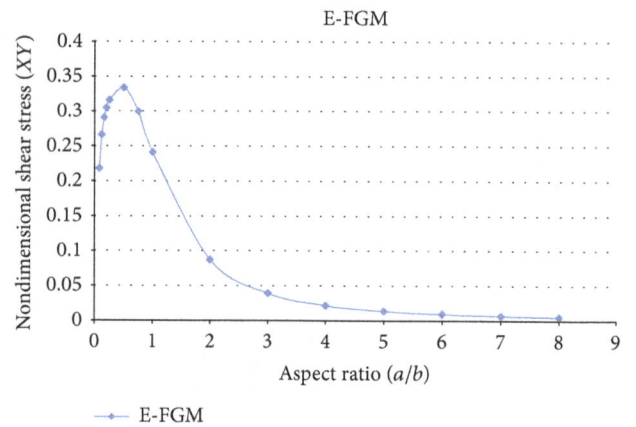

FIGURE 12: Nondimensional shear stress (σ_{xy}) versus aspect ratio (a/b) for simply supported plates under UDL (S-FGM).

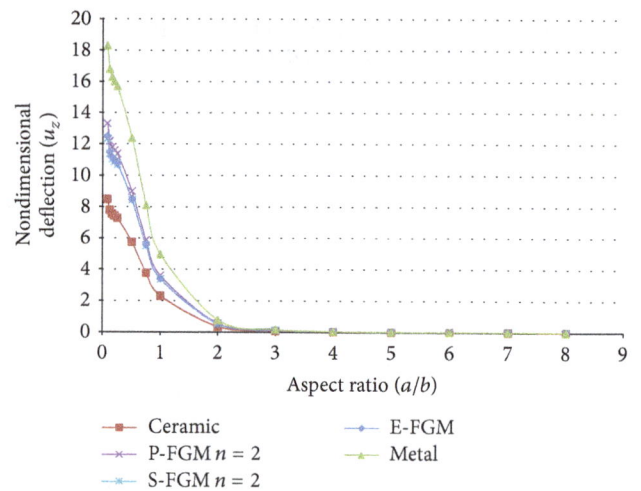

FIGURE 14: Nondimensional deflection (u_z) versus aspect ratio (a/b) for simply supported plates under UDL for various FGMs, ceramic, and metal.

metal, and FGMs following power law, sigmoid, and exponential distribution. Figures 14, 15, and 16 show the comparison graphs for pure ceramic ($n = 0$), pure metal ($n = \infty$), P-FGM ($n = 2$), S-FGM ($n = 2$), and E-FGM.

4.2.1. Nondimensional Deflection ($\overline{u_z}$). See Figure 14.

4.2.2. Nondimensional Tensile Stress ($\overline{\sigma_x}$). See Figure 15.

4.2.3. Nondimensional Shear Stress ($\overline{\sigma_{xy}}$). See Figure 16.
It is observed that

(a) the characteristics of P-FGM and E-FGM are closer to each other as compared to that of S-FGM. The nondimensional tensile stress, shear stress, nondimensional deflection, transverse strain, and shear strain for the three FGMs are in between that of ceramic and metal.

shear stress (σ_{xy}) is maximum for the case of pure metal ($n = \infty$) and minimum for the case of pure ceramic ($n = 0$). The nondimensional shear stress (σ_{xy}) for S-FGM remains closer for various values of "n" as compared to that of the P-FGM.

4.2. Comparison of P-FGM, S-FGM, E-FGM, Ceramic, and Metal. It is also interesting to see the comparison of various parameters like nondimensional deflection, tensile stress, shear stress, transverse strain, and shear strain for ceramic,

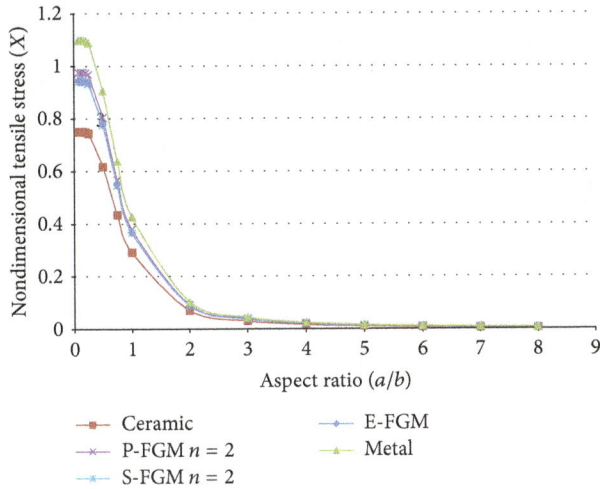

FIGURE 15: Nondimensional tensile stress (σ_x) versus aspect ratio (a/b) for simply supported plates under UDL for various FGMs, ceramic, and metal.

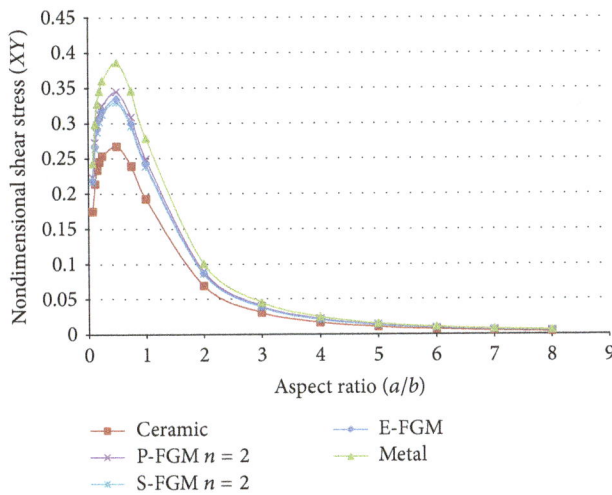

FIGURE 17: Nondimensional deflection (u_z) versus aspect ratio (a/b) for simply supported plates under point load (P-FGM).

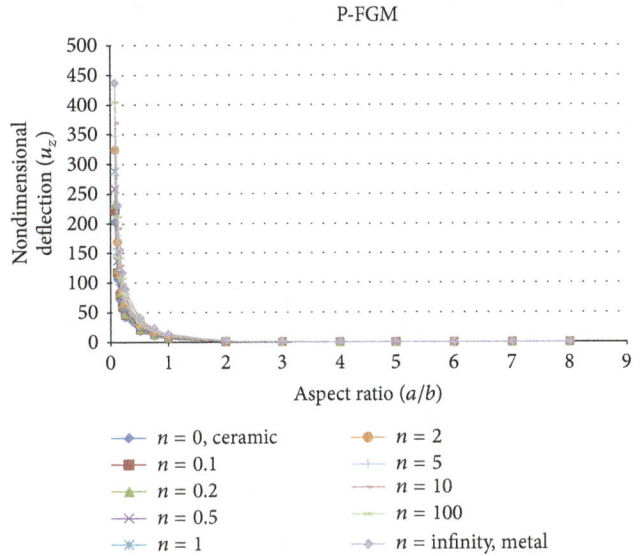

FIGURE 16: Nondimensional shear stress (σ_{xy}) versus aspect ratio (a/b) for simply supported plates under UDL for various FGMs, ceramic, and metal.

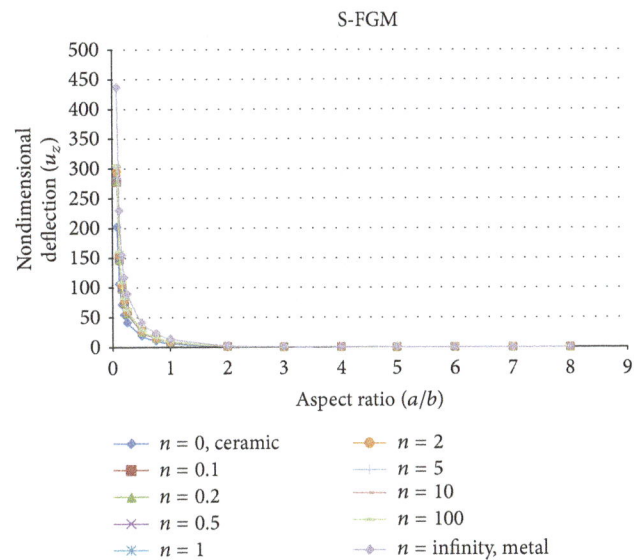

FIGURE 18: Nondimensional deflection (u_z) versus aspect ratio (a/b) for simply supported plates under point load (S-FGM).

4.3. Variation of Aspect Ratio (a/b) with Mechanical Point Load.

This section discusses the results of the analyses performed on FGM plate with varying aspect ratio and subject to constant point load acting at the geometric center of the plate. The results are presented in terms of nondimensional parameters, that is, nondimensional deflection ($\overline{u_z}$), nondimensional tensile stress ($\overline{\sigma_x}$), and nondimensional shear stress ($\overline{\sigma_{xy}}$).

4.3.1. Nondimensional Deflection ($\overline{uz}$).

Figures 17, 18, and 19 show the variation of nondimensional deflection (u_z) with aspect ratio (a/b) for simply supported plates under point load for P-FGM, S-FGM, and E-FGM, respectively. In case of P-FGM and S-FGM the comparison of various values of volume fraction exponent (n) has been presented. In case of E-FGM a single graph is obtained.

It can be observed that

(a) the nondimensional deflection reduces as the aspect ratio increases up to 3 and it becomes constant as the aspect ratio is increased beyond the value 3. The nondimensional deflection reduces steeply up to aspect ratio 3;

(b) the nondimensional deflection is maximum for the case of pure metal ($n = \infty$) and minimum for the case of pure ceramic ($n = 0$). As the "n" increases the nondimensional deflection increases. This is due

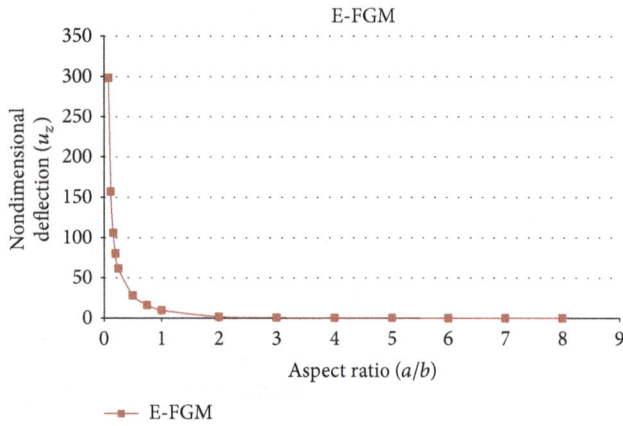

FIGURE 19: Nondimensional deflection (u_z) versus aspect ratio (a/b) for simply supported plates under point load (E-FGM).

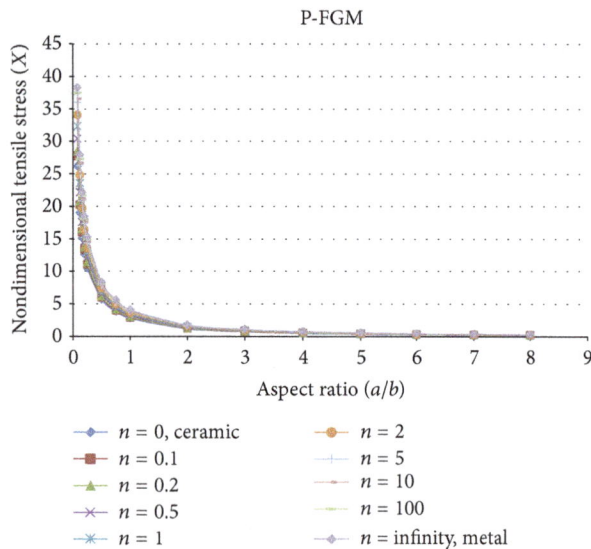

FIGURE 20: Nondimensional tensile stress (σ_x) versus aspect ratio (a/b) for simply supported plates under point load (P-FGM).

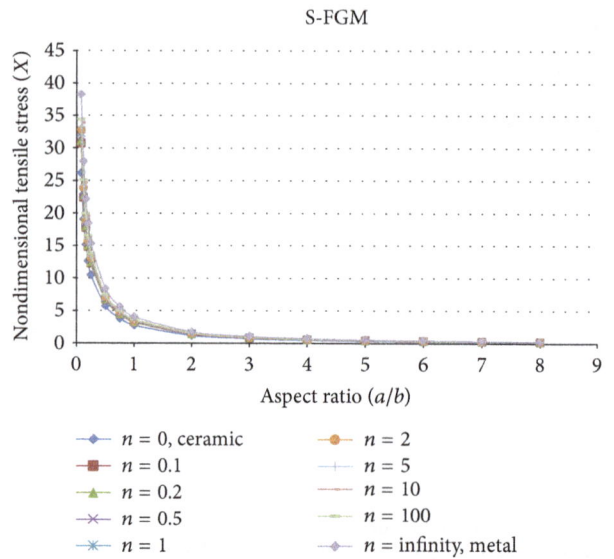

FIGURE 21: Nondimensional tensile stress (σ_x) versus aspect ratio (a/b) for simply supported plates under point load (S-FGM).

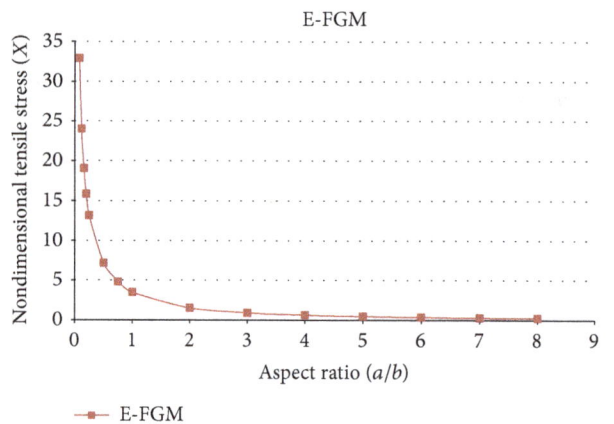

FIGURE 22: Nondimensional tensile stress (σ_x) versus aspect ratio (a/b) for simply supported plates under point load (E-FGM).

to the fact that the bending stiffness is maximum for ceramic plate, while minimum for metallic plate, and degrades continuously as n increases. The nondimensional deflection for S-FGM remains closer for various values of "n" as compared to that of the P-FGM.

4.3.2. Nondimensional Tensile Stress ($\overline{\sigma_x}$).

Figures 20, 21, and 22 show the variation of nondimensional tensile stress ($\overline{\sigma_x}$) with aspect ratio (a/b) for simply supported plates under point load for P-FGM, S-FGM, and E-FGM, respectively. In case of P-FGM and S-FGM the comparison of various values of volume fraction exponent (n) has been presented. In case of E-FGM a single graph is obtained.

It can be observed that

(a) the nondimensional tensile stress reduces steeply up to aspect ratio 1. The nondimensional tensile stress reduces as the aspect ratio increases and it becomes constant as the aspect ratio is increased beyond the value 6;

(b) the nondimensional tensile stress is maximum for the case of pure metal ($n = \infty$) and minimum for the case of pure ceramic ($n = 0$). The nondimensional tensile stress for S-FGM remains closer for various values of "n" as compared to that of the P-FGM.

4.3.3. Nondimensional Shear Stress ($\overline{\sigma_{xy}}$).

Figures 23, 24, and 25 show the variation of nondimensional shear stress ($\overline{\sigma_{xy}}$) with aspect ratio (a/b) for simply supported plates under point load for P-FGM, S-FGM, and E-FGM, respectively. In case of P-FGM and S-FGM the comparison of various values of volume fraction exponent (n) has been presented. In case of E-FGM a single graph is obtained.

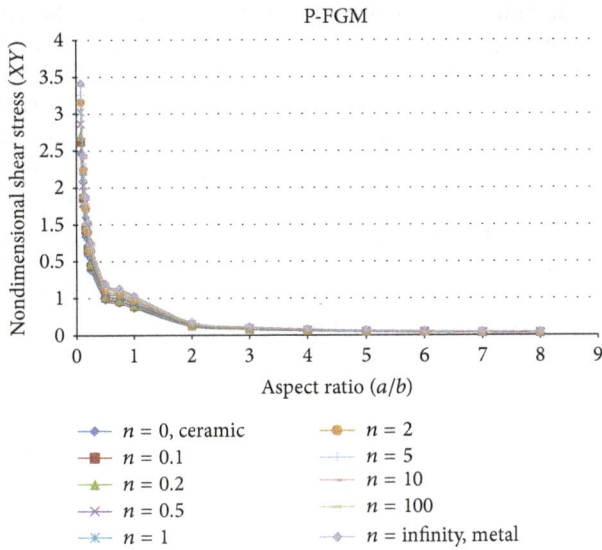

FIGURE 23: Nondimensional shear stress (σ_{xy}) versus aspect ratio (a/b) for simply supported plates under point load (P-FGM).

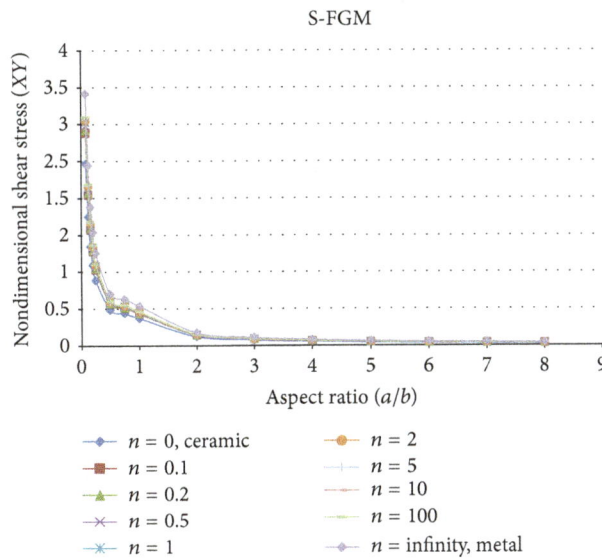

FIGURE 25: Nondimensional shear stress (σ_{xy}) versus aspect ratio (a/b) for simply supported plates under point load (E-FGM).

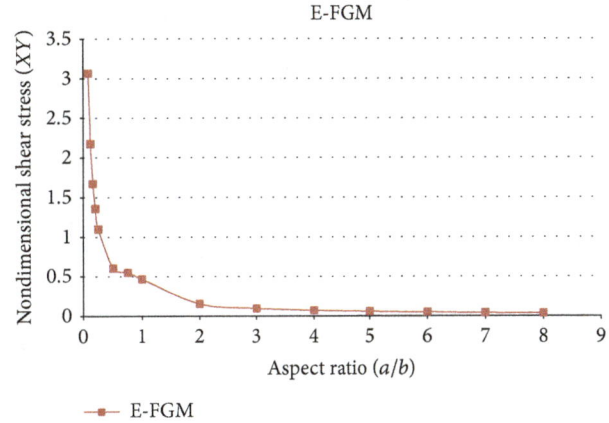

FIGURE 24: Nondimensional shear stress (σ_{xy}) versus aspect ratio (a/b) for simply supported plates under point load (S-FGM).

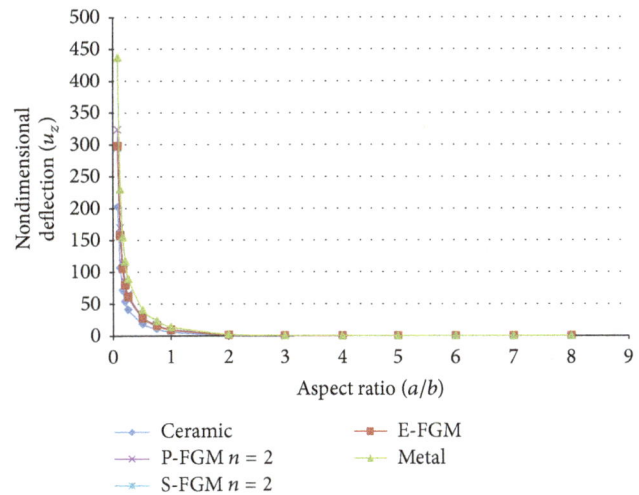

FIGURE 26: Nondimensional deflection (u_z) versus aspect ratio (a/b) for simply supported plates under point load for various FGMs, ceramic, and metal.

It can be observed that the nondimensional shear stress (σ_{xy}) has a steep decline up to aspect ratio 1 and a gradual decline after aspect ratio 1. The nondimensional shear stress (σ_{xy}) is maximum for the case of pure metal ($n = \infty$) and minimum for the case of pure ceramic ($n = 0$). The nondimensional shear stress (σ_{xy}) for S-FGM remains closer for various values of "n" as compared to that of the P-FGM.

4.4. Comparison of P-FGM, S-FGM, E-FGM, Ceramic, and Metal.
It is also interesting to see the comparison of various parameters like nondimensional deflection, tensile stress, shear stress, transverse strain, and shear strain for ceramic,

metal, and FGMs following power law, sigmoid, and exponential distribution. Figures 26, 27, and 28 show the comparison graphs for pure ceramic ($n = 0$), pure metal ($n = \infty$), P-FGM ($n = 2$), S-FGM ($n = 2$), and E-FGM.

4.4.1. Nondimensional Deflection ($\overline{u_z}$). See Figure 26.
It is observed that the characteristics of P-FGM and E-FGM are closer to each other as compared to that of S-FGM. Also the nondimensional tensile stress, shear stress, nondimensional deflection, transverse strain, and shear strain for the three FGMs are in between that of ceramic and metal.

4.4.2. Nondimensional Tensile Stress ($\overline{\sigma_x}$). See Figure 27.

4.4.3. Nondimensional Shear Stress ($\overline{\sigma_{xy}}$). See Figure 28.

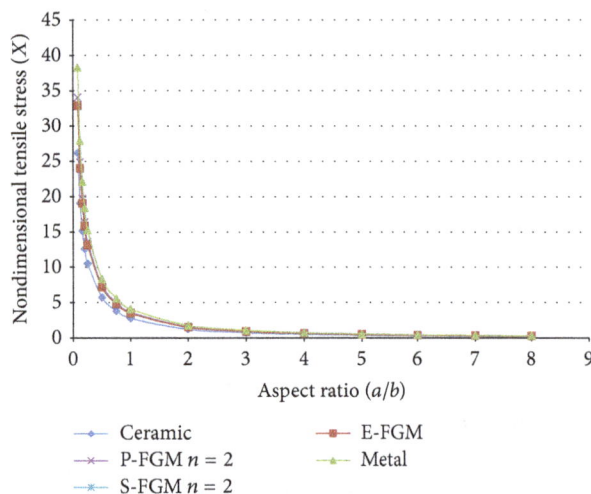

FIGURE 27: Nondimensional tensile stress (σ_x) versus aspect ratio (a/b) for simply supported plates under point load for various FGMs, ceramic, and metal.

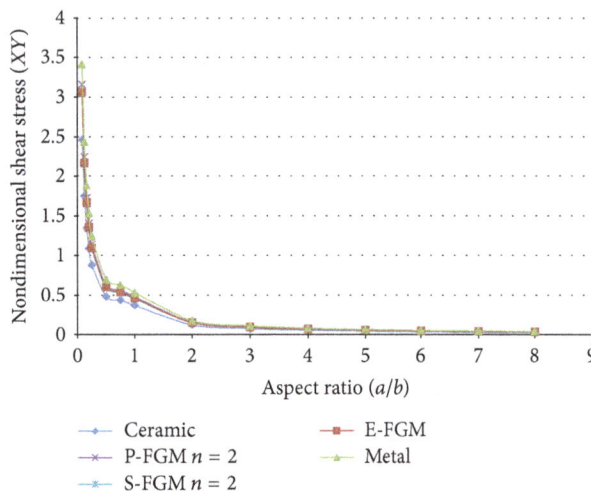

FIGURE 28: Nondimensional shear stress (σ_{xy}) versus aspect ratio (a/b) for simply supported plates under point load for various FGMs, ceramic, and metal.

5. Conclusions and Future Scope

Mechanical deformation of functionally graded ceramic-metal plates with varying aspect ratio is analysed. It is observed that the bending response of the functionally graded plate is intermediate to those of the metal and the ceramic plates. The nondimensional deflection is maximum for the case of pure metal ($n = \infty$) and minimum for the case of pure ceramic ($n = 0$). As the "n" increases the nondimensional deflection increases. The nondimensional tensile stress reduces as the aspect ratio increases and it becomes constant as the aspect ratio is increased beyond the value 6. The nondimensional shear stress (σ_{xy}) increases as the aspect ratio is increased, it reaches maximum value at aspect ratio 1, and it reduces as the aspect ratio increases beyond 1. The bending response for S-FGM remains closer for

various values of "n" as compared to that of the P-FGM. The bending response of E-FGM is nearer to the behavior of P-FGM. The work can be extended for variation in load, loading pattern, and other ceramic metal combinations. Also thermal environment may be imposed in addition to the mechanical loading.

Conflict of Interests

The authors declare that there is no conflict of interests regarding the publication of this paper.

References

[1] G. N. Praveen and J. N. Reddy, "Nonlinear transient thermoelastic analysis of functionally graded ceramic-metal plates," *International Journal of Solids and Structures*, vol. 35, no. 33, pp. 4457–4476, 1998.

[2] J. N. Reddy, "Thermomechanical behavior of functionally graded materials," Final Report for AFOSR Grant F49620-95-1-0342, CML Report 98–01, 1998.

[3] Z. Q. Cheng and R. C. Batra, "Three-dimensional thermoelastic deformations of a functionally graded elliptic plate," *Composites B: Engineering*, vol. 31, no. 2, pp. 97–106, 2000.

[4] J. N. Reddy, "Analysis of functionally graded plates," *International Journal for Numerical Methods in Engineering*, vol. 47, no. 1–3, pp. 663–684, 2000.

[5] B. V. Sankar and J. T. Tzeng, "Thermal stresses in functionally graded beams," *AIAA Journal*, vol. 40, no. 6, pp. 1228–1232, 2002.

[6] L. F. Qian, R. C. Batra, and L. M. Chen, "Static and dynamic deformations of thick functionally graded elastic plates by using higher order shear and normal deformable plate theory and meshless local Petrov-Galerkin method," *Composite Part B*, vol. 35, no. 6–8, pp. 685–697, 2004.

[7] A. J. M. Ferreira, R. C. Batra, C. M. C. Roque, L. F. Qian, and P. A. L. S. Martins, "Static analysis of functionally graded plates using third-order shear deformation theory and a meshless method," *Composite Structures*, vol. 69, no. 4, pp. 449–457, 2005.

[8] M. Tahani, M. A. Torabizadeh, and A. Fereidoon, "Non-linear response of functionally graded beams under transverse loads," in *Proceedings of the 14th Annual International Techanical Engineering Conference*, Isfahan University of Technology, Isfahan, Iran, May 2006.

[9] S. Chi and Y. Chung, "Mechanical behavior of functionally graded material plates under transverse load—part I: analysis," *International Journal of Solids and Structures*, vol. 43, no. 13, pp. 3657–3674, 2006.

[10] S.-H. Chi and Y.-L. Chung, "Mechanical behavior of functionally graded material plates under transverse load—Part II: numerical results," *International Journal of Solids and Structures*, vol. 43, no. 13, pp. 3675–3691, 2006.

[11] H. Wang and Q.-H. Qin, "Meshless approach for thermomechanical analysis of functionally graded materials," *Engineering Analysis with Boundary Elements*, vol. 32, no. 9, pp. 704–712, 2008.

[12] Y. M. Shabana and N. Noda, "Numerical evaluation of the thermomechanical effective properties of a functionally graded material using the homogenization method," *International Journal of Solids and Structures*, vol. 45, no. 11-12, pp. 3494–3506, 2008.

[13] M. Mahdavian, "Buckling analysis of simply-supported functionally graded rectangular plates under non-uniform in-plane compressive loading," *Journal of Solid Mechanics*, vol. 1, no. 3, pp. 213–225, 2009.

[14] A. M. Zenkour and D. S. Mashat, "Thermal buckling analysis of ceramic-metal functionally graded plates," *Natural Science*, vol. 2, no. 9, pp. 968–978, 2010.

[15] S. S. Alieldin, A. E. Alshorbagy, and M. Shaat, "A first-order shear deformation finite element model for elastostatic analysis of laminated composite plates and the equivalent functionally graded plates," *Ain Shams Engineering Journal*, vol. 2, no. 1, pp. 53–62, 2011.

[16] K.-S. Na and J.-H. Kim, "Comprehensive studies on mechanical stress analysis of Functionally Graded Plates," *World Academy of Science, Engineering and Technology*, vol. 60, pp. 768–773, 2011.

[17] B. C. L. Vanam, M. Rajyalakshmi, and R. Inala, "Static analysis of an isotropic rectangular plate using finite element analysis (FEA)," *Journal of Mechanical Engineering Research*, vol. 4, no. 4, pp. 148–162, 2012.

[18] M. Raki, R. Alipour, and A. Kamanbedast, "Thermal buckling of thin rectangular FGM plate," *World Applied Sciences Journal*, vol. 16, no. 1, pp. 52–62, 2012.

[19] M. Talha and B. N. Singh, "Thermo-mechanical deformation behavior of functionally graded rectangular plates subjected to various boundary conditions and loadings," *International Journal of Aerospace and Mechanical Engineering*, vol. 6, p. 1, 2012.

[20] G. Srinivas and U. Shiva Prasad, "Simulation of traditional composites under mechanical loads," *International Journal of Systems, Algorithms & Applications*, vol. 2, pp. 10–14, 2012.

[21] G. Srinivas and U. Shiva Prasad, "Simulation of traditional composites under thermal loads," *Research Journal of Recent Sciences*, vol. 2, pp. 273–278, 2013.

[22] A. E. Alshorbagy, S. S. Alieldin, M. Shaat, and F. F. Mahmoud, "Finite element analysis of the deformation of functionally graded plates under thermomechanical loads," *Mathematical Problems in Engineering*, vol. 2013, Article ID 569781, 13 pages, 2013.

13

A Preliminary Study on Adhesion on Steel Cylinder Filled with Aluminum Foam

G. Marinzuli,[1] **L. A. C. De Filippis,**[1] **R. Surace,**[2] **and A. D. Ludovico**[1]

[1] *Politecnico di Bari, Dipartimento di Meccanica, Matematica e Management (DMMM), Viale Japigia 126, 70126 Bari, Italy*
[2] *ITIA CNR, Institute of Industrial Technology and Automation, National Research Council, Via Paolo Lembo 38F, 70124 Bari, Italy*

Correspondence should be addressed to G. Marinzuli; gaiamarinzuli@gmail.com

Academic Editor: Luca Tomesani

In the last decades, metallic foams found commercial and industrial interests, thanks to their physical properties combined with good mechanical characteristics. Metal foam structures are very light and they can be used to reduce the weight of machinery without compromising the mechanical behavior. In this work, a study of the direct junction of metal foam with metal massive components was carried out. Aluminium foams were manufactured starting from commercial foamable precursors. First of all, attention was paid to the repeatability of foaming process. Then, a direct connection between the foamed samples and the steel shell elements was pursued. The materials that seemed to facilitate the formation of an intermetallic layer were studied and the geometry of the steel mould and the most useful way to place the precursor in the steel mould and then in the furnace were considered. To evaluate the produced aluminum foam, morphological and mechanical characterizations were done. Results showed that, keeping constant the contour conditions, it was possible to control the process and a first result, in terms of interaction between foam and mould, was obtained using an X210Cr12 steel as mould material. The SEM observation revealed the presence of an intermetallic phase.

1. Introduction

Metal foams are cellular materials generally obtained by the dispersion of a gas in a solid material. There are different methods to produce metal matrix cellular solid and they are classified according to the starting state of the metal processed. Metal foams can be manufactured starting by solid, liquid, and vapor metal of a solution of metal ions. In this work, foamed samples were realized by the so-called "powder metallurgy" method; it was developed at the Fraunhofer Institute in Bremen (Germany) [1] and it leads to foamed structures because it involves the decomposition of particles that release gas in semisolid. Foamable precursors are thermally treated in an oven, so that the foaming process occurs. Foamable powder compacts can be produced indoor; thus, the process begins with the mixing of metal powders with a blowing agent, titanium hydride (TiH_2) typically. Then, the mix is compacted to gain a dense, semifinished product called "precursor." The compaction of the mixed powders must be done by a technique that ensures the blowing agent is embedded into the metal matrix without any notable residual open porosity. In this work, commercially available precursors of AlSi0.6Mg1 and AlSi10 with TiH_2 as blowing agent were used; thus, the production of the metal foamable compacts was avoided; circular and rectangular profiles of precursors were used for the experimental part. Figure 1 shows the scheme of the followed process. The temperature of the oven is fixed at a value greater than the melting point of the aluminium that constitutes the base material of the foam: during the heating of the precursor, at temperature near to $460°C$, the titanium hydride, which is homogeneously distributed within the dense metallic matrix, decomposes. The gas, released from the blowing agent, forces the precursor to expand and the internal structure of the metallic sample becomes highly porous. The time needed for the complete expansion is around few minutes and it depends on the temperature of the oven in which the foaming takes place and on the size of the precursor. The volume yielded by the expanded foam depends mainly on the time that the foam remains at high temperatures. Precursors are inserted in a

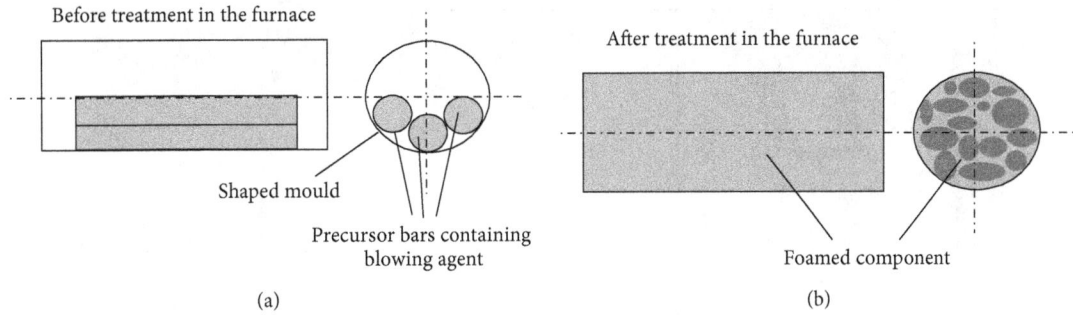

FIGURE 1: Foaming process of compact precursors.

mould, to avoid that the foam is free to expand, and then the ensemble is placed in the oven: the foamed sample gets the shape of the mould once the foaming is completed.

After the foaming occurs, the foamed sample is extracted from the heating chambers and it has to be cooled. There are two main possibilities: air or water cooling. The more time a foam spends in the liquid state, the more drainage can induce a density profile along the height of the foam, with the bottom part being of highest density [1], so, a quick cooling would be better, to avoid that the foam keeps on expanding at room temperature. Generally, if an alloy formed by a solute dissolved in a solvent is heated until the solute is completely dissolved and then is cooled roughly until room temperature, the atoms of the solute are blocked in metastable conditions; in this way, a more ductile and plastic alloy is obtained. Such an aging treatment strongly improves mechanical characteristics of an aluminum alloy, because this process leads to the growth of very fine precipitates that inhibit the movement of the dislocations. The temper prevents a strong diffusion of the elements, so that it can be assumed that the solid solution reaches room temperature without significant variations (in this way an alloy slightly unsaturated at high temperature becomes roughly unsaturated at room temperature).

The aims of this work were the control of process parameters (temperature, means of cooling, and precursor material) and the direct adhesion between metal foam and steel mould during the foaming process. About the former aim, in the literature, there are references to the randomness of the foaming process [2]. Obviously, differences between precursors, uneven temperature distribution during foaming, different thermal expansion between the core and the hollow element, and intrinsic stochastic behavior of the foaming process itself are variables responsible of scarce reproducibility of the foaming process [3]. Also the insertion position of precursors in the mould and then in the heating chamber is a variable that influences the process: a small change in the contact conditions between the mould surface and foamable precursor can change the heat transfer and consequently the starting of foaming [4].

The idea to fill massive hollow metal, such as steel, with aluminum foam is industrially interesting, because the foam gives the core high-energy absorption capacity and flexural stiffness, together with comparatively light weight, while it maintains almost all the other mechanical properties

of the massive metal. Moreover, the cellular structure can be successfully used for producing machine with a strong damping capacity. How to bond massive metal and foam core is an open question; connections between metal foam and massive metal can be obtained in a series of different ways: the two different materials can be welded, glued, or linked by forced elements.

Although the connection between skin and core can be realized with a gluing or a welding of the surfaces, a recent development foresees the adhesion between cellular metals and massive metals already during the foaming process or the direct application of the foam on the massive metal cover. The direct adhesion between metal and aluminum foam should provide a saving in terms of times and costs. In the literature, there are many works about the growth and the development of intermetallic layers between aluminum (not as foam) and massive steel [5–8].

About adhesion between aluminum foam and steel, Neugebauer et al. [9] reported that the factors that influence the adhesion between the two materials are

 (i) the chemical adsorption,

 (ii) the mechanic contraction,

 (iii) the metallurgical interactions.

For what concerns the intermetallic compounds, many literature sources report the formation of a series of intermetallic compounds between aluminum and steel, but among these compounds only Fe_2Al_5 is clearly recognizable at room temperature [10]. Fe_2Al_5 is quite stable and perhaps a surface preparation before proceeding with the trials would be necessary.

2. Experimental

2.1. Setup. The study was carried out in the laboratory of the Department of Mechanics, Mathematics and Management (DMMM) of the Politecnico di Bari. Commercial foamable precursors of AlMg1Si0.6 and AlSi10 (0.8% wt. TiH_2) were used. The aim was to investigate how the components of the solid precursor and the process parameters influenced the foaming process. For foaming process, a thermal treatment station, composed of tempering furnaces and of a quench tank, was used. In particular, tests were conducted in the tempering furnace with a heating chamber of 190×130

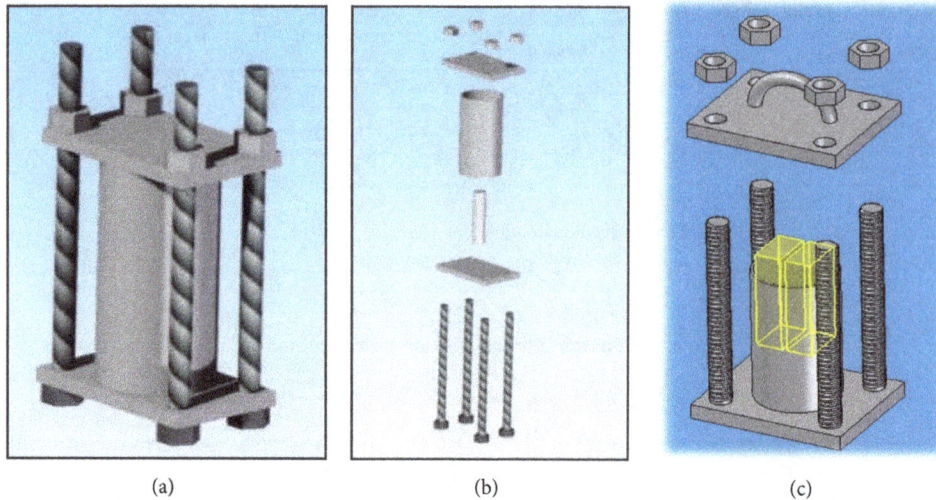

FIGURE 2: (a) Mould model, (b) exploded ensemble of mould, and precursor material and (c) exploded ensemble and rectangular-section precursor material.

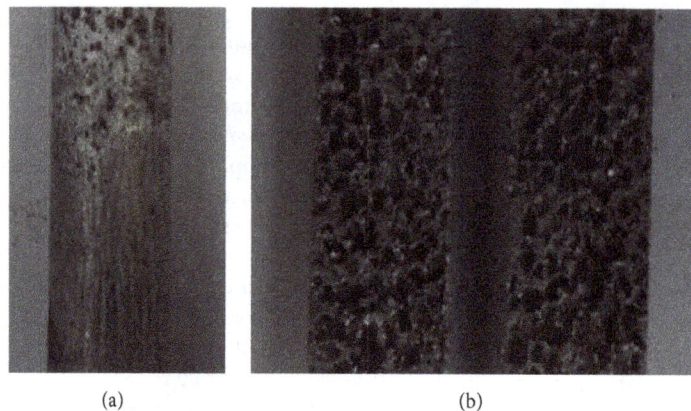

FIGURE 3: (a) View and (b) cross section of a sample made in the mould.

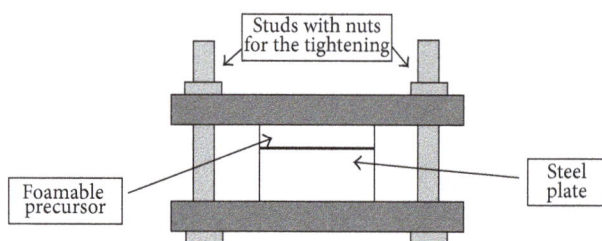

FIGURE 4: New configuration of the mould.

$\times$ 260 mm^3 in which the maximum temperature of 1000°C could be reached. For a correct foaming, the precursor must not be free to expand and it was necessary to use a mould in which it could be inserted; when foaming was performed inside a closed mould and the foaming time was set appropriately, a foam of reproducible volume—that of the mould—and density could be obtained. Foaming in a mould was difficult to control because of several variables involved; thus, the research of a certain control of the process was

needed and pursued to avoid the limits of a free expansion. In this work, steel hollow cylinders were used as moulds, and to prevent the foamed liquid material from coming out from the ends of the cylinder, a closed mould was designed indoor. Figure 2 shows the CAD model of the ensemble. The realized foamed samples showed to reproduce the geometry of the mould in which the foaming process occurred and the cross section of the sample was quite uniform. Figure 3 shows the foamed samples resulting from the use of the mould. The ensemble foam/steel cylinder was interchangeable between the plates of the mould. Insertion position of precursors in the mould was fixed and it is shown in Figure 2(c). Rectangular-section precursors were used during the experimental part. Insertion position of the samples in the mould and then in the oven had influence on foaming; the precursors were placed side by side in the steel cylinder to try to make the junction line almost invisible at the end of the process, and the ensemble precursors steel mould was placed in the middle of the heating chamber to ensure a symmetrical heat diffusion.

Mechanical tests were carried out in the materials test laboratory of the Politecnico di Bari, by means of an INSTRON

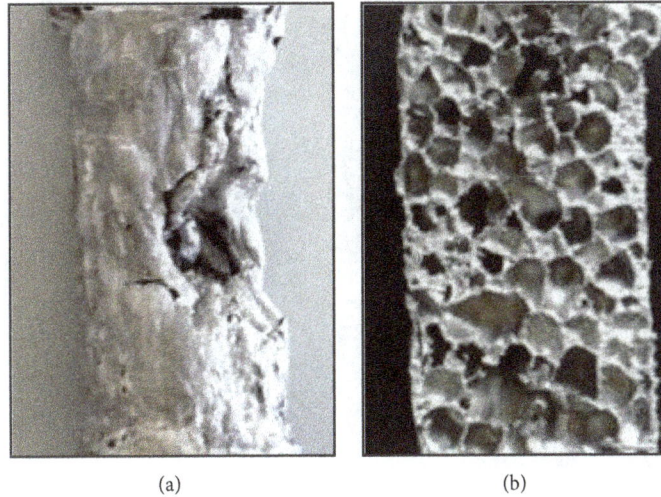

(a)   (b)

FIGURE 5: (a) View of a foamed sample from precursors of different initial length, and (b) cross section of a foamed sample from precursors of the same initial length.

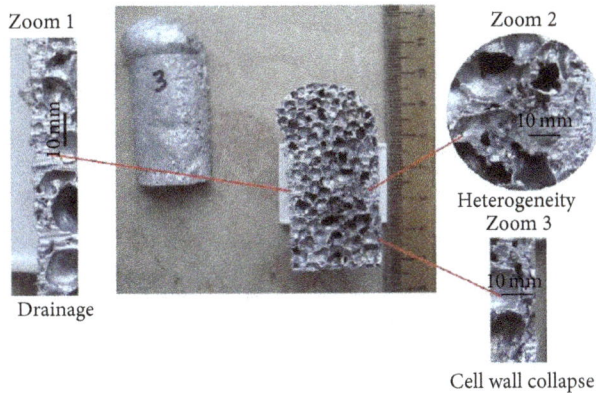

FIGURE 6: First level of morphological analysis: qualitative investigation.

TABLE 1: Parameters and relative levels.

Experimental plan	
Parameters	Levels
T (°C)	700
	800
Means of cooling	Air
	Water
Precursor material	AlSi10
	AlSi0.6Mg1

TABLE 2: Combinations of parameters and levels.

	Full factorial plan		
1	700°C	Air	AlSi10
2	700°C	Air	AlMg1Si0.6
3	700°C	Water	AlSi10
4	700°C	Water	AlMg1Si0.6
5	800°C	Air	AlSi10
6	800°C	Air	AlMg1Si0.6
7	800°C	Water	AlSi10
8	800°C	Water	AlMg1Si0.6

H1015 machine. The compressive force was parallel to the foaming direction of the precursors material and the samples were deformed of the 70% beyond the initial thickness, to point out the densification phase. Each test was interrupted when the load-displacement curve assumed an asymptotic trend. The output data of the compressive machine were the load-displacement curve for each sample. From data couples acquired related stress-strain values were calculated. Compressive strength was evaluated following the statements of the standard test method of the UNI 558-85, on the compressive test on metallic materials at room temperature. According to standard samples, dimensions must respect the following relation: $L_0/D_0 = 1.5$, so that, since the diameter of the samples coincided with the internal diameter of the steel mould, the length is calculated consequently.

About adhesion a new configuration of the mould was realized (Figure 4). Foamed samples, after shearing with a manual saw, stayed connected to the steel part. Little plates of the steel of $L = 15$ mm and $h = 5$ mm were obtained with a thin layer of foam on them. The samples were then embedded in resin and observed at the optical microscope, to verify the existence of a possible intermetallic phase.

2.2. Design of Experiments. The authors assumed three reference main factors and verified their effects on cells area, plateau stress, and adhesion between metal foam and steel mould to ensure repeatability. The chosen parameters were temperature of the heating chamber (°C), means of cooling, and material of the precursors. Each parameter varied on two levels, as indicated in Table 1. Levels and parameters were combined with Minitab software to gain a full factorial plan with 8 combinations, as shown in Table 2. Three replications for each test were made; in this way the trials became 24 and

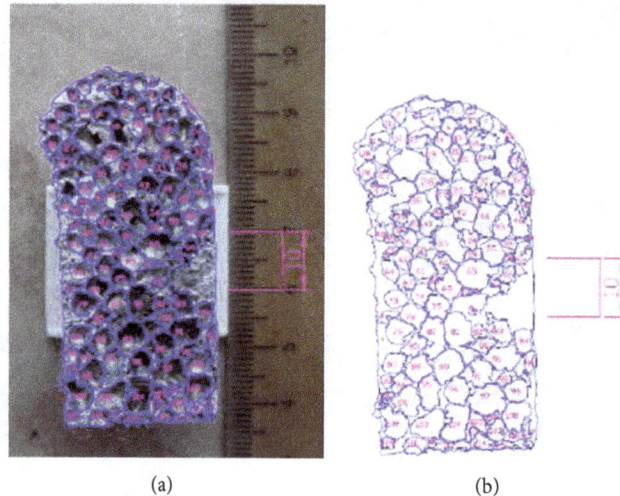

FIGURE 7: (a) Cross section of a foamed sample processed with software AutoCAD and (b) pores area drawing.

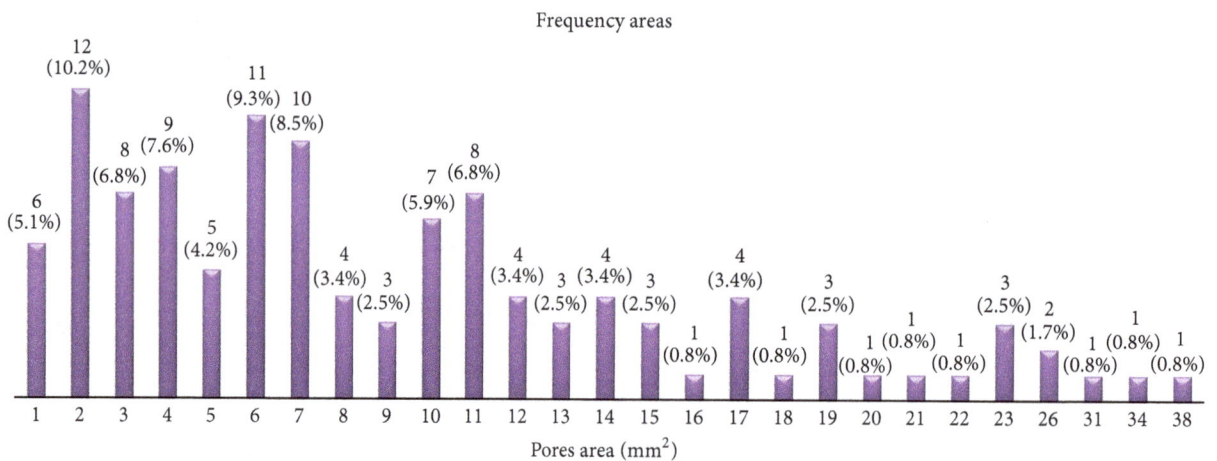

FIGURE 8: Graph of the distribution of the area's dimensions for a foamed sample.

they were randomized to avoid that systematical errors could influence the final results.

The other parameters, not included in the factorial plan, were fixed during screening: starting dimensions of the precursors, which were used to calculate the expansion curve of the foamed samples, and the position of the precursor insertion in the mould. The initial location of the foamable precursor in the mould and its anisotropy had a significant effect on the structure of aluminum foam, which was different due to the heat conduction and pore formation. Symmetric heating of precursors is required to obtain foam with a relatively uniform structure [4]. Time of sample permanence in the oven is a dependent variable and it is a function of the temperature of the heating chamber: this is another value useful for the test plan and it can be known *a posteriori*.

3. Results about Process Control

3.1. Morphological Characterization. Study of the adhesion required a preliminary knowledge of foaming process, to know what happened in fixed conditions of time and temperature. All the samples underwent foaming in a steel mould, to avoid the collapse of the structure of the foam (if it was free to expand without a mould). Then, initial lengths of precursors were the same in each test: the external aspect of the foamed sample depends on the lengths of the starting materials, while the final volume depends on the quantity of used precursor material. Figure 5 shows the differences between the choices just described.

Morphological characterization was carried out in two steps: a qualitative inspection and a quantitative evaluation of pore features. The visual inspection focused on the main defects of the foamed sample: heterogeneity, drainage, coalescence, corrugation, and collapse of the cell walls. Corrugations in the cell walls were observed to be small and they were possibly caused by shrinkage of the cell wall solid during solidification. These defects could be avoided by reducing the stress applied to the foam immediately after solidification (the foam must not be manipulated until the cooling finishes) and by reducing the rate of cooling. Anyway, in the presence of cell wall corrugation both the axial stiffness and the flexural rigidity of the curved or corrugated structural member are reduced [11] up to a 70% drop in the modulus and strength below the values estimated for planar cell walls [12]. In Figure 6, an example of the qualitative evaluation of a sample of AlSi0.6Mg1 foamed at 800°C and cooled in water is shown; the side in which the drainage is indicated corresponded to the bottom of the sample that is the part of the precursor material in contact with the mould in the base of the heating chamber. This part of the precursor in contact with heat zone underwent first the foaming and spent more time at the liquid state.

Another observed feature was the pore shape; in particular, bubble sphericity quantitative evaluation was carried out by measuring the circularity parameter. It is expressed by

$$\text{circularity parameter} = \frac{4\pi A}{P^2}, \tag{1}$$

where A and P are, respectively, area and perimeter of bubbles. The circularity is defined to be shifted from 0 to 1 when the structure of pore is closer to spherical from irregular shape; the values of circularity parameter are determined after pores area measurements.

The second step of morphological characterization was the quantitative evaluation of pores area and perimeter. The quantitative analysis was made by the acquisition and editing of the image of the cross section of the foamed sample to measure area and perimeter of each pore. Each sample was cut longitudinally; even if it was not possible to cut in half each bubble of the sample, it was approximated that the distribution of the pores was realistic and that what resulted in 2D corresponded to the 3D situation.

In cellular material, small bubbles, with uniform size and spherical shape, were desired, because they guarantee better mechanical properties; if the dispersion of the bubbles is low, the foam will perform better. Moreover, improvement of the mechanical properties through control of the cells during manufacturing appears to be particularly challenging. Figure 7 shows an example of elaborated image editing for a sample of AlSi0.6Mg1; for each sample a table was created.

Each table reported the average perimeter of the cells and the standard deviation of the perimeters of each pore, the average area of the cells and the standard deviation of the areas of each pore, the total area of voids in the sample and the total area of the sample, and the total number of pores and the average circularity parameter. Figure 8 represents a plot of data collected to get clearer graphical evidence; on the y-axis the frequencies and the percentages of the cells area were indicated to understand the repartition of the cell according to the areas. Although the trend of the frequency gave a good vision of data distribution, by this graph it was impossible to confront the data obtained with a target value. Thus, a box plot, Figure 9, was also made to put in evidence the central trend, data dispersion and variability, departure for symmetry distribution, and outliers presence.

3.1.1. Data Analysis. After quantitative analysis, from area and perimeter values table and from box plot, the foam was considered homogeneous in the absence of outliers and

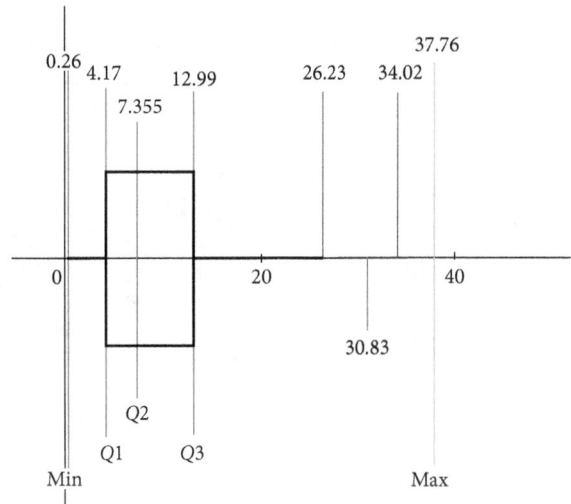

FIGURE 9: Box plot of a foamed sample with a low dispersion of data.

TABLE 3: Scores for cylindrical samples.

Execution order	T (°C)	Means of cooling	Precursor material	Score
1	700	Water	AlSi10	2
2	800	Water	AlSi10	1
3	700	Air	AlSi10	2
4	800	Air	AlSi10	3
5	800	Air	AlSi10	0
6	800	Water	AlMg1Si0.6	1
7	800	Air	AlMg1Si0.6	2
8	700	Air	AlSi10	2
9	800	Water	AlSi10	0
10	700	Water	AlSi10	3
11	800	Water	AlMg1Si0.6	1
12	700	Water	AlMg1Si0.6	2
13	700	Air	AlMg1Si0.6	3
14	700	Water	AlSi10	3
15	700	Water	AlMg1Si0.6	2
16	700	Air	AlMg1Si0.6	2
17	800	Air	AlMg1Si0.6	0
18	800	Air	AlSi10	1
19	800	Water	AlMg1Si0.6	1
20	800	Air	AlMg1Si0.6	1
21	700	Air	AlSi10	2
22	700	Water	AlMg1Si0.6	2
23	800	Water	AlSi10	0
24	700	Air	AlMg1Si0.6	4

anomalous observations. After the two steps of morphological investigation, each sample received a score from 0 to 4 (from sample to discard to very good sample) on the base of cell dimensions, shape, data dispersion, and circularity parameter and on the base of the defects detected. The assigned scores are reported in Table 3. All the cut samples

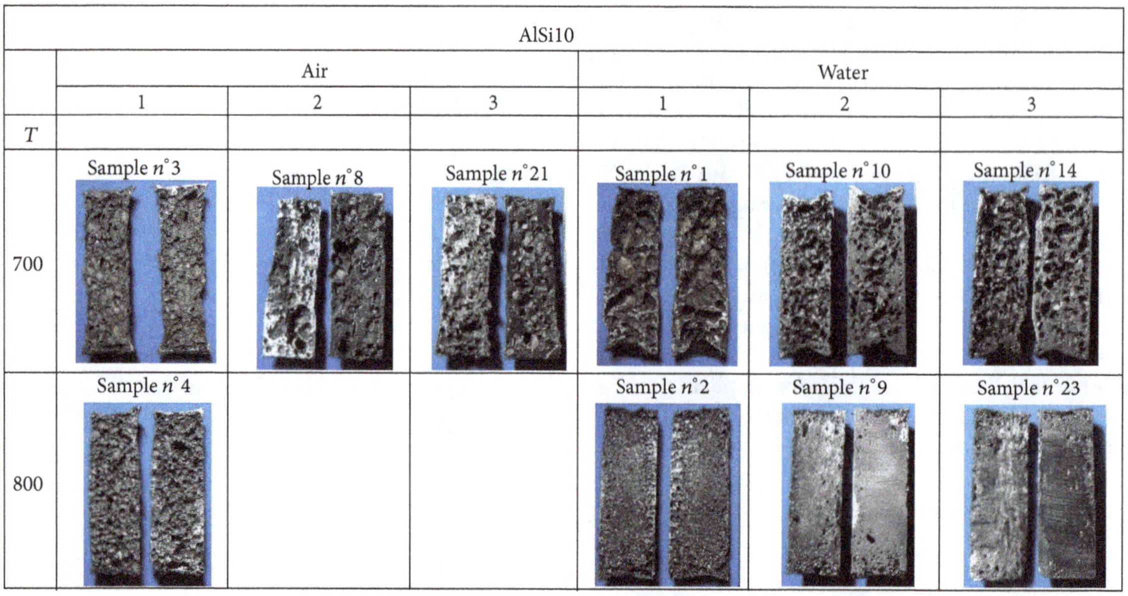

FIGURE 10: Samples of AlSi10 prepared for morphological analysis.

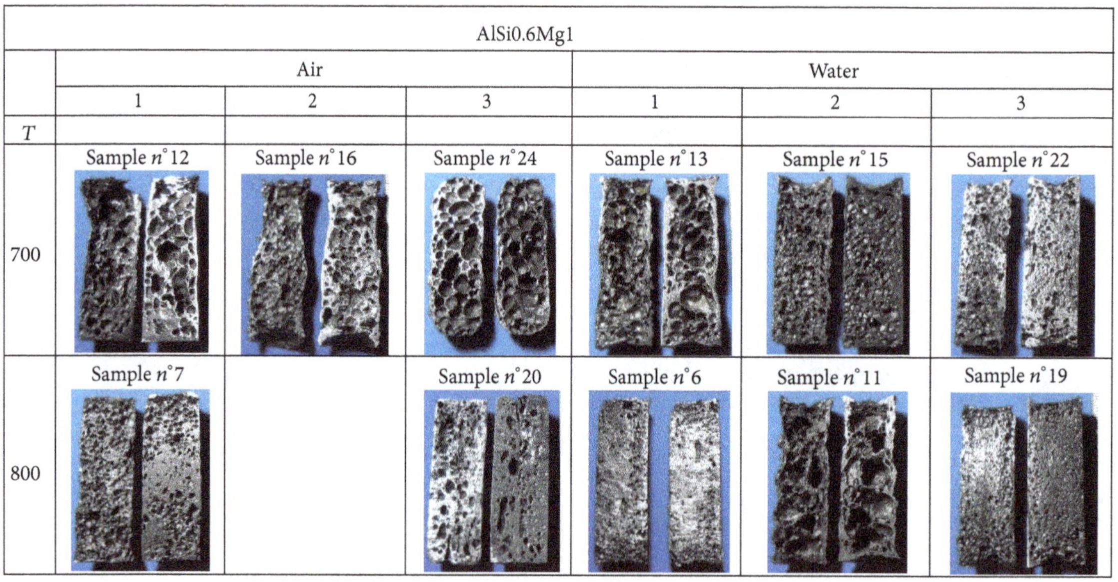

FIGURE 11: Samples of AlSi0.6Mg1 prepared for morphological analysis.

are shown in Figures 10 and 11. Missing samples are those that are damaged in each replication.

The scores assigned to each sample cross section were used to calculate main effect and interaction plot graphs (Figure 12). According to the graphs, the temperature of 700°C gave best results in terms of morphology. It was partially unexpected data because, in preliminary works [13], 800°C gave the best results. Actually, the use of two different precursor materials strongly influenced this result; since the alloy composition influences the melting point of the precursor, a different optimal temperature in comparison to test performed on one precursor material at time was reasonable. From main effects graph it emerged that:

(i) temperature of 700°C increases score results of 250% in comparison to 800°C;

(ii) the use of air as means of cooling increases score results of 15% in comparison to water;

(iii) AlSi0.6Mg1 increases score results of 10% in comparison to AlSi10.

Interaction plot graph shows the interaction of the factors considered in pairs: there was no interaction between temperature and precursor material or between means of cooling and material. The different slope of the curves related to temperature and means of cooling indicated an interaction

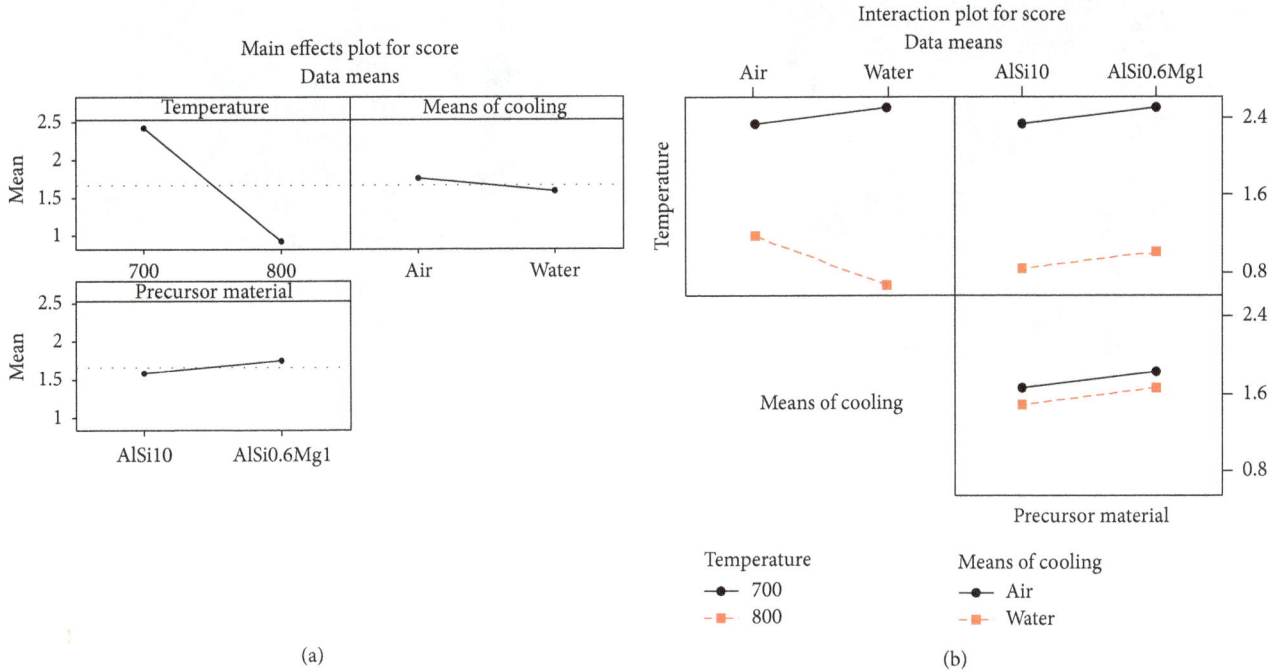

(a)

(b)

FIGURE 12: (a) Main effects and (b) interaction plot for scores.

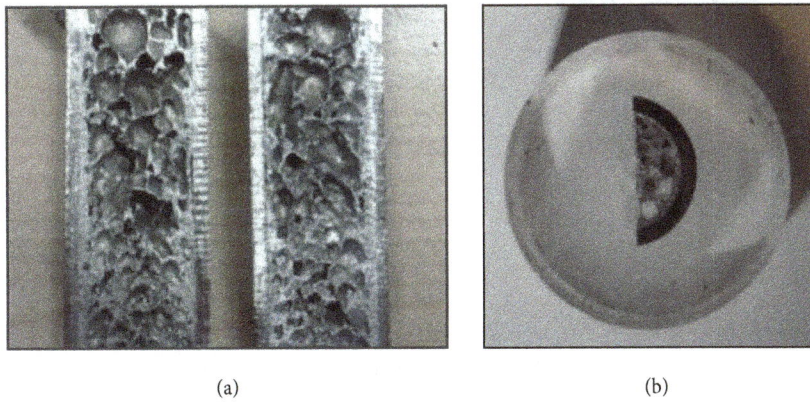

(a)

(b)

FIGURE 13: (a) Section of a steel cylinder filled with aluminum foam. (b) Sample embedded in resin.

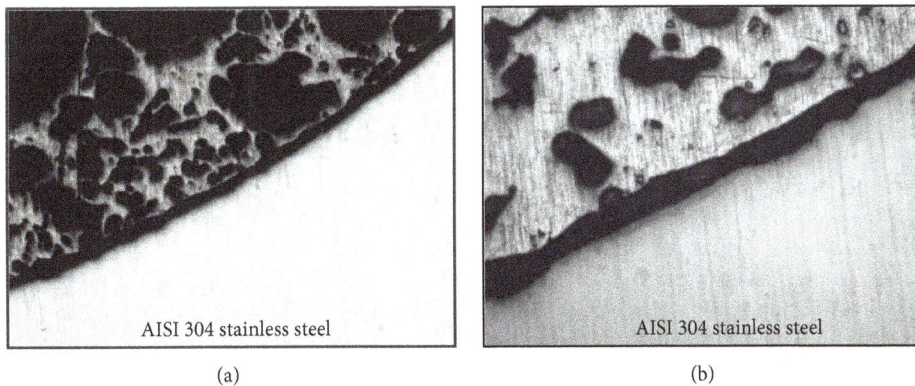

(a)

(b)

FIGURE 14: (a) 50x and (b) 200x micrographic images of the interface aluminum foam/steel mould.

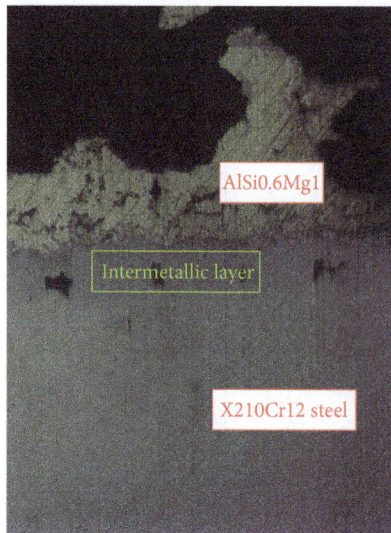

FIGURE 15: Presence of an intermetallic layer at the optical microscope.

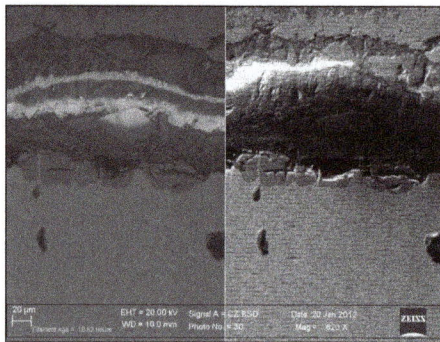

FIGURE 16: Left edge of the sample.

between these two factors, even if their curves had no points of intersection between them.

4. Mechanical Characterization

4.1. Adhesion. Cellular solids are extensively used in structural applications as core materials for loaded sandwich structures, where they experience compressive loading. These materials are also used for packaging and energy absorption applications. So, there is considerable interest in the strength and energy absorption properties under compression. The observed sequence of deformation events is an elastic deflection of the cell elements, followed by a localized deformation in a few cells, the formation of a deformation band, collapse and densification of cells within this band, and gradual spreading of this band through the entire sample. This trend is confirmed by the graphs obtained after the compressive tests. The definition of compression strength for foams, however, is not unambiguous and there are different protocols [14]. In this experimentation, the stress at a certain given deformation (20%) is simply taken. Literature works put in evidence that after foaming a great limit emerged; the

FIGURE 17: Interfacial situation in the central part of the sample.

foamed core had a good bond only with the inferior steel plate but there is almost no bond with the superior one. A growth of the adhesion comes from superficial treatments as sandblasting or nickel electroplating. To gain the adhesion, a mixture of concurrent elements is necessary: there must coexist chemical, physical, and mechanical bonds (not only physical ones, this is just to say, links formed by electrostatic forces that keep together the molecules but do not give any mechanical strength). The existence of a metallurgical bond at the interface aluminum/steel depends on the possibility of a chemical reaction between the molted aluminum and the solid support. This reaction is possible only if the aluminum can wet the metal mold (the superficial oxide that covers the precursor makes this step very difficult) [15]. The superficial oxide reduces roughly the superficial energy of the aluminum and then the contact angle goes up to very high value that corresponds to the nonwettability of the solid by means of the molted aluminum. The roughness of the steel influences positively the adhesion, as it emerges from the literature, because the intermetallic compound starts its growth just in the vales of the steel, so the surface roughness of the samples was improved. The steel mould and the precursor were manually pretreated with coarse sand paper.

A first conclusion, derived from the experimental tests, is that the formation of a layer at the interface aluminum foam/steel required that the foaming is constrained even in a short time and at low temperatures. In the first trials conducted, no adhesion signs are registered; because the carbon reduces the wettability of the steel, steel with a low content of carbon was used.

A hint of adhesion was obtained between precursors of AlSi0.6Mg1 foam and stainless steel hollow cylinder (AISI 304 stainless steel); the ensemble was kept in oven for 7 minutes at 800°C and the cooling is made in water. Once made the longitudinal and the cross sections a very continuous layer at the interface was observed and adhesion in that zone seemed total (Figure 13). The sample was further cut and it was incorporated in resin for the observation under the optical microscope.

What resulted at a macroscopic level was not confirmed at microscopic one. The microscopic investigation revealed a void between the steel and the aluminum foam: no intermetallic phase was formed. Although there was no

(a)

(b)

(c)

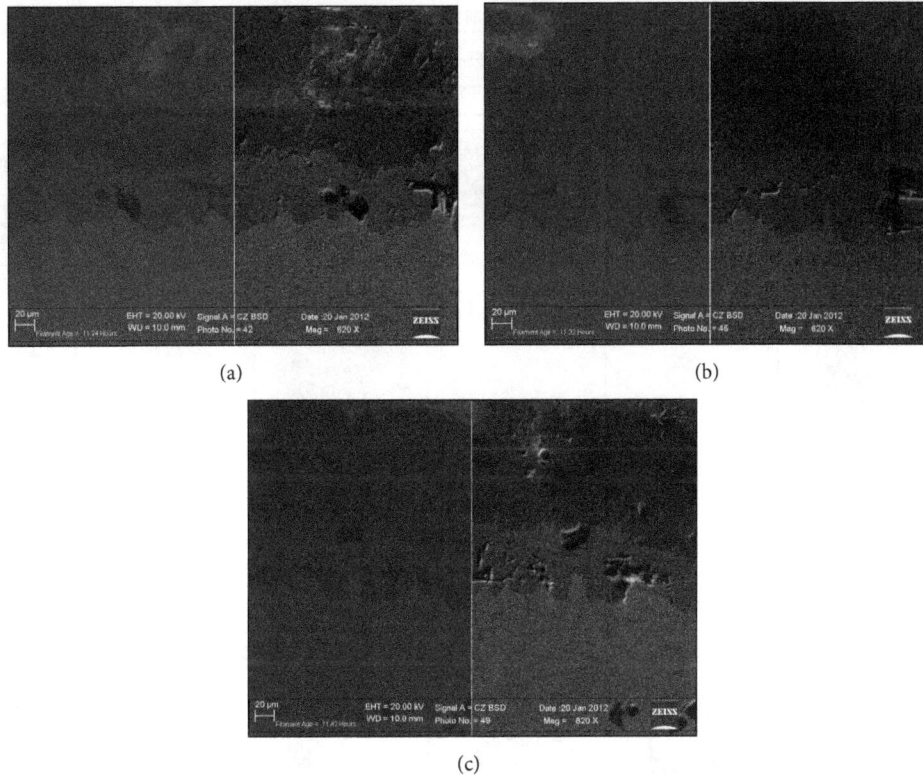

FIGURE 18: Images of compact intermetallic phase at the right side of the sample.

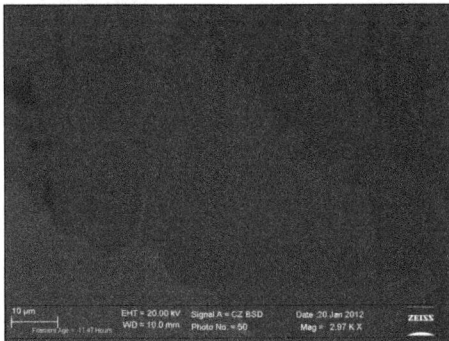

FIGURE 19: Enlargement at 3000x of the interfacial situation to appreciate the homogeneity of the third phase created.

intermetallic formation, the aluminum foam is not divided from the stainless steel; in the obtainment of the adhesion, probably, the role of the stainless steel is fundamental; in particular an AISI 304 stainless steel, austenitic steel with a very low percentage of carbon (<0.08%) is used. From the literature emerged that a low percentage of carbon in the steel facilitates the wettability of the metal and then the formation of intermetallic compounds. This grade of steel is very resistant to chemical agents but at low temperatures its corrosion resistance falls down rapidly, because the acids broke the oxide film. Figure 14 shows two micrographics images (at 50x and at 200x) of the interface situation.

Further trials were conducted by varying the mould material and the configuration mould/precursor to have an idea of the influence of the kind of the mould on the pursuit of direct adhesion. Rectangular-section foamable precursors were used; they were put in contact with an X210Cr12 steel plate and then the ensemble was inserted between the two plates of the mould. In this way, a sort of sandwich configuration was gained. The plates of the mould were tightened to give a force on the ensemble precursor/steel.

Latest articles about this matter refute the statement according to which carbon influences the formation of possible intermetallic layer. So that, steel with a high content of carbon is used for the subsequent attempts, to verify this question: a tool steel, X210Cr12, with 2% of carbon is used. The ensemble remained 6 min. in the furnace at 800°C. The sample, realized with the configuration shown in Figure 4, observed by an optical microscope at 200x revealed the existence of an intermetallic phase, as shown in Figure 15. A further analysis made by a SEM microscope was done. The typical configuration also found in the literature already emerged from optical microscope; the intermetallic layer diffused in the foam with many peaks and valleys and it had the same trend also in the steel part, also if in the steel peaks and valleys were less evident. The SEM observation was made at 620x and from them emerged the presence of an intermetallic phase both in the foam and in the steel (also if this phase had some void zones). Next, some images (Figures 16, 17, 18, and 19) of the interfacial are reported. In particular, Figure 16 shows that more clear grains in different gray scale were evident on the steel. The brighter stripe between foam and steel was due to the charge effect of the resin. Another

AlSi10						
	Air			Water		
	1	2	3	1	2	3
T						
700		8	21	1	10	14
800	4			2	9	23

FIGURE 20: Samples of AlSi10 prepared for compressive tests.

AlSi0.6Mg1						
	Air			Water		
	1	2	3	1	2	3
T						
700	22	16	24	13	15	22
800	7	17		8	21	11

FIGURE 21: Samples of AlSi0.6Mg1 prepared for compressive tests.

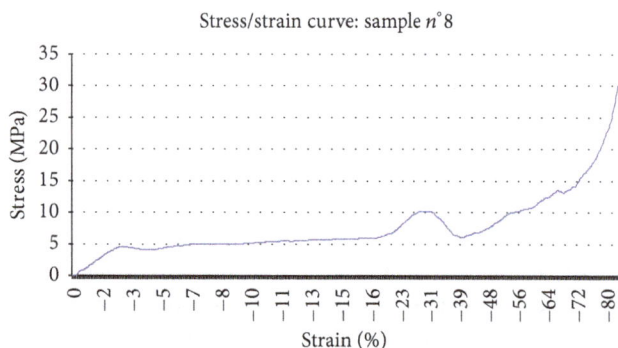

FIGURE 22: Stress/strain curve for the sample of AlSi10 obtained at 700°C and cooled in air (plateau stress = 4.26 MPa).

FIGURE 23: Example of a steel tube filled with aluminium foams cut after compressive test.

As the investigation towards right side of the sample advanced, it is evident that inclusions of a third phase are in the foam and in the steel along all the interface foam/steel, while their consistency and homogeneity varied far from edge in correspondence of the central part. This behavior could be caused by the tightening strength applied on the sample by the mould; this force is clearly stronger on the sides of the sample and it could favor the formation of a continuous and dense intermetallic layer. The compact intermetallic zone starts again as one approached to the right edge of the sample.

Some considerations were done as follows.

(i) All the samples analyzed, at the aluminum foam/ massive steel interface, presented a metallic phase

interesting element was the presence of a grain (the greater one that is in the intermetallic phase) that was composed by two different elements because it was coloured by two different tones of gray.

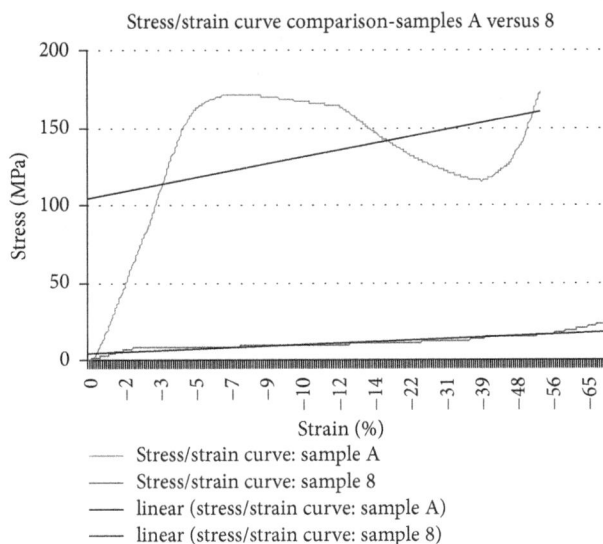

FIGURE 24: Stress/strain curves comparison between sample A (plateau stress = 157.6 MPa) and $n°8$ (plateau stress = 4.26 MPa).

different from metal foam and from steel. This phase was present in the form of grains both in the steel, both in the foam also if the third phase presented some void zone.

(ii) The most homogenous intermetallic layer was observed at the edges of the sample, on which the tightening strength had a greater effect. There was symmetry in what was observed.

4.2. Compressive Test.

How the foam reacts to compressive strength is an important question in terms of adhesion: a possible way to evaluate the interaction between foam and mould is to investigate how the ensemble performs in compressive conditions. So compression tests were carried out on AlSi10 and AlSi0.6Mg1 samples and on the ensemble foam/steel mould. In general damaged samples are not suitable for compressive tests for irregular diameters, fractures, cracks, and carvings. Regular foamed samples were needed for a correct compression test; among the 24 combinations executed, samples damaged after three replications were discarded. A set of 24 other samples was replicated, following the factorial plan of Table 2; the samples prepared for the mechanical characterization were realized with the same process parameters of those prepared for the morphological analysis to verify the correlation of the results. Flat surfaces of foam specimens were machined (Figures 20 and 21). Figure 22 shows the stress/strain curve of a sample of AlSi10 obtained at 700°C and cooled in air.

Hollow stainless steel tubes (AISI 304) filled with AlSi10 and AlSi0.6Mg1 aluminium foam were realized and a set of 8 samples were prepared following the combination of the factorial plan shown in Table 2, without the three replications, because the repeatability was studied yet. Then, samples were cut to view the disposition of the foam in the deformed tube

(Figure 23) and to evaluate the adhesion between the two materials.

It was interesting to make a comparison between stress/strain curves of steel tubes filled with aluminium foam and aluminium foam alone (obtained with the same process parameters), as shown in Figure 24. The following figure shows an example of overlapping of the two curves to put in evidence how the plateau stress is naturally increased and to notice the trend of the curve of the filled element; it had an oscillatory trend and each knee of the curve represented a peak of compression strength. In this work, working with square sample there was only one peak clearly observed. This was in line with other works examined [16].

4.2.1. Data Analysis.

Plateau stress values, calculated for each sample, were used to diagram main effects and interaction plot graphs (Figure 25) and to evaluate which process parameters best influenced such mechanical result. From the analysis of main effects it emerged that

(i) AlSi0.6Mg1 samples increased plateau stress of 55% than AlSi10;

(ii) water cooling gave plateau stress values over 132% than air cooling;

(iii) temperature of 700°C increased plateau stress of 34% than 800°C.

From the analysis of the interaction plot graph, no interactions between couples of factors emerged. There is a great difference between levels but not between factors. Main effects plot for plateau stress results partially agreed with main effect results for score, except for means of cooling. Actually, water cooling improved mechanical response of the foam because it allows freezing of the structure and blocking of the dislocation and it acts like a quench on the foamed structure. By comparing data results it emerged that the effect of a steel skin overshadowed the average effect of the other parameters, as shown in Figure 26.

5. Conclusions

The aims of this work were to reach repeatability of the foaming process and to obtain the adhesion between aluminum foam and steel hollow mould. Some full factorial plans were replicated, considering three parameters that varied on three levels.

About the former aim, on the bases of process parameters and related levels chosen, the results were as follows.

(i) The temperature that gave the best results from a morphological and a mechanical point of view is 700°C.

(ii) The second influent parameter for the morphology is the means of cooling; specimens obtained with air cooling gave a better response. On the contrary samples cooled in water showed a better behavior under the mechanical aspect.

(iii) About the precursor material, AlSi0.6Mg seemed to give better results. It could be caused by the silicon

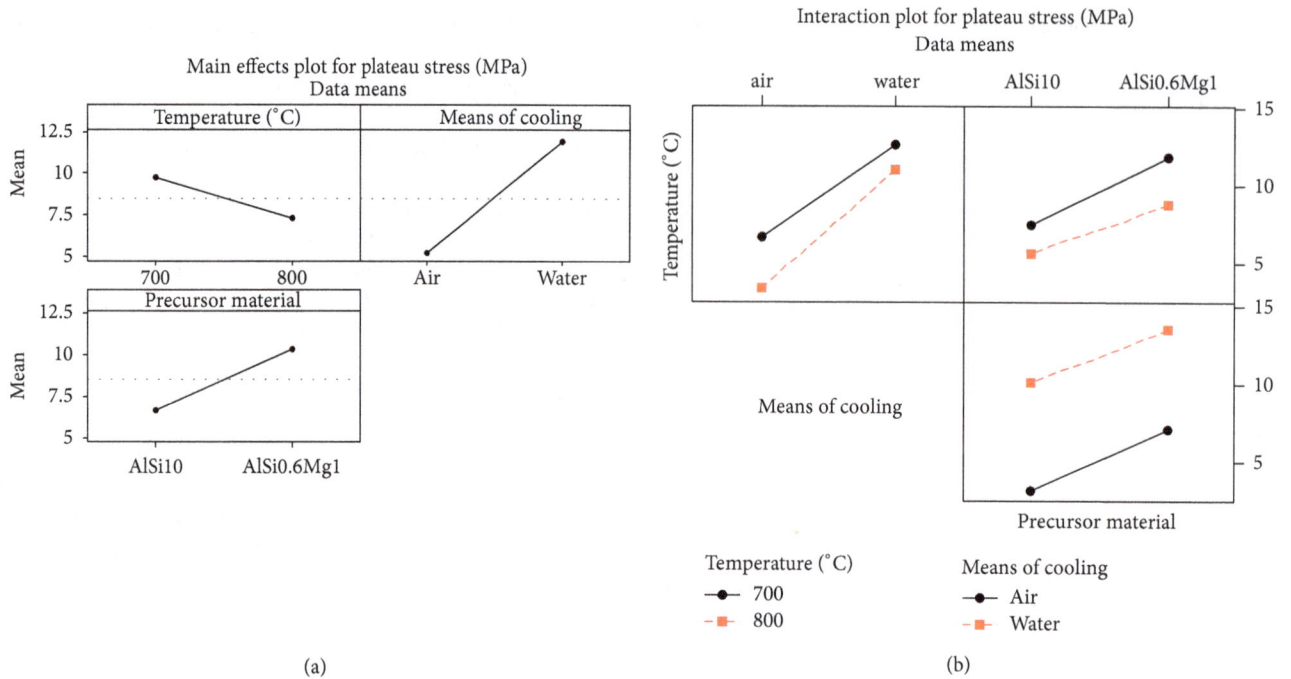

Figure 25: Main effects and interaction plot for plateau stress.

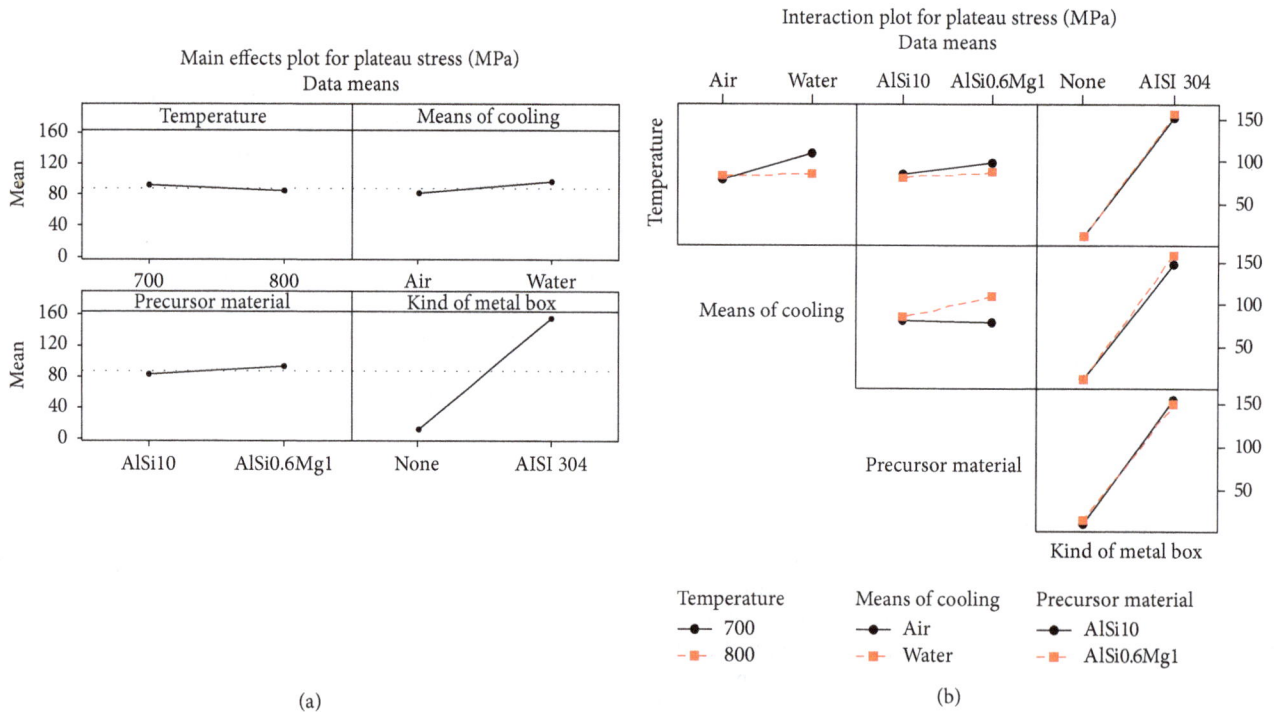

Figure 26: Main Effects and Interaction Plot for the comparison between AISI 304 tubes filled with aluminium foams and aluminium foam alone.

fraction: when it is too high, it derives foam with a great heterogeneity.

Thus, results showed that, keeping constant the contour conditions, it was possible to control the process. The repeatability was preliminary to the study of a direct joining between aluminum foam and steel hollow mould, and the

parameters that gave the best results in terms of morphology and mechanical behavior were used to investigate adhesion with the following results:

(i) the low content of carbon did not seem to influence the direct adhesion: a first result in term of interaction between foam and mould was obtained using an X210Cr12 steel as mould material. The SEM observation revealed the presence of an intermetallic phase between the two different materials.

Moreover, compressive tests on steel tubes filled with aluminium foams were conducted. Even without an intermetallic phase, mechanical features of the ensemble steel mould/foam increased in comparison to the single foam. Future works will be concentrated on the use of an inert atmosphere to investigate how oxygen affects the direct adhesion.

References

[1] J. Banhart, "Manufacture, characterisation and application of cellular metals and metal foams," *Progress in Materials Science*, vol. 46, no. 6, pp. 559–632, 2001.

[2] J. Banhart, "Metal foams: the mystery of stabilization," in *Porous Metal and Metal Foaming Technology (CellMat2005)*, pp. 75–787, 2005.

[3] L. Bonaccorsi, E. Proverbio, and N. Raffaele, "Effect of the interface bonding on the mechanical response of aluminium foam reinforced steel tubes," *Journal of Materials Science*, vol. 45, no. 6, pp. 1514–1522, 2010.

[4] M. Nosko, F. Simančík, and R. Florek, "Reproducibility of aluminum foam properties: effect of precursor distribution on the structural anisotropy and the collapse stress and its dispersion," *Materials Science and Engineering A*, vol. 527, no. 21-22, pp. 5900–5908, 2010.

[5] K. Bouché, F. Barbier, and A. Coulet, "Intermetallic compound layer growth between solid iron and molten aluminium," *Materials Science and Engineering A*, vol. 249, no. 1-2, pp. 167–175, 1998.

[6] V. I. Dybkov, "Interaction of iron-nickel alloys with liquid aluminium," *Journal of Materials Science*, vol. 35, no. 7, pp. 1729–1736, 2000.

[7] K. Barmak and V. I. Dybkov, "Interaction of iron-chromium alloys containing 10 and 25 mass% chromium with liquid aluminium," *Journal of Materials Science*, vol. 38, no. 15, pp. 3249–3255, 2003.

[8] T. Sasaki and T. Yakou, "Features of intermetallic compounds in aluminized steels formed using aluminum foil," *Surface and Coatings Technology*, vol. 201, no. 6, pp. 2131–2139, 2006.

[9] R. Neugebauer, C. Lies, J. Hohlfeld, and T. Hipke, "Adhesion in sandwiches with aluminum foam core," *Production Engineering*, vol. 1, pp. 271–278, 2007.

[10] A. Bouayad, C. Gerometta, A. Belkebir, and A. Ambari, "Kinetic interactions between solid iron and molten aluminium," *Materials Science and Engineering A*, vol. 363, no. 1-2, pp. 53–61, 2003.

[11] A. E. Simone and L. J. Gibson, "The effects of cell face curvature and corrugations on the stiffness and strength of metallic foams," *Acta Materialia*, vol. 46, no. 11, pp. 3929–3935, 1998.

[12] J. L. Grenestedt, "Influence of wavy imperfections in cell walls on elastic stiffness of cellular solids," *Journal of the Mechanics and Physics of Solids*, vol. 46, no. 1, pp. 29–50, 1998.

[13] G. Marinzuli, L. A. C. de Filippis, R. Surace, and A. D. Ludovico, "A preliminary study on adhesion on steel cylinder filled with aluminum foam," in *Proceedings of the 10th AITEM Conference*, Naples, Italy, September 2011.

[14] J. Banhart and J. Baumeister, "Deformation characteristics of metal foams," *Journal of Materials Science*, vol. 33, no. 6, pp. 1431–1440, 1998.

[15] D. Naoi and M. Kajihara, "Growth behavior of Fe_2Al_5 during reactive diffusion between Fe and Al at solid-state temperatures," *Materials Science and Engineering A*, vol. 459, no. 1-2, pp. 375–382, 2007.

[16] F. Campana and D. Pilone, "Effect of heat treatments on the mechanical behaviour of aluminium alloy foams," *Scripta Materialia*, vol. 60, no. 8, pp. 679–682, 2009.

A Comparative Study of the Inhibitory Effect of Gum Exudates from *Khaya senegalensis* and *Albizia ferruginea* on the Corrosion of Mild Steel in Hydrochloric Acid Medium

Paul Ocheje Ameh

Physical Chemistry Unit, Department of Chemistry, Nigeria Police Academy, PMB 3474, Wudil, Kano State, Nigeria

Correspondence should be addressed to Paul Ocheje Ameh; amehpaul99@gmail.com

Academic Editor: Yanqing Lai

A comparative study of the inhibitory potentials of gum exudates from *Albizia ferruginea* (AF) and *Khaya senegalensis* (KS) on the corrosion of mild steel in HCl medium was investigated using weight loss and gasometric method. The active chemical constituents of the gum were elucidated using GC-MS while FTIR was used to identify the bonds/functional groups in the gums. The two gum exudates were found to be good corrosion inhibitors for mild steel in acidic medium. On comparison, maximum inhibition efficiency was found in *Khaya senegalensis* with 82.56% inhibition efficiency at 0.5% g/L concentration of the gum. This may be due to the fact that more compounds with heteroatoms were identified in the GCMS spectrum of KS gum compared to the AF gum. The presence of such compounds may have enhanced their adsorption on the metal surface and thereby blocking the surface and protecting the metal from corrosion. The adsorption of the inhibitors was found to be exothermic and spontaneous and fitted the Langmuir adsorption model.

1. Introduction

Interaction between valuable metals (such as mild steel) and aggressive media (such as acid, base, or salt) is a serious impediment that may risk cost benefit analysis in the operation of some industries [1]. The effects of these on the safe, reliable, and efficient operation of equipment or structures sometimes are often more serious than simple loss of a mass of a metal [2–4]. Several methods have been investigated and implemented to reduce the corrosion process and extend the lifetime of the metals/structures including painting, electroplating, coating, and cathodic protection [5–9]. The use of inhibitors has been found to be one of the best options available for the protection of metals against corrosion [10].

The use of plant products (such as extracts, gums, and latex), as corrosion inhibitors, is also on the increase [11–20]. The greatly expanded interest in these naturally occurring substances is attributed to the fact that they are cheap, readily available, and ecologically friendly and possess no threat to the environment. In addition, they contain compounds that may be aromatic and rich in π-electrons and suitable functional groups (such as C=C, C=O, and -OH) [21]. Plant gum exudates from *Ferula assa-foetida* [22], *Dorema ammoniacum* [22], *Guar gum* [23], *Raphia hookeri* [18], *Pacchylobus edulis* [24], and so forth have been reported as good corrosion inhibitors recently because they are less toxic, green, and eco-friendly.

It is also known that Nigeria is rich in many plant gums species which have not been put into use. Current trends in corrosion inhibition researches are directed towards finding inhibitors (green corrosion inhibitors such as gums) that are eco-friendly, less expensive, and biodegradable.

Hence, the present study is aimed at elucidating the chemical structures of *Khaya senegalensis* and *Albizia ferruginea* and evaluating their corrosion inhibition potentials. From the identified chemical constituents or structures that are inherent in the gums, other industrial potentials of the plant were also investigated. The assessment of the corrosion behaviour was studied using weight loss and gasometric techniques while FTIR measurements were used to study the functional groups associated with the adsorption of the inhibitor.

2. Materials and Methods

2.1. Materials. Corrosion experiments were performed on mild steel specimens with weight percentage composition as follows: Mn (0.6), P (0.36), C (0.15), Si (0.03), and the rest Fe. The sheet was mechanically press-cut into different coupons, each of dimension $5 \times 4 \times 0.11$ cm. These coupons were degreased in absolute ethanol, dried in acetone, and stored in a desiccator free of moisture prior to their use in corrosion studies. The aggressive solutions for gasometric and weight loss studies were 2.5 and 0.1 M, respectively.

Albizia ferruginea (AF) and *Khaya senegalensis* (KS) used as inhibitors were obtained as dried exudates from their parent trees grown at Kanya Babba village in Bubura Local Government Area of Jigawa State, Nigeria, and purified following the method described by Eddy et al. [25]. The concentrations of AF and KS (inhibitors) were prepared and used for the study range from 0.1 g/L to 0.5 g/L.

2.2. GC-MS Analysis. GC-MS analysis was carried out as described by Eddy et al. [25]. Interpretation on mass spectrum GC-MS was conducted using the database of National Institute Standard and Technology (NIST) having more than 62,000 patterns. The spectrum of the unknown component was compared with the spectrum of the known components stored in the NIST library. The name, molecular weight, and structure of the components of the test materials were ascertained. Concentrations of the identified compounds were determined through area and height normalization.

2.3. Corrosion Inhibition Study

2.3.1. Weight Loss Method. The weight loss of the mild steel in 0.1 M HCl with and without the various concentrations of the inhibitors (AF and KS) was determined at 303, 313, 323, and 333 K as described by Oguzie [13]. The coupons were retrieved every 24 hrs for 7 days (168 hrs.) and the difference in weight for a period of 168 hours was taken as total weight loss.

The inhibition efficiency (%I) for each inhibitor was calculated using [25]

$$\%I = \left(1 - \frac{W_1}{W_2}\right) \times 100, \tag{1}$$

where W_1 and W_2 are the weight losses (g/dm³) for mild steel in the presence and absence of inhibitor in HCl solution, respectively. The degree of surface coverage θ is given by [25]

$$\theta = \left(1 - \frac{W_1}{W_2}\right). \tag{2}$$

The corrosion rates for mild steel corrosion in different concentrations of the acid were determined for 168-hour immersion period from weight loss using [26]

$$\text{Corrosion rate (mpy)} = \frac{534W}{DAT}, \tag{3}$$

where W is weight loss (mg), D is density of specimen (g/cm³), A is area of specimen (square inches), and T is period of immersion (hour).

2.3.2. Gasometry Method. The reaction vessel and procedure for determining the corrosion behaviour by this method have been described elsewhere [25]. The experiment was performed at 303 and 333 K for different concentrations of HCl (blank), AF, and KS acting as inhibitors. From the results obtained, the corrosion inhibition efficiency was calculated using the following equation:

$$\%IE = \frac{V_b - V_t}{V_b} \times 100, \tag{4}$$

where V_b is the volume of hydrogen gas evolved by the blank and V_t is the volume of hydrogen gas evolved in the presence of the inhibitor, after time, t.

2.4. FTIR Analysis. FTIR analysis of the corrosion product of mild steel and those of the studied gums was carried out at the National Research Institute of Chemical Technology (NARICT), Zaria, Kaduna State, Nigeria, using Shimadzu FTIR-8400S Fourier transform infrared spectrophotometer. The sample was prepared using KBr and the analysis was done by scanning the sample through a wave number range of 400 to 4000 cm⁻¹.

3. Results and Discussions

3.1. GC-MS Study. Chemical structures of most probable compounds deduced from the GC-MS spectra of KS and AF gums are presented in Figures 1 and 2, respectively. The retention time (RT), IUPAC names of the compounds suggested by reliable spectral library, molecular weight (MW), and charge to mass ratio (m/z) are presented in Tables 1 and 2. Since the area under a GC spectrum is proportional to concentration, normalization of area of the peaks was carried out and used for estimation of percentage concentration (%C) of the respective constituents of the gums. Height normalization was also carried out and the concentrations of constituents (%) based on height normalization were comparable to those obtained from area normalization.

The GC-MS spectra of KS gum revealed 18 peaks. However, only 9 peaks were prominent. From the results presented, it is evident that the most abundant compound in KS gum is nerolidol isobutyrate (peak 15), which constituted about 28%.

This compound undergoes fragmentation into six molecular ions and is characterized with a mass peak value of 41. The compound is a fragrance agent and up to 1 : 10 0000 in the fragrance concentrate is recommended for use [27]. Another major component of KS gum is pinene (line 2: 20.31%), which was separated with characteristic mass peak value of 50 and with about 9 fragment ions. Pinene ($C_{10}H_{16}$) is a bicyclic monoterpene. There are two structural isomers of pinene found in nature: α-pinene and β-pinene. As the name suggests, both forms are important constituents of pine resin; they are also found in the resins of many other

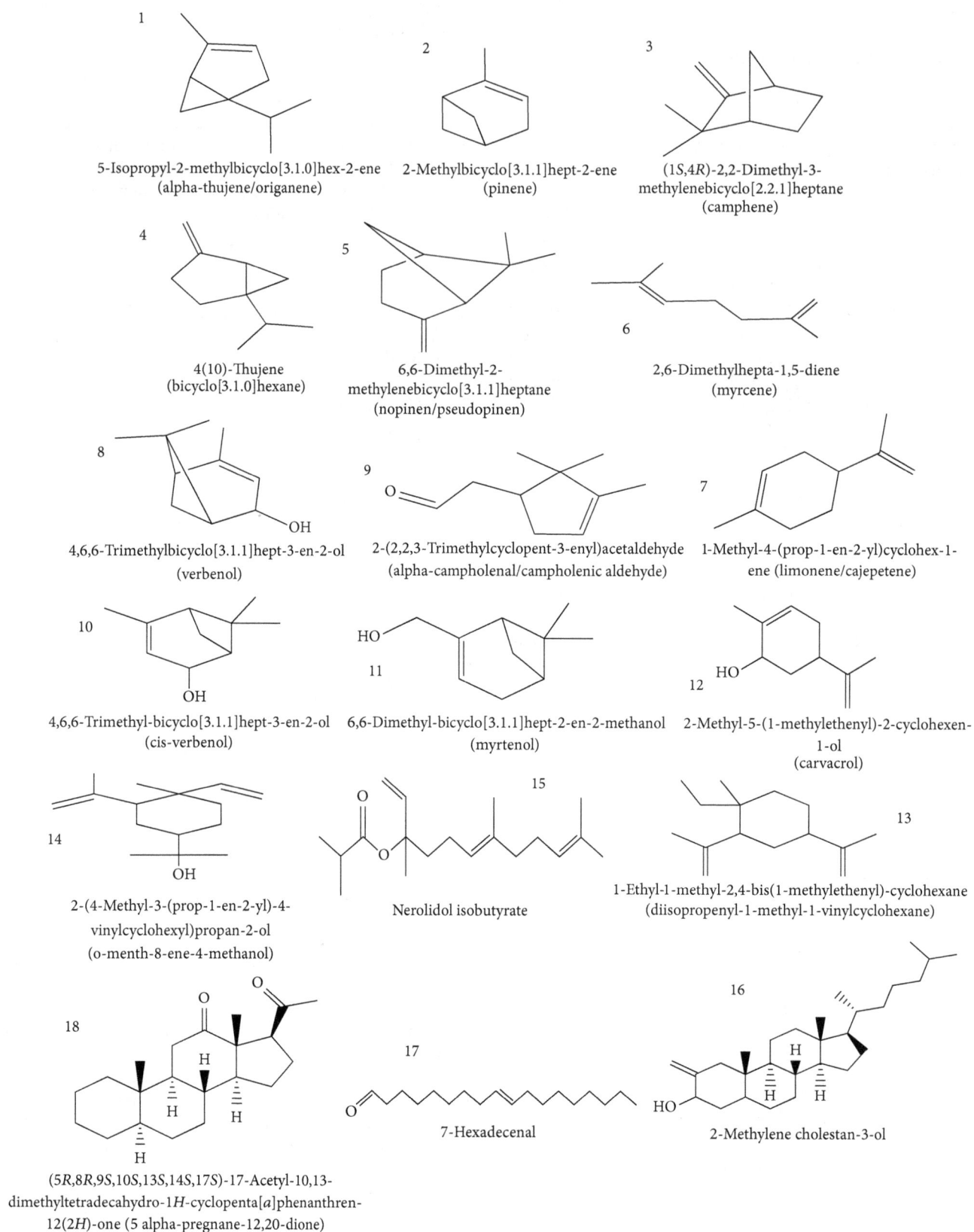

1

5-Isopropyl-2-methylbicyclo[3.1.0]hex-2-ene
(alpha-thujene/origanene)

2

2-Methylbicyclo[3.1.1]hept-2-ene
(pinene)

3

(1*S*,4*R*)-2,2-Dimethyl-3-
methylenebicyclo[2.2.1]heptane
(camphene)

4

4(10)-Thujene
(bicyclo[3.1.0]hexane)

5

6,6-Dimethyl-2-
methylenebicyclo[3.1.1]heptane
(nopinen/pseudopinen)

6

2,6-Dimethylhepta-1,5-diene
(myrcene)

8

4,6,6-Trimethylbicyclo[3.1.1]hept-3-en-2-ol
(verbenol)

9

2-(2,2,3-Trimethylcyclopent-3-enyl)acetaldehyde
(alpha-campholenal/campholenic aldehyde)

7

1-Methyl-4-(prop-1-en-2-yl)cyclohex-1-
ene (limonene/cajepetene)

10

4,6,6-Trimethyl-bicyclo[3.1.1]hept-3-en-2-ol
(cis-verbenol)

11

6,6-Dimethyl-bicyclo[3.1.1]hept-2-en-2-methanol
(myrtenol)

12

2-Methyl-5-(1-methylethenyl)-2-cyclohexen-
1-ol
(carvacrol)

14

2-(4-Methyl-3-(prop-1-en-2-yl)-4-
vinylcyclohexyl)propan-2-ol
(o-menth-8-ene-4-methanol)

15

Nerolidol isobutyrate

13

1-Ethyl-1-methyl-2,4-bis(1-methylethenyl)-cyclohexane
(diisopropenyl-1-methyl-1-vinylcyclohexane)

18

(5*R*,8*R*,9*S*,10*S*,13*S*,14*S*,17*S*)-17-Acetyl-10,13-
dimethyltetradecahydro-1*H*-cyclopenta[*a*]phenanthren-
12(2*H*)-one (5 alpha-pregnane-12,20-dione)

17

7-Hexadecenal

16

2-Methylene cholestan-3-ol

FIGURE 1: Chemical structures of compounds identified in GC-MS spectrum of KS gum (numbering corresponds to the GC line number).

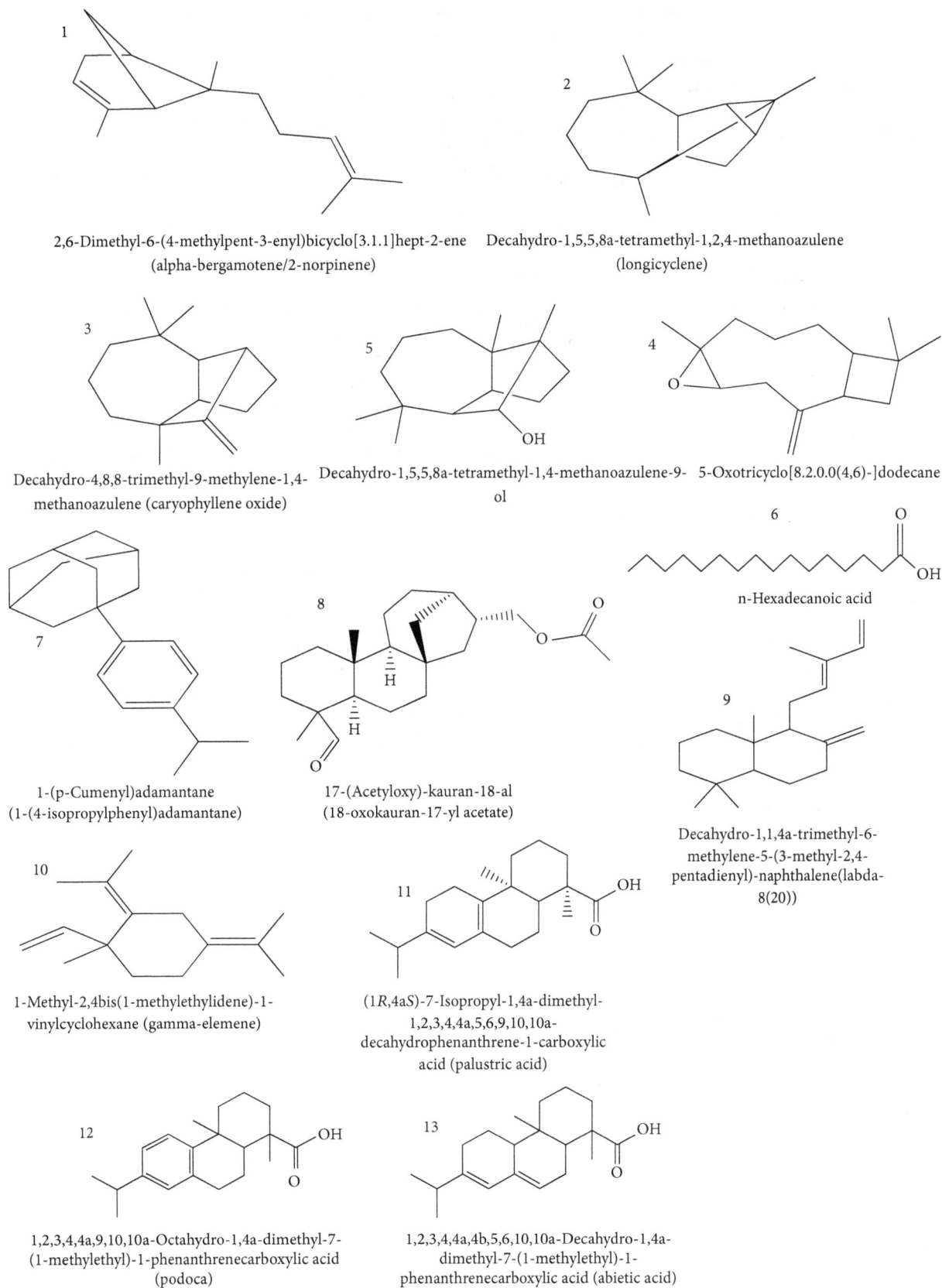

1 2,6-Dimethyl-6-(4-methylpent-3-enyl)bicyclo[3.1.1]hept-2-ene (alpha-bergamotene/2-norpinene)

2 Decahydro-1,5,5,8a-tetramethyl-1,2,4-methanoazulene (longicyclene)

3 Decahydro-4,8,8-trimethyl-9-methylene-1,4-methanoazulene (caryophyllene oxide)

5 Decahydro-1,5,5,8a-tetramethyl-1,4-methanoazulene-9-ol

4 5-Oxotricyclo[8.2.0.0(4,6)-]dodecane

6 n-Hexadecanoic acid

7 1-(p-Cumenyl)adamantane (1-(4-isopropylphenyl)adamantane)

8 17-(Acetyloxy)-kauran-18-al (18-oxokauran-17-yl acetate)

9 Decahydro-1,1,4a-trimethyl-6-methylene-5-(3-methyl-2,4-pentadienyl)-naphthalene(labda-8(20))

10 1-Methyl-2,4bis(1-methylethylidene)-1-vinylcyclohexane (gamma-elemene)

11 (1R,4aS)-7-Isopropyl-1,4a-dimethyl-1,2,3,4,4a,5,6,9,10,10a-decahydrophenanthrene-1-carboxylic acid (palustric acid)

12 1,2,3,4,4a,9,10,10a-Octahydro-1,4a-dimethyl-7-(1-methylethyl)-1-phenanthrenecarboxylic acid (podoca)

13 1,2,3,4,4a,4b,5,6,10,10a-Decahydro-1,4a-dimethyl-7-(1-methylethyl)-1-phenanthrenecarboxylic acid (abietic acid)

FIGURE 2: Chemical structures of compounds identified in GC-MS spectrum of AF gum (numbering corresponds to the GC line number).

TABLE 1: Summary of GC-MS results from peaks in KS gum spectrum.

Line number	%C	Compound	MF	MW	RT	Fragmentation peaks
1	6.22	alpha-Thujene/origanene	$C_{10}H_{16}$	136	6.9	27 (15%), 41 (15%), 43 (15%), 65 (10%), 77 (50%), 93 (100%), 105 (2%), 121 (2%), and 136 (10%)
2	20.31	Myrcene	$C_{10}H_{16}$	136	7.1	27 (15%), 41 (20%), 53 (10%), 67 (100%), 77 (35%), 93 (100%), 105 (10%), 121 (20%), and 136 (10%)
3	1.39	Camphene	$C_{10}H_{16}$	136	7.4	27 (20%), 39 (30%), 53 (20%), 67 (30%), 79 (40%), 93 (100%), 107 (30%), 121 (60%), and 136 (20%)
4	1.90	4(10)-Thujene (bicyclo[3.1.0]hexane)	$C_{10}H_{16}$	136	8.0	27 (25%), 41 (40%), 43 (15%), 69 (15%), 77 (40%), 93 (100%), 105 (2%), 121 (10%), and 136 (20%)
5	2.61	Nopinen/pseudopinen	$C_{10}H_{16}$	136	8.1	27 (20%), 41 (60%), 53 (15%), 69 (40%), 77 (30%), 93 (100%), 105 (2%), 121 (15%), and 136 (15%)
6	1.39	2,6-Dimethylhepta-1,5-diene	$C_{10}H_{16}$	136	8.5	27 (20%), 41 (100%), 53 (20%), 69 (80%), 77 (25%), 93 (100%), 107 (8%), 121 (10%), and 136 (5%)
7	2.37	Limonene/cajepetene	$C_{10}H_{16}$	136	9.6	27 (40%), 39 (60%), 53 (45%), 68 (100%), 79 (40%), 93 (65%), 107 (20%), 121 (20%), and 136 (25%)
8	3.01	Verbenol	$C_{10}H_{16}O$	152	11.6	27 (35%), 41 (65%), 43 (40%), 59 (55%), 79 (60%), 94 (100%), 109 (90%), 119 (30%), and 137 (20%)
9	1.14	alpha-Campholenal	$C_{10}H_{16}O$	152	12.0	27 (15%), 39 (20%), 55 (20%), 67 (25%), 81 (20%), 93 (55%), 108 (100%), 119 (5%), 137 (2%), and 152 (2%)
10	7.24	cis-Verbenol	$C_{10}H_{16}O$	152	12.8	27 (35%), 41 (60%), 43 (40%), 59 (50%), 79 (60%), 94 (100%), 109 (100%), 119 (30%), and 137 (20%)
11	2.61	Myrtenol	$C_{10}H_{16}O$	152	14.3	27 (20%), 41 (40%), 43 (20%), 67 (20%), 79 (100%), 91 (50%), 108 (40%), 119 (30%), 134 (20%), and 152 (10%)
12	0.67	Carvacrol	$C_{10}H_{16}O$	152	14.8	27 (10%), 41 (32%), 55 (35%), 69 (20%), 83 (30%), 84 (55%), 109 (100%), 119 (20%), 137 (10%), and 152 (10%)
13	2.91	Diisopropenyl-1-methyl-1-vinylcyclohexane	$C_{15}H_{24}$	204	31.8	27 (32%), 41 (100%), 53 (60%), 68 (100%), 81 (100%), 93 (80%),107 (40%), 121 (35%), 133 (15%), 147 (20%), 161 (15%), and 189 (15%)
14	8.44	o-Menth-8-ene-4-methanol	$C_{15}H_{26}O$	222	33.2	27 (12%), 39 (18%), 43 (38%), 59 (100%), 81 (50%), 93 (70%), 107 (40%), 121 (30%), 135 (22%), 147 (10%), 161 (40%), 189 (20%), and 204 (10%)
15	28.82	Nerolidol isobutyrate	$C_{19}H_{32}O_2$	292	33.3	41 (42%), 43 (100%), 69 (25%), 71 (50%), 93 (30%), 107 (10%), 121 (40%), 127 (5%), 143 (2%), and 161 (2%)
16	6.23	2-Methylene cholestan-3-ol	$C_{28}H_{48}O$	400	34.0	65 (10%), 69 (100%), 81 (75%), 95 (70%), 105 (35%), 121 (25%), 133 (15%), 149 (15%), and 161 (10%)
17	0.81	7-Hexadecenal	$C_{16}H_{30}O$	238	35.2	41 (80%), 55 (90%), 71 (78%), 85 (50%), 98 (40%), 121 (40%), 135 (20%), and 141 (10%)
18	1.93	5 alpha-pregnane-12,20-dione	$C_{23}H_{36}OS_2$	392	39.9	43 (80%), 67 (25%), 81 (35%), 95 (18%), 105 (20%), 119 (40%), 131 (15%), 145 (20%), 159 (10%), 189 (8%), 203 (5%), 229 (10%), 255 (50%), 299 (10%), 331 (10%), 349 (20%), 350 (4%), 364 (4%), and 392 (100%)

conifers, as well as in nonconiferous plants. In chemical industry, selective oxidation of pinene with catalysts gives many compounds for perfumery, such as artificial odorants.

In line 14, 8.8% of o-menth-8-ene-4-methanol (2-(4-methyl-3-(prop-1-en-2-yl)-4-vinylcyclohexyl)propan-2-ol) was isolated with characteristic mass peak of 128. Eighteen fragmentation ions were identified in the mass spectrum of the sample. In line 10, 7.24% of cis-verbenol (4,6,6-trimethyl-bicyclo[3.1.1]hept-3-en-2-ol) was identified. The mass peak for this fraction was 63 and 8 fragment ions characterized the mass spectrum. Verbenol is one of the common ingredients in flavour and fragrance. An isomer of verbenol was also identified in line 8 (3.01%) of the spectrum. However, the concentration of this isomer was much lower than that found in line 10. About 6.22% of alpha-thujene/origanene (5-isopropyl-2-methylbicyclo[3.1.0]hex-2-ene) was separated

in line 1 of the GC-MS spectrum of KS. Mass peak for the separated compound was 41 and 7 fragment ions were identified in the mass spectrum. Thujene (or α-thujene) is a natural organic compound classified as a monoterpene. It is found in the essential oils of a variety of plants and contributes pungency to the flavour of some herbs such as Summer savory. The term *thujene* usually refers to α-thujene. A less common chemically related double-bond isomer is known as β-thujene (or 2-thujene). Another double-bond isomer is known as sabinene which is one of the chemical compounds that contributes to the spiciness of black pepper and is a major constituent of carrot seed oil. Thujene has long been known for its fragrance and medicinal functions.

Area normalization of the GC-MS spectrum of line 16 indicated the presence of 6.23% of 2-methylene cholestan-3-ol. This compound is a derivative of cholesterol, an

TABLE 2: Summary of GC-MS results from peaks in AF gum spectrum.

Peak number	%C	Compound	MF	MW	RT	MP	Fragmentation peaks
1	0.27	alpha-Bergamotene	$C_{15}H_{24}$	204	18.9	42	69 (40%), 77 (35%), 91 (100%), 107 (40%), 119 (90%), 133 (10%), 147 (5%), 161 (15%), 189 (5%), and 204 (5%)
2	0.65	Longicyclene	$C_{15}H_{24}$	204	19.5	57	27 (20%), 41 (60%), 55 (30%), 69 (25%), 79 (30%), 94 (100%), 105 (60%), 119 (50%), 133 (40%), 147 (35%), 161 (40%), 180 (30%), and 204 (35%)
3	9.15	1,4-Methanoazulene	$C_{15}H_{24}$	204	20.5	89	27 (35%), 41 (100%), 55 (55%), 67 (40%), 79 (70%), 91 (90%), 107 (70%), 119 (55%), 133 (10%), 149 (10%), 161 (5%), and 177 (5%)
4	1.25	Caryophyllene oxide	$C_{15}H_{24}O$	220	26.5	68	27 (25%), 41 (90%), 69 (45%), 81 (40%), 85 (100%), 109 (45%), 119 (45%), 137 (20%), 151 (10%), 161 (15%), 189 (35%), and 204 (40%)
5	0.66	1,4-Methanoazulene-9-ol	$C_{15}H_{26}O$	222	26.9	64	27 (10%), 41 (60%), 55 (40%), 69 (45%), 81 (30%), 85 (100%), 109 (45%), 119 (45%), 137 (20%), 151 (10%), 161 (15%), 189 (35%), and 204 (40%)
6	0.64	n-Hexadecanoic acid	$C_{16}H_{32}O_2$	256	31.7	72	27 (20%), 41 (80%), 43 (100%), 60 (90%), 73 (100%), 85 (25%), 98 (20%), 115 (15%), 129 (40%), 143 (5%), 157 (10%), 171 (10%), 185 (10%), 213 (20%), 227 (5%), and 256 (50%)
7	0.45	1-(p-Cumenyladamantane)	$C_{19}H_{26}$	254	32.5	128	39 (40%), 41 (85%), 67 (15%), 79 (35%), 91 (40%), 105 (20%), 115 (20%), 135 (40%), 145 (10%), 155 (50%), 169 (5%), 197 (20%), 211 (5%), 239 (60%), and 254 (50%)
8	1.38	18-Oxokauran-17-ylacetate	$C_{22}H_{34}O_3$	346	33.3	150	31 (5%), 41 (50%), 43 (100%), 67 (40%), 81 (50%), 91 (35%), 109 (30%), 123 (40%), 135 (10%), 149 (5%), 161 (5%), 187 (5%), 257 (5%), and 286 (10%)
9	1.27	Naphthalene	$C_{20}H_{32}$	272	33.9	132	27 (15%), 41 (80%), 55 (60%), 69 (55%), 81 (65%), 95 (55%), 105 (50%), 119 (40%), 137 (40%), 149 (20%), 161 (40%), 175 (20%), 187 (25%), 257 (100%), and 272 (40%)
10	11.97	Gamma-elemene	$C_{15}H_{24}$	204	34.5	196	28 (15%), 41 (100%), 53 (50%), 67 (60%), 79 (40%), 93 (80%), 107 (50%), 121 (90%), 133 (20%), 147 (10%), 161 (20%), 189 (10%), and 204 (5%)
11	14.77	1-Phenenathrenecarboxylic acid,1,2,3,4,4a,5,6,9,10,10a-decahydro-1,4a	$C_{20}H_{30}O_2$	302	34.9	203	41 (20%), 81 (15%), 91 (30%), 105 (35%), 117 (20%), 13 (35%), 149 (35%), 157 (20%), 171 (10%), 185 (25%), 197 (10%), 213 (35%), 241 (50%), 256 (35%), 287 (80%), and 302 (100%)
12	18.31	1-Phenenathrenecarboxylic acid,1,2,3,4,4a,5,6,9,10,10a-octahydro-1,4a-dimethyl-7-(1-methylethyl)-	$C_{20}H_{28}O_2$	300	35.1	198	41 (10%), 69 (5%), 81 (5%), 91 (10%), 105 (5%), 117 (10%), 129 (15%), 141 (20%), 155 (15%), 169 (10%), 183 (10%), 197 (30%), 211 (5%,) 225 (5%), 239 (90%), and 285 (100%)
13	39.24	Abietic acid, 1-phenenathrenecarboxylic acid	$C_{20}H_{30}O_2$	302	35.6	195	18 (10%), 41 (20%), 67 (10%), 81 (30%), 9 (35%), 105 (40%), 121 (20%), 136 (50%), 143 (20%), 157 (20%), 171 (10%), 185 (20%), 213 (30%), 241 (40%), and 259 (50%)

essential organic compound with numerous biochemical and pharmaceutical applications. In lines 13 and 7, 2.91 and 2.37% of diisopropenyl-1-methyl-1-vinylcyclohexane and limonene/cajepetene were identified through area normalization of the GC-MS spectrum of KS gum. Limonene is a colourless liquid hydrocarbon classified as a cyclic terpene. The more common isomer possesses a strong smell of oranges. Limonene is used in chemical synthesis as a precursor to carvone and as a renewably based solvent in cleaning products.

Other minor components of KS gum include 5 alpha-pregnane-12,20-dione ((5R,8R,9S,10S,13S,14S,17S)-17-acetyl-10,13-dimethyltetradecahydro-1H

cyclopenta[a]phenanthren-12(2H)-one) whose concentration is 1.93% (line 18) isomer of sabinene (line 4; 1.90%), myrcene (line 6; 1.39%), camphene (line 3; 1.39%), alpha-campholenic aldehyde (line 9; 1.14%), 7-hexadecenal (line 17: 0.81%), and carvacrol (line 12: 0.67%).

The characteristics of compounds identified in GC-MS spectrum of AF are presented in Table 2. The results obtained indicate that AF gums have 13 likely compounds and from area normalization of lines in the spectrum, it is found that the most abundant component is alpha-bergamotene. Bergamotene is a sesquiterpenoid that is a component of a number of volatile oils; it functions as an insect repellent in plants. They are usually found as racemic pairs of the cis- and trans-isomer. However, the concentration of alpha-bergamotene in AF gum is very low (0.27%), the least of all its components.

In line 2, longicyclene (0.65%) was identified in the GC-MS spectrum. This compound has been found to be the first tetracycline sesquiterpene and was isolated from *Pinus longifolia* [28], suggesting that this compound has some pharmaceutical values. From the mass spectrum of the compound, 11 fragmentation peaks were identified under a mass peak value of 57.

1,4-Methanoazulene was found to be the most likely compound in lines 3 (9.15%) and 5 (0.66%) of the GC-MS spectrum of AF gum. 1,4-Methanoazulene is also called longifolene and is commercially categorized as polymer/resin and plastic indicating that the gum can be a good source of raw materials for the polymer industries. Present usage of longifolene includes but is not limited to perfumery chemicals, cosmetic products, soaps detergents, deodorants, and fabric products. Twelve (12) fragmentation peaks were identified in the mass spectrum of this compound 9 for both lines 3 and 5 and the mass peak values were 89 and 64, respectively.

In line 4, caryophyllene oxide (1.25%) was identified as the most likely compound (S1 = 95) in the GC-MS spectrum of AF gum. Caryophyllene oxide, an oxygenated terpenoid, well known as preservative in food, drugs, and cosmetics, has been tested in vitro as an antifungal agent against dermatophytes. Its antifungal activity was found to be comparable to that of ciclopiroxolamine and sulconazole, commonly used in onychomycosis treatment and chosen because of their very different chemical structures [29]. Mass peak value for this compound was 68 and 12 different fragmentation peaks characterized its mass spectrum.

Analysis of spectra obtained from AF gum indicated that, in line 6, 0.64% of hexadecanoic acid is the most likely compound that is present. Mass peak values of 72 and 16 fragmentation peaks were identified from these lines. 1-(p-Cumenyl)adamantane at concentration of 0.45% was identified as the most likely component in line 7 of the GC-MS spectrum of AF gum. The mass peak value for this fraction was 128 and 17 fragmentation peaks were also identified. 1-(p-Cumenyl)adamantine is adamantane derivative and has practical application as drugs, polymeric materials, and thermally stable lubricants. Adamantane is a colorless, crystalline chemical compound with a camphor-like odour. It is a cycloalkane and also the simplest diamondoid. Adamantane molecules consist of three cyclohexane rings arranged in the "arm-chair" configuration. It is used in some dry etching masks and polymer formulations.

Kauran-18-al (1.38%) was identified as the likely compound in line 8. Eighteen (18) fragmentation peaks were identified and the mass peak value was 150. In line 9, decahydro-1,1,4a-trimethyl-6-methylene-5-(3-methyl-2,4-pentadienyl)-naphthalene(labda-8(20)) was likely isolated from GC-MS of AF gum.

The most dominant fractions isolated from GC-MS spectrum of AF gum were found to be concentrated in lines 10 to 13. Compounds in lines 10, 11, 12, and 13 were gamma-elemene (11.97%), palustric acid (14.77%), 1-phenanthrene carboxylic acid (18.31%), and abietic acid (39.24%).

TABLE 3: Peaks, wavelength, and assignment of functional group for FTIR adsorption by KS.

Wave number (cm^{-1})	Intensity	Area	Assignments
786.02	45.47	28.83	CH oop
1033.88	22.71	58.13	C-O stretch
1245.09	22.78	56.12	C-O stretch
1377.22	23.52	42.07	C-H scissoring and bending
1457.27	23.43	46.35	CH rock
1704.17	21.57	29.46	C=O stretch: carboxylic group
2870.17	17.08	81.07	C-H stretch
2929.00	11.79	65.06	CH- stretch
3390.97	23.18	3.67	OH stretch

TABLE 4: Peaks, wavelength, peak area, and assignments of functional group for FTIR adsorption by AF.

Wave number (cm^{-1})	Intensity	Area	Assignments
710.79	67.12	7.69	C-H bend
827.49	64.41	7.39	C-H bend
888.25	56.94	15.32	C-H bend
955.76	58.30	17.83	C-H bend
1027.13	60.25	11.39	C-O stretch
1185.30	52.15	10.84	C-N stretch
1276.92	39.14	35.31	NO$_2$ symmetric stretch
1384.94	49.95	18.37	NO$_2$ symmetric stretch
1461.13	45.91	19.79	C-H bend
1694.52	13.64	47.31	C=C stretch
2652.21	56.40	35.53	-OH stretch
2933.83	15.87	178.41	C-H aliphatic stretch
3426.66	54.33	10.71	-OH stretch

3.2. FTIR Study. Wave number and peaks of adsorption deduced from the respective FTIR spectra of KS and AF gums as well as the correlation area and concentrations are presented in Tables 3 and 4. The common features in the adsorption peaks of the gums studied are the appearance of bands and peaks that are typical of polysaccharides. The 2800–3000 cm^{-1} wave number range is associated with the stretching modes of C-H bonds of methyl groups (-CH$_3$). The broad bands around 3400 cm^{-1} are consequence of the presence of -OH groups. However, in AF, -OH groups are is shifted to 3426.68 cm^{-1}, and in KS gum, it is shifted to 3390.97 cm^{-1}. The shifts may be due to dissociating carboxylic acid. The 900–1200 cm^{-1} range represents various vibrations of C-O-C glycosidic and C-O-H bonds.

Of special interest to this study is the wave length range of 1500 and 1800 cm^{-1}, typically used to detect the presence of carboxylic groups. In KS gum, phenyl ring substitution band was found at 1704.17 cm^{-1}. Several adsorption bands were also found in the finger print regions (400–1500 cm^{-1}) for AF.

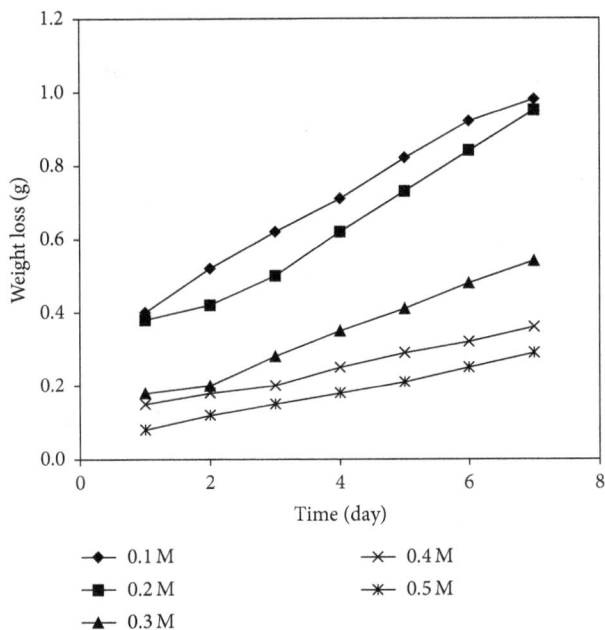

FIGURE 3: Variation of weight loss of mild steel with time for the corrosion of mild steel in various concentrations of HCl.

3.3. Corrosion Studies

3.3.1. Weight Loss Study. According to Eddy et al. [30, 31], the basic requirements for a given compound to be a good corrosion inhibitor are as follows:

(i) Possession of aromatic or long carbon chain that has heteroatom.

(ii) Presence of heteroatom(s) in the compound.

(iii) Presence of suitable functional groups (i.e., π-electron rich functional systems).

(iv) Presence of conjugated system.

From the chemical structures of the studied compounds reported above (Figures 1 and 2), it is evident that all the compounds present in AF and KS gum meet these conditions.

Figure 3 shows the variation of weight loss of mild steel with time for the corrosion of mild steel in various concentrations of HCl while Figures 4 and 5 present the variation of weight loss with time for the corrosion of mild steel at 303 K in 0.1 M HCl containing various concentrations of KS and AF gums, respectively. From the plots, it can be seen that weight loss of mild steel decreases with increase in the concentration of the gums indicating that KS and AF gums retarded the corrosion rate of mild steel in solutions of HCl. However, weight loss was also found to increase with increase in the period of contact. At 333 K (plots not shown), weight loss of mild steel was found to increase with increasing temperature indicating that the mechanism of inhibition of mild steel corrosion by AL gum is by physisorption [15]. Calculated values of corrosion rates of mild steel in various media obtained using (3) are recorded in Table 5. The results also indicate that the corrosion rate of mild steel in solutions of HCl increases with increasing time

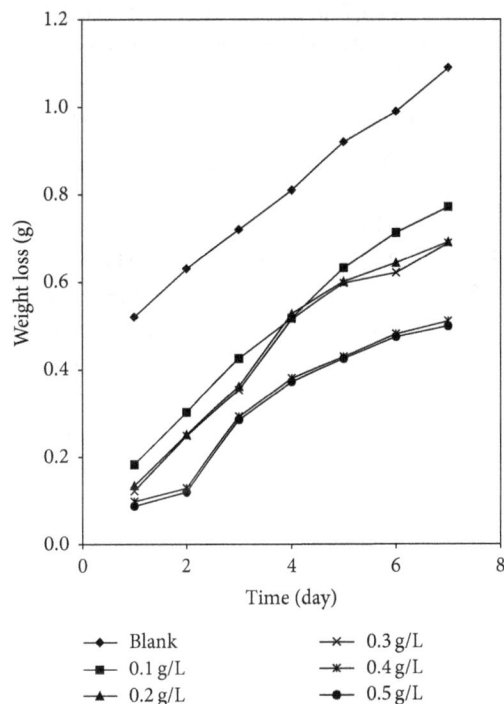

FIGURE 4: Variation of weight loss with time for the corrosion of mild steel in 0.1 M of HCl containing various concentrations of KS at 303 K.

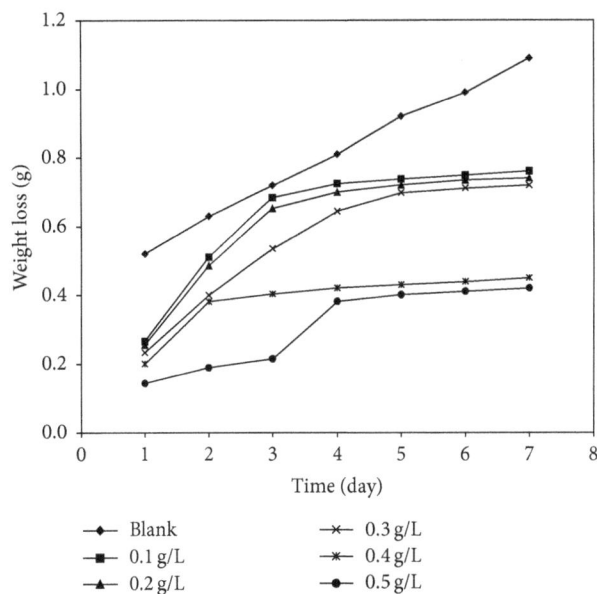

FIGURE 5: Variation of weight loss with time for the corrosion of mild steel in 0.1 M of HCl containing various concentrations of AF at 303 K.

but decreases with increase in the concentration of the gums. These trends confirm that KS and AF gums are inhibitors for the corrosion of mild steel in solutions of HCl. Values of inhibition efficiencies of KS and AF gums for the corrosion of mild steel in solutions of HCl are also presented in Table 5. The result indicates that the inhibition efficiencies of the gums

TABLE 5: Corrosion rates of mild steel and inhibition efficiencies of KS and AF gums at 303 K and 333 K for the corrosion of mild steel in 0.1 M HCl.

System	Inhibition efficiency (%)		Corrosion rate $(gh^{-1}cm^{-2})$	
	KS	AF	KS	AF
Blank at 303 K	—	—	0.000329	0.000329
0.1 g at 303 K	58.75	43.53	0.000201	0.000282
0.2 g at 303 K	62.93	48.05	0.000143	0.000221
0.3 g at 303 K	71.16	53.93	0.000125	0.000219
0.4 g at 303 K	77.37	55.91	0.000122	0.000176
0.5 g at 303 K	82.56	66.80	0.000118	0.000168
Blank at 333 K	—	—	0.001863	0.001863
0.1 g at 333 K	55.53	36.74	0.000917	0.001429
0.2 g at 333 K	58.10	39.85	0.000803	0.001389
0.3 g at 333 K	65.05	43.60	0.000676	0.001333
0.4 g at 333 K	72.62	53.10	0.000648	0.001301
0.5 g at 333 K	76.87	60.36	0.000581	0.001294

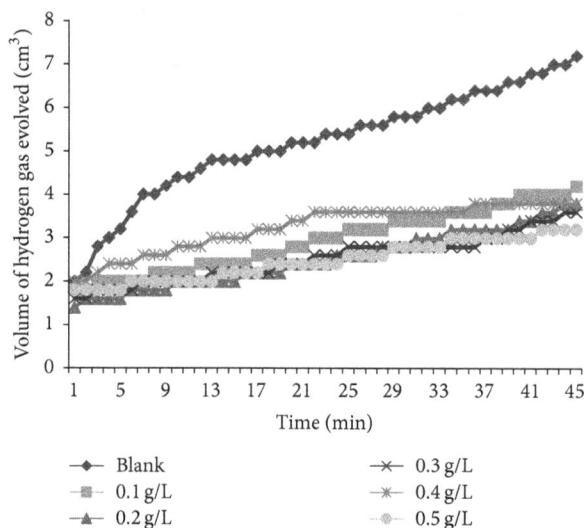

FIGURE 6: Variation of volume of hydrogen gas evolved with time for the corrosion mild steel in 1 M HCl containing various concentrations of AF at 303 K.

studied increase with increasing concentration. However, the inhibition efficiencies values obtained for KS gum were higher than that of the AF gum. This may be due to the fact that more compounds with heteroatoms were identified in the GCMS spectrum of KS gum compared to the AF gum. The presence of such compounds may have enhanced their adsorption on the metal surface and thereby blocking the surface and protecting the metal from corrosion. Inhibition efficiencies obtained at higher temperatures (Table 5) were lower than those for 303 K, indicating that the mechanism of adsorption of the inhibitor is physical adsorption [24, 31].

3.3.2. Gasometric Study. Figures 6 and 7 present plots for the variation of the volume of hydrogen gas evolved during the corrosion of mild steel in solution of HCl containing

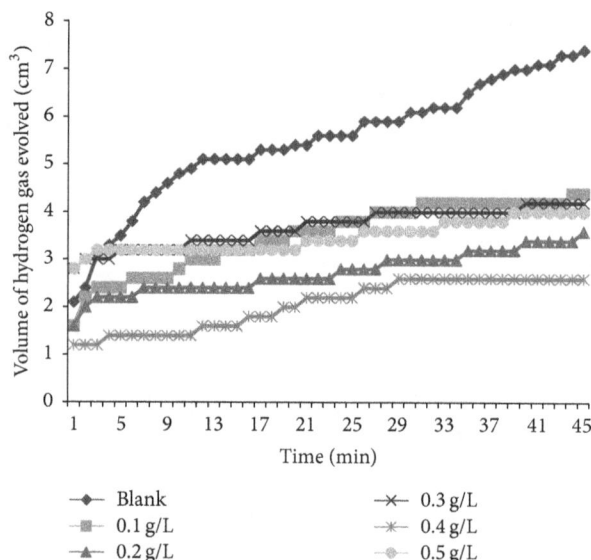

FIGURE 7: Variation of volume of hydrogen gas evolved with time for the corrosion mild steel in 1 M HCl containing various concentrations of KS gum at 303 K.

various concentrations of AF and KS, respectively. From the plots, it is generally evident that the volume of hydrogen gas evolved increases with increase in time but decreases with increase in the concentration of the respective inhibitors. This indicates that the corrosion of mild steel in 0.1 M HCl is inhibited by these inhibitors and that the inhibitors are adsorption inhibitors because their corrosion rates decrease with increase in concentration of the inhibitors [32]. At higher temperatures (333 K), plots obtained for the variation of volume of hydrogen gas with time (figure not shown) also reveal that the volume of hydrogen gas evolved increases with time but decreases with concentration of the added spices, which also indicate that the rate of corrosion of mild steel in 0.1 M HCl increases with time but decreases with increase in the concentration of the added inhibitors.

3.4. Effect of Temperature. In order to understudy the temperature dependence of corrosion rates in uninhibited and inhibited solutions, the corrosion rates calculated from the gravimetric measurements were fitted into the Arrhenius equation as follows [33]:

$$\log \frac{CR_2}{CR_1} = \frac{E_a}{2.303R} \left(\frac{1}{T_1} - \frac{1}{T_2} \right),$$ (5)

where CR_1 and CR_2 are the corrosion rates of mild steel at the temperatures T_1 (303 K) and T_2 (333 K), respectively, E_a is the activation energy, and R is the gas constant. Calculated activation energies are presented in Table 6. These values which ranged from 42.00 to 57.16 kJ/mol were greater than the value of 38.97 kJ/mol obtained for the blank indicating that KS and AF gums retarded the corrosion of mild steel in solutions of HCl. Also the activation energies are within the limits required for the mechanism of physical adsorption [30, 34]. Therefore, the adsorption of KS and AF gums on the surface of mild steel is consistent with the mechanism

TABLE 6: Activation energy and heat of adsorption for the inhibition of the corrosion of mild steel surface by KS and AF gums.

	C (g/l)	E_a (kJ/mol)	Q_{ads} (kJ/mol)
	Blank	38.97	—
	0.1	42.50	−72.30
	0.2	48.31	−63.79
KS	0.3	47.26	−58.86
	0.4	46.75	−64.15
	0.5	44.63	−60.36
	0.1	45.44	−56.65
	0.2	51.47	−52.69
AF	0.3	50.57	−47.85
	0.4	56.01	−75.16
	0.5	57.16	−57.76

of charge transfer from charged inhibitor to charged metal surface, which confirms physical adsorption.

The heat of adsorption (Q_{ads}) of KS and AF gums on mild steel surface was calculated using [35]

$$Q_{ads} = -2.303R \log\left(\frac{\theta_2}{1-\theta_2} - \frac{\theta_1}{1-\theta_1}\right) \times \left(\frac{T_1 \times T_2}{T_2 - T_1}\right). \quad (6)$$

Values of Q_{ads} calculated from (6) are also recorded in Table 6. These values are negative and ranged from −75.16 to −47.85 kJ/mol indicating that the adsorption of KS and AF gums on mild steel surface is exothermic.

3.5. *Thermodynamic/Adsorption Study.* Adsorption isotherms are useful in studying the adsorption characteristics and mechanism of corrosion inhibition. Generally adsorption isotherms are of the general form [36]

$$f(\theta, x) \exp(-2a\theta) = bC, \quad (7)$$

where $f(\theta, x)$ is the configurational factor which depends upon the physical model and the assumptions underlying the derivation of the isotherm, θ, the surface coverage, C, the inhibitor concentration in the electrolyte, x, the size factor ratio, a, and the molecular interaction parameter and b is the equilibrium constant of the adsorption process.

The values of surface coverage θ obtained from weight loss measurement corresponding to different concentrations of KS and AF at 303 and 333 K were fitted into different adsorption isotherms including those of Langmuir, Freundlich, Temkin, Flory Huggins, Frumkin, and El Awardy.

The tests revealed that the adsorption behaviour of the inhibitors is best described by the Langmuir adsorption model, which can be expressed as [35]

$$\log\left(\frac{C}{\theta}\right) = \log b_{ads} - \log C, \quad (8)$$

where b_{ads} is the adsorption equilibrium constant and θ is the degree of surface coverage of the inhibitor. From (6), plots of $\log(C/\theta)$ versus $\log C$ should be linear with intercept equal to $\log(b_{ads})$. Figures 8 and 9 present the Langmuir isotherms

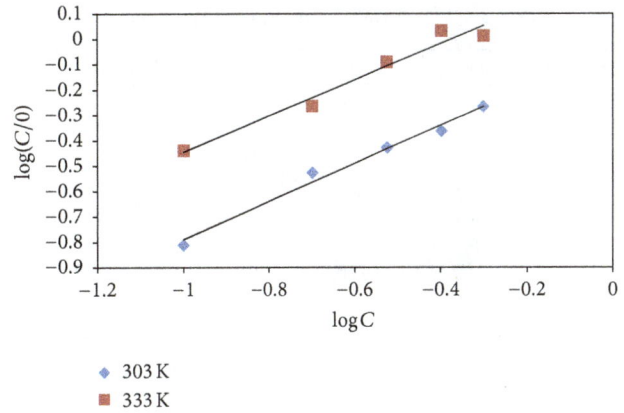

FIGURE 8: Langmuir isotherms for adsorption of KS gum on mild steel surface at 303 and 333 K.

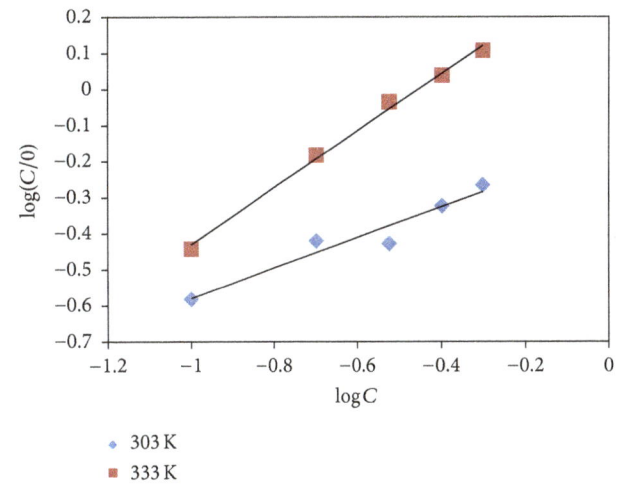

FIGURE 9: Langmuir isotherms for adsorption of AF gum on mild steel surface at 303 and 333 K.

for the adsorption of KS and AF gums on mild steel surface, respectively.

Adsorption parameters deduced from the plots are presented in Table 7. From the results obtained, it is evident that regression coefficient (R^2) values and slopes of the plots are very close to unity confirming that the inhibitors exhibited a single-layer adsorption characteristic [37].

The equilibrium constant of adsorption obtained from the Langmuir adsorption isotherm is related to the standard free energy of adsorption according to

$$b_{ads} = -\frac{1}{55.5} \exp\left(\frac{\Delta G^0_{ads}}{RT}\right), \quad (9)$$

where R is the gas constant in kJ/mol, T is the temperature in Kelvin, b_{ads} is the equilibrium constant of adsorption, and 55.5 is the molar concentration of HCl in water. Calculated values of ΔG^0_{ads} are recorded in Table 7. The negative values of ΔG^0_{ads} indicate the spontaneity of the adsorption process and the stability of the adsorbed layer on the mild steel surface [38]. Generally, the adsorption type is regarded as physisorption if the absolute value of ΔG^0_{ads} is in the range of −40 kJ/mol

TABLE 7: Langmuir parameters for the adsorption of KS and AF gums on mild steel surface.

Inhibitor	T (K)	Slope	$\log b$	ΔG^0_{ads} (kJ/mol)	R^2
KS	303	0.7841	0.0288	−33.95	0.9957
	333	0.7911	0.0642	−34.63	0.9928
AF	303	0.7582	0.1338	−35.96	0.9887
	333	0.7002	0.1627	−36.51	0.9715

TABLE 8: FTIR spectrum of the corrosion product of mild steel in the presence of KS gum as an inhibitor.

Peak (cm^{-1})	Intensity	Assignment (functional group)
722.37	25.108	C-H oop due to aromatic bond
819.77	26.354	C-H oop due to aromatic bond
896.93	26.597	N-H wag due to primary or secondary amines
1026.16	24.095	C-O stretch due to alcohol, carboxylic acids, esters, and ethers
1220.98	24.462	C-O stretch due to alcohol, carboxylic acids, esters, and ethers
1381.08	22.733	C-H rock due to alkane
1600.97	20.221	C=C aromatic stretch
1902.84	20.51	C-H stretch due to phenyl ring substitution
1970.35	20.625	C-H stretch due to phenyl ring substitution
2170.95	19.653	C≡C stretch
2437.14	20.002	OH stretch
2600.13	19.521	OH stretch
3038.95	20.449	=CH stretch due to alkene
3144.07	21.063	OH stretch due to carboxylic acid
3289.7	21.852	C≡C stretch
3474.88	22.834	OH stretch due to alcohol or phenol
3607.97	22.261	OH stretch (free hydroxyl) due to alcohol or phenol

TABLE 9: FTIR spectrum of the corrosion product of mild steel in the presence of AF gum as an inhibitor.

Peak (cm^{-1})	Intensity	Area (cm^2)	Assignment (functional group)
872.82	40.031	36.912	C-H oop due to aromatics
1030.99	38.267	37.875	C-O stretch due to alcohol, carboxylic acids, esters, or ethers
1084.99	38.785	59.234	C-O stretch due to alcohol, carboxylic acids, esters, or ethers
1360.82	39.535	2.330	NO$_2$ symmetric stretch
1466.91	38.277	2.010	C-H scissoring and bending
1635.69	34.353	2.236	-C=C- stretch due to alkene
3277.17	18.327	4.257	OH stretch due to phenols and alcohols
3444.98	17.393	7.301	OH stretch due to phenols and alcohols

or lower. The results obtained show that the free energies are negatively less than the threshold value of −40 kJ/mol specified for physical adsorption [39]. This behaviour is in good agreement with that obtained at 303 and 333 K using weight loss measurements. Therefore the adsorption of KS and AF gums on mild steel surface is spontaneous and supports the mechanism of physical adsorption.

3.6. FTIR Study. Corrosion inhibitors are mostly compounds that have suitable functional groups in addition to the presence of heteroatoms. Based on these principles, almost all the chemical structures of compounds identified in the studied gums are potential corrosion inhibitors. The functional groups associated with the adsorption of the inhibitor onto the metal surface can be studied by comparing the FTIR spectra of the inhibitor before and after adsorption. When this is done, missing functional groups suggest adsorption. Frequencies and peaks of FTIR spectra of the corrosion product of mild steel when KS and AF gums were used as inhibitors are recorded in Tables 8 and 9.

From the adsorption band of KS gum and the corrosion product of mild steel when KS gum was used as an inhibitor, it is evident that the C-H "oop" at 786.02 was shifted to 819.77 cm^{-1}; the C-O stretches at 1033.88 and 1245.09 were shifted to 1026.16 and 1220.98 cm^{-1}, respectively, while the C-H rock at 1457.27 was shifted to 1381.08 cm^{-1}. These shifts indicate that there is interaction between the inhibitor and the metal surface. However, the C-H rocking vibration at 1377.22,

C=O stretch at 1704.17, OH stretch at 3390.97, and C-H stretches at 2870.17 and 2920.00 cm^{-1} were absent in the spectrum of the corrosion product suggesting that KS gum was adsorbed on mild steel surface through these functional groups.

On the other hand, the C-H oop vibration due to aromatics, N-H wagging vibrations due to primary amine, C-H rock due to alkane at 1381.08, C=C aromatic stretch, C-H stretches due to phenyl ring substitutions at 1902.84 and 1970.35, the C≡C stretch at 2170.95, OH stretches at 2437.14, 2600.13, 3144.07, 3474.88, and 3607.97, and =CH stretch at 3038.95 cm^{-1} were found in the spectrum of the corrosion product indicating that these functional groups were used in forming new bonds.

Comparison of the adsorption band of the corrosion product with that of AF gum revealed that the C-O stretch at 1027.13 was shifted to 1030.99 cm^{-1}, the NO$_2$ symmetric stretch at 1384.94 was shifted to 1360.82, the CH bend at 1461.13 was shifted to 1466.91 cm^{-1}, the C=C stretch at 1694.52 was shifted to 1635.69 cm^{-1}, and the OH stretch at 3426.66 was shifted to 3444.98 cm^{-1}, which also indicate that there is an interaction between AF inhibitor and metal surface. However, the CH bends at 710.79, 827.49, 888.25, and 955.76 cm^{-1} as well as the C-N stretch at 1185.30 cm^{-1}, NO$_2$ symmetric stretch at 1276.92 cm^{-1}, OH stretch at 2652.21 cm^{-1}, and CH aliphatic stress at 2933.83 cm^{-1} were missing in the spectrum of the corrosion product indicating that these bonds were probably used in the adsorption of AF gum onto the metal surface. Also, the CH "oop" due to aromatics, the C-O stretch at 1084.99 cm^{-1}, and the OH stretch at 3277.17 cm^{-1} were found in the spectrum of the corrosion product of mild steel indicating that these functional groups were used in forming new bonds between the metal surface and the inhibitor.

The analysis of the FTIR spectra indicates that the formation of multimolecular layers of adsorption between the inhibitor and mild steel is likely. This also supports the mechanism of physical adsorption. It is also possible that the gums inhibited the corrosion of mild steel by forming stable complexes between the metal and the inhibitors.

4. Conclusion

(1) GC-MS study reveals that the studied gums have some potential for use in the polymer and pharmaceutical industries as well as intermediates for other chemicals.

(2) Functional groups identified in the gums were found to be those typical for other carbohydrates.

(3) *Albizia ferruginea* and *Khaya senegalensis* were found to be good corrosion inhibitors for mild steel in acidic medium. However maximum inhibition efficiency was exhibited by *Khaya senegalensis* with 82.56% inhibition efficiency at 0.5% g/L concentration.

(4) The corrosion inhibition efficiencies of the inhibitors are dependent on the period of contact with the corrodent, concentration of the inhibitors, and the temperature.

(5) The inhibitors displayed progressive increase in efficiencies as the concentration increases, but a decrease with increasing temperature, which supported the mechanism of physical adsorption. The adsorption of the inhibitors was found to be exothermic and spontaneous and fitted the Langmuir adsorption model.

Conflict of Interests

The author declares that there is no conflict of interests regarding the publication of this paper.

References

[1] E. A. Noor, "Potential of aqueous extract of *Hibiscus sabdariffa* leaves for inhibiting the corrosion of aluminum in alkaline solutions," *Journal of Applied Electrochemistry*, vol. 39, no. 9, pp. 1465–1475, 2009.

[2] E. H. El Ashry, A. El Nemr, S. A. Essawy, and S. Ragab, "Corrosion inhibitors part III: quantum chemical studies on the efficiencies of some aromatic hydrazides and Schiff bases as corrosion inhibitors of steel in acidic medium," *ARKIVOC*, vol. 11, pp. 205–220, 2006.

[3] E. S. H. El Ashry, A. El Nemr, S. A. Esawy, and S. Ragab, "Corrosion inhibitors part II: quantum chemical studies on the corrosion inhibitions of steel in acidic medium by some triazole, oxadiazole and thiadiazole derivatives," *Electrochimica Acta*, vol. 51, no. 19, pp. 3957–3968, 2006.

[4] S. P. Cardoso, E. Hollauer, L. E. P. Borges, and J. A. D. C. P. Gomes, "QSPR prediction analysis of corrosion inhibitors in hydrochloric acid on 22%-Cr stainless steel," *Journal of the Brazilian Chemical Society*, vol. 17, no. 7, pp. 1241–1249, 2006.

[5] S. Bilgiç and N. Çaliskan, "Investigation of some Schiff bases as corrosion inhibitors for austenitic chromium-nickel steel in H$_2$SO$_4$," *Journal of Applied Electrochemistry*, vol. 31, no. 1, pp. 79–83, 2001.

[6] J. Fang and J. Li, "Quantum chemistry study on the relationship between molecular structure and corrosion inhibition efficiency of amides," *Journal of Molecular Structure*, vol. 593, pp. 179–185, 2002.

[7] P. E. Francis and A. D. Mercer, *Chemical Inhibition for Corrosion Control*, edited by: B. G. Clubly, Royal society of Chemistry, London, UK, 1990.

[8] P. C. Okafor and E. E. Ebenso, "Inhibitive action of *Carica papaya* extracts on the corrosion of mild steel in acidic media and their adsorption characteristics," *Pigment and Resin Technology*, vol. 36, no. 3, pp. 134–140, 2007.

[9] S. Rajendran, M. R. Joany, B. V. Apparao, and N. Palaniswamy, "Synergistic effect of calcium gluconate and Zn^{2+} on the inhibition of corrosion of mild steel in neutral aqueous environment," *Transaction of the SEAST*, vol. 35, no. 3-4, pp. 113–117, 2000.

[10] D. Gopi, K. M. Govindaraju, V. Collins Arun Prakash, V. Manivannan, and L. Kavitha, "Inhibition of mild steel corrosion in groundwater by pyrrole and thienylcarbonyl benzotriazoles," *Journal of Applied Electrochemistry*, vol. 39, no. 2, pp. 269–276, 2009.

[11] A. Y. El-Etre and M. Abdallah, "Natural honey as corrosion inhibitor for metals and alloys. II. C-steel in high saline water," *Corrosion Science*, vol. 42, no. 4, pp. 731–738, 2002.

[12] M. Kliskic, J. Radoservic, S. Gudic, and V. Katalinic, "Aqueous extract of *Rosmarinus officinalis* L. as inhibitor of Al–Mg alloy

corrosion in chloride solution," *Journal of Applied Electrochemistry*, vol. 30, no. 7, pp. 823–830, 2000.

[13] E. E. Oguzie, "Studies on the inhibitive effect of *Occimum viridis* extract on the acid corrosion of mild steel," *Materials Chemistry and Physics*, vol. 99, no. 2-3, pp. 441–446, 2006.

[14] E. E. Oguzie, "Adsorption and corrosion inhibitive properties of *Azadirachta indica* in acid solutions," *Pigment and Resin Technology*, vol. 35, no. 6, pp. 334–340, 2006.

[15] E. E. Oguzie, A. I. Onuchukwu, P. C. Okafor, and E. E. Ebenso, "Corrosion inhibition and adsorption behaviour of *Ocimum basilicum* extract on aluminium," *Pigment and Resin Technology*, vol. 35, no. 2, pp. 63–70, 2006.

[16] E. E. Oguzie, G. N. Onuoha, and E. N. Ejike, "Effect of *Gongronema latifolium* extract on aluminium corrosion in acidic and alkaline media," *Pigment and Resin Technology*, vol. 36, no. 1, pp. 44–49, 2007.

[17] P. C. Okafor, V. I. Osabor, and E. E. Ebenso, "Eco-friendly corrosion inhibitors: Inhibitive action of ethanol extracts of Garcinia kola for the corrosion of mild steel in H_2SO_4 solutions," *Pigment and Resin Technology*, vol. 36, no. 5, pp. 299–305, 2007.

[18] S. A. Umoren and E. E. Ebenso, "Studies of the anti-corrosive effect of *Raphia hookeri* exudate gum-halide mixtures for aluminium corrosion in acidic medium," *Pigment and Resin Technology*, vol. 37, no. 3, pp. 173–182, 2008.

[19] F. Zucchi and I. H. Omar, "Plant extracts as corrosion inhibitors of mild steel in HCl solution," *Surface Technology*, vol. 24, no. 4, pp. 391–399, 1985.

[20] S. Martinez, "Inhibitory mechanism of mimosa tannin using molecular modeling and substitutional adsorption isotherms," *Materials Chemistry and Physics*, vol. 77, no. 1, pp. 97–102, 2003.

[21] N. O. Eddy, "Part 3. Theoretical study on some amino acids and their potential activity as corrosion inhibitors for mild steel in HCl," *Molecular Simulation*, vol. 36, no. 5, pp. 354–363, 2010.

[22] M. Behpour, S. M. Ghoreishi, M. Khayatkashani, and N. Soltani, "The effect of two oleo-gum resin exudate from *Ferula assafoetida* and *Dorema ammoniacum* on mild steel corrosion in acidic media," *Corrosion Science*, vol. 53, no. 8, pp. 2489–2501, 2011.

[23] M. Abdallah, "Guar gum as corrosion inhibitor for carbon steel in sulphuric acid solutions," *Portugaliae Electrochimica Acta*, vol. 22, pp. 161–175, 2004.

[24] S. A. Umoren, I. B. Obot, and E. E. Ebenso, "Corrosion inhibition of aluminium using exudate gum from *Pachylobus edulis* in the presence of halide ions in HCl," *E-Journal of Chemistry*, vol. 5, no. 2, pp. 355–364, 2008.

[25] N. O. Eddy, P. O. Ameh, C. E. Gimba, and E. E. Ebenso, "GCMS studies on *Anogessus leocarpus* (Al) gum and their corrosion inhibition potential for mild steel in 0.1 M HCl," *International Journal of Electrochemistry*, vol. 6, no. 11, pp. 5815–5829, 2011.

[26] A. Yurt, G. Bereket, and C. Ogretir, "Quantum chemical studies on inhibition effect of amino acids and hydroxy carboxylic acids on pitting corrosion of aluminium alloy 7075 in NaCl solution," *Journal of Molecular Structure: THEOCHEM*, vol. 725, no. 1-3, pp. 215–221, 2005.

[27] P. O. Ameh, N. O. Eddy, and C. E. Gimba, *Physiochemical and Rheological Studies on Some Natural Polymers and Their Potentials as Corrosion Inhibitors*, Lambert Academic, 2012.

[28] U. R. Nayak and S. Dev, "Studies in sesquiterpenes—XXXV: longicyclene, the first tetracyclic sesquiterpene," *Tetrahedron*, vol. 24, no. 11, pp. 4099–4104, 1968.

[29] L. J. Yan, L. Niu, H. C. Lin, W. T. Wu, and S. Z. Liu, "Quantum chemistry study on the effect of Cl- ion on anodic dissolution of iron in H_2S-containing sulfuric acid solutions," *Corrosion Science*, vol. 41, no. 12, pp. 2303–2315, 1999.

[30] N. O. Eddy, F. E. Awe, A. A. Siaka, L. Magaji, and E. E. Ebenso, "Chemical information from GC-MS studies of ethanol extract of *Andrographis paniculata* and their corrosion inhibition potentials on mild steel in HCl solution," *International Journal of Electrochemical Science*, vol. 6, no. 9, pp. 4316–4328, 2011.

[31] M. Şahin, G. Gece, F. KarcI, and S. Bilgiç, "Experimental and theoretical study of the effect of some heterocyclic compounds on the corrosion of low carbon steel in 3.5% NaCl medium," *Journal of Applied Electrochemistry*, vol. 38, no. 6, pp. 809–815, 2008.

[32] B. I. Ita, "A study of corrosion inhibition of mild steel in 0.1 M hydrochloric acid by O-vanilinhydrazone," *Bulletin of Electrochemistry*, vol. 20, no. 8, pp. 363–370, 2004.

[33] E. C. Ogoko, S. A. Odoemelam, B. I. Ita, and N. O. Eddy, "Adsorption and inhibitive properties of clarithromycin for the corrosion of Zn in 0. 01 to 0.05 M H_2SO_4," *Portugaliae Electrochimica Acta*, vol. 27, no. 6, pp. 713–724, 2009.

[34] G. Moretti, F. Guidi, and G. Grion, "Tryptamine as a green iron corrosion inhibitor in 0.5 M deaerated sulphuric acid," *Corrosion Science*, vol. 46, no. 2, pp. 387–403, 2004.

[35] S. A. Odoemelam, E. C. Ogoko, B. I. Ita, and N. O. Eddy, "Inhibition of the corrosion of zinc in H_2SO_4 By 9-deoxy-9a-aza9a-methyl-9a-homoerythromycin A (azithromycin)," *Portugaliae Electrochimica Acta*, vol. 27, no. 1, pp. 57–68, 2009.

[36] K. Y. Foo and B. H. Hameed, "Insights into the modeling of adsorption isotherm systems," *Chemical Engineering Journal*, vol. 156, no. 1, pp. 2–10, 2010.

[37] F. Bentiss, M. Lebrini, and M. Lagrenée, "Thermodynamic characterization of metal dissolution and inhibitor adsorption processes in mild steel/2,5-bis(n-thienyl)-1,3,4-thiadiazoles/hydrochloric acid system," *Corrosion Science*, vol. 47, no. 12, pp. 2915–2931, 2005.

[38] M. Bouklah, N. Benchat, B. Hammouti, A. Aouniti, and S. Kertit, "Thermodynamic characterisation of steel corrosion and inhibitor adsorption of pyridazine compounds in 0.5 M H_2SO_4," *Materials Letters*, vol. 60, no. 15, pp. 1901–1905, 2006.

[39] A. Popova, M. Christov, and A. Zwetanova, "Effect of the molecular structure on the inhibitor properties of azoles on mild steel corrosion in 1 M hydrochloric acid," *Corrosion Science*, vol. 49, no. 5, pp. 2131–2143, 2007.

Numerical Investigations on Characteristics of Stresses in U-Shaped Metal Expansion Bellows

S. H. Gawande,[1] N. D. Pagar,[2] V. B. Wagh,[3] and A. A. Keste[1]

[1]Department of Mechanical Engineering, M. E. Society's College of Engineering, Pune, Maharashtra 411001, India
[2]Department of Mechanical & Materials Technology and Department of Technology, S.P. Pune University, Pune 411007, India
[3]Department of Mechanical Engineering, G.S.M. College of Engineering, Pune, Maharashtra 411045, India

Correspondence should be addressed to S. H. Gawande; shgawande@yahoo.co.in

Academic Editor: Yuanshi Li

Metal expansion bellows are a mechanical device for absorbing energy or displacement in structures. It is widely used to deal with vibrations, thermal expansion, and the angular, radial, and axial displacements of components. The main objective of this paper is to perform numerical analysis to find various characteristics of stresses in U-shaped metal expansion bellows as per the requirement of vendor and ASME standards. In this paper, extensive analytical and numerical study is carried out to calculate the different characteristics of stresses due to internal pressure varying from 1 MPa to 2 MPa in U-shaped bellows. Finite element analysis by using Ansys14 is performed to find the characteristics of U-shaped metal expansion bellows. Finally, the results of analytical analysis and finite element method (FEM) show a very good agreement. The results of this research work could be used as a basis for designing a new type of the metal bellows.

1. Introduction

Metal bellows are structural component in which a wavy shape is formed on the surface of a circular tube to introduce elastic property. Expansion joints used as an integral part of heat exchangers or pressure vessels shall be designated to provide flexibility for thermal expansion and also to function as a pressure-containing element. Normally metal bellows are used as an expansion joint in shell and tube heat exchanger. It deals with vibrations, thermal expansion, and angular, radial, and axial displacements of components. Its present applications are in AC equipment, industrial plants, hose pipes, vacuum systems, and aerospace equipment.

Limited amount of research work has been carried out by some researchers working in the area of the expansion joint for shell and tube heat exchanger. Their work has been reported by performing industrial survey (namely, Alfa Laval India Ltd., Pune) and exhaustive literature review through earlier published research work, journal papers, and technical reports. Many design formulae of bellows can be found in ASME code [1]. And the most comprehensive and widely accepted text on bellows design is the Standards of Expansion Joint Manufactures Association, EJMA [2]. Number of pilot and test experiments have been performed for analysis of AM350 steel bellows by Shaikh et al. [3]. As bellows are exposed to marine atmosphere for more than 13 years which leads to pitting effect, hence the determination of dynamic characteristics of beam finite elements by manipulating certain parameters on commercial software was done by Broman et al. [4]. In comparison with semianalytical, methods have potential of considering axial, bending, and torsion degrees of freedom at the same time, and the rest are modeled by finite elements in which experimental results are also verified. The effect of the elliptic degree of Ω-shaped bellows toroid on its stresses is investigated by Li [5]. In addition, Becht IV [6] has investigated the fatigue behavior of expansion joint bellows. The results of Ω-shaped bellows with elliptic toroid calculated stresses correspond to experiments. The elliptic degree of Ω-shaped toroid affects the magnitude of internal pressure-induced stress and axial deflection-induced stress. It especially produces a considerable effect on the pressure-induced stress. To maintain the fatigue life of toroid bellows, during manufacturing process toroid elliptic degree must be reduced. EJMA stresses for unreinforced bellows are evaluated by Becht IV [6]. Using linear axisymmetric shell elements parametric analysis is conducted. Finite element analysis is carried out using commercial code. Meridional stresses due to internal pressure and displacement are accurate. Bellows-forming process is done after evaluating effective parameters by Faraji et al. [7]. FEM results are compared with analytical solutions. Faraji et al. [8] used a commercial FEM code, ABAQUS Explicit, to simulate manufacturing process of metal bellows. Forming of different shapes of tubular bellows using a hydroforming process is proposed by Kang et al. [9]. The conventional manufacturing of metallic tubular bellows consists of four-step process: deep drawing, ironing, tube bulging, and folding. In their study, single-step tube hydroforming combined with controlling of internal pressure and axial feeding was proposed. These reviewed papers show that there is need for rigorous analysis and forming parameters of bellows. It is stated that the Ω-shaped bellows have much better ability to endure high internal pressure than common U-shaped bellows. Metal bellows have wide applications in piping systems, automotive industries, aerospace, and microelectromechanical systems. Kang et al. [10] have developed a microbellows actuator using microstereo lithography technology. Numerous papers have dealt with various aspects of bellows except for forming process. Broman et al. [4] have determined dynamic characteristics of bellows by manipulating certain parameters of the beam finite elements. Jakubauskas and Weaver [11] have considered the transverse vibrations of fluid-filled double-bellows expansion joints. Jha et al. [12] have investigated the stress corrosion cracking of stainless steel bellows of satellite launch vehicle propellant tank assembly. Zhu et al. [13] have investigated the effect of environmental medium on fatigue life for U-shaped bellows expansion joints. However, few papers have shown the manufacturing process of the metal bellows. Wang et al. [14] have developed a new process for manufacturing of expansion joint bellows from Ti-6Al-4V alloys with high degree of spring back. Wang et al. [14] have used gas pressure instead of fluid pressure, because the process was done in high temperature ambient. Kang et al. [10] have investigated the forming process of various shapes of tubular bellows using a single-step hydroforming process. Lee [15] has carried out parametric study on some of the forming process parameters of the metal bellows by finite element only. He has mentioned that, in general, metal bellows are manufactured in four stages: deep drawing, ironing, tube bulging, and folding.

From the literature survey, it is seen that a number of researchers have worked on study and applications of different types of bellows under various working conditions, their comparison, and manufacturing processes, and few are working on fatigue life enhancement. But investigations on need for selection of proper material of bellows for given application, their proper design, stresses induction, fatigue life analysis, and prediction of failure and investigations on various characteristics of different bellows and vibration effect are essential.

2. Problem Formulation and Objective

As per literature and industrial survey, it is seen that bellows are one of the most important elements in the expansion joint

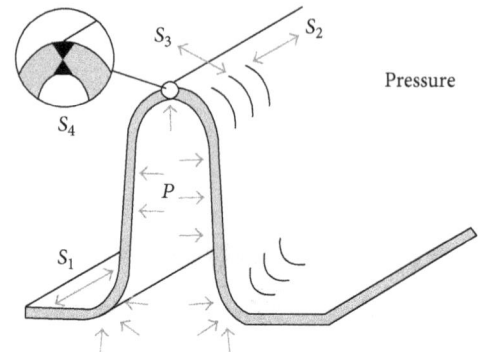

FIGURE 1: Stress directions in bellows.

and have the function to absorb regular as well as irregular expansion and contraction of the system. Bellows require high strength and good flexibility, which can be achieved by good design and proper manufacturing method. The design referred to from EJMA requires proper configuration selection which makes it difficult. The metal bellows are manufactured with different methods like forming, hydroforming, bulging, drawing, and deep drawing, which depend on applications. The materials used for bellows are normally stainless steel; in rare cases Inconel and aluminum are also used. Different shapes of bellows are U-shaped, semitoroidal, S-shaped, flat, stepped, single sweep, and nested ripple. As per discussion with experts working in the same field, it is observed that the concept of study in this paper needs detailed understanding of proper design and investigations on selection of materials, shapes, vibration effect, joining of bellows to shell, stresses, flow analysis, fatigue life analysis, and prediction of failure. Hence this work focuses on selection of materials of bellows for the given application, their proper design, and determination of characteristics of stresses of bellows, fatigue life analysis, and prediction of failure.

3. Determination of Characteristics of Stresses of Bellows by Analytical Analysis

Metal expansion bellows are a very distinctive component of a piping system. They must be designed strong enough to accommodate the system design pressure as well as flexible enough to accept the design deflections for a calculated number of occurrences, with a minimum resistive force. In order to understand the static and dynamic behavior of metal expansion bellows as shown in Figure 5, it is necessary to study the selection of materials of bellows for the given application, basic fundamental, their proper design, and working. The different mechanical properties and design parameters for bellows under consideration are shown in Table 1.

The design and analytical analysis of metal expansion bellows is performed as per ASME standards. Figure 1 shows the direction of different stresses induced in metal expansion bellows. According to ASME standards, the circumferential

TABLE 1: Different design parameters.

Design parameters	Notations	Specifications
Expansion joint material		SA-240 321
Material UNS number		S32100
Bellows design allowable stress	S	129.65 N/mm^2
Bellows ambient allowable stress	S_a	137.89 N/mm^2
Bellows yield stress	S_y	157.39 N/mm^2
Bellows elastic modulus at design temp.	E_b	183090 N/mm^2
Bellows elastic modulus at ambient temp.	E_o	195121 N/mm^2
Poisson's ratio	v_b	0.300
Bellows material condition		Formed
Design cycle life, required number of cycles	N_{req}	7000
Design internal pressure	P	1.099 N/mm^2
Design temperature for internal pressure		190°C
Bellow type		U-shaped
Bellows inside diameter	D_B	131.000 mm
Convolution depth	w	8.000 mm
Convolution pitch	Q	8.000 mm
Expansion joint opening per convolution	ΔQ	0.2985 mm
Total number of convolutions	N	10
Nominal thickness of one ply	t	0.300 mm
Total number of plies	n	3
End tangent length	L_T	13.000 mm
Fatigue strength reduction factor	K_g	1.500

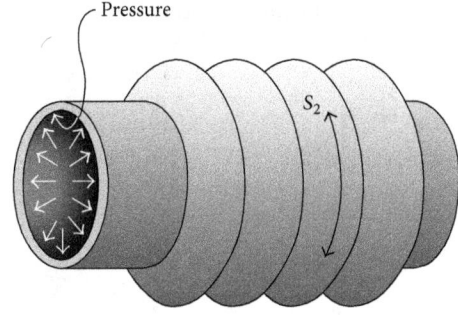

FIGURE 2: Deflection stresses acting on bellows.

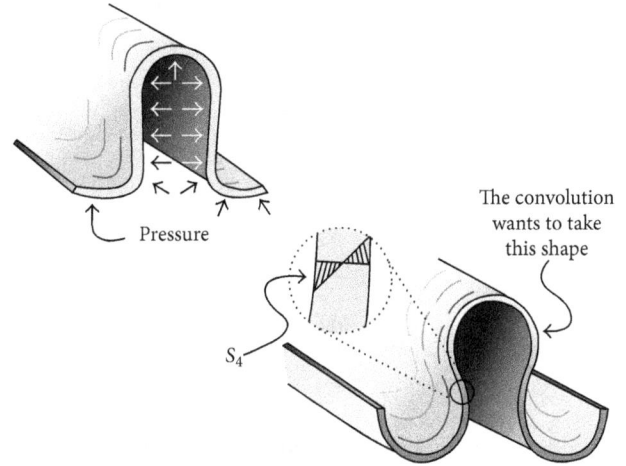

FIGURE 3: Meridional bending stress due to internal pressure.

membrane stress (S_1) in bellows tangent due to internal pressure is given as per

$$S_1 = \frac{1}{2} \left\{ \frac{L_t \times E_b \times K \times P \left(D_b + n \times t\right)^2}{\left[n \times t \times \left(D_b + n \times t\right) \times L_t \times E_b + t_c \times D_c \times L_c \times E_c \times K\right]} \right\}. \quad (1)$$

The end convolution circumferential membrane stress (S_2) due to internal pressure based on the equilibrium considerations is as shown in Figure 2. Equation (2) represents the end convolution circumferential membrane stress:

$$S_{2,E} = \frac{1}{2} \left\{ \frac{\left[q \times D_m + L_t \times \left(D_b + n \times t\right)\right] \times P}{\left(A + n \times t_p \times L_t + t_c \times L_c\right)} \right\}, \quad (2)$$

where D_m is mean diameter of bellows convolution and it is given as

$$D_m = D_b + w + n \times t. \quad (3)$$

The intermediate convolution circumferential membrane stress ($S_{2,I}$) due to internal pressure is calculated by using the following equation:

$$S_{2,I} = \frac{1}{2} \left\{ \frac{P \times q \times D_m}{A} \right\}. \quad (4)$$

The bellows meridional membrane stress (S_3) due to internal pressure is calculated based on the component of pressure in axial direction acting on the convolution divided by the metal area of root and crown by using the following equation:

$$S_3 = \frac{1}{2} \left\{ \frac{W \times P}{n \times t_p} \right\}. \quad (5)$$

The bellows meridional bending stress (S_4) due to internal pressure as represented in Figure 3 is given by (6). Figure 4 shows the variation of meridional bending stresses induced in bellows:

$$S_4 = \left\{ \frac{1}{2 \times n} \right\} \times \left\{ \frac{W}{t_p} \right\}^2 \times P \times C_p. \quad (6)$$

The bellows meridional membrane stress (S_5) and meridional bending stress (S_6) due to deflection are given by (7). Figure 4

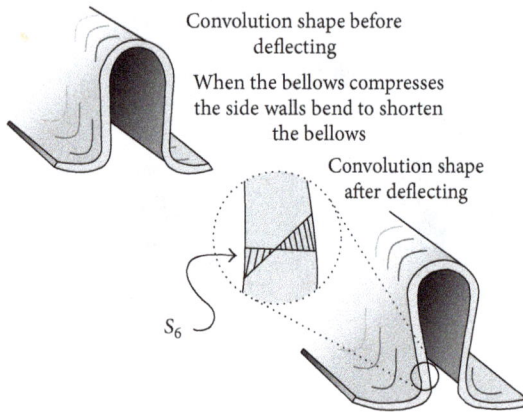

FIGURE 4: Meridional bending stress due to deflection.

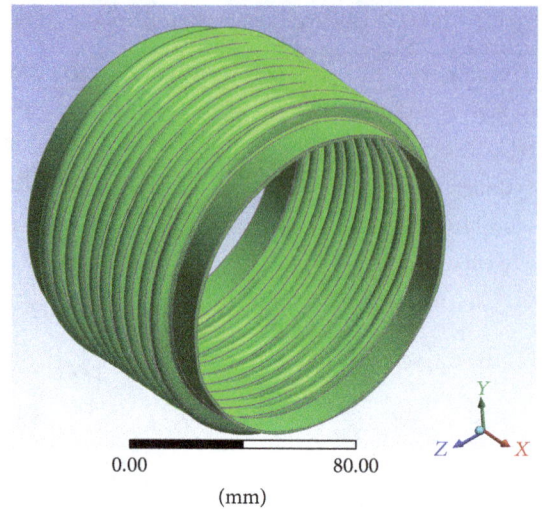

FIGURE 5: Geometry of metal expansion bellows.

shows the representation of meridional bending stress due to deflection. Consider

$$S_5 = E_b \times \left(t_p\right)^2 \times \frac{\Delta Q}{\left(2 \times w^3 \times C_f\right)},$$

$$S_6 = 5 \times E_b \times t_p \times \frac{\Delta Q}{\left(3 \times w^2 \times C_d\right)},$$

$$(7)$$

where C_p, C_f, and C_d are the factors for calculating S_4, S_5, S_6 respectively. E_b is modulus of elasticity for bellows. Figure 5 shows the metal expansion bellows under consideration in this paper.

4. Numerical Simulation

In order to perform numerical simulation, it is necessary to develop solid model of metal expansion bellows. Hence metallic expansion bellows is first modeled in Creo2.0 as shown in Figure 6, which is latest CAD software and makes modeling easy and user friendly. The model is then transferred in IGES format and geometry is imported for analysis to Ansys14.0 software. Then metal expansion bellows is analyzed in Ansys14.0 software.

4.1. Finite Element Procedure and Mesh Generation. Numerical simulation includes three stages of analysis as shown in Figure 7. First is preprocessing which involves modeling, geometric clean-up, element property definition, and meshing. Second step is solution of problem, which involves applying

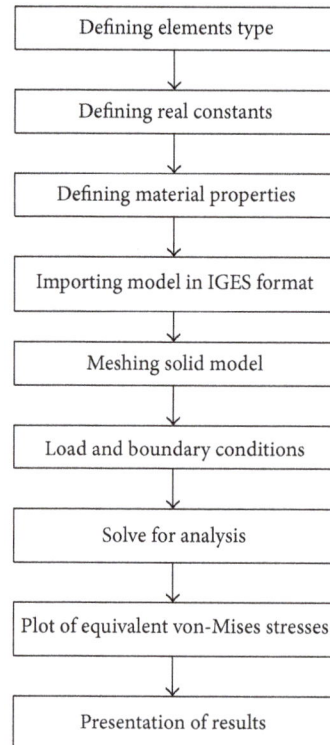

FIGURE 6: Solid model of metal expansion bellows.

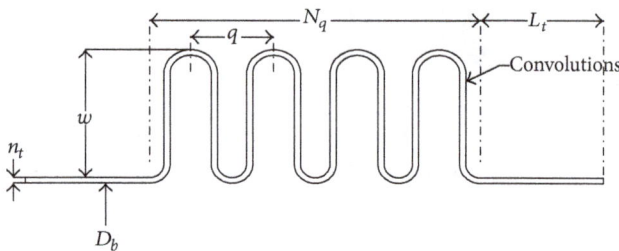

FIGURE 7: Stages of analysis.

boundary conditions on the model and then solution. Third step is postprocessing, which involves analyzing the results plotted with different parameters like stresses and deformation. The objective in creating a solid model is to mesh that model with nodes and elements. After completing the solid model, set element attributes and establishing meshing controls, which turn the Ansys program to generate the finite element mesh. For defining the elements attributes, the user has to select the correct element type.

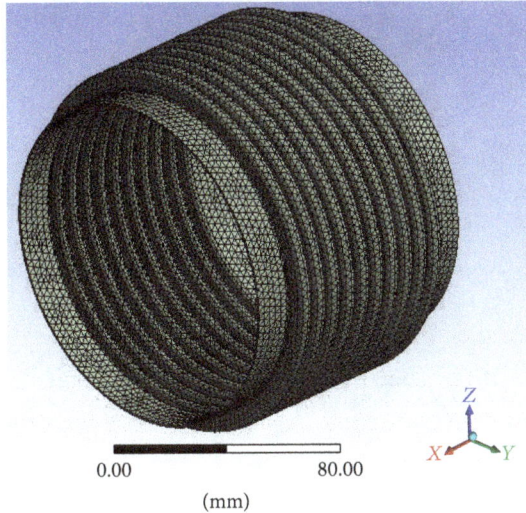

FIGURE 8: Meshed model of metal expansion bellows.

TABLE 2: Analytic and FEA stresses due to internal pressure.

Stress	Source	Internal pressure (MPa)			
		1	1.12	1.5	2
S_1	ASME	73.26	82.05	109.89	146.52
	FEA	43.83	53.02	70.26	80.53
S_{2I}	ASME	31.28	35.03	46.92	62.56
	FEA	34.55	32.12	48.63	57.67
S_{2E}	ASME	48.52	54.35	81.52	163.05
	FEA	42.36	34.44	61.46	86.41
S_3	ASME	4.89	5.14	6.56	9.19
	FEA	30.39	33.26	45.41	61.85
S_4	ASME	82.18	92.04	123.28	164.37
	FEA	33.12	31.73	47.48	54.68

Figure 8 shows the meshed model of metal expansion bellows. In this work, structural solid element 20 node plane 183 element was used as element type. Elastic analyses were carried out on full convolutions of the bellows with axisymmetric model. The computational domain is divided into 10 elements in thickness and 200 elements in length. Therefore, the model with elements 10 × 200 is used in all analyses. In the present analysis, a U-shaped bellow named VLC Shell Dia. 129 mm is picked. The bellows inside diameter is 131 mm with outside diameter of 147 mm, thickness of 0.9 mm, pitch of 8.00 mm, and height of the convolution is 8.00 mm. The bellows is made of stainless steel SA-240 321 with the modulus of elasticity of 195 GPa and Poisson's ratio of 0.3. In this work, the internal pressure in applied by applying the constraints.

5. Results and Discussions

5.1. Numerical Validations. Comparison test is performed for verification of the results obtained by numerical method. For the given solid element, FEM stresses are evaluated. The circumferential membrane stress at bellows tangent, intermediate and end convolution membrane stress, meridional membrane stress, and meridional bending stress due to internal pressure of U-shaped bellows are calculated. The applied internal pressures are 1 MPa, 1.12 MPa, 1.5 MPa, and 2 MPa, respectively. In Table 2, the results obtained from analytical approach and numerical simulations are presented. The meridional membrane stress and meridional bending stress for various internal pressures are presented in Table 2. After comparing the results, it is observed that the obtained stresses by two approaches for U-shaped bellows are in good agreement and show very closed match.

5.2. Comparison of Induced Design and Simulated Stresses of Metal Expansion Bellows. In the present work, numerical values of stresses are used for evaluation of characteristics of metallic bellows. Initially, the circumferential membrane stress is simulated for various internal pressures. As per the requirement, the internal pressures selected were 1 MPa, 1.12 MPa, 1.5 MPa, and 2 MPa, respectively. Figure 9 shows comparison of circumferential membrane stress induced in bellows tangent due to internal pressure of 2 MPa. Similar plots are obtained for various pressures as 1 MPa, 1.12 MPa, and 1.5 MPa. Comparison of different stresses for various pressures is explained in Figures 10–14. From Figure 10, it is seen that the circumferential membrane stress obtained by both approaches shows considerable variation in induced stress, but, as per design criterion, this is within acceptable agreement. This is an important membrane stress that runs circumferentially around the bellows. For safety, the value must be lower than the allowable stress for the bellows material multiplied by the bellows longitudinal weld joint efficiency. Figure 11 shows variation of intermediate convolution circumferential membrane stress due to internal pressure. From Figure 11, it is observed that the intermediate convolution circumferential membrane stress obtained by both approaches shows very closed match. This means that the stresses obtained by both approaches are in good agreement.

From Figure 12, it is seen that the end convolution circumferential membrane stress obtained by both approaches shows considerable variation as pressure varies from 1.12 MPa to 2 MPa, but as per design criterion this is within acceptable limit. The end convolution circumferential membrane stress obtained by both approaches shows much closed match for pressure of 1 MPa. This means that the stress obtained by both approaches is in good agreement.

Figure 13 shows the variation of meridional membrane stress due to internal pressure. It is seen that the meridional membrane stress obtained by both approaches shows considerable variation in induced stresses, but as per design criterion this is within acceptable limit. From Figure 13, it is observed that the calculated meridional membrane stress as per ASME standard almost remains constant as pressure varies from 1 MPa to 2 MPa, but the simulated meridional membrane stress increases significantly as pressure increases from 1 MPa to 2 MPa.

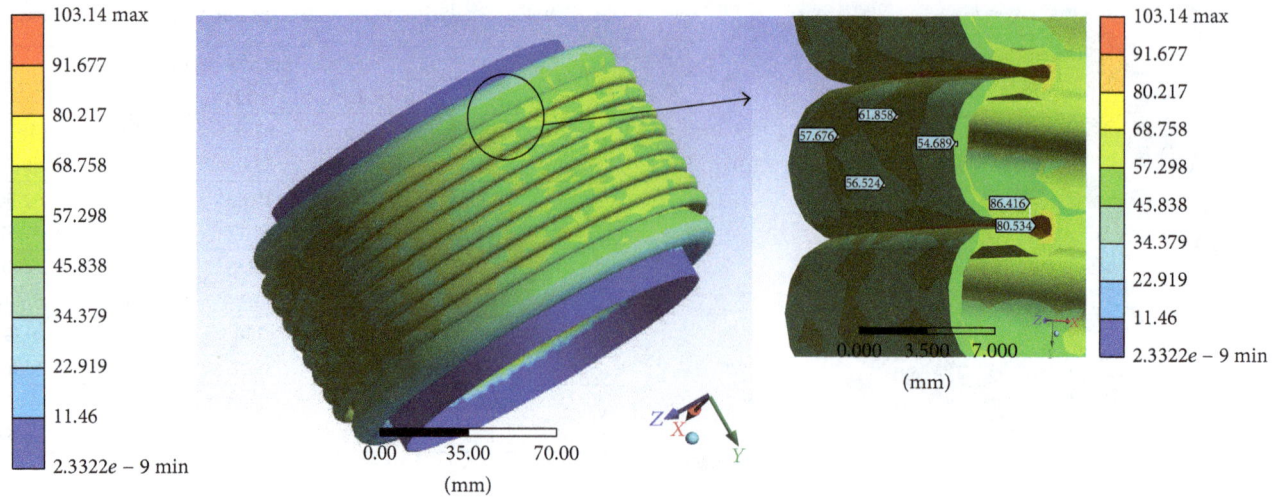

FIGURE 9: Simulated model of metal expansion bellows for internal pressure of 2 MPa.

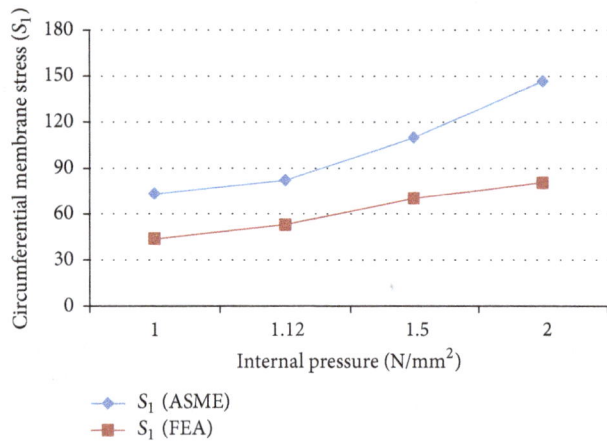

FIGURE 10: Circumferential membrane stresses in bellows tangent.

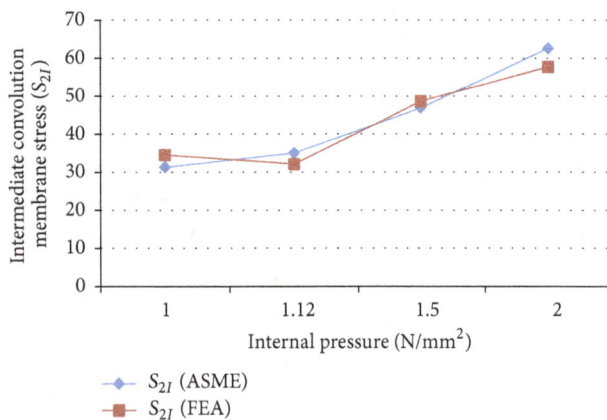

FIGURE 12: End convolution circumferential membrane stress.

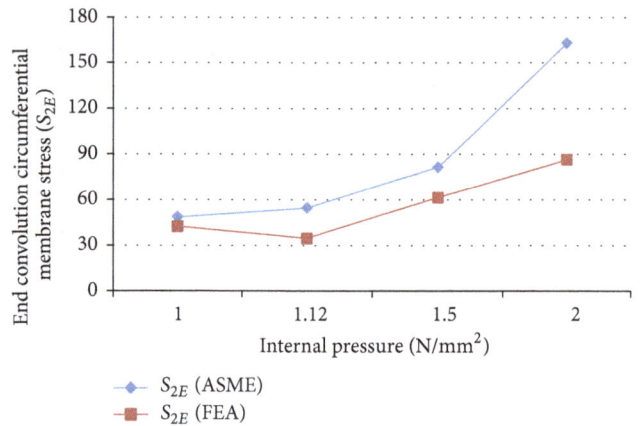

FIGURE 11: Intermediate convolution circumferential membrane stress.

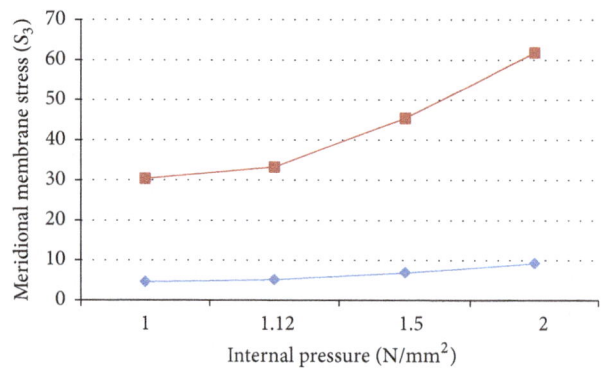

FIGURE 13: Meridional membrane stress.

Figure 14 shows the variation of meridional bending stress due to internal pressure. It is seen that the meridional bending stress obtained by both approaches shows considerable variation in induced stresses, but as per design criterion this is within acceptable limit. From Figure 14 again, it is found that the calculated meridional bending stress as per ASME standard almost remains constant as pressure varies from 1 MPa to 2 MPa, but the simulated meridional bending stress

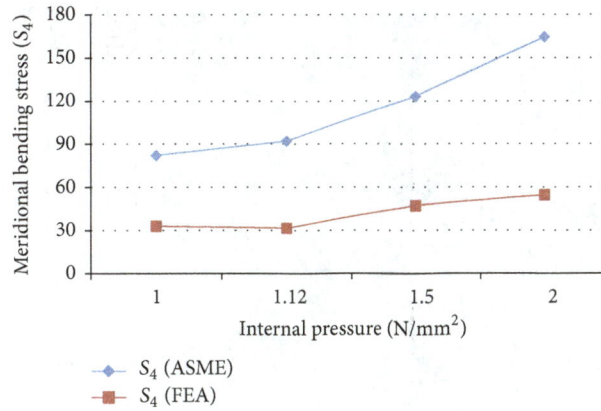

FIGURE 14: Meridional bending stress.

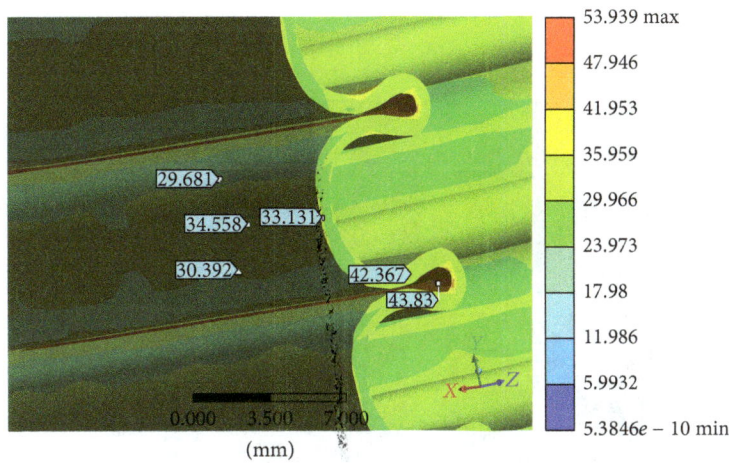

FIGURE 15: Stress distribution due to internal pressure of 1 MPa.

increases significantly as pressure increases from 1 MPa to 2 MPa.

5.3. Stress Distribution due to Internal Pressure. The internal pressure varying from 1 MPa to 2 MPa was applied on considered metal expansion bellows with boundary condition, as no (fixed) displacement of the six sides of the bellows. Figures 15–18 show stress distribution in the metal expansion bellows under consideration for internal pressure of 1 MPa, 1.12 MPa, 1.5 MPa, and 2 MPa.

6. Conclusions

In this paper, analytical and simulation study for characteristics of U-shaped metallic bellows is conducted. The results obtained as per ASME standards are compared with the FEA for stress distribution. The design stresses and distributions are compared for U-shaped bellows. The main conclusion is that the most destructive stress in bellows due to internal pressure is meridional bending stress and circumferential membrane stress. The circumferential membrane stress is an important membrane stress that runs circumferentially around the bellows. For bellows functional safety, this value must be lower than the allowable stress.

Notations

S: Bellows design allowable stress
S_a: Bellows ambient allowable stress
S_y: Bellows yield stress
E_b: Bellows elastic modulus at design temperature
E_o: Bellows elastic modulus at ambient temperature
ν_b: Poisson's Ratio
N_{req}:   Design cycle life, required number of cycles
P: Design internal pressure
D_B: Bellows inside diameter
w: Convolution depth
Q: Convolution pitch
ΔQ: Expansion joint opening per convolution
N: Total number of convolutions
t: Nominal thickness of one ply
n: Total number of plies
L_T: End tangent length
K_g: Fatigue strength reduction factor.

Conflict of Interests

The authors declare that there is no conflict of interests regarding the publication of this paper.

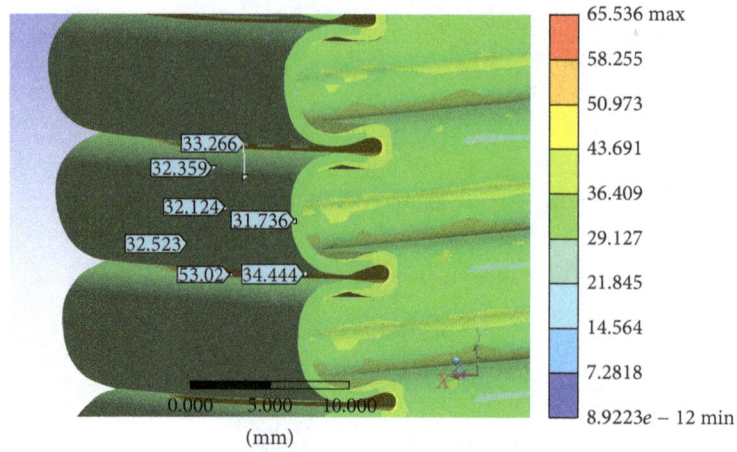

FIGURE 16: Stress distribution due to internal pressure of 1.12 MPa.

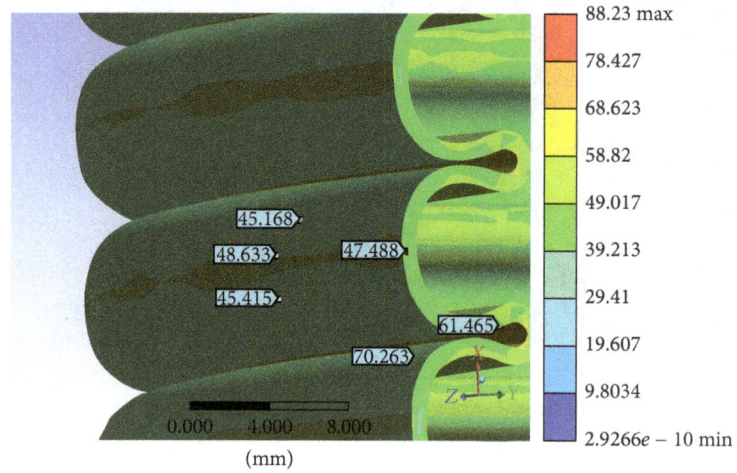

FIGURE 17: Stress distribution due to internal pressure of 1.5 MPa.

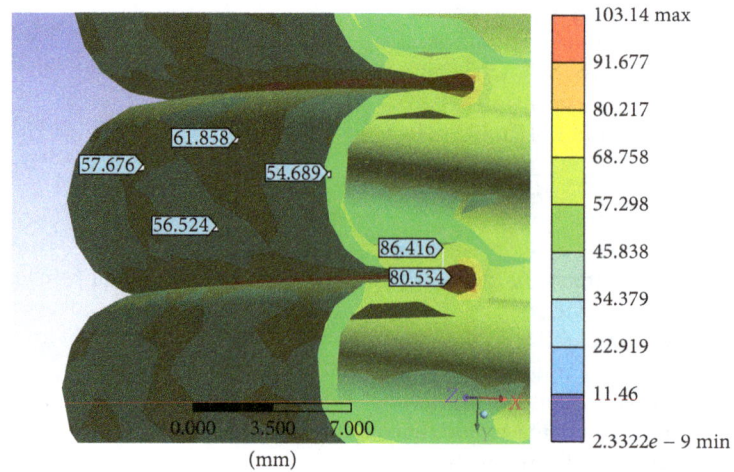

FIGURE 18: Stress distribution due to internal pressure of 2 MPa.

Acknowledgment

Authors would like to thank Mr. Umesh Ubarhande, *Senior Manager (R&D, PEM)* of Alfa Laval India Pvt. Ltd., Pune, for helping them in formulating the problem and providing the necessary input and guidance to achieve the objective.

References

[1] ASME, "ASME boiler and pressure vessel code-section VIII, division 1," in *Appendix 26—Pressure Vessel and Heat Exchanger Joints*, ASME, New York, NY, USA, 2000.

[2] EJMA, *Standards of Expansion Joint Manufacturers Association*, Expansion Joint Manufacturers Association, New York, NY, USA, 9th edition, 2008.

[3] H. Shaikh, G. George, and H. S. Khatak, "Failure analysis of an AM 350 steel bellows," *Engineering Failure Analysis*, vol. 8, no. 6, pp. 571–576, 2001.

[4] G. I. Broman, A. P. Jönsson, and M. P. Hermann, "Determining dynamic characteristics of bellows by manipulated beam finite elements of commercial software," *International Journal of Pressure Vessels and Piping*, vol. 77, no. 8, pp. 445–453, 2000.

[5] T. Li, "Effect of the elliptic degree of Ω-shaped bellows toroid on its stresses," *International Journal of Pressure Vessels and Piping*, vol. 75, no. 13, pp. 951–954, 1998.

[6] C. Becht IV, "Fatigue of bellows, a new design approach," *International Journal of Pressure Vessels and Piping*, vol. 77, no. 13, pp. 843–850, 2000.

[7] G. H. Faraji, M. M. Mashhadi, and V. Norouzifard, "Evaluation of effective parameters in metal bellows forming process," *Journal of Materials Processing Technology*, vol. 209, no. 7, pp. 3431–3437, 2009.

[8] G. H. Faraji, M. K. Besharati, M. Mosavi, and H. Kashanizadeh, "Experimental and finite element analysis of parameters in manufacturing of metal bellows," *The International Journal of Advanced Manufacturing Technology*, vol. 38, no. 7-8, pp. 641–648, 2008.

[9] B. H. Kang, M. Y. Lee, S. M. Shon, and Y. H. Moon, "Forming various shapes of tubular bellows using a single-step hydro-forming process," *Journal of Materials Processing Technology*, vol. 193, no. 1-3, pp. 1–6, 2007.

[10] H.-W. Kang, I. H. Lee, and D.-W. Cho, "Development of a micro-bellows actuator using micro-stereolithography technology," *Microelectronic Engineering*, vol. 83, no. 4–9, pp. 1201–1204, 2006.

[11] V. Jakubauskas and D. S. Weaver, "Transverse natural frequencies and flow induced vibrations of double bellows expansion joints," *Journal of Fluids and Structures*, vol. 13, no. 4, pp. 461–479, 1999.

[12] A. K. Jha, V. Diwakar, and K. Sreekumar, "Stress corrosion cracking of stainless steel bellows of satellite launch vehicle propellant tank assembly," *Engineering Failure Analysis*, vol. 10, no. 6, pp. 699–709, 2003.

[13] Y. Z. Zhu, H. F. Wang, and Z. F. Sang, "The effect of environmental medium on fatigue life for u-shaped bellows expansion joints," *International Journal of Fatigue*, vol. 28, no. 1, pp. 1–5, 2006.

[14] G. Wang, K. F. Zhang, D. Z. Wu, J. Z. Wang, and Y. D. Yu, "Superplastic forming of bellows expansion joints made of titanium alloys," *Journal of Materials Processing Technology*, vol. 178, no. 1-3, pp. 24–28, 2006.

[15] S. W. Lee, "Study on the forming parameters of the metal bellows," *Journal of Materials Processing Technology*, vol. 130-131, pp. 47–53, 2002.

16

Material Properties of Wire for the Fabrication of Knotted Fences

Dirk J. Pons, Gareth Bayley, Christopher Tyree, Matthew Hunt, and Reuben Laurenson

Department of Mechanical Engineering, University of Canterbury, Private Bag 4800, Christchurch 8020, New Zealand

Correspondence should be addressed to Dirk J. Pons; dirk.pons@canterbury.ac.nz

Academic Editor: Francisca Caballero

This paper describes the materials properties of galvanised fencing wire, as used in the fabrication of knotted wire fences. A range of physical properties are investigated: tensile strength, ductility in tension, Young's modulus, three-point bending, and bending span. A range of commercially available wire products were tested. The results show that most, but not all, high tensile wire samples met the minimum tensile and ductility requirements. Young's modulus results failed to provide any meaningful insights into wire quality. Flexural modulus results also failed to provide any insight into wire quality issues, with no statistically significant differences existing between acceptable and problematic wire batches. The implications are that premature fence failures are unlikely to be caused solely by reduced tensile properties. Existing test methods, including tensile strength and ductility, are somewhat incomplete, perhaps even unreliable, as measures of wire quality.

1. Introduction

Knotted wire fencing, despite being agriculturally ubiquitous, has a limited research literature. There are several aspects to the problem, including the manufacturing of the fencing, its erection, and its in-field use. The focus of the present paper is on the manufacturing, and the particular area under examination is the physical properties of wire.

2. Background

Knotted wire fencing, which is differentiated from single strands, diamond (chain link), welded mesh, and hexagonal mesh (chicken mesh), is a rectangular knotted mesh that is used for livestock retention, for example, sheep and deer. The fencing is an integral feature of the stock farming landscape worldwide and serves its purpose for years and even decades totally exposed to the environment. It is such a common product as to be overlooked. Yet the fabrication of this product is a sophisticated task that requires specialist machines [1–4]. Modern machines are numerically controlled to permit different configurations, for example, sizes of the rectangular openings. There are other challenges too: the steel wire used for fences tends to have high tensile strength, and this makes it particularly difficult to form the knotted features, since plastic deformation is required. Consequently high localized forces are required to bend the wire, and the wire needs to have the necessary material properties to accommodate this without fracture. Breakages of wire inside the fence knotting machine is highly problematic from a production perspective, since the high speed of operation causes wire to be jammed and tangled in the machine, and the high tensile strength of the material makes it difficult to extract. The machines also become damaged in these events. This results in the machine being out of production for an extended period, which adversely affects the production economics. Consequently the quality of the input wire is an important factor in the production process. The wire quality is also important in that it affects the fence erection processes. Similarly breakages of fencing in-field can have significant repercussions for the contractors who erect the fence or the agricultural operations of live-stock retention. This is an unsatisfactory customer experience. The fence manufacturer therefore wants to ensure that the raw wire stock is suitable for both fabrication and erection. Against this is the opposing pressure to reduce the cost of the input wire, which has negative effects on wire quality. The issue from the engineering perspective is therefore to obtain a wire feedstock that is fit for purpose: has sufficient material properties to withstand the fabrication and erection processes, without properties that are excessive (and therefore unnecessarily costly) to the process.

Standards do exist to control the quality of the input wire, and their foci are tensile strength and ductility; for example, see [5–9]. Corrosion resistance, hence also coating properties, is the other main strand of research [10–18] but is peripheral to the breakage problem considered here.

At present wire quality is only controlled by tensile strength and coating properties. However, these metrics are poor predictors of the performance of the wire in the manufacturing process in that wire that passes these tests can still fail in the machines [19]. The problem is that the production capability of the machines has increased such that wire failure is now a significant event, and the standards, which are mostly based on a previous generation of technology, are not providing the necessary quality control. Consequently there is a need to identify critical wire properties that can be used to differentiate between acceptable and poor wire quality, before it is input to the knotting machines.

3. Purpose and Approach

The two conventional materials properties used in the fencing industry are tensile strength and percent elongation (ductility) [6–9]. However the role of tensile strength and elongation, and their possible role in premature fence failures, is unclear. The objective of this particular paper was to apply a basket of physical-material tests to a set of wire samples of known history and infer the efficacy of the various tests in distinguishing wire quality. The background to the project is described in [19] along with a summary of some of the other tests.

The approach taken was first to obtain samples of manufactured fences and coils of input wire, of known acceptable and problematic product. These were commercially available products. The quality attributes of the wire were identified by the manufacturer (South Fence Machinery Ltd, New Zealand) and categorised as "acceptable" if it had not broken during fence fabrication or erection or "problematic" if it had failed in one or both situations. The quality categorisation thus supplied to the researchers at the outset was as follows: Acceptable: all Onesteel wire batches, Bekaert wire batches; Problematic: pacific wire breaking rolls 1, 2, and HiSPAN and Hurricane wire batches. To reduce external variability, all fence panels were produced by the same fence fabricator ("Producer") on machines from the manufacturer.

Samples were of complete fence panels and coils and covered a variety of input wire batches and brands. Assay pieces were cut from these fences or coils and subject to a variety of material tests. These were then statistically examined and compared to the fate of the fence/coil (where known), to determine the sensitivity of the test. Several tests were conducted, and the method for each is given preceding the results.

4. Results

4.1. Tensile Testing. Many materials properties can be calculated from undertaking tensile tests on wire specimens.

FIGURE 1: Clamps for elongation test.

Firstly, the ultimate tensile strength (UTS) of wire can be determined by measuring the maximum force required to fracture the specimen in tension. Secondly, the addition of gauge marks before a tensile test allows the percentage elongation (measure of ductility) to be evaluated. The wire stiffness, measured by Young's modulus (E), can also be inferred from the stress-strain curve. The tensile results may then be compared against the relevant wire standards, to assess compliance.

4.1.1. Experimental Method. Tensile tests were undertaken in accordance with the procedures set by ASTM A370-05 [20]. Wire specimens to be tested were cut to 380 mm lengths. A gauge length of 250 mm was used to enable ductility calculations to be conducted. Special-purpose wire clamps were used to grip the wire without introducing stress concentrations, as shown in Figure 1. These clamps ensured that the final fracture position would occur inside the gauge length. However, since these clamps worked on a wedge tightening principle, the strain values (and hence Young's modulus) calculations were significantly affected. In some cases, separate tests were conducted using standard clamps to find Young's modulus. At least three samples per wire batch were tested, to gain a representative average for the batch.

Tensile tests were also conducted on line wires cut from fabricated fence samples. For these samples, the knots and stay wires were cut off, and a sample containing two crimps and one knot was tested. Due to the presence of the crimps, ductility measurement was not possible.

The wire specimens were mounted in a MTS 810 Materials Testing System. The rate of cross-head displacement was set at 6 mm/minute until failure. After failure two parts of the specimen were matched back together and the new gauge length was measured to obtain ΔL, the change in gauge length after fracture. The elongation of the specimen is then $\Delta L/L$ where original specimen length $L = 250$ mm in this case. The ultimate tensile strength (UTS) was calculated as the maximum tensile force relative to the average wire diameter before testing. Young's modulus was calculated from the slope of the elastic region of the stress-strain curve.

TABLE 1: Statistical high tensile UTS results for ANOVA: single factor test.

Source of variation	SS	df	MS	F	P value	F_{crit}
Between groups	709486.1	12	59123.8	18.661	$2.193E-15$	1.930
Within groups	177417.4	56	3168.2			
Total	886903.6	68				

From the multiple tests carried out for each supplier, averages were found and statistical analyses performed: box plots, single factor ANOVA tests, and multiple comparisons of means.

4.1.2. Statistical Method. The one-way ANOVA technique tests the null hypothesis that two or more groups are drawn from the same population. For this investigation the null hypothesis is that no difference in tensile properties exists across the different wire batches (all batches are from the same "population" of wire). To confirm/deny the null hypothesis, the ANOVA test produces an F statistic (ratio of the variance calculated among the means to the variance within the samples). If the group means are drawn from the same population, then the central limit theorem infers that the variance between the group means should be lower than the variance of the samples. Therefore, a higher F statistic (ratio) implies that the samples are from different populations, and thus the null hypothesis can be rejected. If $F > F_{crit}$, then the null hypothesis can be rejected at the α% confidence level. ANOVA cannot be used to determine which groups are "statistically different." For this, a multiple comparison of means study was carried out, using confidence intervals for mean differences.

4.1.3. Wire Samples. From discussions with the producer, it was evident that no significant problems had been encountered with Onesteel wire, either during fabrication or in service. The majority of the wire quality issues had arisen around the use of certain Pacific Steel wire batches, namely, the Pacwire-HiSPAN, -breaking roll 1, and -breaking roll 2 samples. Later on in the year, a fencing contractor from Matawai (Gisborne, New Zealand) was experiencing wire breakages during tie off procedures when using Hurricane wire. Two different roll samples were obtained for testing purposes. In the accompanying figures and tables, the identified known acceptable wire is indicated in green whilst wire from failed fences is labelled problematic and indicated in red. Any other wire tested is for comparative purposes and is indicated in a neutral (blue) colour.

4.2. Results for Tensile Strength. The UTS results for each grade of wire are compiled below. Tests are named via the tag supplied with the wire coil. Tests performed on the fence after manufacture (i.e., wire cut from a fabricated fence) are named "after" and thus any other tests can be assumed to be on coiled wire.

4.2.1. UTS for High Tensile Wire. NZS 3471:1974 stipulates that 2.5 mm HT wire must have a minimum UTS of 1235 MPa (or minimum breaking force of 6060 N). As shown in

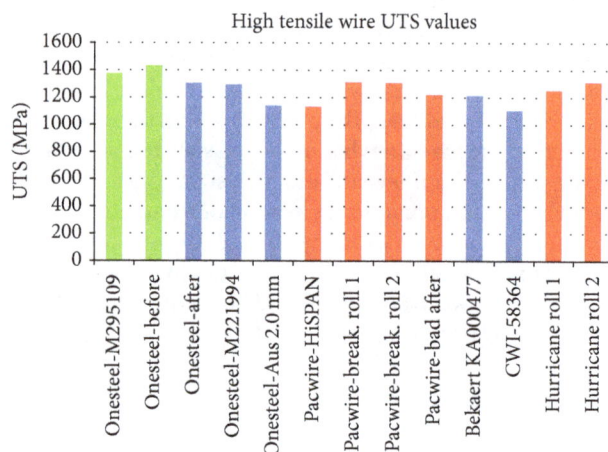

FIGURE 2: Average UTS results for a range of high tensile wire batches.

Figure 2, several wire batches exceeded this standard, which is good. The Onesteel "before" wire batch had the greatest UTS at 1433 MPa (average over the 8 different batches tested). The Pacific Steel HiSPAN batch failed to meet the minimum standard, scoring an average UTS of 1131 MPa. The CWI batch registered the lowest UTS, with an average of 1106 MPa being recorded. Interestingly, both of the problematic Hurricane wire rolls surpassed the minimum standard.

The accompanying boxplot in Figure 3 is used to show the variance of the wire and to look for statistical differences between batches. Of most importance are the results given by Pacific wires HiSPAN batch. Here the wire is observed to have a high variance as shown by the long box and whiskers. This shows that the UTS of this batch is highly variable and cannot be relied on to consistently meet the standard of 1235 MPa. Furthermore the upper range of UTS for HiSPAN wire is still below that of any Onesteel brand tested. This indicates a clear difference in tensile strength between the different product batches.

The statistical ANOVA analysis gives a statistical test of whether the means of several groups are all equal (the null hypothesis). For this investigation, a significance level of 5% (α = 0.05) was chosen. Table 1 shows that the F value for the test (18.661) is greater than F_{crit} (1.930). This gives a strong indication that the UTS values from the different wire brands/batches are not the same (i.e., null hypothesis can be rejected). Since the ANOVA test cannot determine which particular pairs are "significantly different," a multiple comparison procedure was carried out, using the results from the ANOVA test. Figure 4 shows that there are 34 significantly different mean pairs, the largest being between Onesteel "before" and Pacific Steel HiSPAN batches.

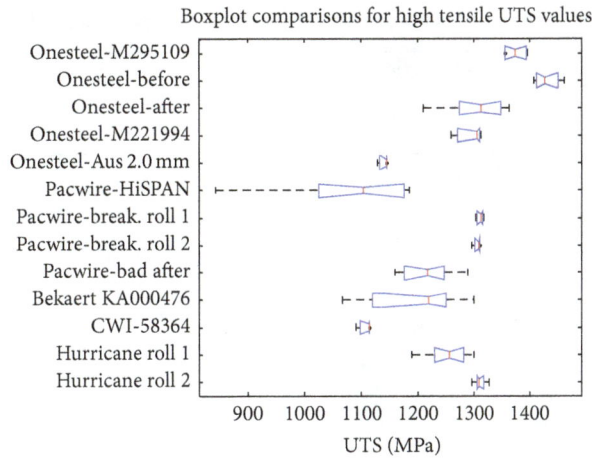

FIGURE 3: Boxplot comparison between the various high tensile wire UTS batches.

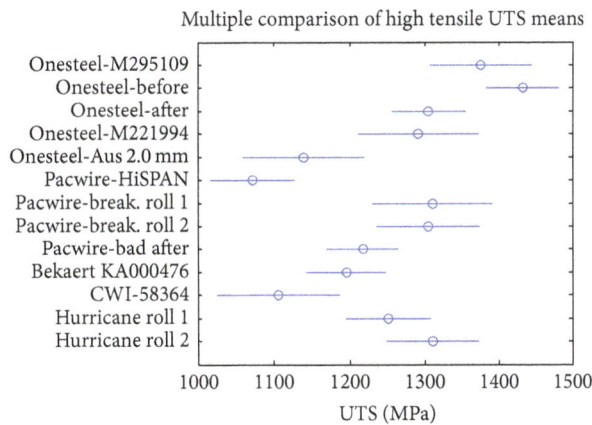

FIGURE 4: Multiple comparison of UTS means for high tensile wire batches.

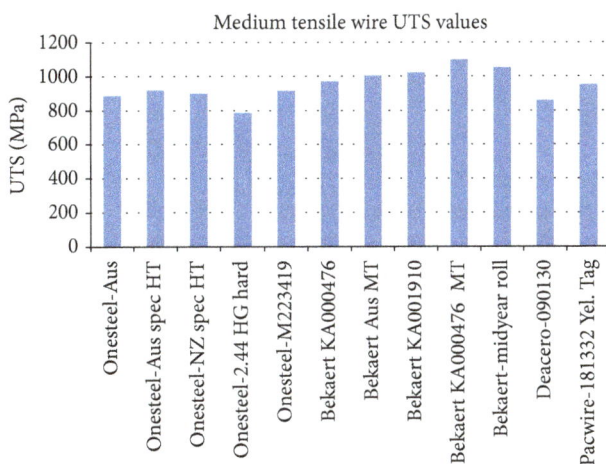

FIGURE 5: Average UTS results for a range of medium tensile wire batches.

4.2.2. UTS for Medium Tensile Wire. Figure 5 shows that the medium tensile wire all performed well over the minimum value of 600 MPa for 2.5 mm diameter wire [21]. That standard also states the following: *"Tensile strength range from any*

1 batch within a type of wire shall not be more than 200 Mpa." From the boxplot in Figure 3, it is evident that only the Pacific steel HiSPAN batch failed to meet this criterion.

4.2.3. Soft Wire. NZS 3471:1974 stipulates that 2.5 mm mild steel (soft) wire should exceed UTS of 430 MPa. Figure 6 shows that this standard is exceeded by all batches, except for some 4.0 mm wire intended for producing staples that were included for interest.

4.3. Percent Elongation (Ductility). The percentage elongation (ductility) results for each grade of wire are assembled below. Tests are named via the tag supplied with the wire coil. Batches match those tested in the UTS tests.

4.3.1. Ductility of High Tensile Wire. Figure 7 shows the percent elongation (ductility) results from the high tensile wire batches tested. NZS 3471:1994 stipulates that the ductility (over a 250 mm gauge) should not fall below 4%. A minimum amount of ductility is required to prevent brittle failures, but too much ductility makes the wire prone to stretching (resulting in baggy fences). Figure 6 shows that the problematic Pacific steel batches feature the lowest ductility. Indeed,

TABLE 2: Statistical high tensile % elongation results for ANOVA: single factor test.

Source of variation	SS	df	MS	F	P value	F_{crit}
Between groups	46.3	10	4.631	41.394	$1.811E - 15$	2.142
Within groups	3.58	32	0.112			
Total	49.89	42				

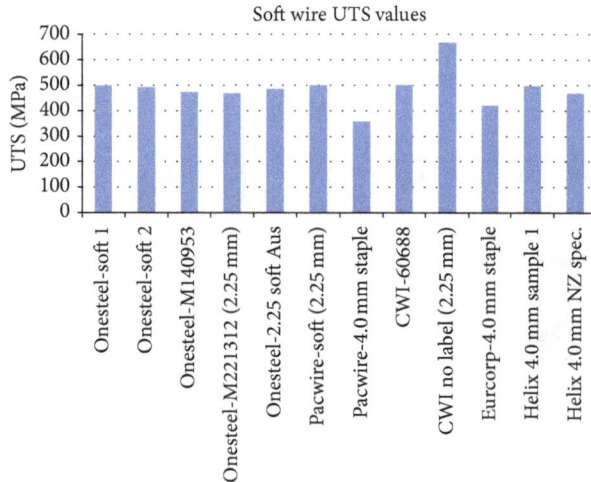

FIGURE 6: Average UTS results for a range of soft wire batches.

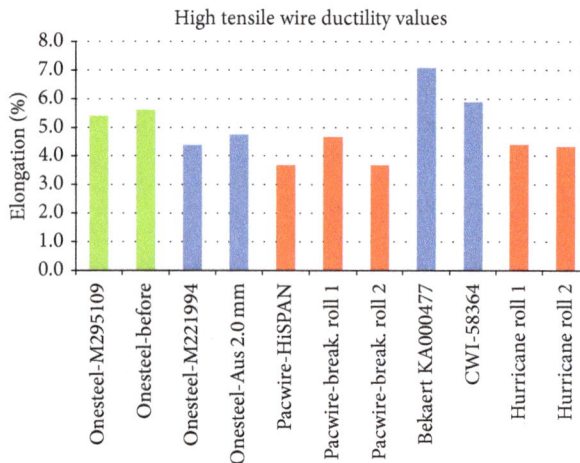

FIGURE 7: Average percent elongation results for a range of high tensile wire batches.

both the HiSPAN and breaking roll 2 batches fail to meet the 4% minimum ductility threshold. The acceptable Onesteel batches both have superior ductility, at approximately 5.5%. Bekaert and CWI which had both performed averagely in the UTS test showed large amounts of elongation with 7% and 5.9%, respectively. The results for Hurricane wire batches show that despite the user's complaint of "brittleness," both have a ductility of 4.35% which is above the required minimum.

The accompanying boxplot in Figure 8 shows that there is less variation between tests, as indicated by narrower whiskers. The highest variations were observed within the Onesteel "before" and Hurricane roll 2 wire batches. Table 2

shows that the F value for the test (41.394) is greater than F_{crit} (2.142). This gives a strong indication that the ductility values from the different wire brands/batches are not the same (i.e., null hypothesis can be rejected). Since the ANOVA test cannot determine which particular pairs are "significantly different," a multiple comparison procedure was carried out, using the results from the ANOVA test. Figure 9 shows that there are 29 significantly different mean pairs, the largest being between Bekaert KA000467 and Pacific steel HiSPAN/breaking roll 2 batches.

4.3.2. Ductility of Medium Tensile Wire. For the medium tensile wire shown in Figure 10, the majority of the brands were found to have around 6% ductility. However, two batches were found to fall below the 4% threshold. The first batch is the Bekaert KA001910 with 3.7% ductility, while the worst is the Pacific steel 181332 yellow tag batch with 1.13% ductility. In the case of the Bekaert batch, it has a relatively high UTS which could explain the reduction in ductility. However, the yellow tag Pacific steel batch is likely to suffer from microstructural defects (i.e., inclusions or grain size effects). The Deacero batch is interesting, displaying a 10.3% ductility. Its UTS is 860 MPa, suggesting that this batch suffers no compromise in ductility or tensile strength.

4.3.3. Ductility of Soft Wire. As expected, the soft wire batches of Figure 11 had very high ductility. For the standard 2.5 mm soft, the average ductility was between 10 and 14%. This was further exceeded by the 2.25 mm wire reaching 16.7%.

Ductility emerges as an unreliable indicator of wire quality, since some known problematic batches nonetheless had adequate ductility.

4.4. Young's Modulus. Young's modulus tests were only able to be performed on a small range of batches due to material (small quantities of wire) and time constraints. Figure 12 reveals that no significant trends exist with wire stiffness (Young's modulus). All of the Young's moduli tested fall within 140–160 GPa. These values are low compared to the usual modulus of steel of about 200 GPa. The reason for this apparent discrepancy could be due to the clamping system and/or the strain measurement system on the MTS 810 tensile machine. Further validation of these results is necessary if they are to be relied on. The fact that no trends exist between acceptable and problematic wire batches indicates that stiffness is not likely to be an important discriminator of wire quality.

4.5. Flexural Modulus. The flexural modulus property measures the ratio of stress to strain in *bending*. From a failure perspective, this property describes the stress that a member will experience for a given deflection. It is similar to the *elastic*

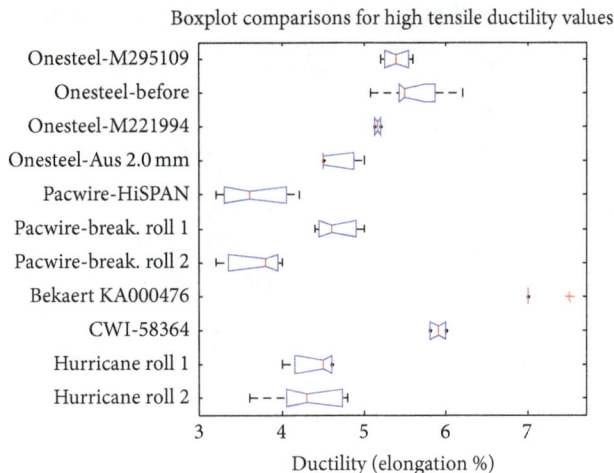

FIGURE 8: Boxplot comparison of ductility values for the various high tensile batches.

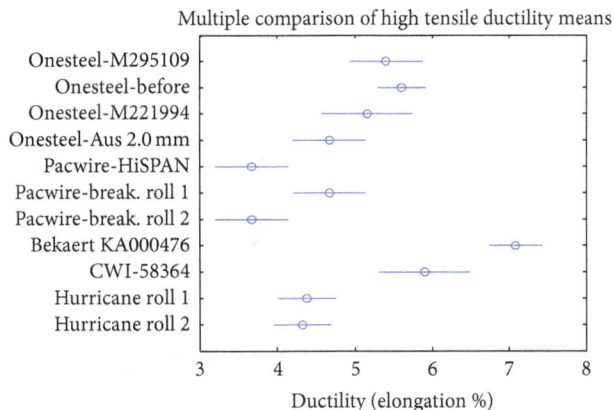

FIGURE 9: Multiple comparison of ductility means for high tensile wire batches.

modulus (or Young's modulus that measures the ratio of stress to strain in *tension*). However an important difference is that the flexural modulus includes the beam geometry and so is not so much a material property as a material-and-beam-size variable.

4.5.1. Method for Three-Point-Bending Test. To better simulate the conditions under which line wires are subjected in the knotting process, a three-point bending test was set up to determine the flexural modulus, E_f, of a wire sample in bending. The objective was to replicate the process whereby the line wire is bent around the vertical stay wires in an extreme loading situation. For a circular section the flexural modulus is given by the formula

$$E_f = \frac{4L^3 m}{3\pi d^4}, \qquad (1)$$

where L is the span between supports (mm), m is the slope of the tangent to the initial straight-line portion of the load-deflection curve (N/mm), d is the diameter of wire (mm), and E_f is the flexural modulus in bending (MPa).

The three-point bend test was based on the standard: *ASTM D790: test methods for flexural properties of plastics* [22]. Since plastics were not being tested, only relevant

sections were implemented. The tests were carried out using a custom built bending rig fitted to a MTS 810 Material Testing System. The base supports were fully adjustable, allowing the bending span to be altered. Figures 13 and 14 show the bending rig layout and dimensions.

For this investigation, a support span of 26 mm was used. This was chosen based on measuring the crimp dimensions of fabricated stiff-stay fences. A thin metal strip was used to apply the midspan force. Wire samples were cut to a length of 50 mm before being tested. Any knotted samples first had their knots carefully removed and were loaded "knot down" to try and exploit the presence of the notch. The MTS 810 Material Testing System was configured into compression mode, with a cross-head displacement rate of −2 mm/minute. As a minimum, four wire samples from any one batch were tested. Plotting the load-deflection results allowed the flexural modulus values to be calculated. ANOVA (analysis of variance) statistical analysis techniques were then used to determine if "statistically significant" results existed.

4.5.2. Results for Flexural Modulus. The flexural modulus results from all tests are shown in Figure 15. A Boxplot showing a summary of the statistical results is shown in Figure 16.

TABLE 3: Statistical results for ANOVA: single factor test.

(a) Summary

Groups	Count	Sum	Average E_f	Variance
Pacwire-HiSPAN	4	143.254	35.814	1.677
Pacwire-breaking 1	4	130.653	32.663	45.283
Pacwire-breaking 2	4	132.910	33.227	20.576
Onesteel B# 295109	5	188.727	37.745	10.868
Bekaert B# KA00476/0101	4	140.213	35.053	7.232
Bekaert B# KA004770101	4	106.384	26.596	4.369
Deacero B# 090130-009 & 014	4	92.908	23.227	9.081

Significance level, α = 5% (0.05).

(b) ANOVA

Source of variation	SS	df	MS	F	P value	F_{crit}
Between groups	691.26	6	115.21	8.226	$9.616E - 05$	2.549
Within groups	308.12	22	14.01			
Total	999.38	28				

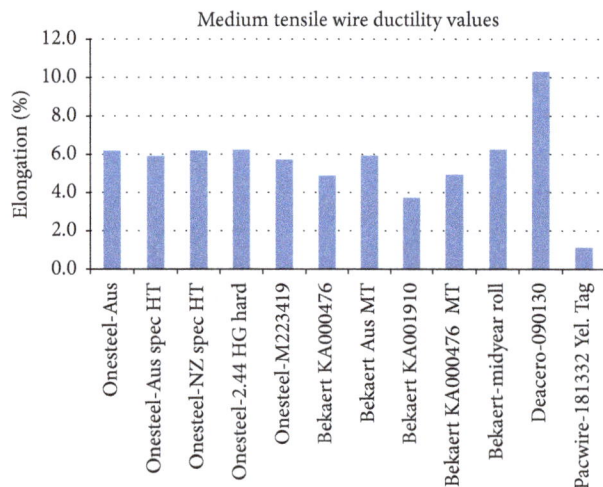

FIGURE 10: Average percent elongation results for a range of medium tensile wire batches.

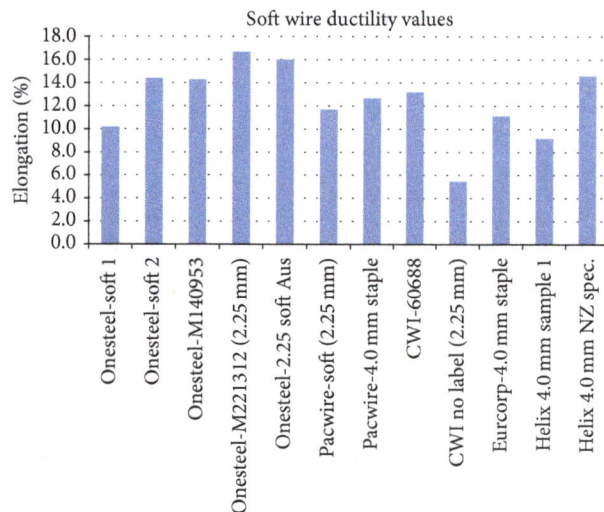

FIGURE 11: Average percent elongation results for a range of soft wire batches.

The statistical ANOVA (analysis of variance) results are depicted in Table 3. To compare each of the wire batches against one another, a multiple comparisons investigation using confidence intervals was conducted. Two results can be said to be "statisitcally different" if there is no overlap of confidence intervals. Figure 17 depicts these results for a 95% confidence level.

4.5.3. Discussion of Three-Point Bending. The three-point bending test results in Figure 15 show a high degree of variability, again confirmed by the boxplot in Figure 16. The source of this variability is unknown, since all known variables were maintained constant (such as span, cross-head displacement, etc.). One possible reason is that the MTS 810 Material Testing System was not sensitive enough for the testing of these wire samples in bending. The resulting load-deflection graphs did contain load fluctuation, which made

curve fitting to the initial portion of the graph more difficult. Since the flexural modulus is highly sensitive to variations in the initial slope, it is possible that the fluctuations are the source of the observed variability. The only way to test this would be to replicate the tests on a smaller, more sensitive compression testing setup.

The statistical ANOVA analysis gives a statistical test of whether the means of several groups are all equal (the null hypothesis). For this investigation, a significance level of 5% (α = 0.05) was chosen. Table 3 shows that the F value for the test (8.226) is greater than F_{crit} (2.549). This gives a strong indication that the flexural moduli from the different wire brands/batches are not the same (i.e., null hypothesis can be rejected). An F statistic as extreme as the observed F would occur by chance one in 10,400 times if the means were truly equal.

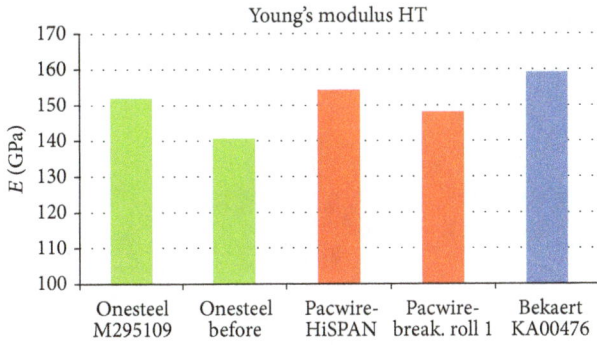

FIGURE 12: Average Young's modulus (E) results for a range of high tensile wire.

FIGURE 13: Three-point bending rig fitted to MTS 810 *Testing System*.

Since the ANOVA test cannot determine which particular pairs are "significantly different," a multiple comparison procedure was carried out, using the results from the ANOVA test. Figure 17 shows that there are seven significantly different mean pairs, the largest being between high tensile Onesteel and medium tensile Deacero wire batches.

Three-point bending is also an unreliable indicator of wire quality, since some known problematic batches nonetheless had high flexural modulus. Also, some known good batches had low modulus. The test thus has poor specificity and sensitivity.

4.5.4. Effect of Span on Flexural Modulus.

During the three-point bend testing, it was noted that many Pacific wire tests resulted in the wire sample fracturing when bent through an acute angle, while the majority of Onesteel, Bekaert, and Deacero wire did not. It was postulated that there might be a critical support spacing that caused all problematic Pacific wire to crack but leave all acceptable wire intact.

For the second part of this investigation, a pass/fail test was adopted, whereby Fail is any visible cracking and Pass is no visible cracking. Wire samples from different batches were again cut to 50 mm lengths and loaded into the bending rig. The cross-head was manually controlled to bend the wire to the point of fracture, or as far as possible. This process was completed for varying support spans.

The bending span investigation results are as shown in Table 4. Bold font indicates fracture, italic font not fracture, and m-dash not tested.

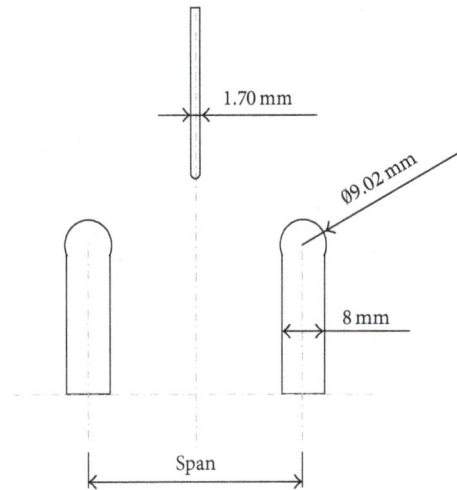

FIGURE 14: Rig dimensions for three-point-bending.

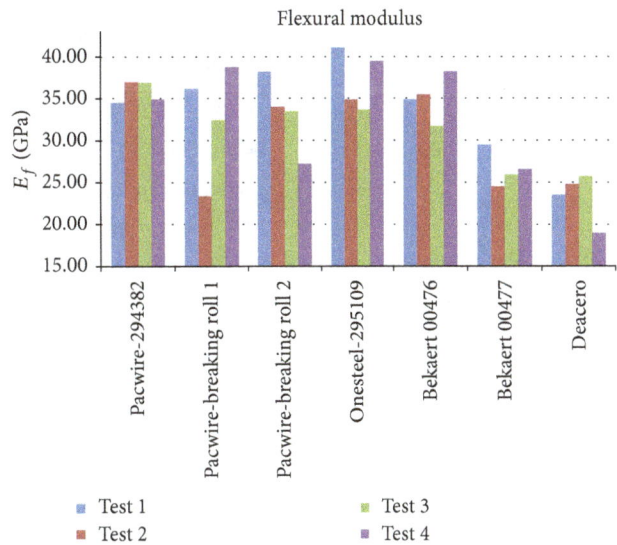

FIGURE 15: Results from all flexural modulus tests from a range of different batches/manufacturers.

The results show a number of interesting trends about the different wire brands/batches. They are as follows.

(i) All HiSPAN and breaking roll 2 Pacific wire batches fail for spans smaller than 22 mm.

(ii) The majority of Onesteel samples fail for spans smaller than 21 mm.

(iii) The majority of Bekaert samples remain intact even for the smallest span (20.44 mm).

(iv) Pacific wire breaking roll 1 begins failing at a span of 21 mm.

These results do confirm the presence of critical support spacings. In general, the problematic Pacific wire batches tend to fracture at larger spans than the acceptable Onesteel and Bekaert wire batches.

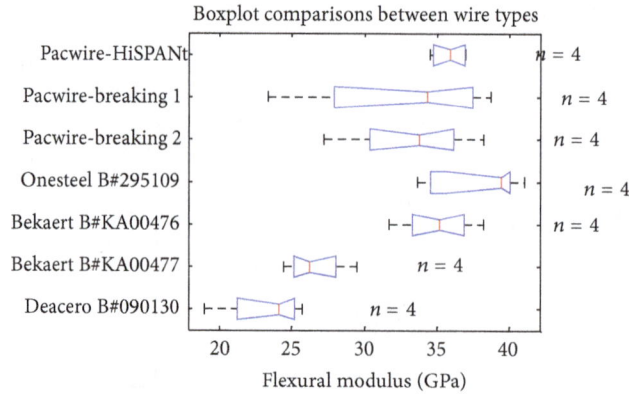

FIGURE 16: Boxplot comparisons between the various wire types.

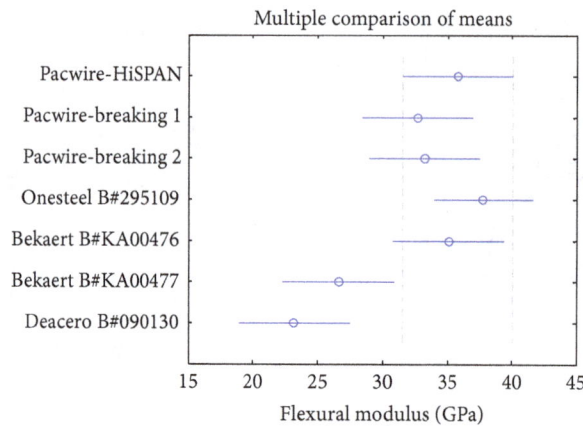

FIGURE 17: Multiple comparison of means between wire batches at 95% confidence level.

This trend is violated by the Pacific wire breaking roll 1 specimens, which appear to have a better fracture tolerance than the acceptable Onesteel specimens do. The medium tensile Bekaert wire proved to be the best performer, only beginning to fracture at the smallest rig span of 20.44 mm.

Span has some potential to be further developed into a test of wire quality, since the results were broadly consistent with the known quality of the wire (other than Pacwire-breaking roll 1). However it is not a quantitative test and the threshold is uncertain (it appears to be around 23 mm). This is a potentially significant finding since the knotting of wire, for example, in the industrial fabrication of fences, or in-field tie-off, involves bending the wire to a tight radius of curvature. Often wire is turned back and wrapped around itself when knotted. Thus wire that passes other tests might fail in the knotting situation because of sensitivity to tight bending.

5. Discussion

5.1. Outcomes: What Has Been Achieved? A batch of wire samples, of different origins, has been subjected to a battery of tests for physical material-properties. These tests have included the standard tensile strength and elongation as presently used in the wire industry, as well as other tests that currently are not: Young's modulus, flexural modulus, and bending-radius (span) effects.

5.1.1. Comparison Across Brands. The results are generally based on multiple rather than single specimens, and this permitted various statistical tests to be performed. These show that there is indeed a statistically significant difference (at the 5% significance level) between many of the brands for tensile strength, ductility, and flexural modulus.

5.1.2. Comparison to Known Wire Quality. Results were obtained for various material properties, for various wire brands. The quality of the wire was also known, and this permitted the sensitivity of the various tests to be determined.

Generally the multiple comparison analyses showed unreliable discrimination of the tests to known good and poor quality wire, in that the distribution of breaking wire overlapped the good wire. This problem was observed for tensile strength and to a lesser extent ductility and three-point bending. This is an issue because it means that the tests are not very powerful: they may let through wire that turns out to fail during production or fence erection (type II error).

5.1.3. Novel Contribution. The novel contribution of this work is first the publishing of multiple material properties

TABLE 4: Bending span investigation results for range of wire batches.

Sample	SPAN 1 = 26.06 mm	SPAN 2 = 24.01 mm	SPAN 3 = 21.96 mm	SPAN 4 = 21.00 mm	SPAN 5 = 20.44 mm (max.)
Pacwire-breaking roll 1					
Test 1	*No fracture*	*No fracture*	*No fracture*	**Fracture, no pop**	**Fracture, no pop**
Test 2	*No fracture*	*No fracture*	*No fracture*	*No fracture*	*No fracture*
Test 3	*No fracture*	*No fracture*	*No fracture*	*No fracture*	*Only galv. crack*
Test 4	*No fracture*	*No fracture*	*No fracture*	**Fracture, pop**	**Fracture, no pop**
Test 5	*No fracture*	*No fracture*	*No fracture*	**Fracture, pop**	**Fracture, pop**
Test 6	—	—	—	*No fracture*	**Fracture, pop**
Pacwire-breaking roll 2					
Test 1	**Fracture**	**Fracture, pop**	**Fracture, pop**	**Fracture, pop**	**Fracture, pop**
Test 2	**Fracture, pop**	*No fracture*	**Fracture, no pop**	**Fracture, pop**	**Fracture, pop**
Test 3	**Fracture, pop**	**Fracture, pop**	**Fracture, pop**	**Fracture, pop**	**Fracture, pop**
Test 4	*No fracture*	*No fracture*	**Fracture, no pop**	**Fracture, pop**	**Fracture, pop**
Test 5	*No fracture*	*No fracture*	**Fracture, pop**	**Fracture, pop**	**Fracture, pop**
Test 6	—	**Fracture, pop**	—	—	—
Pacwire-HiSPAN					
Test 1	**Fracture, pop**	**Fracture, pop**	**Fracture, pop**	**Fracture, pop**	**Fracture, pop**
Test 2	*No fracture*	**Fracture, pop**	**Fracture, pop**	**Fracture, pop**	**Fracture, pop**
Test 3	**Fracture, pop**	**Fracture, pop**	**Fracture, pop**	**Fracture, pop**	**Fracture, pop**
Test 4	*No fracture*	**Fracture, pop**	**Fracture, pop**	**Fracture, pop**	**Fracture, no pop**
Test 5	**Fracture, pop**	**Fracture, pop**	**Fracture, pop**	**Fracture, pop**	**Fracture, pop**
Onesteel line 7					
Test 1	*No fracture*	*No fracture*	**Fracture, pop**	**Fracture, pop**	*No fracture*
Test 2	*No fracture*	*No fracture*	*No fracture*	**Fracture, loud pop**	**Fracture, pop**
Test 3	*No fracture*	*No fracture*	**Fracture, pop**	**Fracture, pop**	**Fracture, pop**
Test 4	*No fracture*	*No fracture*	*No fracture*	**Fracture, loud pop**	**Fracture, pop**
Test 5	*No fracture*	**Fracture, pop**	*No fracture*	**Fracture, pop**	**Fracture, pop**
Onesteel line 8					
Test 1	*No fracture*	*No fracture*	**Fracture, pop**	**Fracture, pop**	**Fracture, pop**
Test 2	*No fracture*	*No fracture*	**Fracture, pop**	**Late Fracture, pop**	**Fracture, pop**
Test 3	*No fracture*	*No fracture*	*No fracture*	*No fracture*	**Fracture, pop**
Test 4	*No fracture*	*No fracture*	*No fracture*	**Fracture, pop**	**Fracture, pop**
Test 5	*No fracture*	*No fracture*	*No fracture*	**Fracture, pop**	*No fracture*
Test 6	—	—	—	—	**Fracture, pop**
Onesteel 295109					
Test 1	*No fracture*	*No fracture*	*No fracture*	**Fracture, pop**	**Late fracture, pop**
Test 2	*No fracture*	*No fracture*	*No fracture*	**Late fracture, pop**	**Fracture, pop**
Test 3	*No fracture*	*No fracture*	*No fracture*	**Fracture, pop**	**Fracture, no pop**
Test 4	*No fracture*	*No fracture*	*No fracture*	**Fracture, pop**	**Fracture, pop**
Test 5	*No fracture*	*No fracture*	*No fracture*	**Fracture, pop**	**Fracture, loud pop**
Bekaert KA00476-0101					
Test 1	*No fracture*	*No fracture*	*No fracture*	*No fracture*	*No fracture*
Test 2	*No fracture*	*No fracture*	*No fracture*	*No fracture*	*No fracture*
Test 3	*No fracture*	*No fracture*	*No fracture*	*No fracture*	*No fracture*
Test 4	*No fracture*	*No fracture*	*No fracture*	*No fracture*	**Beginnings of crack (in galv.)**
Test 5	*No fracture*	*No fracture*	*No fracture*	*No fracture*	**Fracture, no pop**

for fencing wire beyond only tensile strength and ductility, secondly the comparison of different batches of wire, and thirdly the categorisation of results by known wire quality in production.

5.2. Implications for Practitioners. The tensile testing results suggest that premature fence failures are unlikely to be caused solely by tensile properties. In general, the acceptable wire batches have displayed higher strengths, but the majority of the wire batches were above the minimum standard. However, neither the UTS nor ductility results have shown any clear distinction between acceptable and problematic wire quality. Ductility is the more powerful of the two, but this property alone cannot be relied on as a means of screening wire compatibility and quality.

For those practitioners who are specifying wire for input into a fence-knotting machine, the implication is that not all wire is created equal. The first issue is that not all wire on the market necessarily meets the existing standards for tensile strength and ductility. This is a production quality issue. The second issue is that the existing standards for tensile strength and ductility cannot be fully relied on to discriminate between good and poor wire quality. It is possible for wire to meet these tests and yet still fail in production [19]. Fence producers who are concerned about wire-quality would want to insist that their suppliers at least meet the existing tensile strength and ductility standards. Ductility appears to be the harder property to meet. There are also other tests, not yet codified into standards, that could be used [19].

Overall, it must be remembered that tensile specifications are fundamental to the wire classification system. Tensile properties therefore should still continue to be evaluated at all stages in the fabricated fence production sequence. Flexural modulus evaluations, on the other hand, appear to have no significant production implications.

5.3. Limitations. The work done here was based on samples of wire and fence. Due to the nature of the wire production process, it is to be expected that there will be differences in wire batches over time. Therefore the data and conclusions apply only to the batches tested over that window of time and not necessarily to the brands as a whole.

Sample sizes were constrained due to the quantity of wire provided. Some batch samples only contained enough wire to permit a bare minimum of three tensile tests, while other samples were much larger.

Also, the power of the results is only as good as the correct assignment of acceptable and problematic quality to wire. For example, the Pacific steel breaking roll 1 batch, despite being labelled as problematic wire by the end user, performed very well in the tensile tests. Not only was its tensile strength comparable with the acceptable batches, but so too was its ductility. It must also be noted that this particular batch performed acceptably in other tests, for example, LTD, three-point bending, and impact energy [19], which indicates that the issues with this batch may have had more to do with the user than the wire itself.

5.4. Implications for Further Research. The first implication is that existing metrics of tensile strength and ductility are unreliable discriminators of wire quality, at least according to the supplied samples. One avenue for further research could be to repeat the tests with a bigger sample of known good and bad wire. This would be useful to quantify the statistical power of the various tests. However the results from the existing work already show that existing tests have poor power, and doing more work to know this with greater certainty might not be the most useful way to move the field forward. It would add greater value to have fundamentally better understanding of the failure mechanism for wire during fence-production and better tests of incoming wire quality. Nonetheless the point is that future work needs to ensure a ready supply of known good and bad material.

In general is has been observed that there is some inverse correlation between tensile strength and ductility. Medium and soft tensile grades have lower UTS values but benefit from having increased ductility. There does appear to be exception to this rule, with some of the high tensile Onesteel batches having excellent UTS values combined with respectable ductility. It is this combination of properties which is most suited to the fabrication of high quality fences. Therefore, some form of test could be developed that can reliably identify these wire batches.

As discussed above, ductility gives the more powerful measure of wire quality. It therefore may be that ductility can be combined into another test, which enhances the discriminating power. One method may include dropping the temperature below zero, thus inducing more brittle behaviour from the poorer quality wire batches (i.e., ductile-to-brittle transition phenomenon exhibited in metals). Furthermore there may be more suitable temperature dependant tests available such as an impact energy experiment, and indeed other work conducted by the research group addresses this and low-temperature effects.

6. Conclusions

The objective of this investigation was to investigate the material properties of various bathes of wire. The results showed that the majority of the high tensile wire samples met the NZS 3471:1974 minimum tensile requirements. Only the Pacific Steel HiSPAN and CWI batches failed to meet the UTS standard. In terms of ductility, again the Pacific steel HiSPAN and breaking roll 2 batches failed to meet the minimum 4% ductility. Young's Modulus results revealed no major differences. Three-point bending and span effects showed significant differences between batches.

However, the bigger question is the power of the tests to discriminate between good and poor wire quality. While there was significant variability between wire batches, this did not correlate reliably with the known wire quality. Therefore it is concluded that existing test methods, including tensile strength and ductility, are somewhat incomplete, perhaps even unreliable, as measures of wire quality.

Conflict of Interests

The authors declare that there is no conflict of interests regarding the publication of this paper.

Authors' Contribution

All authors contributed to the work and writing of the paper.

References

[1] Anon, "Ring-lock wire-fence making machine," *Engineering*, vol. 133, no. 3452, p. 309, 1932.

[2] S. S. DeForest, "Development of fence erecting machine," *Agricultural Engineering*, vol. 42, no. 2, pp. 66–67, 1961.

[3] C. Hamann, "Wire crimping," *Draht*, vol. 12, no. 4, pp. 155–160, 1961.

[4] C. E. Hann, D. Aitchison, D. Kirk, and E. Brouwers, "Modelling and system identification of a stiff stay wire fence machine," *Proceedings of the Institution of Mechanical Engineers, Part B (Journal of Engineering Manufacture)*, vol. 224, no. B7, pp. 1069–1083, 2010.

[5] Australian Standards, *AS 1650-1989 Hot-Dipped Galvanized Coatings on Ferrous Articles*, Australian Standards, Sydney, Australia, 1650.

[6] SNZ, "NZS 3471:1974A1—Specification for Galvanised Steel fencing wire," NZS 3471, Standards New Zealand, Wellington, New Zealand, 1974.

[7] BSI, *Specification for Fences*, Part 10, British Standards Institution—British Standard, 1963.

[8] ASTM, *Annual Book of ASTM Standards, 1981, Part 3: Steel Plate, Sheet, Strip And Wire; Metallic Coated Products; Fences*, Annual Book of ASTM Standards, Part 3, ASTM, 1981.

[9] Australian Standards, *AS 1650–1989 Hot-Dipped Galvanized Coatings on Ferrous Articles*, Australian Standards, 1650.

[10] Anon, "Atmospheric corrosion tests on wire and wire products begun by A.S.T.M," *Steel*, vol. 99, no. 25, p. 5154, 1936.

[11] W. H. Bleecker, "Choice of fencing material affected by atmospheric conditions," *Oil and Gas Journal*, vol. 36, no. 51, p. 48, 1938.

[12] C. C. Crane, "New electro-galvanizing process for round wire," *Steel*, vol. 100, no. 19, pp. 71–72, 1937.

[13] B. A. Jennings, "Fence exposure tests," *Agricultural Engineering*, vol. 25, no. 4, pp. 140–141, 1944.

[14] F. M. Reinhart, "Twenty-year atmospheric corrosion investigation of zinc-coated and uncoated wire and wire products," Special Technical Publications 290, American Society for Testing and Materials, 1961.

[15] J. L. Schueler, "New process for making zinc coated farm fencing more durable," *Agricultural Engineering*, vol. 14, no. 12, pp. 339–340, 1933.

[16] O. W. Storey, *The Corrosion of Fence Wire*, American Electrochemical Society, 1917.

[17] R. Nuenninghoff and K. Sczepanski, "Galfan—an improved corrosion protection for steel wire. Part 2: applications of aluminium-zinc coated steel wires and tests with coated products," *Wire*, vol. 37, no. 4, pp. 321–324, 1987.

[18] R. Fabien, M. Robertson, and A. V. Nguyen, "Study on parameters influencing the corrosion of metallic coatings on wire exposed to marine environments," *Wire Journal International*, vol. 40, no. 7, pp. 94–98, 2007.

[19] D. Pons, G. Bayley, R. Laurenson, M. Hunt, C. Tyree, and D. Aitchison, "Wire fencing (part 1): determinants of wire quality," *The Open Industrial & Manufacturing Engineering Journal*, vol. 5, no. 5, pp. 19–27, 2012.

[20] ASTM, *ASTM A370-05 Standard Test Methods and Definitions for Mechanical Testing of Steel Products*, A370, American Society for Testing and Materials (ASTM), 2005.

[21] British Standards Institute (BSI), "BS EN 10223-5:1998-steel wire and wire products for fences. Steel wire woven hinged joint and knotted mesh fencing," Tech. Rep. BS EN 10223, British Standards Institute, London, UK, 1998.

[22] ASTM, *STM D790-07e1 Standard Test Methods for Flexural Properties of Unreinforced and Reinforced Plastics and Electrical Insulating Materials*, American Society for Testing and Materials (ASTM), 2007.

17

The Elastic Constants of the Single Crystal of the Mg-Zn-Zr-REM Alloy from the Data of the Elastic Anisotropy and the Texture of the Polycrystalline Sheet

S. V. San'kova, N. M. Shkatulyak, V. V. Usov, and N. A. Volchok

South Ukrainian National Pedagogical University, 26 Staroportofrankovskaya Street, Odessa 65020, Ukraine

Correspondence should be addressed to V. V. Usov; valentin_usov50@mail.ru

Academic Editor: Manoj Gupta

The measuring of the constants of single-crystals requires the availability of crystals of relatively big size. In this paper the elastic constants of the single crystals of magnesium alloy with zinc, zirconium, and rare earth metals (REM) were determined by means of the experimental anisotropy of Young's modulus and integral characteristics of texture (ICT), which were found from pole figures. Using these constants the anisotropy of Young's modulus of alloy sheet ZE10 was calculated. Deviation of calculated values from experimental values did not exceed 2%.

1. Introduction

Magnesium alloys attracted attention in recent years because they are the lightest among available metallic commercial materials. Due to the low density ($1740\,\text{kg/m}^3$) and a relatively high specific strength, magnesium alloys enable the reduction of weight of construction by replacing the steel and aluminum parts in the aerospace and the transport industry. High resistance to the formability due to unfavorable texture, which was formed in the process of its production, impedes wide use of magnesium alloys in practice. Magnesium alloys with zinc, modified by rare earth metals (REM), such as cerium, neodymium, yttrium, were created with the aim to weaken the undesirable texture by different thermomechanical treatments [1]. However, many questions of formation of texture and anisotropy of properties (in particular, elastic) at different types of heat treatment and deformation of magnesium alloys with zinc, zirconium, and rare earth metals have not yet been sufficiently studied.

Elastic modules and constants of compliances of single crystals are included in all equations of physics and mechanics of deformable solids. Elastic constants are determined of appropriate experiments on single crystals. Growing the artificial single crystals of the desired size is associated with technical difficulties. Using the elastic constants of a single crystal of pure metal for calculating properties of alloys in polycrystalline state can lead to significant errors.

The aim of this work is to determine the elastic constants of a single crystal of magnesium alloy with zinc, zirconium, and rare earth metals (REM) from the data of the measurements of the anisotropy of Young's modulus and integral characteristics of texture (ICT).

2. Materials and Methods

Magnesium alloy ZE10 containing 1.3% Zn, 0.15% Zr, and 0.2% of REM (mainly cerium) of 1 mm thickness was used as material for studies. Before the study the alloy was treated according to industrial technology [2]. From these sheets of alloy we cut out seven rectangular specimens of size 100 × 10 mm through every 15° from the rolling direction (RD) up to the transverse direction (TD) of sheet for measuring of anisotropy of Young's modulus. Young's modulus was measured by a dynamic method on the frequency of flexural oscillations of a flat sample [3]. Samples were processed in a package to reduce measurement errors due to geometrical dimensions. The error did not exceed 1%.

The elastic anisotropy of policrystal is determined by the properties of single crystal and crystallographic texture of the policrystal. The grain structure of metal may also play a certain role there. If the relationship between elastic properties of polycrystalline material and its texture is known then it is possible to find the elastic constants of the single crystal or a combination of them. The orientation of a crystal relatively the system of coordinates of the sample is given by means of three variables. The most complete description of the texture, taking into account all its details, is given by the three-dimensional orientation distribution function (ODF) of the crystals. Experimental pole figures (PF) contain the data for calculation of ODF in nonobvious kind. Pole figures are functions of two variables because the PF are the result of the convolutions of three-dimensional ODF [4]. Information about texture in ODF is redundant for calculation of anisotropy of properties in the approximation of continuum mechanics [4]. Therefore for the estimation of the anisotropy of the properties on the texture analysis data it is sufficient to use not the complete ODF but some characteristics of ODF, which contain sufficient information about the anisotropy of the properties. Such characteristics are the certain combinations of directional cosines, averaged according to the law of distribution of crystals on orientations, and named integral characteristics of texture (ICT) [5, 6]:

$$I_1 = \langle \alpha_{13}^2 \rangle; \qquad I_2 = \langle \alpha_{23}^2 \rangle; \qquad I_3 = \langle \alpha_{33}^2 \rangle;$$
$$I_4 = \langle \alpha_{13}^4 \rangle; \qquad I_5 = \langle \alpha_{33}^4 \rangle; \qquad I_6 = \langle \alpha_{13}^2 \cdot \alpha_{23}^2 \rangle. \tag{1}$$

Here α_{ik} are cosines of angles that set the orientation of the crystal relative to the system of coordinates of the sample; the brackets denote the averaging on all possible orientations of the crystals. The ICT of the orthorhombic polycrystalline objects with a hexagonal lattice may be found [5, 6] by averaging of combinations of the direction cosines of the axis C of hexagonal crystal relative to the coordinate system of sample. The coordinate axes of this system are the rolling direction (RD), the transverse direction (TD), and the normal direction (ND) to the sheet plane. Only five ICT are independent for textures of sheets of hexagonal polycrystalline materials, because $I_1 + I_2 + I_3 = 1$. Averaging can be performed using direct PF of isotropic plane of the crystal. For hexagonal metal this may be PF {0002}. Then (1) takes the form [5, 6]:

$$I_1 = \frac{1}{2\pi} \int_0^{\pi/2} \int_0^{2\pi} \sin^3\alpha\cos^2\beta P_{(0002)}(\alpha, \beta)\, d\alpha\, d\beta,$$

$$I_2 = \frac{1}{2\pi} \int_0^{\pi/2} \int_0^{2\pi} \sin^3\alpha\sin^2\beta P_{(0002)}(\alpha, \beta)\, d\alpha\, d\beta,$$

$$I_3 = \frac{1}{2\pi} \int_0^{\pi/2} \int_0^{2\pi} \sin\alpha\cos^2\beta P_{(0002)}(\alpha, \beta)\, d\alpha\, d\beta,$$

$$I_4 = \frac{1}{2\pi} \int_0^{\pi/2} \int_0^{2\pi} \sin^5\alpha\cos^4\beta P_{(0002)}(\alpha, \beta)\, d\alpha\, d\beta, \tag{2}$$

$$I_5 = \frac{1}{2\pi} \int_0^{\pi/2} \int_0^{2\pi} \sin^5\alpha\sin^4\beta P_{(0002)}(\alpha, \beta)\, d\alpha\, d\beta,$$

$$I_6 = \frac{1}{2\pi} \int_0^{\pi/2} \int_0^{2\pi} \sin^5\alpha\cos^2\beta P_{(0002)}(\alpha, \beta)\, d\alpha\, d\beta.$$

Here $P_{(0002)}(\alpha, \beta)$ is the pole density in the corresponding points of PF {0002}; α, β are azimuth and meridian angles of the exit of the pole [0002] on the projection sphere. Anisotropy of Young's modulus in the plane of sheet takes the form [7]:

$$E(\varphi) = \left[s_{11} + 2a\psi_2^T(\varphi) + b\psi_4^T \right]^{-1}. \tag{3}$$

Here

$$\psi_2^T(\varphi) = I_1^{(h)}\cos^4\varphi + I_2^{(h)}\sin^4\varphi + \frac{1}{4}\left(I_1^{(h)} + I_2^{(h)}\right)\sin^2 2\varphi, \tag{4}$$

$$\psi_4^T = I_6^{(h)}\cos^4\varphi + I_4^{(h)}\sin^4\varphi + 1.5I_5^{(h)}\sin^2 2\varphi, \tag{5}$$

$$a = s_{13} - s_{11} + \frac{1}{2}s_{44}, \tag{6}$$

$$b = s_{11} + s_{33} - 2s_{13} - s_{44}. \tag{7}$$

Young's modulus in the ND will be found according to the formula

$$(E_{HH})^{-1} = s_{11} + aI_3^{(h)} + b\left(1 - 2I_1^{(h)} - 2I_2^{(h)} + I_4^{(h)} + I_5^{(h)} + 2I_6^{(h)}\right). \tag{8}$$

Anisotropy of shear modulus in the plane of quasisingle crystal of orthorhombic symmetry (sheet) is expressed as

$$G(\varphi) = \left[c + d\psi_2^T(\varphi) + 2b\psi_4^T \right]^{-1}. \tag{9}$$

Here b is defined by (7), and c and d are expressed as

$$c = \frac{1}{2}s_{44} + s_{11} - s_{12},$$
$$d = s_{11} - 2s_{33} - 4s_{13} - \frac{3}{2}s_{44} + s_{12}. \tag{10}$$

The functions $\psi_2^T(\varphi)$ and $\psi_4^T(\varphi)$ do not depend on properties of the individual crystals. They define only the texture of sheets. Together with single-crystal characteristics they are present in the expression for the anisotropy of properties that are accessible to the tensor description. These functions describe distribution of certain texture characteristics depending on the direction in the plane of sheet or in directions, which form some angles with the plane of sheet. In fact they describe the anisotropy of the orthorhombic sample with hexagonal structure of its inner-element. The anisotropy of the elastic properties in the textured sheet can be represented by a Fourier series with the even harmonics with sufficient accuracy as [7]

$$E^{-1}(\varphi) = A_0 + A_2 \cos 2\varphi + A_4 \cos 4\varphi. \tag{11}$$

Amplitudes of harmonics (Fourier coefficients) of the series (11) in terms of ICT and elastic constants for hexagonal metals are expressed as [8]

$$A_0 = s_{11} + a(I_1 + I_2) + \frac{3}{8}b(I_4 + 2I_5 + I_6),$$

$$A_2 = \frac{1}{2}\left[a(I_1 - I_2) + b(I_6 - I_4)\right], \tag{12}$$

$$A_4 = \frac{1}{8}b(I_4 + I_6 - 6I_5),$$

a and b are defined in (6) and (7).

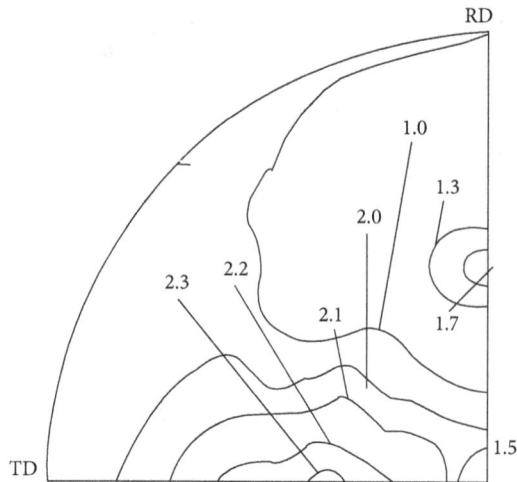

FIGURE 1: Pole figure {0002} for the sheet of the alloy ZE10.

TABLE 1: Integral characteristics of texture of alloy ZE10.

ICT					
I_1	I_2	I_3	I_4	I_5	I_6
0.267466	0.327686	0.404848	0.150986	0.191116	0.04711

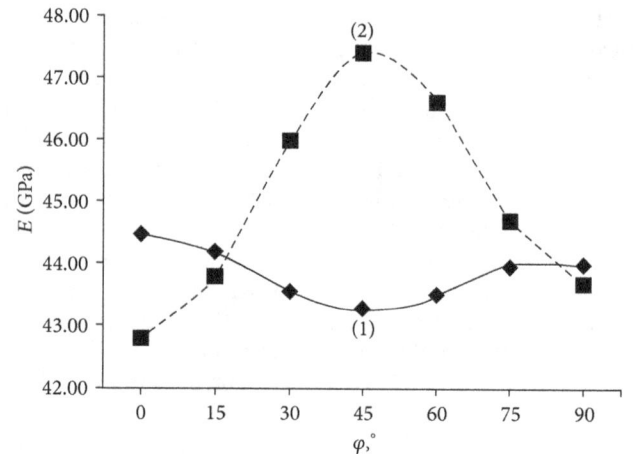

FIGURE 2: Anisotropy of Young's modulus in the sheet of the alloy ZE10. (1) Experiment; (2) calculation by elastic constants of Mg single crystal.

If experimental anisotropy of $1/E$ is represented as a Fourier series and is defined experimentally, then we can find constants s_{11} of alloy and some combinations of them:

$$s_{11} = A_0 - \frac{A_2 (I_1 + I_2)}{I_1 - I_2} + 4\frac{A_4 (I_6 - I_4)(I_1 + I_2)}{(I_1 - I_2)(I_4 + I_6 - 6I_5)}$$

$$- 3\frac{A_4 (I_4 + I_6 + 2I_5)}{(I_4 + I_6 - 6I_5)},$$

$$a = \frac{2A_2}{I_1 - I_2} - 8A_4 \frac{I_6 - I_4}{(I_1 - I_2)(I_4 + I_6 - 6I_5)},$$

$$b = \frac{8A_4}{(I_4 + I_6 + 6I_5)}.$$

(13)

Similar data on the anisotropy of a shear modulus will allow to write three equations which relate the characteristics of the single crystal with the amplitudes of the harmonics Fourier series via the ICT and to find all the constants of compliances.

Above mentioned ICT were calculated from (2) using data of the pole figure {0002}. X-ray technique was used for constructing this PF [9]. Pole density in a particular point of the PF was found from the pole density curves normalized by means of standard sample without the texture (powder sample). Such sample has been prepared from recrystallized saw dusts ZE10 alloy.

3. Results and Discussions

The PF {0002} for the sheet of the alloy ZE10 is presented in Figure 1.

The texture is characterized by two maxima of pole density at the angular distances from RD and TD in 45°. The texture of alloy ZE10 sheets differs considerably from the texture of central base type of pure magnesium [1].

Table 1 shows the ICT calculated from (2) and PF data (Figure 1).

Table 2 shows the elastic constants of a single crystal of magnesium, taken from [10].

Table 3 shows the experimental anisotropy of Young's modulus that was calculated according to the relation (11). It is easy to see that the anisotropy of Young's modulus expressed as a Fourier series (harmonic model) with 4 even harmonics describes the experimental anisotropy of alloy ZE10 with a deviation not more than 1%. We have calculated the anisotropy of Young's modulus in the alloy ZE10 sheet, using ICT (Table 1) and the elastic constants of a single crystal of pure magnesium (Table 2).

The comparison of the theoretical anisotropy with corresponding experimental data showed satisfactory consent. Results are shown in Figure 2.

Figure 2 shows that the calculated curve does not coincide to the experimental data not only quantitatively but also qualitatively. The experimental curve exhibits a minimum at 45° to the RD, while the calculated curve shows the maximum here. Thus, the use of the elastic constants of single crystal Mg in calculation of Young's modulus anisotropy in the alloy ZE10 leads to significant errors. We have calculated the constant of compliances s_{11} and combinations of a, b for the alloy ZE10 by formulas (12) using Fourier coefficients found earlier (Table 3) and ICT (Table 1). We obtained the following values:

$$(s_{11} = 2.287; a = -0.100; b = 0.128) \cdot 10^{-11}\text{Pa}^{-1}. \quad (14)$$

Values of a, b were defined by means (6) and (7).

Next we used the values of the elastic constants obtained from (14) and ICT (Table 1) for estimation of Young's modulus anisotropy in the alloy ZE10 sheet using relationship reverse

TABLE 2: The elastic constants and some of its combinations for Mg single crystal [10].

Single crystal	The elastic constants $S_{ij} \cdot 10^{-11}$, Pa^{-1}					Combinations $(a, b) \cdot 10^{-11}$, Pa^{-1} according to (6) and (7)	
	S_{11}	S_{12}	S_{44}	S_{33}	S_{13}	a	b
Magnesium	2.213	−0.771	6.024	1.975	−0.491	0.308	−0.854

TABLE 3: Experimental and Fourier model of anisotropy of Young's modulus in sheet of alloy ZE10.

φ,°	$E^{-1} \cdot 10^{-11}$, Pa^{-1}		$(\Delta E^{-1})/E^{-1}$, %	Fourier coefficients, Pa^{-1}		
	Experiment	Model		$A_0 \cdot 10^{-11}$	$A_2 \cdot 10^{-14}$	$A_4 \cdot 10^{-13}$
0	2.25	2.27	−0.89			
15	2.26	2.27	−0.44			
30	2.30	2.29	0.43			
45	2.31	2.30	0.43	2.29	−3.64	−1.52
60	2.30	2.29	0.43			
75	2.27	2.28	−0.44			
90	2.27	2.27	0.00			

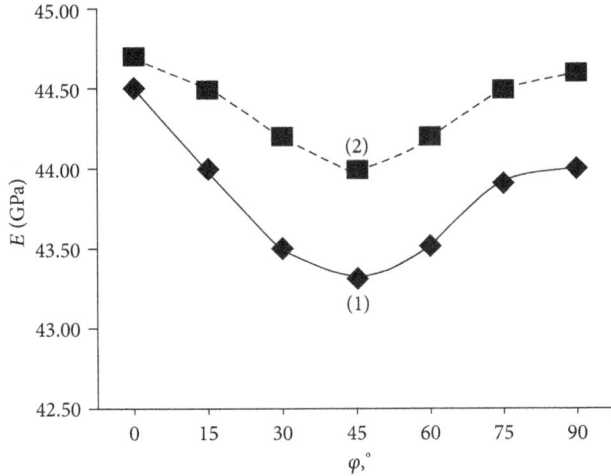

FIGURE 3: Anisotropy of Young's modulus in the sheet of the alloy ZE10. (1) Experiment; (2) calculation by elastic constants of alloy ZE10 single crystal.

to (3) and compared with experiment. The results are shown on Figure 3. Analysis showed that the deviation of the calculated and experimental data does not exceed 2%.

Additionally we have calculated also the value of Young's modulus in ND (END) of the sheet using the relationship reverse to (8), ICT (Table 1), and combination of constants of single crystal (14) of alloy ZE10. We obtained that END = 43.84 GPa.

In principle, even if the number of grains is large enough, Young's modulus depends on the orientation of each grain of the textured material.

In order to estimate Young's modulus of such structures, it is important to identify the local properties of different grains relatively to their orientation. Once the crystal orientation of each grain or its statistical distribution is identified, Young's modulus of the aggregate can be evaluated by averaging Young's modulus of each grain based on a geometrical

assumption. However, the geometrical condition (uniform local strain or stress) in a polycrystal is not obvious due to the complexity of the geometrical structure of a crystal grain. Furthermore, the complex geometry causes nonuniform stresses at the microstructural level even under a uniform remote stress condition. Deformation of one grain becomes difficult due the presence of neighboring grains. It leads to additional stresses. The variation of Young's modulus from one grain to another as well induces the large stress (or strain) near the grain boundary. Such nonuniform stress may affect the macroscopic Young's modulus [11]. In order to quantify the effect of complex geometry and the local stress distribution in the polycrystalline material, it is necessary to use a numerical approach such as the finite element method [11]. Numerical simulations have shown that the crystal orientation makes the main contribution to the value of Young's modulus, when the number of grains is large enough. The error due to ignoring of the structure is the average of less than 0.3 GPa [11]. For Young's modulus of alloy ZE10 investigated by us this error amounted to 0.7%. This is included in above mentioned interval 2% deviation of our results of calculations from experimentally measured values (Figure 3). Thus, the traditional analytic theory of averaging properties of the crystals may be used to find a link between the (anisotropic) properties of the individual grains and the effective macroscopic elastic behavior of polycrystalline materials for large volumes of material. A similar conclusion is also confirmed by experimental measurements of barium titanate at different temperatures in [12].

Sheets of alloy ZE10 are anisotropic even in the initial state, as it follows from the results of measurement and calculation of elastic anisotropy. The coefficient of experimental anisotropy of Young's modulus is equal to 2.8%. For calculated anisotropy of Young's modulus this coefficient is 1.6%. We calculated the anisotropy coefficient of Young's modulus E by means of relation

$$A = \frac{E_{\max} - E_{\min}}{E_{\min}} \cdot 100\%. \qquad (15)$$

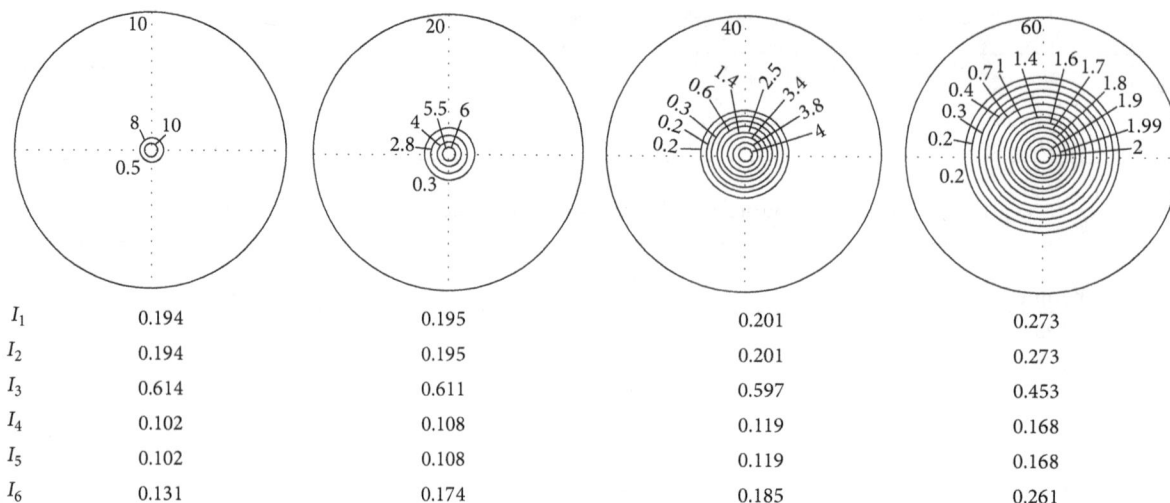

I_1	0.194	0.195	0.201	0.273
I_2	0.194	0.195	0.201	0.273
I_3	0.614	0.611	0.597	0.453
I_4	0.102	0.108	0.119	0.168
I_5	0.102	0.108	0.119	0.168
I_6	0.131	0.174	0.185	0.261

FIGURE 4: Change of ICT with increasing scattering basis poles [0001]. Scattering angle is shown in the upper-portion of each PF.

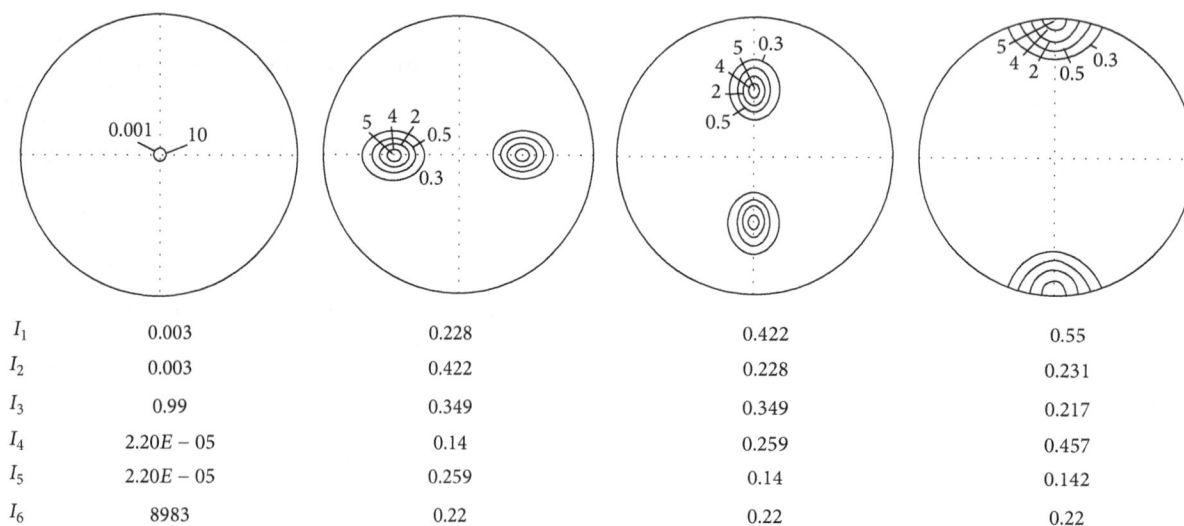

I_1	0.003	0.228	0.422	0.55
I_2	0.003	0.422	0.228	0.231
I_3	0.99	0.349	0.349	0.217
I_4	$2.20E-05$	0.14	0.259	0.457
I_5	$2.20E-05$	0.259	0.14	0.142
I_6	8983	0.22	0.22	0.22

FIGURE 5: Change of ICT with increasing tilt of hexagonal axis [0001] to the plane of sheet.

The maximum of Young's modulus takes place in the RD. The minimal value is observed at the angle 45° to the RD. On Figure 3 it is visible also that Young's modulus in the RD is greater than in the TD. This is contrasted to the anisotropy of Young's modulus in sheets of magnesium alloy AZ31 [13].

It should be noted that the "effect of texture" is rarely taken into account in engineering designing due to the complexity of description of textures by conventional methods. Method of the representation of texture in terms of the ODF is complicated and requires special training of engineers. Method of ideal orientations does not allow directly estimating the anisotropy of properties caused by the texture. The approach described here, which is based on the set of the integral characteristics of texture, uses a minimal number of parameters that is sufficient to predict the properties in various directions of a sheet or product. These characteristics may be used for certification of the texture state of sheet metal. For example, by modeling of ICT for different types of textures

it may be seen [8] that the simultaneous growth of the ICT corresponds to an increase of scattering of poles [0001] by angles of azimuth and meridian (Figure 4).

The increasing of the difference between these characteristics is responsible for development of additional peaks removed from the center of the PF. For example, if $I_2 > I_1$, and $I_4 > I_5$, the texture is corresponded to the deviation from the center of the PF to RD, and vice versa, if the $I_2 < I_1$ and $I_4 < I_5$, then the texture is corresponded to the deviation from the center of the PF to TD (Figure 5). Naturally, there are possible more complicated cases (Figure 1, Table 1).

4. Conclusion

(1) Texture of sheet of magnesium alloy ZE10 (1.3% Zn, 0.15% Zr, and 0.2% of REM) after industrial processing is characterized by two pole density maxima, spaced from the center of the pole figure {0002} in the

RD and TD on angles $45°$. The corresponding integral characteristics of texture (I_i) make up $I_1 = 0.267466$; $I_2 = 0.327686$; $I_3 = 0.404848$; $I_4 = 0.150986$; $I_5 = 0.191116$; $I_6 = 0.04711$.

(2) Integral characteristics of texture include the necessary information to predict tensor properties in any direction and may be recommended for certification of the sheet materials on the textural characteristics

(3) Using the results of the Fourier analysis of the experimental anisotropy of Young's modulus of alloy sheet ZE10 and the integral characteristics of the texture, obtained by means of X-ray analysis, the value of the s_{11} and the certain combinations of the constants compliances of the single crystal of alloy ZE10 were determined, which amounted to $s_{11} = 2.29 \cdot 10^{-11} \mathrm{Pa}^{-1}$; $a = s_{13} - s_{11} + (1/2)s_{44} = -0.1 \cdot 10^{-11} \mathrm{Pa}^{-1}$; $b = s_{11} + s_{33} - 2s_{13} - s_{44} = 0.128 \cdot 10^{-11} \mathrm{Pa}^{-1}$. Calculated and measured curves of the anisotropy of Young's modulus in the plane of the sheet alloy ZE10 differ by no more than 2%.

Conflict of Interests

The authors declare that there is no conflict of interests regarding the publication of this paper.

References

[1] Y. Chino, K. Sassa, and M. Mabuchi, "Texture and stretch formability of Mg-1.5 mass%Zn-0.2 mass%Ce alloy rolled at different rolling temperatures," *Materials Transactions*, vol. 49, no. 12, pp. 2916–2918, 2008.

[2] "Magnesium-based alloy for wrought applications," http://www.sumobrain.com/patents/wipo/Magnesium-based-alloy-wrought-applications/WO2011146970A1.html.

[3] "Elastic moduli: overview and characterization methods. Technical Review ITC-ME/ATCP," http://www.atcp-ndt.com/images/products/sonelastic/articles/RT03-ATCP.pdf.

[4] H.-J. Bunge, *Mathematische Methoden der Teksturanalyse*, Akademie, Berlin, Germany, 1969.

[5] A. A. Bryukhanov and A. R. Gokhman, "Integral characteristics of the texture of cubic and hexagonal metals," *Izvestiâ vysših učebnyh zavedenij. Fizika*, no. 9, pp. 127–131, 1985 (Russian).

[6] A. A. Bryukhanov and A. R. Gokhman, "Calculation method for determining of texture parameters of the tensor properties of cubic and hexagonal metals," *Zavodskaya Laboratoriya*, vol. 53, no. 3, pp. 572–578, 1987 (Russian).

[7] F. Koutný, "Elementary numerical methods & Fourier analysis," http://www.koutny-math.com/.

[8] A. A. Bryukhanov, N. A. Volchok, and T. S. Sovkova, "Effect of cold rolling on the characteristics of texture and anisotropy properties of alpha-alloy Ti-3Al-1,5V," *Materialy*, no. 4, pp. 9–14, 2010 (Russian).

[9] G. Wassermann and J. Greven, "Texturen metallischer Werkstoffe," in *Zweite Neubearbeitet und Erweiterte Auflage*, Springer, Berlin, Germany, 1962.

[10] H. P. R. Frederikse, "Elastic constants of single crystals," http://www.docstoc.com/docs/45454109/Elastic-Constants-of-Single-Crystals.

[11] M. Kamaya, "A procedure for estimating Young's modulus of textured polycrystalline materials," *International Journal of Solids and Structures*, vol. 46, no. 13, pp. 2642–2649, 2009.

[12] J. M. J. den Toonder, J. A. W. dan Dommelen, and F. P. T. Baaijens, "The relation between single crystal elasticity and the effective elastic behaviour of polycrystalline materials: theory, measurement and computation," *Modelling and Simulation in Materials Science and Engineering*, vol. 7, no. 6, pp. 909–928, 1999.

[13] A. A. Bryukhanov, Y. Zil'berg, M. Schaper et al., "Effect of cold straightening the texture and anisotropy properties of magnesium alloy AZ31 sheets," *Deformatsiya i Razrushenie Materialov*, no. 8, pp. 34–41, 2010 (Russian).

18

Synthesis and Characterization of $Ca_xSr_yBa_{1-x-y}Fe_{12-z}La_zO_{19}$ by Standard Ceramic Method

Rohit K. Mahadule,[1] Purushottam R. Arjunwadkar,[2] and Megha P. Mahabole[3]

[1] Smt. Radhikatai Pandav College of Engineering, Nagpur, Maharashtra 411204, India
[2] Institute of Science, Nagpur, Maharashtra 440001, India
[3] School of Physical Sciences, S.R.T.M. University, Nanded, Maharashtra 431606, India

Correspondence should be addressed to Rohit K. Mahadule; nodoubt.mahadule@gmail.com

Academic Editor: Koppoju Suresh

The polycrystalline compounds with chemical formula $Ca_xSr_yBa_{1-x-y}Fe_{12-z}La_zO_{19}$ (CSBFLO) were synthesized via standard ceramic method. The chemical phase analysis was carried out by X-ray powder diffraction (XRD) method, which confirmed the formation of the magnetoplumbite phase belonging to ferrite structure. The frequency dependence of AC conductivity and dielectric constant was studied in the frequency range of 10 Hz to 2 MHz. The experimental results revealed that AC conductivity increases with increasing frequency, which is in agreement with Koop's phenomenological theory. However, variation in dielectric constant required explanation in light of dielectric polarization. Magnetic characterization included studies of parameters such as Ms, Mr, Hc, and Tc, and results were explained via magnetic dilution and canting spin structure.

1. Introduction

M-hexaferrite is a hard ferrimagnetic material possessing magnetoplumbite phase of hexagonal structure which is widely used in various industrial applications. Along with M-hexaferrites, various other members like W, Z, Y, X, and U belong to this family and they can be distinguished as per their stoichiometry. However, M-hexaferrites have gained more attention because of their special characteristic of being magnetoplumbite in nature which leads to greater structural stability compared to other members. The general chemical formula by which M-hexaferrites are represented is $MeFe_{12}O_{19}$ where Me is the divalent alkaline metal cations and can be replaced by a suitable cation or their combinations. Among various hexaferrites, M-hexaferrite is preferred as permanent magnetic material due to its cost effectiveness, reasonable magnetic performances, and wide availability of raw materials needed for synthesis [1].

It finds numerous applications in diverse fields like high-density magnetic recording, microwave absorption devices, high power transmitters, high permeability ferrite components for digital switching equipment for the telecom requirement, high frequency microwave ferrites for VHF/UHF communication sets, defense radar requirement, and transmitter and receiver application in railway projects and can be used as building blocks for hexaferrite isolators [2–4]. The ferrimagnetic oxides with hexagonal crystal structure were first synthesized at the Philips laboratory in 1950 and were called hexagonal ferrites in order to distinguish them from the ferrimagnetic oxides with Spinel and Garnet structure [5]. The basic crystallographic and magnetic properties of the main hexagonal ferrites have been reviewed by Smit and Wijn [6]. The literature survey shows that a lot of works on various combinations of Sr^{2+}, Ca^{2+}, Ba^{2+}, Pb^{2+}, and La^{3+} as alkali earth metals cations have been carried out, which revealed the evolution of the uniaxial anisotropy in M-hexaferrite [7–17]. Furthermore, combination of Sr-La is found to be affecting positively the magnetic properties of the M-hexaferrites in comparison to Ba-La and Ca-La combination [18–21].

In addition to magnetic properties, the exhaustive work on dielectric properties of M-hexaferrites has also been reported by various researchers. Perieria and his group have shown that combined substitution of Sr^{2+} and Ba^{2+} in

TABLE 1: The structural properties of mixed M-hexaferrite samples.

Code	x, z			a (Å)	c (Å)	V	D_x	D_M	Mol. Wt.	Porosity	a/c
	Ca	Sr	Ba								
R1	0.1	0.2	0.7	5.8569	23.1070	686.45	5.32	4.13	1100.09	22.33	0.25
R2	0.1	0.4	0.5	5.8157	22.7836	667.35	5.42	4.02	1090.14	25.85	0.25
R3	0.1	0.6	0.3	5.8170	22.7823	667.61	5.37	3.99	1080.20	25.66	0.25
R4	0.25	0.2	0.55	5.8278	22.7442	668.97	5.43	3.96	1093.80	27.01	0.25
R5	0.25	0.4	0.35	5.8158	22.7633	666.78	5.39	3.87	1083.85	28.13	0.25
R6	0.25	0.6	0.15	5.8079	22.7410	664.32	5.36	3.83	1073.92	28.57	0.25
R7	0.4	0.2	0.4	5.8959	23.0741	694.63	5.19	3.97	1087.52	23.57	0.25
R9	0.4	0.6	0.0	5.8956	23.1369	696.45	5.09	3.87	1067.64	23.88	0.25
R8	0.4	0.4	0.2	5.8958	23.1108	695.71	5.14	3.81	1077.58	25.91	0.25
R10	0.4	0	0.6	5.9028	23.0910	696.76	5.23	3.89	1097.47	25.63	0.25

M-hexaferrite possesses high value of dielectric constant with low loss in radio frequency range [22, 23]. High values of complex relative permittivity and low loss tangent for pure Ba-M hexaferrites have been reported by Mallick [24]. Debnath et al. have studied dielectric properties of Sr, La-M hexaferrite wherein high value of loss tangent is observed at lower frequency side [25].

From the literature review, it is depicted that very meager work is carried out on the simultaneous combinational effect of these cations, namely, Ca^{2+}, Ba^{2+}, Sr^{2+}, and La^{3+}, on electric, magnetic, and dielectric properties of the M-hexaferrites. Hence, attempt has been made to study the electric, dielectric, and magnetic properties of M-type hexaferrite with combined substitution of divalent ions Ca, Sr, and Ba (CSB) along with trivalent La with compositional formula $Ca_xSr_yBa_{1-x-y}Fe_{12-z}La_zO_{19}$ synthesized using standard ceramic method.

2. Experimental Details

2.1. Synthesis. The preparation of polycrystalline compounds with chemical formula $Ca_xSr_yBa_{1-x-y}Fe_{12-z}La_zO_{19}$ (CSBFLO) (with $x = 0.1, 0.25, 0.4$; $z = 0.1, 0.2, 0.3$, and $y = 0.2, 0.4, 0.6$) was carried out via standard ceramic method. The molecular concentration (x and y) of substituted cations in the chemical formula was chosen that the stoichiometry of the compound remains unaffected. The AR grade oxides Fe_2O_3, La_2O_3, CaO, SrO, and BaO (Merck grade) were used as starting precursors for the synthesis of present series of compounds. The preparation process involved the mixing of oxides with respective stoichiometry and grounded together in agate mortar in an acetone medium. The synthesis was divided into two steps. Initially the mixture was calcined at 773 K for 8 h in air followed by further mixing and rigorous grinding and final thermal treatment at 1430 K for 72 hr. The compounds, thus formed, were coded as R-1 to R-10 concerning different combination of substituted cations (Table 1), were characterized by XRD technique, and were used as sample to carry out further studies.

2.2. Characterization. X-ray diffraction patterns of $Ca_xSr_yBa_{1-x-y}Fe_{12-z}La_zO_{19}$ hexagonal ferrites, under investigation, were obtained using Cu-Kα radiation on a Philips X-ray diffractometer (Model PW1732) within scanning range from 10° to 90°. Dielectric parameters were measured by using samples in the pellet form (13 mm diameter) with the help of QuadTech LCR meter in the frequency range of 10 Hz–2 MHz. The study on magnetic behavior at room temperature of the prepared M-hexaferrite samples was carried out at magnetic field of 1T with the help of vibrating sample magnetometer (VSM).

3. Results and Discussion

3.1. Structural Analysis. The lattice parameters, X-ray density, bulk density, and porosity are calculated for each sample and are presented in Table 1. It is observed that not only the values of lattice parameter a vary from 5.8079 to 5.9208 Å but also the lattice parameter c changes from 22.7442 to 23.2143 Å. The XRD profiles of the standard M-hexaferrite are presented in Figure 1 along with the recorded X-ray diffraction patterns of all the samples. In comparison, the presence of reflection planes (006), (107), (114), (201), (108), (220), and (304) corresponding to pure magnetoplumbite phase of hexaferrite family which belongs to the space group P63/mmc was found (no. 194) [14]. The recorded values of lattice parameters also strengthen the results, as the values lie within the lattice parameter range ($a = 5.8$–5.9 Å and $c = 22$-23 Å) of pure magnetoplumbite phase of hexaferrite. Moreover, due to high sintering temperature, the intensity of the peaks becomes stronger and narrower, indicating a better structural quality of materials.

However, it seems that substitution of La^{3+} for Sr^{2+} ion leads to decreasing lattice parameters "a," since La^{3+} ion (1.13 Å) has ionic radii less than those of Sr^{2+} (1.27 Å) ion. Hence, it was concluded that the lattice expansion is higher for the sample having lowest amount of La^{3+} ion. Whereas overall variations of lattice parameter can be attributed to average ionic radius of substituted cations, as the ratio of a/c has remained fairly constant. Similar behavior was

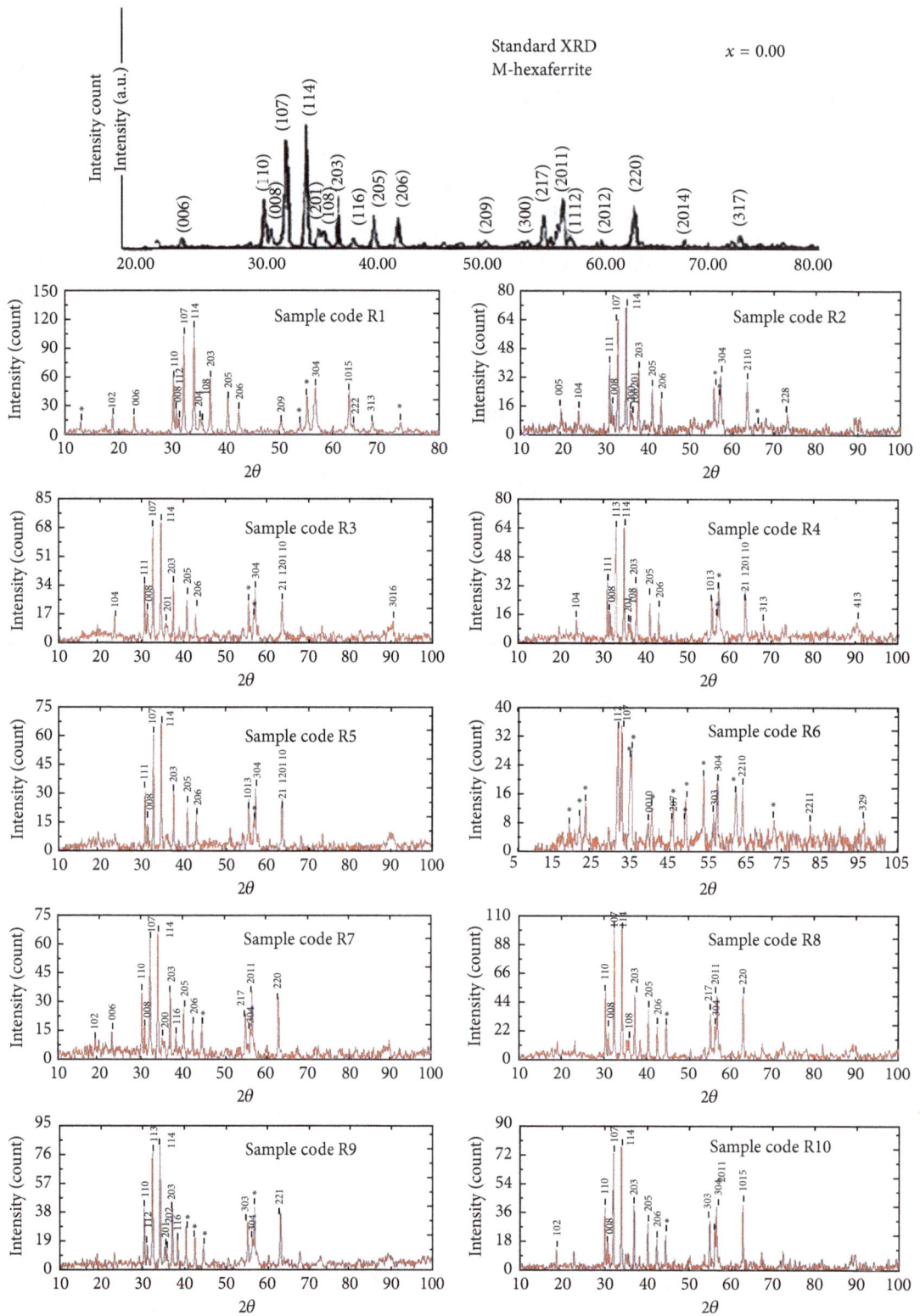

FIGURE 1: XRD pattern for the samples belonging to different composition (sample code).

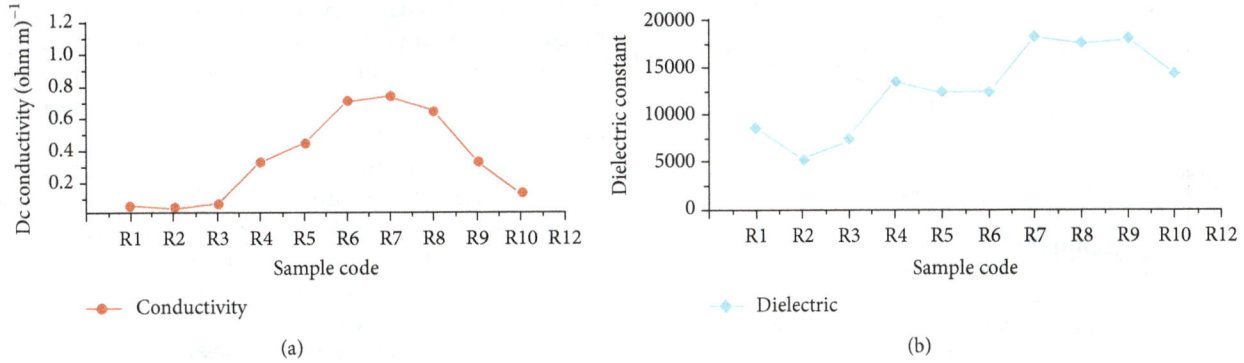

FIGURE 2: Variation of electrical conductivity and dielectric constant as a function of composition (sample code).

TABLE 2: Electric, dielectric, and magnetic parameters of mixed M-hexaferrite samples.

Code	x, z			E_A (eV)	σ_{DC} $(\Omega\,m)^{-1}$	$\sigma_{AC} \times 10^{-3}$	ε	Ms (emu/g)	Mr (emu/g)	Hc (G)	Tc (K)
	Ca	Sr	Ba								
R1	0.1	0.2	0.7	0.34	0.05537	2.5	8.54 k	49.298	24.339	2307.5	700–750
R2	0.1	0.4	0.5	0.31	0.04442	1.11	5.15 k	57.88	30.440	2427.3	700–750
R3	0.1	0.6	0.3	0.35	0.07205	1.49	7.36 k	68.471	35.569	2904.6	700–750
R4	0.25	0.2	0.55	0.279	0.32859	3.89	13.50 k	55.610	23.671	1170.8	670–700
R5	0.25	0.4	0.35	0.262	0.44474	2.59	12.27 k	48.143	23.043	2074.14	670–700
R6	0.25	0.6	0.15	0.268	0.70824	4.48	12.45 k	52.98	23.297	2327.67	670–700
R7	0.4	0.2	0.4	0.31	0.73797	6.58	18.02 k	54.811	22.200	1074.9	650–670
R8	0.4	0.4	0.2	0.248	0.64729	7.54	17.47 k	54.140	26.934	2101.8	650–670
R9	0.4	0.6	0.0	0.244	0.32852	5.31	14.26 k	43.381	19.910	1657.1	650–670
R10	0.4	0.0	0.6	0.265	0.1369	4.76	17.47 k	52.350	21.670	1091.1	650–670

reported in La^{3+} substituted M-type strontium ferrites [15–17]. Hence, the behavior confirmed that interaction and solubility between Sr^{2+} ion and La^{3+} ion is higher than other divalent ions with La^{3+} ion in the compounds.

The variation in the densities shows general behavior; that is, the X-ray density is higher than the apparent density. The densification of samples depends on oxygen ions which diffuse through the material during sintering process. The variation in porosity attributes to function of lattice parameters; it is reported that variation in porosity is inverse to variation in effective cross sectional area of grain-to-grain contact. This concludes that if densification increases, the volume of unit cell and lattice constant ultimately decreases and vice-versa [23, 26]. This showed a good agreement with our results.

3.2. Compositional Variation of Electrical Conductivity and Dielectric Constant.
Electrical conductivity and dielectric constant, both are basically electrical properties and it has been recognized that the same mechanism, namely, exchange of electron between $Fe^{2+} \leftrightarrow Fe^{3+}$, is responsible for variation in both properties. Figure 2 represents variation in electrical conductivity and dielectric constant with respect to compositions or sample codes (R-1 to R-10). An increase is observed along with the increasing substitution of Sr^{2+} and La^{3+} ion till compound R-7, which is further followed by a decrease

up to R-10. The maximum value is obtained for electrical conductivity and dielectric constant for sample R7 (Table 2) having proportion of $Ca^{2+} = 0.4$, $Sr^{2+} = 0.2$, $Ba^{2+} = 0.4$, and $La^{3+} = 0.3$.

These results can be explained with small polaron hopping mechanism and Maxwell Wagner interfacial polarization. Both of the mechanisms deal with the production of Fe^{2+} ions resulting from the partial substitution of $Fe^{2+} \leftrightarrow Fe^{3+}$ at octahedral site $4f_1$ or 2b and volatilization of substituted ions during sintering process. As the structure possesing cations and anions separately in tetrahedral site and octahedral site surrounded by oxygen ions (excluding only trigonal bipyramidal site) can be treated isolated from each other. Thus the localized electron model, namely, hopping mechanism, is more appropriate to discuss the condition mechanism rather than the band model. It is expected that till R7 the interaction and solubility of $Sr^{2+} \leftrightarrow La^{3+}$ ion dominate semiconducting phenomenon, to compensate the charge neutrality at $4f_1$ or 2b site and produce electron transfer between Fe^{2+} and Fe^{3+} [27]. Moreover, in compound R-7, high value of electrical conductivity reflects transfer of maximum number of Fe^{2+} ions which are involved in the phenomenon of exchange interaction between Fe^{2+} and Fe^{3+}, giving rise to maximum conduction process. It may be due to high activation energy ($E_A = 0.31$ eV) among the compounds having the high concentration of La^{3+} ion along with Ba^{2+}

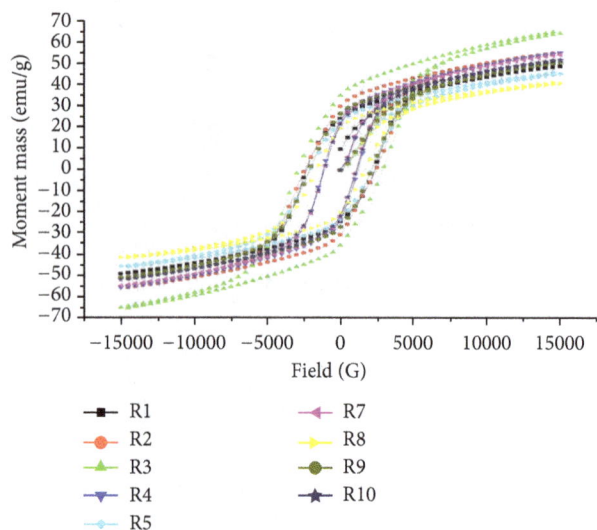

FIGURE 3: Variation in M-H curve for the samples.

and Sr^{2+}. As reported, the transition energy between Fe^{2+} and Fe^{3+} is 0.2 eV and if the activation energy of resistivity is greater than 0.28 eV, the energy is mainly utilized in moving the charges, not for the production of further charge carriers [26]. Hence, higher concentration of La^{3+} ion, which resides at the grain boundaries, results in high mobility of charges and the decrease of Fe^{3+} ion concentration, which ultimately enhances conductivity and reflects semiconducting behavior [28, 29]. However, further decrease in conductivity can be attributed to migration of Fe^{3+} ions to 12k site due to higher substitution of Ba^{2+} and Sr^{2+}, separately, along with La^{3+} ion in M-hexaferrites, resulting in weakening the hopping mechanism and increasing resistivity [30]. Whereas, the high value of dielectric constant for the series of synthesized compounds is due to high conductivity. As it was already reported that sintered ferrites with high conductivity at low frequencies have a high dielectric constant [30].

3.3. Magnetic Measurements

3.3.1. Compositional Variation of Magnetic Parameters. In general magnetic moment found in the range between 43 and 70 emu/g, retentivity between 19 and 36 emu/g, and the coercivity between 1074 and 2905 G for all the samples confirming the good quality of samples are shown in Figure 3 and Table 2.

Taking these results into account, we can conclude that sample code R3 (having composition Ca = 0.1, Sr = 0.6, Ba = 0.3, and La = 0.1) has shown the maximum value for magnetic saturation (M_s), retentivity (M_r), and coercivity (H_c). The observed increase in saturation magnetic moment, retentivity, and coercivity showed high solubility and interaction between La^{3+} and Sr^{2+} as compared to Ba^{2+} among the substitution. The increase in value of magnetization reflects the substitution of ions at spin down—sublattice at octahedral site [31, 32]. Whereas increase in coercivity can be explained

by the loss of magnetocrystalline anisotropy and growth of large shape anisotropy in the magnetic particles [33]. However, substitution of nonmagnetic ion La^3 increases for further samples having Sr^{2+} contents constant, reflecting the decrease in magnetic parameters. This variation concerned with phenomenon known as canting spin structure plays its role, when there is the substitution of divalent ion by trivalent ion associates with a valency change of one Fe^{3+} to Fe^{2+} which reduces the strength of interaction. This results into the shift from collinear to non-collinear of magnetically hard axis, for example, c-axis in spin structure. This showed a relevant support to our results and also led to strengthening the assumption that exchange of Fe^{3+} to Fe^{2+} referred to octahedral sites and Fe^{2+} anisotropy on the octahedral site could be dominant in all M-hexaferrites.

3.3.2. Curie Temperature. The variation of the Curie temperature Tc (K) with composition for all the samples is shown in Table 2. It is observed that the Curie temperature Tc (K) was higher in sample having lower amount of La^{3+} ion. This trend may be attributed to the exchange interactions between different magnetic ions, concentration of these ions, and their magnetic moments. It is therefore expected that a greater amount of energy will be required to offset the effects of exchange interactions in the material having a larger number of magnetic ions. As the magnetic moment of La^{3+} ion is 2.78 μB compared to the magnetic moment of 10 μB for the two Fe^{3+} ions [34], this concludes that the replacement of Fe^{3+} ions by lower amount of La^{3+} ion and Sr^{2+} is likely to increase hard magnetic properties and the Curie temperature. However, lower values for other samples can be explained on the basis of the number of magnetic ions present in the two sublattices and their mutual contraction. As Fe^{3+} ions are gradually replaced by rare earth La^{3+} ions, the number of magnetic ions begins to decrease at both sites, thus leading to a decrease of exchange interaction of the type Fe^{3+}-O^{2-}-Fe^{3+} [35]. As the Curie temperature Tc (K) is determined by the overall strength of the exchange interactions, the weakening of exchange interactions results in a decrease in the Curie temperature [29], which is in good agreement with our result.

4. Conclusions

The structural study of samples reveals that Ca^{2+} does not replace Ba^{2+} in proper quantity as compared to Sr^{2+} and Fe ions. The substitution of La^{3+} for Ba^{2+} and Sr^{2+} in M-type hexaferrite is associated with a valance change of $Fe^{2+} \leftrightarrow Fe^{3+}$ at octahedral 2a or $4f_2$ site. Among the samples having higher concentration of Sr^{2+} with suitable concentration of La^{3+} in M-hexaferrite shows largest variation in AC. Conductivity and dielectric polarization along with conduction enable us to conclude excess formation of Fe^{2+} ion and dually supported by negative value of thermoelectric power. Due to such a high value of dielectric parameters, it can be very useful for application discussed like RAMs. Furthermore, in the magnetic studies we observed high values of magnetic parameters for La-substituted Ca, Sr, Ba, and M-hexaferrite

(R3) are concerned with high interaction between La^{3+} and Sr^{2+} ions and also support the formation of single domain particles as the magnetization process takes place by spin rotation instead of domain wall displacement. These samples can be selected for the production of permanent magnets.

References

[1] J. I. Kraschwit and M. Howe-grami, "Explosive and propellant to flame retardants for textiles," in *Encyclopedia of Chemical Technology*, vol. 10, p. 381, 4th edition, 1993.

[2] D. Ravinder and P. V. B. Reddy, "High-frequency dielectric behaviour of Li-Mg ferrites," *Materials Letters*, vol. 57, no. 26–27, pp. 4344–4350, 2003.

[3] K. Iwauchi, "Dielectric properties of fine particles of Fe_3O_4 and some ferrites," *Japanese Journal of Applied Physics*, vol. 10, pp. 1520–1528, 1971.

[4] C. G. Koops, "On the dispersion of resistivity and dielectric constant of some semiconductors at audiofrequencies," *Physical Review*, vol. 83, no. 1, pp. 121–124, 1951.

[5] J. J. Went, G. W. Rathenau, E. W. Gorter, and G. W. van Oosterhout, "Ferroxdurc, a class of new permanent magnet materials," *Philips Technical Review*, vol. 13, pp. 194–208, 1951.

[6] J. Smit and H. P. J. Wijn, *Ferrites*, Philips Technical Library, Eindhoven, The Netherlands, 1959.

[7] P. B. Braun, "Ferrites," *Journal of Chemical Education*, vol. 37, p. 380, 1960.

[8] J. Smith and H. P. J. Wijn, "Ferrites," *Journal of Chemical Education*, vol. 37, no. 7, p. 380, 1960.

[9] E. P. Wohlforth, "Transport properties of ferromagnets," in *Ferromagnetic Materials*, vol. 3, chapter 9, North-Holland, Amsterdam, 1982.

[10] R. Atkinson, "Optical and magneto-optical properties of Co-Ti-substituted barium hexaferrite single crystals and thin films produced by laser ablation deposition," *Journal of Magnetism and Magnetic Materials*, vol. 138, no. 1-2, pp. 222–231, 1994.

[11] G. Asghar and M. Anis-ur-Rehman, "Structural, dielectric and magnetic properties of Cr-Zn doped strontium hexa-ferrites for high frequency applications," *Journal of Alloys and Compounds*, vol. 526, pp. 85–90, 2012.

[12] R. Grössinger, M. Küpferling, M. W. Pieper, M. Müller, J. F. Wang, and R. Harris, "The effect of substituting rare earth on M-Type hard magnetic ferrites," in *Proceedings of the 9th International Conference on Ferrites (ICF '04)*, pp. 573–578, San Francisco, Calif, USA, 2004.

[13] F. L. Wei, "Magnetic properties of $BaFe_{12-2x}Zn_xZr_xO_{19}$ particles," *Journal of Applied Physic*, vol. 87, no. 12, Article ID 8636, 2000.

[14] H. W. Starkweather, P. Avakian, J. J. Fontanella, and M. C. Wintersgill, "Dielectric properties of polymers based on hexafluoropropylene," *Journal of Thermal Analysis*, vol. 46, no. 3-4, pp. 785–794, 1996.

[15] H. Ismael, "Dielectric behavior of hexaferrites $BaCo_{2-x}Zn_xFe_{16}O_{27}$," *Journal of Magnetism and Magnetic Materials*, vol. 150, no. 3, pp. 403–408, 1995.

[16] D. Autissier, A. Podembski, and C. Jacquiod, "Microwaves properties of M and Z type hexaferrites," *Journal de Physique IV France*, vol. 7, no. C1, pp. C1-409–C1-412, 1997.

[17] Y. K. Hong, 1901 5th Avenue East, Unit 1322, Tuscaloosa, Ala, USA, 35401, US.

[18] X. Liu, "Research on La^{3+}-Co^{2+}-substituted strontium ferrite magnets for high intrinsic coercive force," *Journal of Magnetism and Magnetic Materials*, vol. 305, no. 2, pp. 524–528, 2006.

[19] H. Yamamoto and H. Seki, "Magnetic properties of Sr-La system M-type ferrite fine particles prepared by controlling the chemical coprecipitation method," *Japan IEEE Transactions on Magnetics*, vol. 35, pp. 3277–3279, 1999.

[20] A. Grusková and J. Lipka, "La-Zn substituted hexaferrites prepared by chemical method," in *Hyperfine Interact*, vol. 164, pp. 27–33, Springer Science+Business Media, Dordrecht, The Netherlands, 2005.

[21] N. K. Dung and N. T. L. Huyen, " Signficantly improving magnetic properties of Sr-La-Co hexagonal ferrite," *VNU Journal of Science, Mathematics*, vol. 25, pp. 199–205, 2009.

[22] F. M. M. Pereira and M. R. P. Santos, "Magnetic and dielectric properties of the M-type barium strontium hexaferrite ($Ba_xSr_{1-x}Fe_{12}O_{19}$) in the RF and microwave (MW) frequency range," *Journal of Materials Science*, vol. 20, no. 5, pp. 408–417, 2009.

[23] K. G. Rewatkar, "Synthesis and the magnetic characterization of iridium-cobalt substituted calcium hexaferrites," *Journal of Magnetism and Magnetic Materials*, vol. 316, no. 1, pp. 19–22, 2007.

[24] K. K. Mallick, "Magnetic and structural properties of M-type barium hexaferrite prepared by co-precipitation," *Journal of Magnetism and Magnetic Materials*, vol. 311, no. 2, pp. 683–692, 2007.

[25] N. Debnath, M. M. Rahman, F. Ahmed, and M. A. Hakim, "Study of the effect of rare-earth oxide addition on the magnetic and dielectric properties of Sr-hexaferrites," *International Journals of Engineering and Sciences*, vol. 12, no. 5, pp. 49–52, 2012.

[26] D. Seifert, "Synthesis and magnetic properties of La-substituted M-type Sr hexaferrites," *Journal of Magnetism and Magnetic Material*, vol. 321, no. 24, pp. 4045–4051, 2009.

[27] A. A. Sattar, "Temperature dependence of the electrical resistivity and thermoelectric power of rare earth substituted Cu-Cd ferrite," *Egyptian Journal of Solids*, vol. 26, no. 2, pp. 113–121, 2003.

[28] C. L. Khobragade, "Structural, mechanical, electrical & magnetic properties of Mn-Zn substituted Ca-hexaferrite," *Journal Materials & Metallurgical Engg*, vol. 1, pp. 1–9, 2011.

[29] G. Litsardakis, I. Manolakis, and K. Efthimiadis, "Structural and magnetic properties of barium hexaferrites with Gd-Co substitution," *Journal of alloys compounds*, vol. 427, no. 1-2, pp. 194–198, 2007.

[30] F. G. Brockman, "Anomalous behavior of the dielectric constant of a ferromagnetic ferrite at the magnetic curie point," *Physical Review*, vol. 75, no. 9, pp. 1440–1448, 1949.

[31] F. K. Lotgering, "Magnetic anisotropy and saturation of $LaFe_{12}O_{19}$ and some related compounds," *Journal of Physics and Chemistry of Solids*, vol. 35, no. 12, pp. 1633–1639, 1974.

[32] O. Kubo, T. Ido, H. Yokoyama, and Y. Koike, "Particle size effects on magnetic properties of $BaFe_{12-2x}Ti_xCo_xO_{19}$ fine particles," *Journal of Applied Physics*, vol. 57, no. 8, Article ID 4280, 1985.

[33] K. G. Rewatkar, "Synthesis and the magnetic characterization of iridium-cobalt substituted calcium hexaferrites," *Journal of*

Magnetism and Magnetic Materials, vol. 316, no. 1, pp. 19–22, 2007.

[34] O. Kubo, T. Ido, H. Yokoyama, and Y. Koike, "Particle size effects on magnetic properties of $BaFe_{12-2x}Ti_xCo_xO_{19}$ fine particles," *Journal of Applied Physics*, vol. 57, no. 8, Article ID 4280, 1985.

[35] M. N. Giriya, "Structural analysis and magnetic properties of substituted Ca-Sr Hexaferrites," *International Journal of Scientific and Engineering Research*, vol. 3, pp. 30–36, 2012.

19

An Analytical Model Approach for the Dissolution Kinetics of Magnesite Ore Using Ascorbic Acid as Leaching Agent

Nadeem Raza, Zafar Iqbal Zafar, and Najam-ul-Haq

Institute of Chemical Sciences, Bahauddin Zakariya University, Multan 60800, Pakistan

Correspondence should be addressed to Nadeem Raza; nadeemr8@hotmail.com

Academic Editor: Chi Tat Kwok

Ascorbic acid was used as leaching agent to investigate the dissolution kinetics of natural magnesite ore. The effects of various reaction parameters such as acid concentration, liquid-solid ratio, particle size, stirring speed, and temperature were determined on dissolution kinetics of the magnesite ore. It was found that the dissolution rate increased with increase in acid concentration, liquid-solid ratio, stirring speed, and temperature and decrease in the particle size of the ore. The graphical and statistical methods were applied to analyze the kinetic data, and it was evaluated that the leaching process was controlled by the chemical reaction, that is, $1 - (1 - x)^{1/3} = 1.256 \times 10^5 e^{-57244/RT} t$. The activation energy of the leaching process was found to be 57.244 kJ mol^{-1} over the reaction temperature range from 313 to 343 K.

1. Introduction

Magnesium is the third most commonly used structural metal after iron and aluminum. The applications of magnesium involve aerospace, automobiles, flash photography, flares, pyrotechnics, Grignard reagent, refractory materials, food, fertilizers, medicinal products, paper, textile, alloy formation, to remove sulfur in the production of iron and steel, fireproof, and so forth as described by Jones et al. [1].

With increase in population, there is continuous increase in demand of magnesium and its compounds. To overcome the increasing demand of magnesium and its compounds there is a need to explore ores of magnesium (magnesite, dolomite, etc.). These ores may vary in composition from deposit to deposit resulting differences in acidulation processes. These rocks generally contain impurities such as calcium, iron, and silica which can cause adverse effects on the applications of magnesium and its compounds.

The leaching and dissolution studies of different ores with different leaching agents are available in the literature [2–11]. On industrial scales different leaching agents are used to leach ores of various compositions to get different metals and their compounds. Inorganic/organic acids or bases and their salts can be used for leaching of magnesite rocks to get magnesium

and its compounds. From the dissolution studies of magnesite rocks by inorganic acids such as HCl, H$_2$SO$_4$, it was found that the dissolution reaction was chemically controlled [12, 13]. The inorganic acids are regarded as good leaching agents when relatively fast reaction rates are required. However, the use of inorganic acids as leaching agents affords certain limitations, like less selectivity, scaling problems, high CO$_2$ pressure, corrosion, environmental problems, froth formation, and pH control of reaction medium [14]. On contrary to inorganic acids, organic acids may be more selective as compared to inorganic acids and can be used for dissolution of specific ores where relatively low-acid concentrations are favorable. The reaction mediums involving organic acids as leaching agents have various advantages like low risk of corrosion and froth accumulation and biodegradability of organic acids. Furthermore, their corrosion effect can be reduced by the addition of corrosion inhibitors such as benzoic acid and salicylic acid [15]. However, organic acids may have less ability to leach some ores at higher temperatures due to their low boiling points and decomposition problems. The leaching kinetics of low-grade phosphate rock involving dilute organic acids such as succinic acid and lactic acid was carried out [16, 17], and it was investigated that the leaching process was chemically controlled. The literature concerning the dissolution studies of magnesite in citric acid and in gluconic acid is also available [18, 19]. In these research studies, it was evaluated that the dissolution process was controlled by chemical reaction. Lacin et al. [20, 21] carried out data analysis using shrinking core models for fluid solid systems in the dissolution kinetics study of natural magnesite in acetic acid and lactic acid solutions and found that the dissolution rate was controlled by chemical reaction.

Large deposits of magnesite are present in Khuzdar area of Balochistan (Pakistan). These deposits have different compositions from ore to ore. Studies of leaching reaction kinetics at different conditions of various reaction parameters of these deposits have not been carried out. Therefore, in the present research work indigenous magnesite ore has been taken to investigate the dissolution kinetics at different conditions of various reaction parameters.

2. Methods and Materials

2.1. Sample Preparation and Analysis. The magnesite ore used in the present research work was obtained from Khuzdar area of Balochistan (Pakistan). Khuzdar area is widely endowed with magnesite ore deposits. Samples of natural magnesite ore were collected and crushed with ball mill and mortar grinder. ASTM standard sieves were used to obtain the desired particle size fractions. Örgül and Atalay [22] found that, when minerals are fractionated, the chemical composition of each fraction with definite particle size is usually changed. All the magnesite samples were dried in an electric oven at 100°C, cooled to room temperature, and stored in dry plastic bottles. EDX was used for the analysis of the magnesite rock fractions along with the other conventional analytical techniques [23]. Analytical results indicating the composition of magnesite ore have been shown in Tables 1 and 2. The EDX pattern indicating the elemental composition of the raw magnesite ore has been shown in Figure 1. Different chemicals used in dissolution studies of magnesite ore were of reagent grade.

The leaching agent used in this research work was ascorbic acid [(5R)-[(1S)-1,2-dihydroxyethyl]-3,4-dihydroxyfuran-2(5H)-one] commonly called vitamin C, abundantly found in citrus fruits. It is a naturally occurring organic compound with antioxidant properties. It is a white solid very soluble in water to give mildly acidic solutions. Ascorbic acid and its salts with sodium, potassium, and magnesium usually act as an antioxidant and are used in curing of different diseases like hypertension. Metal salts of ascorbic acid typically react with oxidants of the reactive oxygen species, such as the hydroxyl radical formed from hydrogen peroxide, and can terminate chain radical reactions.

2.2. Detection Measurement. Scanning electron microscope (Hitachi S-3000H) was used to observe the magnesium contents in the magnesite ore.

2.3. Experimental Procedure. Different size fractions (150–590 μm) were used in a 500 mL well-mixed spherical glass batch reactor, equipped with a mechanical stirrer, digital controller unit, timer, and thermostat. Various experiments were carried out with known amount of ascorbic acid having different concentrations at various L/S ratios. Each time a known amount of the ascorbic acid was added slowly to the reaction vessel containing 5 g of magnesite ore. The vessel contents were stirred at a certain speed along with different times and temperatures. At the end of each experiment, an ice bath was used to stop the reaction in reaction vessel. The contents of reaction vessel were filtered using suitable filter paper. The filtrate solution was analyzed volumetrically for magnesium contents to evaluate the degree of conversion.

3. Mechanism of Leaching

The leaching process of the magnesite ore with ascorbic acid can be represented as follows.

(a) Ionization of $C_6H_8O_6$

$$C_6H_8O_6 \longrightarrow 2H^+ + C_6H_6O_6{}^{-2} \tag{1}$$

(b) Diffusion of H^+ ions to the exposed surface of the magnesite particles.

(c) H^+ ions attack on the magnesite particles in the rock:

$$2H^+ + MgCO_3 \longrightarrow H_2CO_3 + Mg^{2+} \tag{2}$$

The H^+ ions taking part in these reactions may come from the ascorbic acid as well as from the carbonic acid formed in the medium.

(d) Reaction between Mg^{2+} and $C_6H_6O_6{}^{-2}$

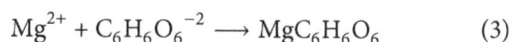

$$Mg^{2+} + C_6H_6O_6{}^{-2} \longrightarrow MgC_6H_6O_6 \tag{3}$$

The leaching process can be represented by using the following general equation:

$$H_2Y_{(aq)} + MCO_3 \longrightarrow CO_2 + MY + H_2O \tag{4}$$

Solubility product constant for $MgCO_3$ is 7.46 at 25°C, and the ionization constants for ascorbic acid are $pK_1 = 4.10$, $pK_2 = 11.6$ at 25°C. Dissociation constants for carbonic acid are $pK_1 = 6.35$, $pK_2 = 10.33$ at 25°C. The equilibrium direction for the above reaction (4) remains in forward direction and may be considered as an irreversible reaction because one of the products (CO_2) produced during the reaction conditions is evacuated from the reaction mixture.

4. Results and Discussion

4.1. Morphology of Magnesite. The SEM micrograph of the raw magnesite ore has been shown in Figure 2, which indicates the morphology of magnesite ore. The material seems to be nongranular with surface roughness. The surface roughness is due to the evolution of volatiles which in this case might be CO_2.

TABLE 1: Chemical analysis of natural magnesite ore.

Component	[Wt %]
MgO	45.4
CaO	1.18
Fe$_2$O$_3$	0.8
SiO$_2$	0.52
Loss on ignition [at 950°C]	52.1

TABLE 2: EDX analysis of natural magnesite ore.

Element	[Wt %]	[Atomic %]
C	14.90	20.79
O	56.907	59.50
Mg	27.24	19.02
Si	0.242	0.14
Ca	0.836	0.35
Fe	0.56	0.17

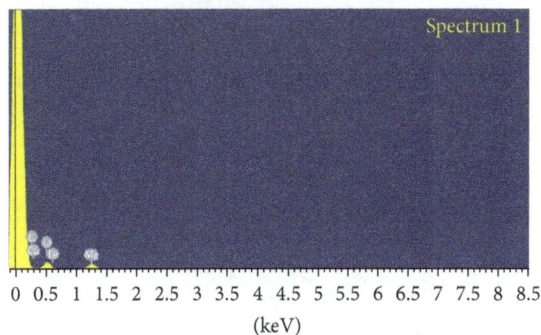

FIGURE 1: EDX pattern of natural magnesite ore.

FIGURE 2: SEM image of natural magnesite ore.

4.2. Effect of Reaction Temperature. The effect of temperature (40°C to 70°C) on rate of conversion of magnesite ore was investigated at different experimental conditions (178 micrometer particle size, 10% ascorbic acid with liquid/solid ratio of 10 : 1, and 350 rpm) as shown in Figure 3. It was observed that the rate of conversion of the magnesite ore increased with an increase in reaction temperature. It was also observed that increase in temperature reduced the reaction time required to attain the equilibrium in reaction medium. Furthermore, higher temperature (above 70°C) can cause contamination of CO$_2$ gas stream with ascorbic acid and water vapors. The experimental results indicated that the reaction temperature was the most effective parameter in the dissolution kinetics of magnesite ore. From these experimental observations it was also evaluated that below 40°C the ascorbic acid was not good leaching agent due to its lower solubility.

4.3. Effect of Acid Concentration and Liquid Solid Ratio. Different experiments were carried out to find the effect of concentration of ascorbic acid and liquid solid ratio on leaching kinetics of magnesite ore under various experimental reaction conditions as given in Figures 4 and 5. The experimental results indicated that an increase in acid concentration caused an increase in magnesium content. However, after certain optimum value of acid concentration, the increase in acid concentration did not have an appreciable effect. It might be considered that when the acid concentration exceeded its maximum required value, the hydrogen ions in the medium might decrease due to decrease in water contents. During the leaching study of colemanite ore with acetic acid, Özmetin et al. [24] found that higher acid concentration in reaction medium increased the rate of appearance of product by attaining the saturation value along with the formation of sparingly solid film layer resulting in a decrease in dissolution process.

From Figure 4, it was found that the acid concentration of 10% was good for leaching kinetics study of magnesite

ore with liquid solid ratio of 10 : 1. The pH of reaction medium depends on the ascorbic acid concentration and its degree of ionization at a particular temperature. The pH decreased as the concentration of ascorbic acid was increased. Figure 5 showed that the rate of dissolution of magnesite ore increased with an increase in the liquid solid ratio. From the experimental results, it was found that the liquid-solid ratio also had a significant effect on dissolution rate of magnesite ore. It may be attributed to the fact that a relatively higher liquid solid ratio may provide a medium of liquid phase to facilitate the mobility of reactive species produced in the reaction medium.

4.4. Effect of Particle Size. In order to investigate the effect of particle size on the leaching of magnesite ore different experiments were carried out. Four different size fractions of magnesite ore (150, 178, 297, and 590 μm) at 60°C were used to find the effect of particle size as shown in Figure 6. The leaching curves indicated that the rate of dissolution process increased as the particle size was decreased. This situation might be attributed to the fact that the surface area for reaction becomes more available with decreasing particle size resulting an increase in the efficiency of the leaching process. In separate experiments, it was observed that the effect of stirring speed on the leaching reaction rate was not appreciable as compared to the other parameters. This

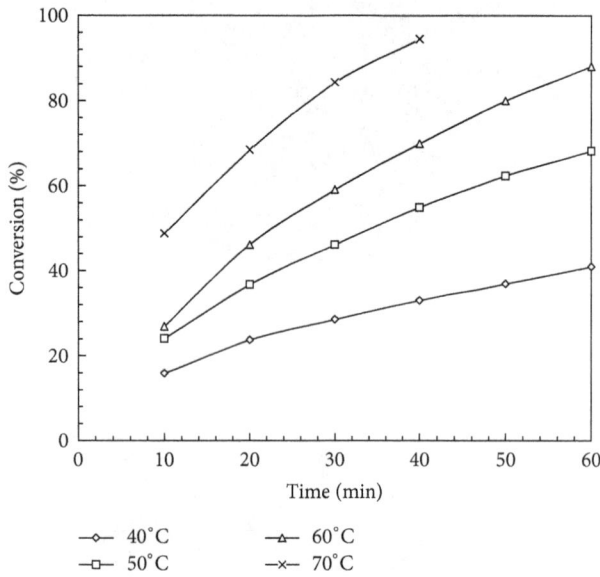

FIGURE 3: Effect of temperature on leaching of magnesite ore.

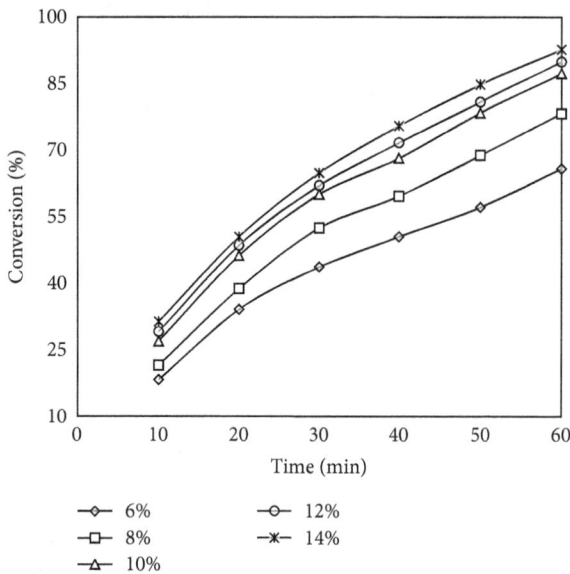

FIGURE 4: Effect of ascorbic acid concentration on leaching of magnesite ore.

FIGURE 5: Effect of liquid solid ratio on leaching of magnesite ore.

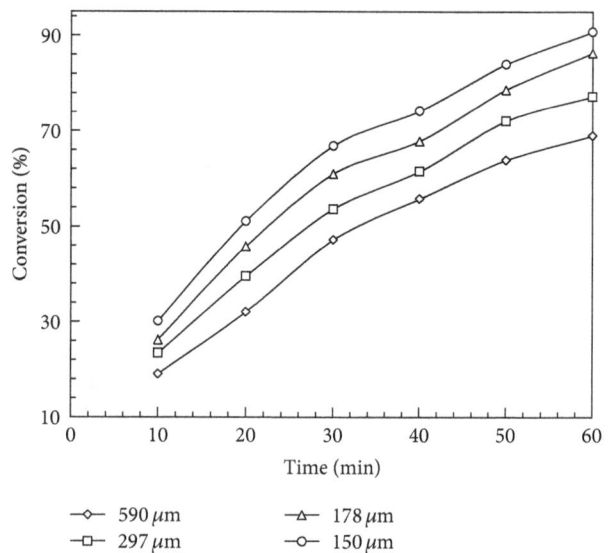

FIGURE 6: Effect of particle size on leaching of magnesite ore.

situation indicated that the leaching of magnesite ore was not product or ash layer controlled process.

5. Kinetic Analysis

Fluid solid heterogeneous reaction systems are usually involved in chemical and hydrometallurgical processes. In fluid solid reaction systems, reaction rate may be controlled by one of the following mechanisms: diffusion through the fluid films, diffusion through ash/product layer, or the chemical reaction at the surface of the core of unreacted materials [25]. The experimental data was analyzed on the basis of shrinking core model to find rate controlling step and

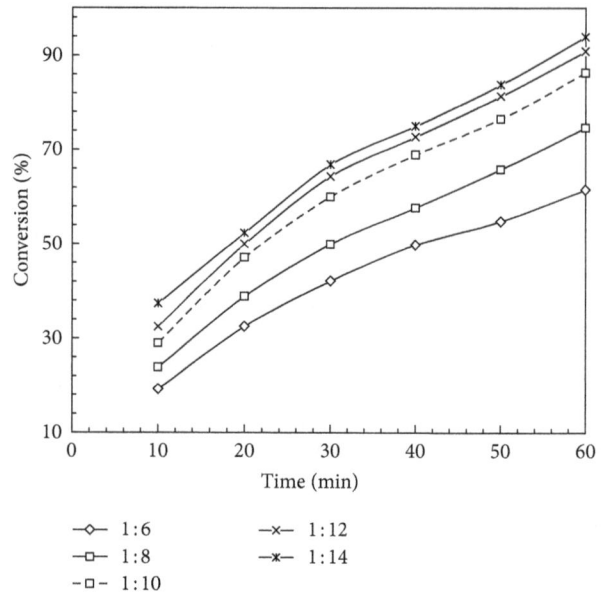

kinetic parameters. The reaction between a solid and fluid can be represented as

$$A_{(fluid)} + bB_{(Solid)} \longrightarrow Products. \tag{5}$$

If no ash/product layer over unreacted core is formed, then two controlling steps may be fluid film diffusion or chemical reaction. If the time of completion of the leaching process is k_o, the fractional conversion of magnesite is x and at any time t the integrated equations for fluid-solid heterogeneous reactions may be represented as follows.

For film diffusion control,

$$t = k^* \left[1 - (1 - x) \right]. \tag{6}$$

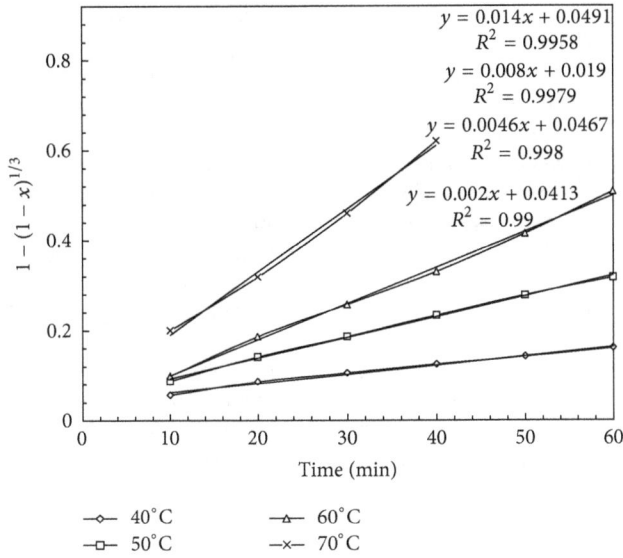

FIGURE 7: $1 - (1 - x)^{1/3}$ at different reaction temperatures.

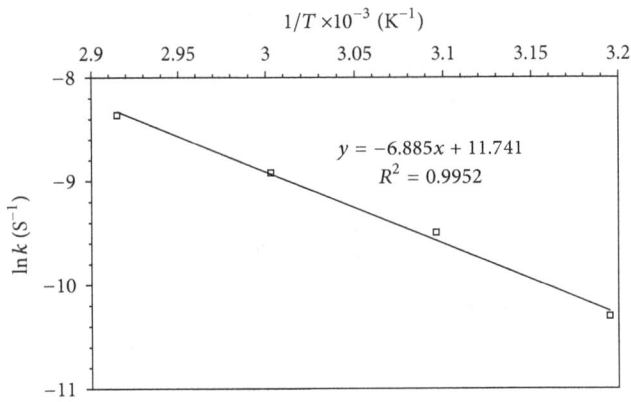

FIGURE 8: Arrhenius plot for leaching of magnesite ore.

For chemical reaction control,

$$t = k^* \left[1 - (1 - x)^{1/3} \right]. \tag{7}$$

The value of k^* may vary with reaction parameters according to the kinetic models. For example, according to the chemical reaction controlled model (7), k^* is

$$k^* = \frac{\rho_B R_o}{b K_s C_A}, \tag{8}$$

where k^* is the time for complete dissolution (min), ρ_B is the molar density of the solid reactant (mol m^{-3}), R_o is the radius of the solid particle (m), b is the stoichiometric coefficient of the solid, k_s is the surface reaction rate constant (m min^{-1}), and C_A is the leaching agent concentration (mol dm^{-3}). The validity of the experimental data into the integral rate was tested by statistical and graphical methods. The kinetic analysis results for the dissolution process were found to be consistent with a chemically controlled reaction and the

integral rate expression was determined to obey the following rate equation:

$$1 - (1 - x)^{1/3} = kt. \tag{9}$$

Using the conversion values for various reaction temperatures, liquid solid ratio, stirring speed, particle size fractions, and acid concentration applied in leaching kinetics of magnesite ore, the apparent rate constants k can be evaluated by plotting $1 - (1 - x)^{1/3}$ versus t as shown in Figure 7. Using the Arrhenius equation, the above equation may be expressed as

$$1 - (1 - x)^{1/3} = k_o e^{-E_a/RT} t. \tag{10}$$

Arrhenius plot for the leaching of magnesite ore in ascorbic acid solutions was obtained by plotting the values of slopes of the straight lines (apparent rate constant) versus $\ln(1/T)$ as shown in Figure 8 and the following values were calculated:

$$1 - (1 - x)^{1/3} = 1.256 \times 10^5 e^{-57244/RT} t. \tag{11}$$

The value of activation energy indicates that the leaching of magnesite with ascorbic acid solutions is controlled by chemical reaction, and this value agrees with the values obtained in the similar research work of fluid solid reaction system [26]. Abdel-Aal [27] described that the activation energy of a diffusion controlled process is characterized to be from 4.18 to 12.55 kJ mol^{-1}, and, for a chemically controlled process, value of activation energy is usually greater than 41.84 kJ mol^{-1}.

6. Conclusions

(i) The experimental results show that the ascorbic acid can be used as leaching agent to extract magnesium contents from the magnesite ore.

(ii) Analysis of the kinetic data by different kinetic models shows that the leaching of magnesite ore in ascorbic acid solutions follows a chemically controlled process with activation energy of 57.244 kJ mol^{-1}.

(iii) In the leaching of magnesite ore with ascorbic acid, the product, that is, magnesium ascorbate obtained is an important medical material and can be used in curing of different diseases like hypertension.

(iv) Nontoxic techniques in terms of environmental pollution and human safety are the major reasons in using environment friendly leaching agents like ascorbic acid for the dissolution studies of magnesite ores.

Explanation of Symbols

E_a: activation energy (J mol^{-1})
x: dissolved fraction of Mg^{2+}
L/S: liquid/solid ratio (cm^3 g^{-1})
t: reaction time (min)
T: reaction temperature (K)
k: reaction rate constant (min^{-1})
EDX: energy dispersive X-ray analysis.

Acknowledgments

The authors thank the Institute of Chemical Sciences, BZU, Multan, and Institute of Chemical Engineering and Technology, NFC, Multan, for providing the facilities.

References

[1] P. T. Jones, B. Blanpain, P. Wollants, R. Ding, and B. Hallemans, "Degradation mechanisms of magnesia-chromite refractories in vacuum-oxygen decarburization ladles during production of stainless steel," *Ironmaking and Steelmaking*, vol. 27, no. 3, pp. 228–237, 2000.

[2] H. T. Doğan and A. Yartaşi, "Kinetic investigation of reaction between ulexite ore and phosphoric acid," *Hydrometallurgy*, vol. 96, no. 4, pp. 294–299, 2009.

[3] N. Demirkýran and A. Künkül, "Dissolution kinetics of ulexite in perchloric acid solutions," *International Journal of Mineral Processing*, vol. 83, pp. 76–80, 2007.

[4] A. Ekmekyapar, N. Demirkiran, and A. Künkül, "Dissolution kinetics of ulexite in acetic acid solutions," *Chemical Engineering Research and Design*, vol. 86, no. 9, pp. 1011–1016, 2008.

[5] N. Demirkiran, "A study on dissolution of ulexite in ammonium acetate solutions," *Chemical Engineering Journal*, vol. 141, no. 1–3, pp. 180–186, 2008.

[6] N. Demirkiran, "Dissolution kinetics of ulexite in ammonium nitrate solutions," *Hydrometallurgy*, vol. 95, no. 3-4, pp. 198–202, 2009.

[7] N. Habbache, N. Alane, S. Djerad, and L. Tifouti, "Leaching of copper oxide with different acid solutions," *Chemical Engineering Journal*, vol. 152, no. 2-3, pp. 503–508, 2009.

[8] S. Kuşlu, F. Ç. Dişli, and S. Çolak, "Leaching kinetics of ulexite in borax pentahydrate solutions saturated with carbon dioxide," *Journal of Industrial and Engineering Chemistry*, vol. 16, no. 5, pp. 673–678, 2010.

[9] A. Mergen and M. H. Demirhan, "Dissolution kinetics of probertite in boric acid solution," *International Journal of Mineral Processing*, vol. 90, no. 1-4, pp. 16–20, 2009.

[10] S. A. Awe, C. Samuelsson, and Å. Sandström, "Dissolution kinetics of tetrahedrite mineral in alkaline sulphide media," *Hydrometallurgy*, vol. 103, no. 1-4, pp. 167–172, 2010.

[11] S. Zhang and M. J. Nicol, "Kinetics of the dissolution of ilmenite in sulfuric acid solutions under reducing conditions," *Hydrometallurgy*, vol. 103, no. 1-4, pp. 196–204, 2010.

[12] L. Chou, R. M. Garrels, and R. Wollast, "Comparative study of the kinetics and mechanisms of dissolution of carbonate minerals," *Chemical Geology*, vol. 78, no. 3-4, pp. 269–282, 1989.

[13] Y. Abali, M. Çopur, and M. Yavuz, "Determination of the optimum conditions for dissolution of magnesite with H_2SO_4 solutions," *Indian Journal of Chemical Technology*, vol. 13, no. 4, pp. 391–397, 2006.

[14] S. Hausmanns, G. Laufenberg, and B. Kunz, "Rejection of acetic acid and its improvement by combination with organic acids in dilute solutions using reverse osmosis," *Desalination*, vol. 104, no. 1-2, pp. 95–98, 1996.

[15] S. Bilgiç, "The inhibition effects of benzoic acid and salicylic acid on the corrosion of steel in sulfuric acid medium," *Materials Chemistry and Physics*, vol. 76, no. 1, pp. 52–58, 2002.

[16] M. Ashraf, Z. I. Zafar, and T. M. Ansari, "Selective leaching kinetics and upgrading of low-grade calcareous phosphate rock in succinic acid," *Hydrometallurgy*, vol. 80, no. 4, pp. 286–292, 2005.

[17] Z. I. Zafar and M. Ashraf, "Selective leaching kinetics of calcareous phosphate rock in lactic acid," *Chemical Engineering Journal*, vol. 131, no. 1–3, pp. 41–48, 2007.

[18] F. Demir, B. Dönmez, and S. Çolak, "Leaching kinetics of magnesite in citric acid solutions," *Journal of Chemical Engineering of Japan*, vol. 36, no. 6, pp. 683–688, 2003.

[19] B. Bayrak, O. Laçin, and H. Saraç, "Kinetic study on the leaching of calcined magnesite in gluconic acid solutions," *Journal of Industrial and Engineering Chemistry*, vol. 16, no. 3, pp. 479–484, 2010.

[20] O. Laçin, B. Dönmez, and F. Demir, "Dissolution kinetics of natural magnesite in acetic acid solutions," *International Journal of Mineral Processing*, vol. 75, no. 1-2, pp. 91–99, 2005.

[21] F. Bakan, O. Laçin, B. Bayrak, and H. Saraç, "Dissolution kinetics of natural magnesite in lactic acid solutions," *International Journal of Mineral Processing*, vol. 80, no. 1, pp. 27–34, 2006.

[22] S. Örgül and Ü. Atalay, "Reaction chemistry of gold leaching in thiourea solution for a Turkish gold ore," *Hydrometallurgy*, vol. 67, no. 1–3, pp. 71–77, 2002.

[23] N. H. Furmann, *Standard Methods of Chemical Analysis*, D. Van Nostrand, Princeton, NJ, USA, 6th edition, 1963.

[24] C. Özmetin, M. M. Kocakerim, S. Yapici, and A. Yartaşi, "A semiempirical kinetic model for dissolution of colemanite in aqueous CH_3COOH solutions," *Industrial and Engineering Chemistry Research*, vol. 35, no. 7, pp. 2355–2359, 1996.

[25] O. Levenspiel, *Chemical Reaction Engineering*, John Wiley & Sons, New York, NY, USA, 2nd edition, 1972.

[26] Z. I. Zafar, "Determination of semi empirical kinetic model for dissolution of bauxite ore with sulfuric acid: parametric cumulative effect on the Arrhenius parameters," *Chemical Engineering Journal*, vol. 141, no. 1–3, pp. 233–241, 2008.

[27] E. A. Abdel-Aal, "Kinetics of sulfuric acid leaching of low-grade zinc silicate ore," *Hydrometallurgy*, vol. 55, no. 3, pp. 247–254, 2000.

20

Accumulative Roll Bonding of Pure Copper and IF Steel

Saeed Tamimi,[1,2] **Mostafa Ketabchi,**[1] **Nader Parvin,**[1] **Mehdi Sanjari,**[3] **and Augusto Lopes**[4]

[1] *Mining and Metallurgical Engineering Department, Amirkabir University of Technology, Tehran 15875-4413, Iran*
[2] *Department of Mechanical Engineering, TEMA, University of Aveiro, 3810-193 Aveiro, Portugal*
[3] *Mining and Materials Engineering Department, McGill University, Montreal, QC, Canada H3A 0E8*
[4] *Departamento de Engenharia de Materiais e Ceramica, CICECO, University of Aveiro, 3810-193 Aveiro, Portugal*

Correspondence should be addressed to Saeed Tamimi; saeed.tamimi@gmail.com

Academic Editor: Mohammad Reza Toroghinejad

Severe plastic deformation is a new method to produce ultrafine grain materials with enhanced mechanical properties. The main objective of this work is to investigate whether accumulative roll bonding (ARB) is an effective grain refinement technique for two engineering materials of pure copper and interstitial free (IF) steel strips. Additionally, the influence of severely plastic deformation imposed by ARB on the mechanical properties of these materials with different crystallographic structure is taken into account. For this purpose, a number of ARB processes were performed at elevated temperature on the materials with 50% of plastic deformation in each rolling pass. Hardness of the samples was measured using microhardness tests. It was found that both the ultimate grain size achieved, and the degree of bonding depend on the number of rolling passes and the total plastic deformation. The rolling process was stopped in the 4th cycle for copper and the 10th cycle for IF steel, until cracking of the edges became pronounced. The effects of process temperature and wire-brushing as significant parameters in ARB process on the mechanical behaviour of the samples were evaluated.

1. Introduction

Recently, much attention has been directed to ultragrain refining of metallic materials, where the grain size is reduced to less than one micrometre. According to Hall-Petch relationship, it is expected that ultrafine grain (UFG) structure would result in higher strength [1, 2]. Producing high strength materials, particularly without alloying, is very important in economical point of view. Severe plastic deformation (SPD) techniques have been known in the last decades as effective methods to produce UFG materials.

The efficiency of traditional SPD techniques has been carried out such as equal channel angular pressing (ECAP) and high-pressure torsion (HPT) for grain refinement of a number of metallic materials, for example, [3, 4]. It has been shown that the UFG microstructure can be achieved using these methods; however, the typical sizes of the samples deformed by ECAP and HPT are small [5]. Furthermore, these types of SPD processes require special and/or expensive equipment. In recent years, a number of alternative SPD technologies have been developed, including equal channel angular rolling, cyclic bending, and accumulative roll bonding (ARB) in which the mentioned limitations were partially omitted [5, 6]. These SPD processes have potential to be adopted by the industry to produce UFG materials in the form of large sheets, due to their possibility as continuous processes. Saito et al. developed the ARB process for the first time in which SPD strategies are applied in a simple rolling process [7, 8]. This SPD process has then been used to fabricate UFG microstructures in various engineering materials and the effect of UFG structure on their mechanical properties have been studied, for example, [9–12]. However, there is a lack of study on the comparison of efficacy of ARB on the microstructure and its induced mechanical response of two different crystallographic structure materials. This study focuses on the ARB process in IF steel and pure copper sheets. The process parameters such as plastic deformation, processing temperature, and sample preparation factor are

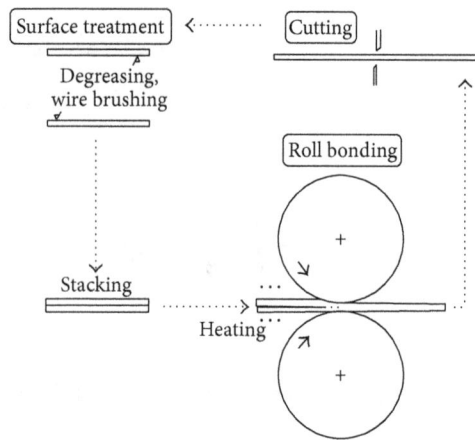

FIGURE 1: A schematic illustration of ARB process [9].

taken into consideration. The work deals with mechanical properties and microstructure observations after ARB process to investigate the influence of the process temperature and wire-brushing on the materials' properties.

2. Technical Work Preparation

2.1. Materials and Samples Preparation. The chemical analyses of pure copper and IF steel are given in Table 1 in which their initial mean grain sizes are 37 and 18 micrometre, respectively.

A schematic illustration of ARB technique is represented in Figure 1. The process is conducted under the conditions that the reduction in thickness per cycle is 50% (equivalent strain of 0.8) [8]. In order to remove the oxide layers and scale on the surface and also to increase the bonding capability, the surfaces of the strips were roughened using a metallic wire brush. After brushing, the surfaces were cleaned using acetone. The strips were then joined on the roughened surfaces by a couple of rivets. Rolling at elevated temperature is advantageous for join-ability and workability, though very high temperature would cause recrystallization and omit the accumulated strain effects. Therefore, the samples were placed in the furnace and preheated to a particular temperature for each case. Subsequently, the preheated samples were ARB-processed at a circumferential velocity of 0.47 m/s. No-lubrication condition was applied during the rolling passes. The principle of the ARB process has been reported previously, for example, [13].

The preheating condition in each pass was holding in the furnace at 550°C and 400°C for 300 s for IF steel and pure copper sheets, respectively. It can be considered a 50°C for thermal losing after taking the samples out of furnace as well as rolling process for both cases. Ten ARB cycles were carried out for IF steel sample. However, due to the earlier pronounced cracks in the edges of the sheet of copper after four passes, the process was stopped after the forth cycle.

2.2. Materials Characterization. Bonding between different metallic layers after fracture was studied by scanning electron

microscopy (SEM) using back scatter electron probe. The microstructure of samples was investigated by an optical microscopy and scanning electron microscopy (SEM) using secondary electron probe for samples. The microstructures were observed on the planes perpendicular to the transverse direction (TD). The etch solution was natal for IF steel, and pure copper samples were etched by 50 mL HCl, 5 gr $FeCl_3$, 100 mL H_2O after mechanical polishing. Furthermore, an electron backscattering diffraction (EBSD) was used to study the effect of ARB on the dislocation organization as well as the misorientation in various directions in the rolling plane. The mean value of hardness in each sample was measured using a Vickers microhardness with the load of 15 gr load for the copper and 50 gr for the IF steel samples. Additionally, the influence of process temperature was investigated on the mechanical properties of samples. To this end, the metallic sheets were processed up to four ARB cycles in various temperatures. The copper sheets were taking place in the furnace in the temperature range of 250–400°C for one hour, and the range of 400–550°C for one hour was also considered for preheating of IF steel before ARB process. Mean values of hardness of these specimens were measured. Finally, the effect of metallic wire-brushing on surface of the samples, as a preparation step, was observed. For this goal, two pieces of strip, a wire-brushed and a not-wire-brushed sheets, were joined with rivets and then followed by ARB process. The variation of mean value of hardness at each area throughout the sheet thickness was measured.

3. Results and Discussion

3.1. Interfaces and Microstructure

3.1.1. Interfaces. The ARB process was successfully performed up to ten cycles without any shape defects in IF specimens and 4 cycles for copper samples. SEM macrographs of ARB-processed IF steel after fracture for different numbers of ARB passes are shown in Figure 2. The number of interfaces for each ARB pass is

$$N = 2^n - 1, \tag{1}$$

where n is the number of passes and N indicates the number of interfaces across the sheet thickness. In case of the specimen of ten cycles, 1023 interfaces exist across the thickness (Figure 2(c)). Figure 2 shows that unbonded regions were decreased in a higher number of ARB passes. In other words, the subsequent rolling improves the quality of bonds between the surfaces introduced in the previous cycles. A similar observation has also been reported by Saito et al. [8].

3.1.2. Microstructure Evolution. Figures 3 and 4 illustrate the optical microstructures observed on the TD plane of the specimens. Microstructural evolution of pure copper through ARB in Figure 3 shows that by increasing the number of cycles, grain shapes become more elongated and finer. Mean grain thickness grew a little in cycle four in comparison to previous cycles. The interface can be clearly seen in

TABLE 1: Chemical analysis of IF steel and pure copper.

	C	Mn	Mg	Cr	Ti	Ni	V	Mo	Ta	Co	N	Nb	P	Pb	Zn	Fe	Cu	Al
IF steel	0.004	0.06	—	0.015	0.040	0.017	0.001	0.002	0.002	0.002	Trace	Trace	—	—	—	Remained	—	—
Pure copper	—	0.003	—	0.002	—	0.03	—	—	—	0.02	—	—	0.001	0.002	0.05	0.07	99.74	0.02

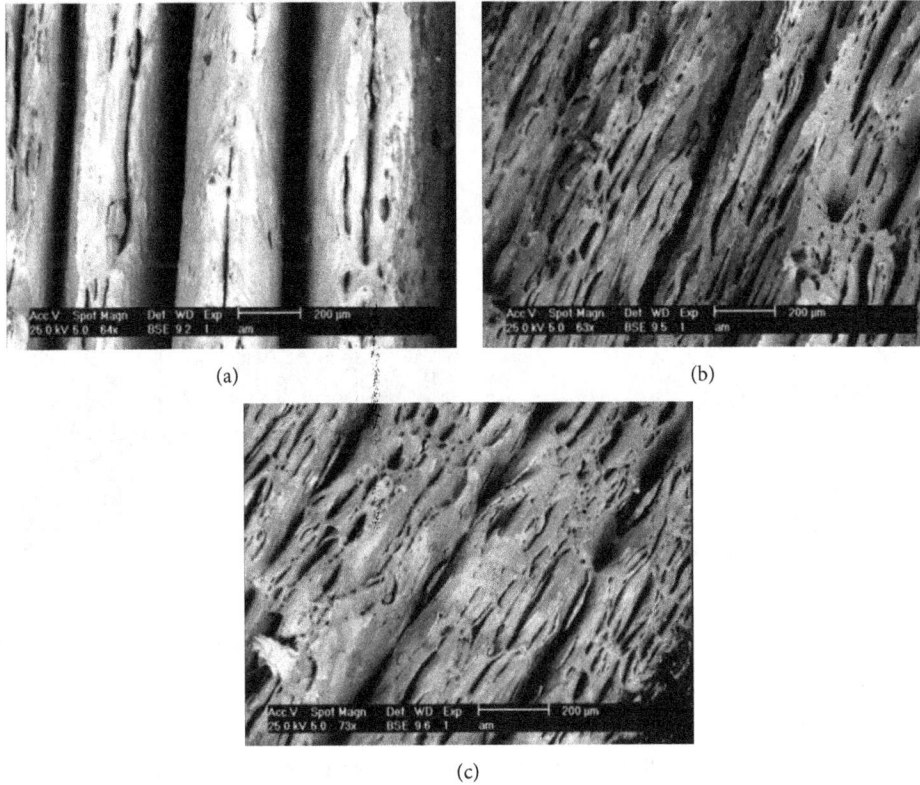

(a)

(b)

(c)

FIGURE 2: Interfaces of IF samples after (a) 4, (b) 8, and (c) 10 cycles ARB, preheating of 550°C.

Figure 3(d). Additionally, Figure 4 shows the microstructure of annealed (as received sample) IF steel as well as the samples produced by one and three ARB cycles. The microstructure of the specimen of one ARB-processed cycle showed relatively large grains and elongated to the rolling direction with clear grain boundaries. As the number of ARB cycles increases, the microstructures become finer and more complicated. After 3 cycles it is difficult to detect any grains and their boundaries by optical microscopy.

Figure 5 presents the SEM micrographs of the copper sheets after ARB process. The grains become finer and more complicated by number of cycles. The microstructure grew very slightly at the last pass of ARB.

Additionally, SEM micrographs of the ARB-processed IF steel after different amounts of strain are shown in Figure 6. New ultrafine grains can be seen in the specimens. The fraction of these ultrafine grains increased with the number of ARB cycles. In Figure 6(c) the specimen after six cycles was covered with small grains with less than 500 nm thickness, surrounded by clear boundaries. With increasing the number of ARB cycles, the mean grain size reduces and reaches to

a minimum value at 8th pass. The grain size of IF steel increases slightly in pass 9 and 10 of ARB.

Figure 7 presents an EBSD analysis of the IF steel sample deformed by the 7th pass of ARB. The ultrafine grains can be seen in this figure. The grain size distribution of this sample indicates that seven passes of ARB process could successfully decrease the grain size down to one micrometre.

3.1.3. Discussion. Microstructure evolution during ARB has been investigated for different engineering materials in which the formation of UFG with the number of ARB cycles is similar (e.g., [14–16]). Generally, with increasing the extent of plastic deformation, the generated dislocations move to the direction of the applied load with subsequent interlocking to the neighbouring dislocations. Dislocation tangles are then formed which contain regions with high dislocation density. The mechanism of grain refining in ARB process has been suggested by Tsuji et al. [17] using geometrical necessary (g-n) dislocations. It has been suggested that the continuous changes in misorientation are converted into the planer boundaries by rearrangement of the g-n dislocations.

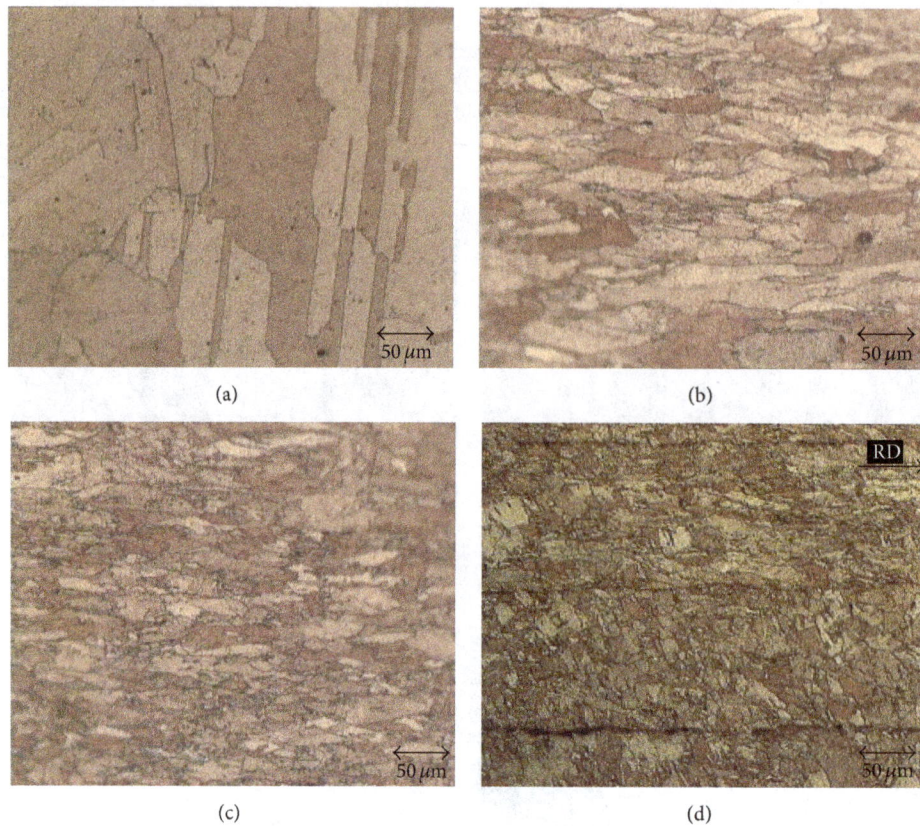

FIGURE 3: Optical images of TD plane in pure copper samples, (a) annealed, (b) two, (c) three, and (d) four ARB cycles, preheated at 400°C.

There exist two types of boundaries in a sample processed by ARB: the extended lamellar boundaries parallel to the rolling plane and the short transverse boundaries interconnecting the lamellar boundaries. Figure 8 shows the misorientation profiles along two directions of TD and RD of the 7th ARB-processed IF steel which are indicated in Figure 7, with line A-A and line B-B, respectively. The misorientation profiles indicate that boundaries parallel to the RD possess higher misorientation (high angle boundaries) in comparison to those boundaries which are perpendicular to the RD (low angle boundaries). This is in agreement with the work of Huang et al. [18]. A comparison between the EBSD result and the SEM results of the samples with roughly equal strain (Figures 6(c) and 7) shows that SEM analysis can only detect the lamellar boundaries (i.e., high angle boundaries).

The results above indicate that the rate of grain refinement decreases at higher cycles. It should be noted that maximum grain refinement rate is related to the first cycle and then the rate decreases at higher amount of plastic deformation. This observation can be explained by saturation phenomenon where the formation of new dislocation in the deformed grains is difficult [19, 20]. The results show slightly grain growth in high cycles for both materials (see Figures 5 and 6). The larger amount of energy accumulated in the material as a consequence of the higher reduction in thickness per pass might constitute a driving force for grain growth. Preheating the sample before each ARB pass, as well as the

heat generated during the plastic deformation, may accelerate this phenomenon. The results indicate that the pure copper experiences the coarsening after the first pass of ARB whereas IF steel sample shows around the 8th pass of ARB process. This shows that the pure copper is more sensitive to amount of accumulated strain compared to the IF steel sample for microstructure growing. That might be attributed to easy motion of dislocations resulted from its relatively higher stacking fault energy of pure copper. In case of pure copper, therefore, boundaries are gradually diminished and grains become larger after the first ARB pass and in case of IF steel coarsening occurs after the 8th ARB pass.

3.2. Mechanical Properties. Figure 9 illustrates the changes in Vickers microhardness with ARB cycles. In IF samples the hardness increased rapidly at initial stage. After 7 cycles, the rate of increasing hardness reduces. Maximum hardness was reported after 8 cycles being 247 HV, at which the trend is reversed. The copper samples present a similar trend. The hardness increased considerably during after first cycle and decreased after second rolling pass.

3.2.1. Temperature Effects. The temperature of the process and the accumulated strain determines the amount of recrystallization. The influence of process temperature on the material properties was observed systematically by measuring

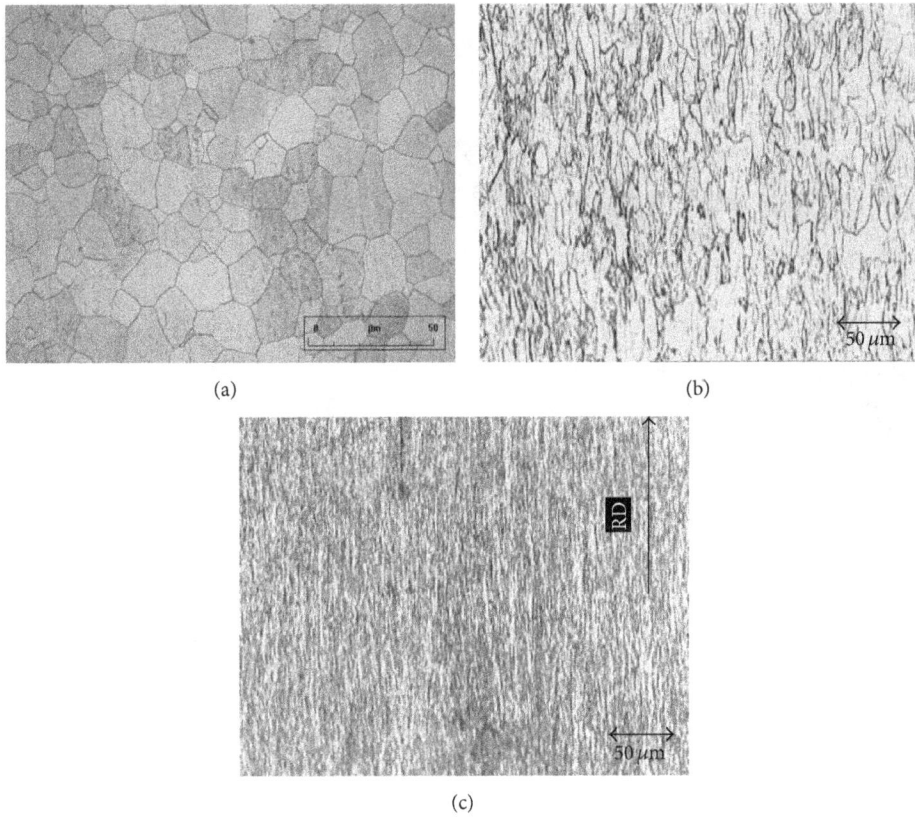

(a)

(b)

(c)

Figure 4: Optical images of TD plane in IF steel samples, (a) annealed, (b) after one cycle, and (c) after three ARB cycles, preheated at 550°C.

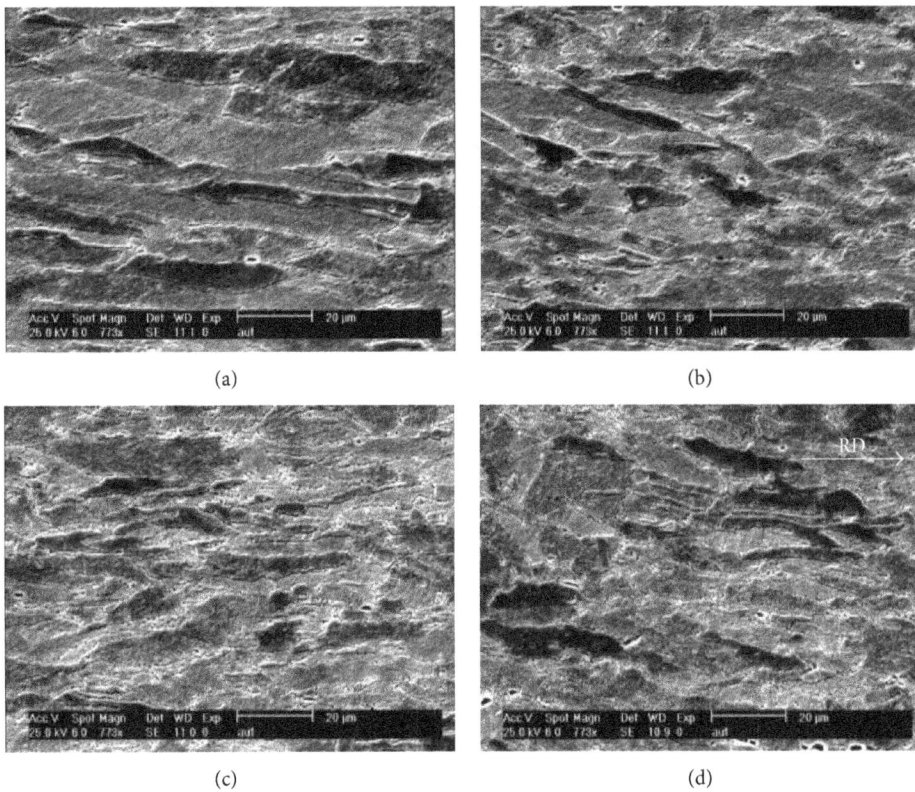

(a)

(b)

(c)

(d)

Figure 5: SEM micrographs in TD plane of pure copper, (a) one, (b) two, (c) three, and (d) four passes of ARB, preheated at 400°C.

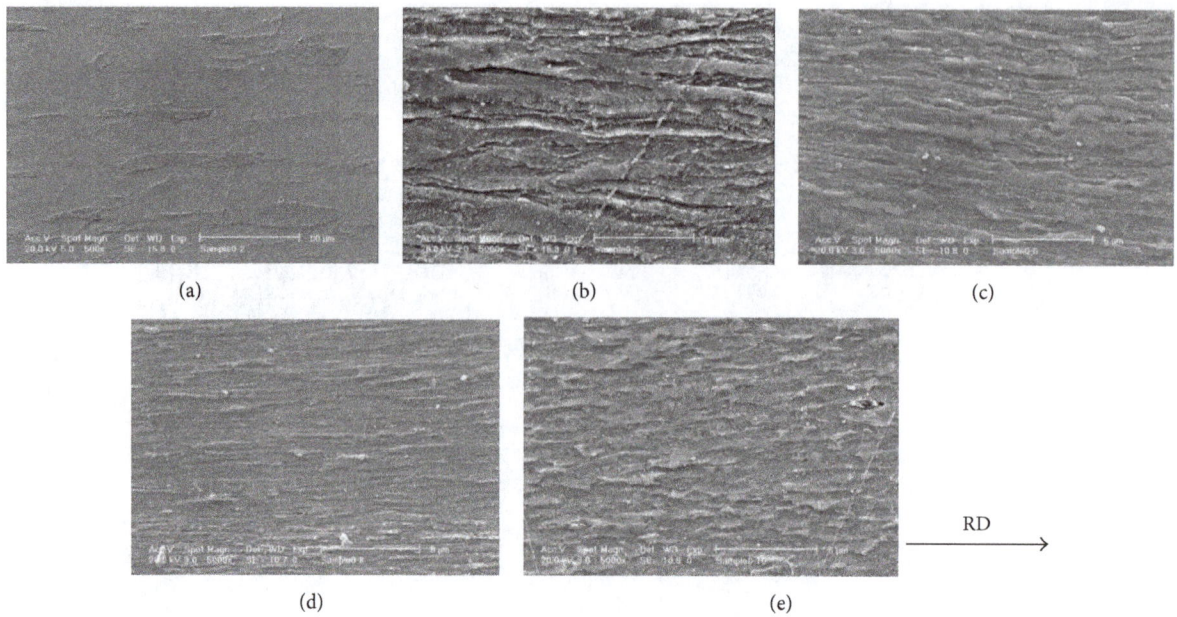

FIGURE 6: SEM micrographs in TD plane of IF steel, (a) one, (b) three, (c) six, (d) eight, and (e) ten ARB cycles; preheated at 550°C.

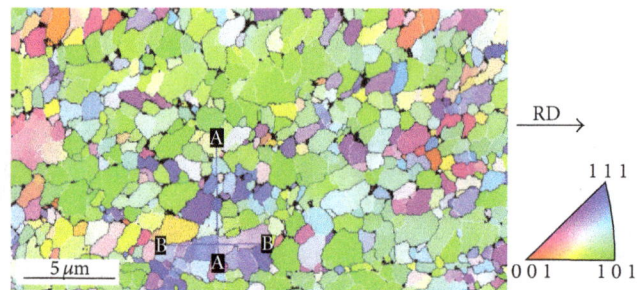

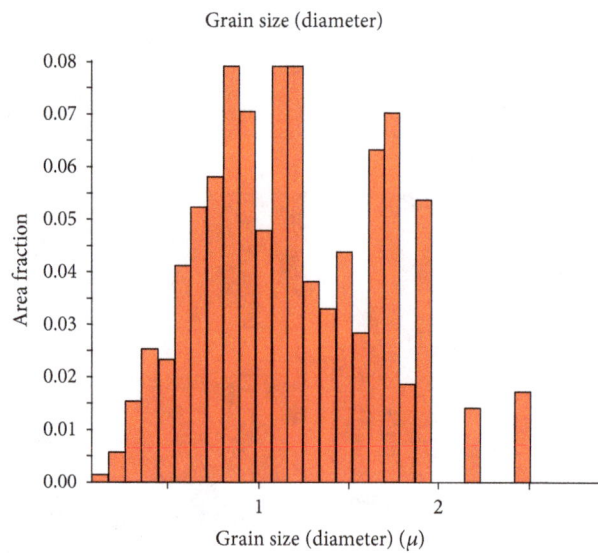

FIGURE 7: EBSD image and grain size distribution of IF steel from rolling plane after the 7th pass of ARB; preheated at 550°C.

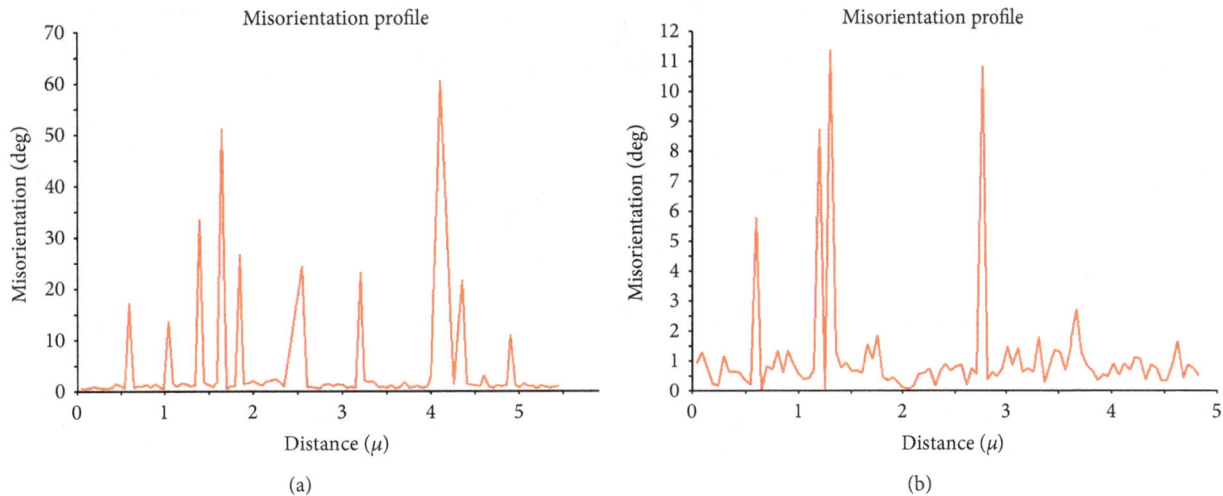

FIGURE 8: Misorientation profile along (a) transvers (line A-A) and (b) rolling (line B-B) directions in Figure 7.

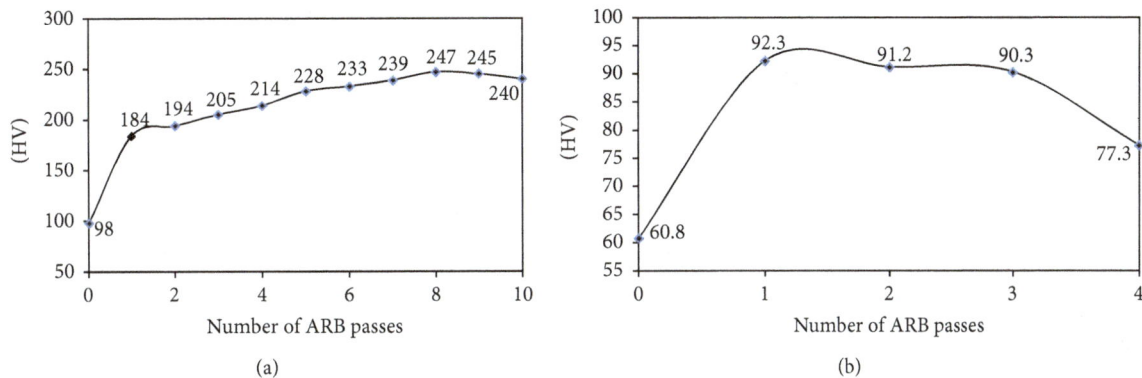

FIGURE 9: Variation of microhardness after various ARB cycles: (a) IF steel preheated at 550°C and (b) pure copper preheated at 400°C.

the hardness evolution for different preheating conditions. Figure 10 shows the effect of temperature during rolling on hardness in IF steel and pure copper samples. Both materials were processed up to four ARB cycles in various temperatures. According to the graphs, at higher ARB temperatures, hardness of the samples decreases to lower values.

With the aim of enhanced mechanical properties, the results of Figure 10 suggest lower temperatures for ARB process; however, the quality of the bonding between layers is modified at higher temperatures. Furthermore, since formability of the materials is affected by process temperature, the cracks during the ARB passes have been detected at lower temperatures.

3.2.2. Wire-Brushing Effects. Figure 11 shows the effect of wire-brushing on microhardness across the sheet thickness for both materials. In each case, a single pass of ARB was carried out on two sheets in which only one sheet was wire-brushed. It was found that the difference on hardness of brushed and not-brushed sheets was in about 6 HV for IF

steel and 3 HV for pure copper. This difference is attributed with work hardening caused by metallic wire-brushing on the surface regions of the material, although it is too small.

3.2.3. Discussion. Investigations have been conducted to find out whether the dislocation theories could describe the mechanical behaviour of ultrafine grain materials [21, 22]. In general, Hall-Petch relationship may explain that fine grains produced by ARB increase the material hardness. The ultrafine structure achieved by ARB in both materials (Figures 5 and 6) leads to increase in their hardness (Figures 9(a) and 9(b)). In addition, Xing et al. claimed that the plastic deformation produces a large number of dislocations inside the grains. This may make the dislocation movement difficult, leading to increased hardness [23]. By increasing the number of ARB cycles up to final stages, the hardness decreased a little in the observed materials. These phenomena could be due to different reasons. In larger accumulated strains at higher passes during the ARB, the saturation of

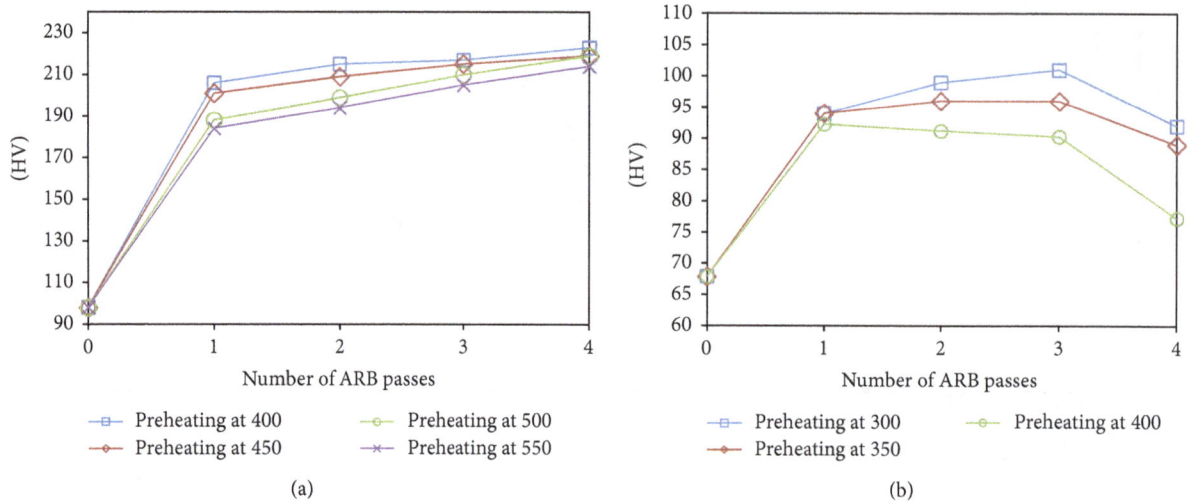

FIGURE 10: Variation of microhardness with ARB temperature up to four cycles (a) IF steel and (b) pure copper.

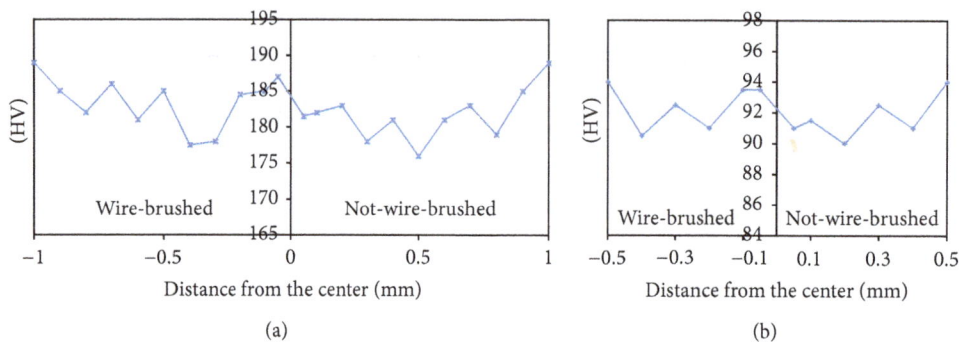

FIGURE 11: Wire-brushing effect on hardness in first ARB cycle in (a) IF steel (preheated at 550°C) and (b) pure copper (preheated at 400°C).

dislocation may diminish the value of hardness. High value of dislocation density in the grains makes it difficult to produce new dislocations. In this level of plastic deformation, misorientation of the boundaries reaches to the threshold angle of 36° for boundaries parallel to the rolling direction, reported in previous observations [19, 20, 24]. Moreover, recrystallization, both types of dynamics during the process and static in the furnace (through preheating), may partially occur during the process. In thermodynamic point of view, a higher total plastic strain produced by ARB process results in a premature recrystallization. The mechanisms of the mechanical behaviour seen in Figure 9 (i.e., a dramatic increase of hardness at the beginning and a slight decrease at higher ARB cycles) are similar for both observed materials. However, each material shows this behaviour at certain strains. These particular points are 6.4 and 0.8 as equivalent to plastic strain for IF steel and pure copper, respectively. After the establishment of accumulated strain in both materials, the variation of hardness becomes reverse; that is, the softening occurs. This observation is in agreement with the grain size variations of each material at the certain strain presented in Figures 5 and 6 (i.e., Hall-Petch theory).

It is well known that the temperature of the process has a great impact on the mechanical behaviour of ARB samples,

which can be due to dynamic recrystallization through the process as well as static one in the holding time when the sample was placed in the furnace. At higher temperatures the dislocations annihilation leads to reduced number of dislocations. The change in dislocation density varies by stacking fault energy of each material. Figure 10(b) indicates that the maximum hardness variations of copper ARB-processed in lower temperatures reach to maximums in the higher amount of accumulated strains; for instance, the maximum hardness for the case of preheating at 300°C was seen after the third ARB pass. This indicates that recrystallization at lower temperature required higher accumulated strain whereas only 0.8 of strain is sufficient for recrystallization at higher temperature (400°C).

4. Conclusion

Present work deals with the influence of severely plastic deformation imposed by ARB process on the properties of two prominent engineering materials: pure copper and IF steel. The results indicate that UFG structures were produced in both materials through ARB. Additionally, the results of hardness test of the samples increase with the number of ARB cycles. The high rate of increase of hardness at the beginning passes, followed by decreasing the rate and then saturation, all occur in different amounts of strain in the materials. The IF steel specimen after eight cycles achieved highest hardness of 247 HV, which is about 2.5 times the initial value and the pure copper samples after the first cycle achieved highest hardness of 92 HV, which is about 1.5 times the initial value. It can be concluded that mean value of hardness distribution along the thickness of the samples was directly proportional to the mean grain size.

Furthermore, the influence of process temperature on the hardness of the samples in different passes of the ARB was studied. The results show that ARB process at lower temperature results in higher hardness in copper and steel sheets; however, it may lead to appear as unbonded area which is undesirable. Additionally, low temperature ARB process causes the quick fracture of the samples at higher strains. Moreover, the presented results indicate that using metallic wire-brushing before each ARB pass causes work hardening on the surface of samples and raises the surface hardness for small values.

Conflict of Interests

The authors declare that there is no conflict of interests regarding the publication of this paper.

References

[1] E. O. Hall, "The deformation and ageing of mild steel: III discussion of results," *Proceedings of the Physical Society Section B*, vol. 64, no. 9, pp. 747–752, 1951.

[2] N. L. Petch, "The cleavage strength of polycrystals," *The Journal of the Iron and Steel Institute*, vol. 174, p. 25, 1953.

[3] V. M. Segal, V. I. Reznikov, A. E. Drobyshevskiy, and V. I. Kopylov, "Plastic working of metals by simple shear," *Russian Metallurgy*, vol. 1, pp. 99–105, 1981.

[4] Z. Horita, D. J. Smith, M. Furukawa, M. Nemoto, R. Z. Valiev, and T. G. Langdon, "An investigation of grain boundaries in submicrometer-grained Al-Mg solid solution alloys using high-resolution electron microscopy," *Journal of Materials Research*, vol. 11, no. 8, pp. 1880–1890, 1996.

[5] N. Tsuji, Y. Saito, S.-H. Lee, and Y. Minamino, "ARB (accumulative roll-bonding) and other new techniques to produce bulk ultrafine grained materials," *Advanced Engineering Materials*, vol. 5, no. 5, pp. 338–344, 2003.

[6] H. Utsunomiya, K. Hatsuda, T. Sakai, and Y. Saito, "Continuous grain refinement of aluminum strip by conshearing," *Materials Science and Engineering A*, vol. 372, no. 1-2, pp. 199–206, 2004.

[7] Y. Saito, N. Tsuji, H. Utsunomiya, T. Sakai, and R. G. Hong, "Ultra-fine grained bulk aluminum produced by accumulative roll-bonding (ARB) process," *Scripta Materialia*, vol. 39, no. 9, pp. 1221–1227, 1998.

[8] Y. Saito, H. Utsunomiya, N. Tsuji, and T. Sakai, "Novel ultra-high straining process for bulk materials development of the accumulative roll-bonding (ARB) process," *Acta Materialia*, vol. 47, no. 2, pp. 579–583, 1999.

[9] N. Tsuji, Y. Saito, H. Utsunomiya, and S. Tanigawa, "Ultra-fine grained bulk steel produced by accumulative roll-bonding

(ARB) process," *Scripta Materialia*, vol. 40, no. 7, pp. 795–800, 1999.

[10] M. T. Pérez-Prado, J. A. Del Valle, and O. A. Ruano, "Grain refinement of MG-Al-Zn alloys via accumulative roll bonding," *Scripta Materialia*, vol. 51, no. 11, pp. 1093–1097, 2004.

[11] H. W. Höppel, J. May, and M. Göken, "Enhanced strength and ductility in ultrafine-grained aluminium produced by accumulative roll bonding," *Advanced Engineering Materials*, vol. 6, no. 9, pp. 781–784, 2004.

[12] K. S. Suresh, S. Sinha, A. Chaudhary, and S. Suwas, "Development of microstructure and texture in Copper during warm accumulative roll bonding," *Materials Characterization*, vol. 70, pp. 74–82, 2012.

[13] S. H. Lee, Y. Saito, T. Sakai, and H. Utsunomiya, "Microstructures and mechanical properties of 6061 aluminum alloy processed by accumulative roll-bonding," *Materials Science and Engineering A*, vol. 325, no. 1-2, pp. 228–235, 2002.

[14] N. Tsuji, Y. Ito, Y. Saito, and Y. Minamino, "Strength and ductility of ultrafine grained aluminum and iron produced by ARB and annealing," *Scripta Materialia*, vol. 47, no. 12, pp. 893–899, 2002.

[15] K. Inoue, N. Tsuji, and Y. Saito, "Ultra Grain Refinement of 36%Ni Steel by Accumulative Roll-Bonding (ARB) Process," in *Proceedings of the International Symposium on Ultrafine Grained Steels (ISUGS '01)*, pp. 126–129, ISIJ, Fukuoka, Japan, 2001.

[16] N. Tsuji, Y. Saito, H. Utsunomiya, and T. Sakai, "Ultra-fine grained ferrous and aluminum alloys produced by accumulative roll-bonding," in *Ultrafine Grained Materials*, p. 207, TMS, 2000.

[17] N. Tsuji, Y. Saito, H. Utsunomiya, and T. Sakai, *The Proceeding of the Fourth International Conference on Recrystallization and Phenomena*, vol. 13, The Japan Institute of Metals, Sendai, Japan, 1999.

[18] X. Huang, N. Tsuji, N. Hansen, and Y. Minamino, "Microstructural evolution during accumulative roll-bonding of commercial purity aluminum," *Materials Science and Engineering A*, vol. 340, no. 1-2, pp. 265–271, 2003.

[19] Y. Iwahashi, Z. Horita, M. Nemoto, and T. G. Langdon, "The process of grain refinement in equal-channel angular pressing," *Acta Materialia*, vol. 46, no. 9, pp. 3317–3331, 1998.

[20] K.-T. Park, H.-J. Kwon, W.-J. Kim, and Y.-S. Kim, "Microstructural characteristics and thermal stability of ultrafine grained 6061 Al alloy fabricated by accumulative roll bonding process," *Materials Science and Engineering A*, vol. 316, no. 1-2, pp. 145–152, 2001.

[21] R. Z. Valiev, N. A. Krasilnikov, and N. K. Tsenev, "Plastic deformation of alloys with submicron-grained structure," *Materials Science and Engineering A*, vol. 137, pp. 35–40, 1991.

[22] M. Furukawa, Z. Horita, M. Nemoto, R. Z. Valiev, and T. G. Langdon, "Microhardness measurements and the hall-petch relationship in an Al-Mg alloy with submicrometer grain size," *Acta Materialia*, vol. 44, no. 11, pp. 4619–4629, 1996.

[23] Z. P. Xing, S. B. Kang, and H. W. Kim, "Structure and properties of AA3003 alloy produced by accumulative roll bonding process," *Journal of Materials Science*, vol. 37, no. 4, pp. 717–722, 2002.

[24] K. Nakashima, Z. Horita, M. Nemoto, and T. G. Langdon, "Development of a multi-pass facility for equal-channel angular pressing to high total strains," *Materials Science and Engineering A*, vol. 281, no. 1-2, pp. 82–87, 2000.

21

Modification of Magnesium Alloys by Ceramic Particles in Gravity Die Casting

Urs Haßlinger,[1] Christian Hartig,[1] Norbert Hort,[2] and Robert Günther[1]

[1] *Institute of Materials Physics and Technology, Hamburg University of Technology, Eißendorfer Straße 42, 21073 Hamburg, Germany*
[2] *Helmholtz-Zentrum Geesthacht, Magnesium Innovation Centre (MagIC), Max-Planck-Straße 1, 21502 Geesthacht, Germany*

Correspondence should be addressed to Urs Haßlinger; hasslinger@tuhh.de

Academic Editor: Manuel Vieira

A critical drawback for the application of magnesium wrought alloys is the limited formability of semifinished products that arises from a strong texture formation during thermomechanical treatment. The ability of second phase particles embedded into the metal matrix to alter this texture evolution is of great interest. Therefore, the fabrication of particle modified magnesium alloys (particle content 0.5–1 wt.-%) by gravity die casting has been studied. Five different types of micron sized ceramic powders (AlN, MgB_2, MgO, SiC, and ZrB_2) have been investigated to identify applicable particles for the modification. Agglomeration of the particles is revealed to be the central problem for the fabrication process. The main factors that influence the agglomerate size are the particle size and the intensity of melt stirring. Concerning handling, chemical stability in the Mg-Al-Zn alloy system, settling and wetting in the melt, and formation of the microstructure in most cases, the investigated powders show satisfying properties. However, SiC is chemically unstable in aluminum containing alloys. The high density of ZrB_2 causes large particles to settle subsequent to stirring resulting in an inhomogeneous distribution of the particles over the cast billet.

1. Introduction

Due to their low density and their high specific strength magnesium wrought alloys reveal great potential for light weight applications in the transportation industry. Nevertheless, magnesium wrought alloys still exhibit a few drawbacks which make their application difficult. One is the limited formability of semifinished products like sheets that arises from a strong texture formation during the thermomechanical treatment. To overcome this drawback a significant part of current research on magnesium wrought alloys aims at designing alloys that show an altered recrystallization behavior. For this purpose those recrystallization mechanisms in magnesium wrought alloys have to be identified and fully understood that can change the texture evolution. A possible recrystallization mechanism that is mainly known from other metals is the so called particle stimulated nucleation (PSN) [1]. Although PSN in magnesium wrought alloys is not fully understood, Laser [2] has shown that basically it works. Typically, the starting material for the thermomechanical treatment is produced by casting and particles are created by adding alloying elements to the melt that form precipitates. However, precipitates exhibit several disadvantages that complicate the systematic investigation of the influence of particles on texture formation. Precipitates usually change the solid solution content of an alloy and thereby its properties. Depending on the amount of deformation particles should at least have a size of a few microns to act as nucleation sites [1]. However, the size distribution of precipitates is bound to precipitation kinetics and they tend to fragment during deformation. This can be overcome if, instead of precipitate forming elements, chemical stable ceramic particles are added to the melt in a comparable content. Normally, magnesium alloys can be easily cast by gravity die casting. Only a tilting furnace with a standard agitator to homogenize the alloying elements and a protective gas system that prevents the melt from oxidation is required. Therefore, this technique is comparatively inexpensive and highly available which makes it favored also for the fabrication of alloys modified by ceramic particles. The content of ceramic particles in wrought alloys that try to benefit from the influence of the particles on the texture formation will in general be different than for other applications of ceramic particles like grain refinement or the fabrication of metal matrix composites. Therefore, findings to the latter applications can only be applied to a limited extent. This makes further investigations on the fabrication of magnesium wrought alloys modified by ceramic particles in gravity die casting necessary. One important aspect of these investigations is to deal with processing issues like the addition of the particles to the melt or the homogenization of the particles in the melt by stirring. Another aspect is to identify applicable ceramic particles that reveal sufficient chemical stability and wetting in the melt as well as a good connection to the matrix in the cast material.

1.1. Selection of Ceramic Particles. For the mechanical properties of the particle modified alloys a good connection, that is, a mechanical stable interface, between the particles and the magnesium matrix is important. The mechanical stability of the interface is expected to increase with decreasing interface energy between particle and matrix. Low interface energy between the particle and the matrix is also desirable for grain refinement by heterogeneous nucleation during solidification of the melt. Therefore, an obvious way to determine suitable ceramic particles for the modification of magnesium alloys is to use the knowledge developed to identify appropriate inoculation particles for heterogeneous nucleation. Turnbull and Vonnegut [3] have shown that the misfit of the lattice parameter in close packed planes between nucleus and substrate is a suitable parameter to identify such appropriate particles. They observed grain refinement for a misfit up to ~15 %. This approach has successfully been applied to find potent grain refiners for magnesium alloys [6]. On this basis 5 ceramic phases with a low misfit (see Table 1) have been selected for this study. The interfaces that form between the selected particles and the matrix will in general at most be partial coherent.

2. Experimental Procedure

2.1. Fabrication of the Alloys. For the investigation of the as cast microstructure of the particle modified alloys, in particular for the analysis of the particle distribution, billets of a diameter of about 10 cm and a weight of 4–6 kg (minimal height 30 cm) have been cast using a tilting furnace from Nabertherm (Lilienthal, Germany) with a steel melting crucible. The melt has been protected against oxidation by a protective atmosphere consisting of SF_6 and argon. First the matrix material was molten and heated up to the casting temperature (typically $T_L = 988$ +/− 15 K (715 +/− 15°C)). As matrix material four different compositions from the well-known Mg-Al-Zn alloy system have been chosen: pure magnesium (Mg), Mg with 1 wt.-% zinc (Z1), Mg with 3 wt.-% aluminum (A3), and Mg with 3 wt.-% aluminum and 1 wt.-% zinc (AZ31). The former three are model alloys which in particular allow for the investigation of the chemical stability of the ceramic particles in different alloy systems. Subsequently, up to two stirrers were inserted into the melt. The agitators (Eurostar power control-visc from IKA, Staufen, Germany) exhibit a maximum power output of about 100 W.

Three different types of impellers shown in Figure 1 have been used in this study (flat blade disc turbine, propeller, and dissolver) to investigate the influence of stirring on the particle distribution in the cast billet. The rotational speed of the stirrers has been adapted depending on the type of impeller in the range of about 150–500 rpm, which for all impellers was in the region of turbulent flow. After the start of stirring the ceramic particles have been added. Due to the protective layer on top of the melt it is not possible to add the particles directly. Therefore, two different methods of adding particles to the melt have been tested as follows.

(i) One is the addition of loose powder wrapped in aluminum foil in the shape of a ball. The balls have been produced manually under argon atmosphere and were stored in argon until the transfer into the melt.

(ii) The other way of adding particles to the melt is the application of a master alloy. The production of master alloy tablets via milling as used for grain refinement has already been described in a previous publication [6]. The particle content in the master alloy usually was 10 vol.-%. The size of the ceramic particles is reduced by the milling process. For example, the size of the AlN particles used in this study (cf. Table 1) is reduced to about $d_{p,50} \approx 5\,\mu m$.

The desired particle content in the cast material was in the range of 0.5–1.0 wt.-% (equals ~0.3 vol.-%). After addition of the preheated aluminum foil balls or master alloy tablets ($T \approx 373$ K (100°C)) the melt is stirred for typically about 10 min. In order to prevent leftovers from the melt from burning, the stirrers have to be removed before the melt is cast into the preheated steel die ($T \approx 473$ K (200°C)). Therefore, the time between switching off the stirrers and beginning to cast was at least 30 s. The billets have been air cooled to room temperature with a typical cooling rate of about 1 K/s [7].

2.2. Metallography. The specimens for optical and electron microscopy were sectioned from the cast billets, grinded, and polished. In order to investigate the grain structure via optical microscopy using polarized light (see Figure 6) some homogenized (20 h at $T = 673$ K (400°C)) specimens were etched by a solution of picric acid similar to that used by Kree et al. [8]. Grain sizes have been determined using the line intercept method on the basis of ASTM E 112-96. The grain sizes given here are the mean value of all completed line segments. To investigate the agglomerate distribution macroscopic images have been made by a stereo microscope (Leica M205 C) with a ring light, which makes the particles appear bright and the matrix appears dark (see Figure 4). Particle and agglomerate sizes have been determined from microscopic images using the particle detection function of the Scandium Software from Olympus. The software is able to identify the particle or agglomerate area by the grey scale contrast to the matrix.

3. Results and Discussion

3.1. Addition of Particles to the Melt. Adding the loose particles to the melt in form of aluminum foil balls proved

TABLE 1: Properties of the selected ceramic particles: misfit of the lattice parameter in closed packed planes calculated according to Turnbull and Vonnegut [3] using the crystallographic data of Villars and Calvert [4], density, and particle size according to manufacturer's specifications.

Material	Misfit [%]	Density [g/cm^3]	Manufacturer	Powder Denotation	Size [μm]
AlN	2.8	3.26	H.C. Starck[1]	Grade A	$d_{p,50} = 7$–11
MgB$_2$	3.9	2.57	H.C. Starck[1]	Grade A	$\overline{d}_p = 2$–5
MgO	7.3	3.58	H.C. Starck[1]	−325 mesh; 98%	$d_p < 44$
α-SiC(6H)	4.0	3.23	ESK[2]	Gruen Mikro F600-D	$d_{p,50} = 9.3$
				Gruen Mikro F1200-D	$d_{p,50} = 3.0$
ZrB$_2$	1.3	6.10	ESK[2]	−400 mesh	$d_p < 37$

[1] H.C. Starck GmbH, Goslar, Germany.
[2] ESK CERAMICS GMBH & CO. KG, Kempten, Germany.

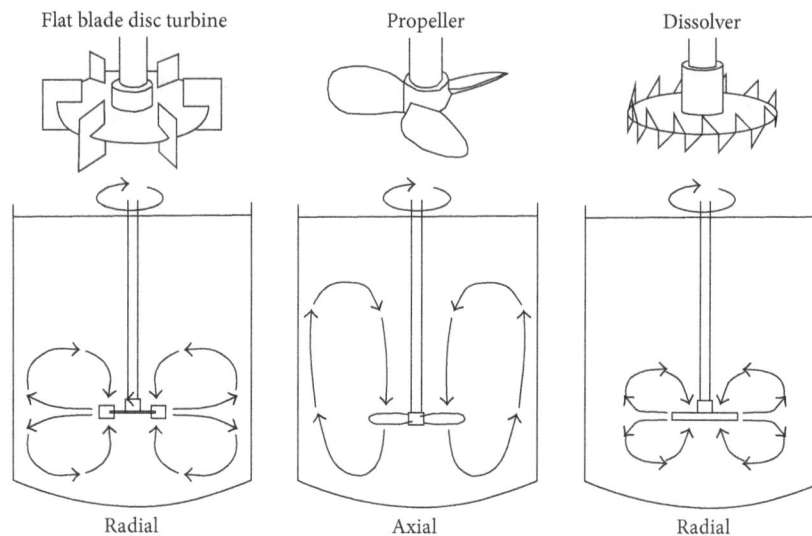

FIGURE 1: Shape, flow pattern, and primary direction of flow for the impeller systems used in this study according to [5].

to be difficult to handle. In order to introduce the particles to the melt the ball has to sink. Therefore, it has to have a higher density than the melt. The density of the ball depends on the amount of enclosed gas which depends on the ball's packing density and the porosity of the particles. While the former is difficult to adjust in the manual fabrication process of the balls the latter varies from one ceramic powder to another. For instance, balls of the MgO powder used in this study never sank due the high porosity of the magnesium oxide. Thus it is generally difficult to produce balls manually, which reproducibly sink into the melt. In addition, the balls are exposed to air when transferred into the furnace and thereby can introduce undesired oxygen to the melt.

Due to the low porosity and high density of the tablets a master alloy proved to be a better way of adding the particles to the melt. Nevertheless, in both cases, either addition of the powder in a ball or application of a master alloy, most of the particles form agglomerates when introduced to the melt, as Figure 2 shows. Therefore, the spacing between the particles within the master alloy is not large enough to avoid agglomeration. Longer milling can increase the distance between the particles in the master alloy by distributing the

particles more homogeneously. However, it also leads to a stronger size reduction of the particles. The spacing between the particles in the master alloy can also be enlarged by lowering the volume content of the particles. Yet, this rapidly increases the amount of master alloy in the cast alloy making it not insignificant anymore.

3.2. Homogenization of Particles in the Melt. The ceramic powders exhibit a higher density than the magnesium melt. Therefore, they tend to settle within the melt causing a macroscopic inhomogeneous distribution of the particles over the cast billet. This could be avoided for the ceramics with moderate density (AlN, MgB$_2$, MgO, and SiC, cf. Table 1). In contrast a macroscopic homogeneous distribution of the ZrB$_2$ particles could not be achieved, as Figure 3 shows. The inverted X-ray image of an AZ31 cast billet which contains 1 wt.-% ZrB$_2$ particles reveals a region of higher density at the bottom, which can be ascribed to the settling of the ZrB$_2$ particles.

In gravity die casting the particles can settle in the melt during stirring if the stirring is not sufficient. Qian et al. [9] described the settling of particles in an unstirred melt using

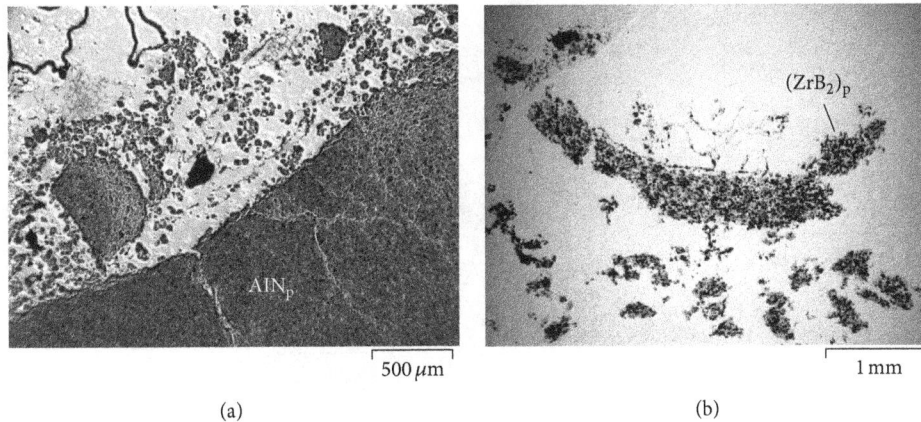

FIGURE 2: (a) Agglomerates detach from a ball of AlN particles; (b) agglomerates of ZrB_2 particles originating from a master alloy tablet below the image.

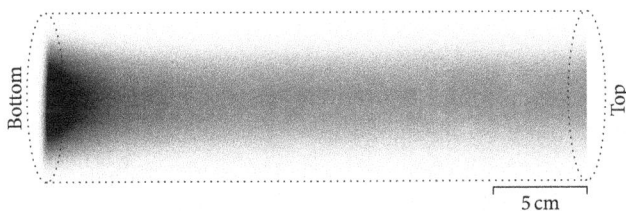

FIGURE 3: Inverted X-ray image of an AZ31 cast billet containing 1 wt.-% ZrB_2 particles revealing a region of higher density at the bottom.

a constant settling velocity that results from equilibrium of gravity, buoyancy, and Stokes drag force for a sphere in laminar flow. This description is also applicable to the particles used in this study. The equilibrium velocity for particles of a few microns is reached in much less than a second. The assumption of laminar flow is sufficiently fulfilled even for 50 μm sized ZrB_2 particles (Reynolds number Re~ 0.4). The Brownian movement of particles in the micrometer regime is negligible. It has to be taken into account for much smaller particles (d_p < 0.1 μm). Stirring is supposed to counteract the settling of the particles. An important parameter for the layout of the stirrer is the settling power of the particles which is the time derivative of the potential energy of the particles and can be calculated using the equilibrium settling velocity. According to [5] the agitator power is supposed to be a multiple of the settling power to counteract settling. This condition is easily fulfilled for the alloy billet shown in Figure 3 since the settling power is only about a few mW. However, the influence of agglomeration on the settling of the particles is difficult to predict exactly. The application of Stokes drag force for a sphere in laminar flow implies that the particles do not interact (infinite dilution). According to Richardson and Zaki [10] the interaction of the laminar flow fields of the particles can lead to a decrease in settling velocity. If the agglomerates act like single particles of larger size, the settling velocity increases. In practice an increase of the settling velocity of about 50% can be observed that can be ascribed to the clustering of particles [11]. But even

if the agglomerates in the alloy that is shown in Figure 3 are assumed to act as 100 μm sized single particles, the settling power is only about 25 mW. Therefore, the stirring system used in this study is even sufficient to counteract the settling of agglomerated particles and the settling visible in Figure 3 is caused by particles which do not settle during but subsequent to stirring.

In order to cast the melt into the die the stirring has to be stopped. As soon as the flow caused by the stirrer ceases, the particles start to settle. Therefore, the time between stopping the stirrer and casting of the melt into the die (typically about 30 s) is critical for the settling of the particles. Within that time, according to the equilibrium settling velocity, 40 μm sized ZrB_2 particles have already settled about 10 cm. Due to their lower density the other investigated ceramics (AlN, MgB_2, MgO, and SiC) settle more than two times slower. In order to achieve a homogeneous distribution of coarse particles with a high density like the ZrB_2 powder used in this study, faster casting techniques, for example, using a melt pump, have to be used.

Figure 4 exemplary shows the agglomerates in three different alloys that have been produced using master alloy tablets. Using a flat blade disc turbine as impeller and fine powder (SiC F1200-D, see Table 1) leads to a relatively large mean agglomerate size of $\overline{d}_{a,ECD}$ = 56 μm (Figure 4(a)). Smaller agglomerate sizes ($\overline{d}_{a,ECD}$ = 31 μm, Figures 4(b) and 4(c)) can be achieved using a propeller together with a dissolver as impellers and coarser powder (AlN, see Table 1). Some of the agglomerates that detach from the master alloy tablet during dissolution (see Figure 2) are much bigger than those in the cast alloy. Therefore, the size of the agglomerates is reduced by the stirring process. In the melt the size of the agglomerates is determined by two factors: the strength of the agglomerates and the forces the stirred melt exerts on the agglomerates. As described by Pietsch [12] the strength of the agglomerates can be associated with the coordination points between neighboring particles, where adhesive forces occur. Assuming that the forces at the coordination points between the different particles are of similar magnitude in the examined systems (Mg, Z1, A3, and AZ31) the agglomerate's

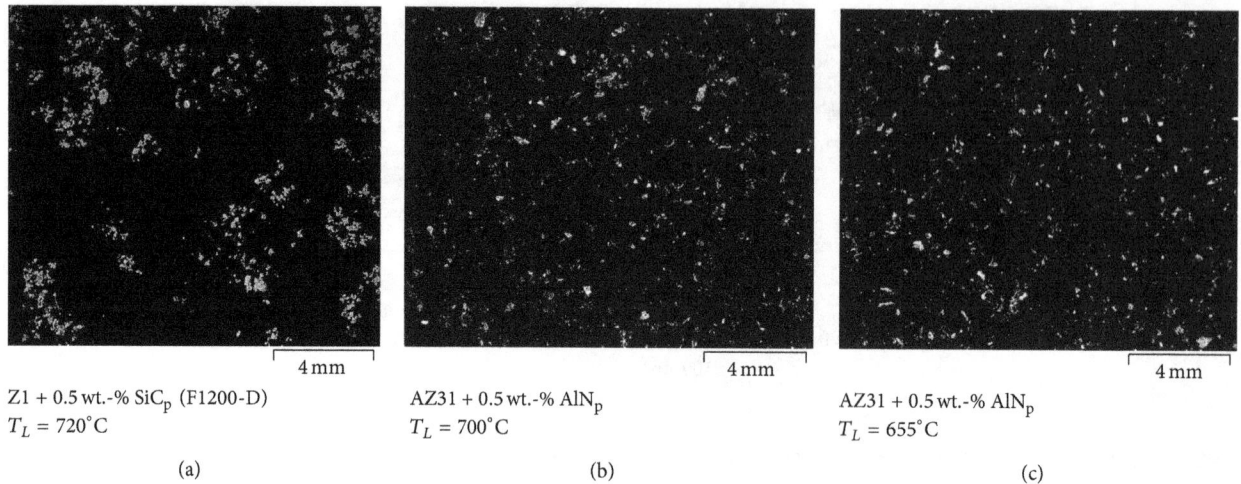

FIGURE 4: Macroscopic images (matrix dark, particles bright) of three-particle-modified-alloy cast at different temperatures T_L using different stirrers ((a) flat blade disc turbine and ((b) and (c)) propeller and dissolver stirrer).

FIGURE 5: Images of ceramic particles and precipitates in different alloys.

strength mainly depends on the number of coordination points per volume. The smaller the particles and the wider the size distribution of the particles the higher the number of coordination points per volume and thereby the strength of the agglomerate. The small particles play an important role for the strength of the agglomerate since they fill up the space between larger particles and create a large amount of coordination points [12]. Small particles could be removed from the powder by sieving making the agglomerates mechanically less stable. However, during fabrication of the master alloy new small particles are created by milling. The higher the strength of the agglomerates is the more intense the stirring has to be in order to reduce the size of the agglomerates. The stirring causes a turbulent flow with microvortices. Since the agglomerates are not sheared inside the microvortices they cannot be smaller than the size of the vortices λ. The size of the vortices is decreased with increasing stirring power and with decreasing melt viscosity η_{Mg} ($\lambda \sim \eta_{Mg}^{3/4}$ [5]). The change of the viscosity with temperature according to Qian et al. [9] is too small to cause a significant change in the size of the agglomerates. Therefore, the reduction of the melt

temperature from 973 K to 928 K (700°C to 655°C) does not influence the agglomerate size (cf. Figures 4(b) and 4(c)).

In addition to the particle size the smaller agglomerate size in Figures 4(b) and 4(c) compared with Figure 4(a) can be ascribed to the use of two independent agitators, one that primarily counteracts settling and one that reduces the size of the agglomerates. The flat blade disc turbine or the propeller creates a large amount of flow that counteracts the settling of the particles. Due to the low viscosity of the magnesium melt they also cause the melt to rotate. The rotation of the melt can create a funnel at the melt surface that perils the integrity of the protection layer. In addition, the melt's rotation pushes the particles towards the wall of the crucible and reduces the relative velocity between impeller and melt. The latter reduces the power that is introduced into the melt by the agitator. Therefore, the ability of the flat blade disc turbine and the propeller to reduce the agglomerate size turns out to be limited. This can be overcome by using an additional independent stirrer that causes only little flow but creates a large local shear gradient in the melt that breaks the agglomerates apart like the dissolver used in this study. The

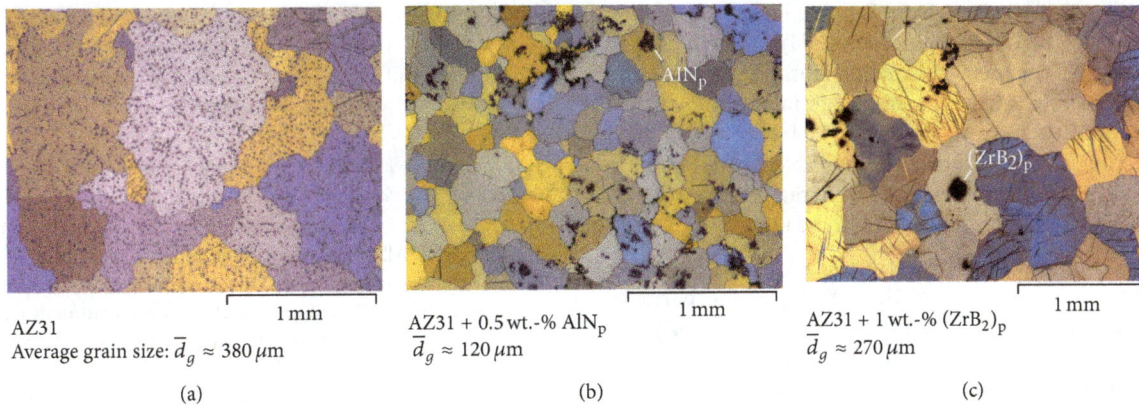

FIGURE 6: Images (polarized light) of the etched microstructures of different homogenized alloys.

rotational speed of this additional stirrer will in general be much higher than that of the stirrer that counteracts settling. In order to further reduce the agglomerate size more complex setups have to be used. For example, baffles can be installed into the crucible that disturb the rotation of the melt or more elaborate stirring devices like being used for intensive melt shearing by Fan et al. [13] can be applied.

3.3. Chemical Stability of the Particles in the Melt.

The only ceramic which proved to be chemical unstable in this study was SiC in the alloy systems containing aluminum (A3 and AZ31). As Figure 5(a) shows, only a small number of SiC particles can be found in the microstructure of these alloys. Instead, a large amount of Mg_2Si precipitates occur. Most of the SiC reacts forming Mg_2Si and aluminum containing carbides. Due to the metallographic preparation the water soluble carbides are difficult to detect in the microstructure. In contrast, SiC is stable in aluminum free Z1 (see Figure 5(b)). Thermodynamical calculations by Schmid-Fetzer et al. [14] show that SiC will begin to stabilize in AZ31 for an initial SiC content of more than about 4 wt.-% due to the aluminum depletion of the melt. The chemical instability of SiC in the MgAl alloy system is not disadvantageous for the application as grain refiner for cast material, since the aluminum containing carbides that form from the dissolution of SiC can also be suitable for heterogeneous nucleation during solidification [15]. However, our experiments show that the kinetics of the SiC dissolution is too fast to use as a stable particle for the investigation of the particle stimulated nucleation during thermomechanical treatment in the aluminum containing alloys of this study.

Figure 5(c) shows that Al-Mn-phases (likely Al_8Mn_5) form on the AlN particles in AZ31. Clear evidence for a decomposition of AlN resulting in Al-Mn phases and magnesium nitride was not found. Due to the highly positive reaction enthalpy such a reaction is not likely.

3.4. Formation of the As Cast Microstructure.

Most of the particles form agglomerates that are randomly distributed over the microstructure. The agglomerates exhibit a compact outer shape with a low aspect ratio. The particles inside the agglomerates are loosely packed and the space in between

the particles is completely filled by the magnesium matrix. A small number of particles are found as single particles in between the agglomerates. In general, the particles exhibit a good connection to the matrix. In the casting experiments an increased deposition of the selected ceramics in pores or at the melt surface due to poor wetting was not observed. In the simple model alloys (Mg, Z1, and A3) often cracks occur along the agglomerates (see Figure 5(b)) which can be associated with the thermal shrinkage of the billet in solid state during cooling. The thermal contraction of the ceramic particles is negligible compared to that of the matrix. Therefore, the missing contraction of the particles has to be compensated by the matrix. This leads to complex stress states within the agglomerates. In contrast to AZ31 the model alloys do not exhibit the required toughness to compensate those stress states by plastic deformation without the initiation of cracks.

The particle modified alloys often reveal a refined grain structure (see Figures 6(b) and 6(c)) in comparison with an unmodified reference alloy (see Figure 6(a)). This can qualitatively be understood by heterogeneous nucleation during solidification of the melt caused by the ceramic inoculants. The effect of a grain refiner is depending not only on the interfacial energies between the inoculant and the melt or nucleus but also on the size distribution of nucleation sites in the melt, as Gosslar et al. [16] have shown, for example. Therefore, despite agglomeration the AlN powder used in this study shows a grain refinement that is comparable to known potent magnesium grain refiners like SiC and more pronounced than that reported before by Lee [17] for AlN in A1. The inhomogeneous distribution of the ZrB_2 particles caused by the settling can lead to a local variation of the grain size, as already has been shown [18].

4. Conclusions

The investigation of the modification of magnesium alloys by ceramic particles in gravity die casting revealed the following results.

(i) All ceramic powders used in this study tend to agglomerate. Most of the particles already pass in form of agglomerates from the powder or master alloy into the melt. The agglomerate size is reduced by stirring of the melt. The main factors determining the agglomerate size in the cast material are the particle size and the stirring intensity. However, a standard stirring system is normally not able to dissolve the agglomerates completely. For that purpose more elaborate casting techniques have to be used.

(ii) The standard stirring system used in this study is sufficient to counteract the settling of the particles in the melt. Settling of the particles only takes place subsequent to stirring. While this is not critical for the ceramics of this study with moderate density (AlN, MgB_2, MgO, and SiC) the settling of the large ZrB_2 particles leads to an inhomogeneous distribution of the particles over the cast billet.

(iii) All ceramics used in this study except SiC reveal sufficient chemical stability for the content range of about 0.5–1 wt.-% in the investigated alloy systems (Mg, Z1, A3, and AZ31). SiC proved to be unstable in the aluminum containing alloys (A3 and AZ31).

In general the investigated particles exhibit adequate wetting in the melt and a good connection to the matrix in the solid state.

The basic casting techniques used in this study are sufficient to produce magnesium wrought alloys modified by ceramic particles that at least enable the systematic investigation of particle stimulated nucleation (PSN) during thermomechanical treatment as [19] shows. This is an important requirement for using particles in a targeted manner to design magnesium wrought alloys that reveal a weakened texture formation and, thereby, an improved formability.

Conflict of Interests

The authors declare that there is no conflict of interests regarding the publication of this paper.

Acknowledgments

This paper is dedicated to Professor R. Bormann. The work presented here was conducted under his supervision. Unexpectedly, he passed away in spring 2013. G. Meister and W. Punessen are acknowledged for their technical assistance during the casting experiments. Parts of this work have been funded by the German Research Foundation (DFG) in the priority program SPP 1168 "InnoMagTec." The financial support by the DFG is gratefully acknowledged.

References

[1] F. J. Humphreys and M. Hatherly, *Recrystallization and Related Annealing Phenomena*, Elsevier, Amsterdam, The Netherlands, 2nd edition, 2004.

[2] T. Laser, *Einfluss von intermetallischen Phasen in der Magnesiumknetlegierung AZ31 auf Rekristallisation, Texturausbildung und mechanische Eigenschaften [Ph.D. thesis]*, Hamburg University of Technology, Hamburg, Germany, 2008.

[3] D. Turnbull and B. Vonnegut, "Nucleation Catalysis," *Industrial & Engineering Chemistry*, vol. 44, no. 6, pp. 1292–1298, 1952.

[4] P. Villars and L. D. Calvert, *Pearson's Handbook of Crystallographic Data for Intermetallic Phases*, ASM International, Metals Park, Ohio, USA, 2nd edition, 1991.

[5] P. Hentrich, *Handbook of Mixing Technology*, EKATO GmbH, Schopfheim, Germany, 2000.

[6] R. Günther, T. Ebeling, U. Haßlinger, N. Hort, C. Hartig, and R. Bormann, "All purpose grain refinement of Mg alloys by inoculation based on ceramic particles: simulation and experiments," in *Magnesium: Proceedings of the 8th International Conference on Magnesium Alloys and Their Applications*, K. U. Kainer, Ed., pp. 1268–1275, Wiley-VCH, New York, NY, USA, 2009.

[7] R. Günther, C. Hartig, and R. Bormann, "Grain refinement of AZ31 by $(SiC)_p$: theoretical calculation and experiment," *Acta Materialia*, vol. 54, no. 20, pp. 5591–5597, 2006.

[8] V. Kree, J. Bohlen, D. Letzig, and K. U. Kainer, "The metallographical examination of magnesium alloys," *Practical Metallography*, vol. 41, no. 5, pp. 233–246, 2004.

[9] M. Qian, L. Zheng, D. Graham, M. T. Frost, and D. H. StJohn, "Settling of undissolved zirconium particles in pure magnesium melts," *Journal of Light Metals*, vol. 1, no. 3, pp. 157–165, 2001.

[10] J. F. Richardson and W. N. Zaki, "The sedimentation of a suspension of uniform spheres under conditions of viscous flow," *Chemical Engineering Science*, vol. 3, no. 2, pp. 65–73, 1954.

[11] J. Happel and H. Brenner, *Low Reynolds Number Hydrodynamics*, Martinus Nijhoff Publishers, The Hague, The Netherlands, 1983.

[12] W. Pietsch, *Agglomeration in Industry*, Wiley-VCH, Weinheim, Germany, 2005.

[13] Z. Fan, Y. B. Zuo, and B. Jiang, in *Proceedings of the 5th International Light Metals Technology Conference*, H. Dieringa, N. Hort, and K. U. Kainer, Eds., vol. 690, pp. 141–144, Materials Science Forum, 2011.

[14] R. Schmid-Fetzer, J. Gröbner, and H. Chen, *TU Clausthal*, Personal Communication, Institute of Metallurgy, Thermochemistry & Microkinetics, 2010.

[15] Y. Huang, K. U. Kainer, and N. Hort, "Mechanism of grain refinement of Mg-Al alloys by SiC inoculation," *Scripta Materialia*, vol. 64, no. 8, pp. 793–796, 2011.

[16] D. Gosslar, R. Günther, U. Hecht, C. Hartig, and R. Bormann, "Grain refinement of TiAl-based alloys: the role of TiB_2 crystallography and growth," *Acta Materialia*, vol. 58, no. 20, pp. 6744–6751, 2010.

[17] Y. C. Lee, *Grain refinement of magnesium [Ph.D. thesis]*, University of Queensland, Brisbane, Australia, 2002.

[18] R. Günther, C. Hartig, N. Hort, and R. Bormann, "On the influence of settling of $(ZrB_2)_p$ inoculants on grain refinement of Mg-alloys: experiment and theoretical calculation," in *Magnesium Technology 2009*, E. A. Nyberg, S. R. Agnew, N. R. Neelameggham, and M. O. Pekguleryuz, Eds., pp. 309–313, TMS, Warrendale, Pa, USA, 2009.

[19] U. Haßlinger, *Einfluss von keramischen Partikeln auf die Texturausbildung von Magnesiumknetlegierungen [Ph.D. thesis]*, Hamburg University of Technology, Hamburg, Germany, 2013.

22

Algae Mediated Green Fabrication of Silver Nanoparticles and Examination of Its Antifungal Activity against Clinical Pathogens

Shanmugam Rajeshkumar, Chelladurai Malarkodi, Kanniah Paulkumar, Mahendran Vanaja, Gnanadas Gnanajobitha, and Gurusamy Annadurai

Environmental Nanotechnology Division, Sri Paramakalyani Centre for Environmental Sciences, Manonmaniam Sundaranar University, Alwarkurichi, Tamilnadu 627412, India

Correspondence should be addressed to Gurusamy Annadurai; gannadurai@hotmail.com

Academic Editor: Yuanshi Li

Algae extract has the great efficiency to synthesize the silver nanoparticles as a green route. Brown seaweed mediates the synthesis of silver nanomaterials using extract of *Sargassum longifolium*. For the improved production of silver nanomaterials, some kinetic studies such as time incubation and pH were studied in this work. 10 mL of algal extract was added into the 1 mM AgNO$_3$ aqueous solution. The pH and reaction time range were changed and the absorbance was taken for the characterization of the nanoparticles at various time intervals, and the high pH level shows the increased absorbance due to the increased nanoparticles synthesis. The synthesized silver nanoparticles were characterized by Scanning Electron Microscope (SEM) showing that the shape of the material is spherical, and X-Ray Diffraction value obtained from range of (1 1 1) confirmed synthesized silver nanoparticles in crystalline nature. TEM measurement shows spherical shape of nanoparticles. The Fourier Transmittance Infrared spectrum (FT-IR) confirms the presence of biocomponent in the algae extract which was responsible for the nanoparticles synthesis. The effect of the algal mediated silver nanoparticles against the pathogenic fungi *Aspergillus fumigatus*, *Candida albicans*, and *Fusarium* sp. *S. longifolium* mediated synthesized silver nanoparticles shows cheap and single step synthesis process and it has high activity against fungus. This green process gives the greater potential biomedical applications of silver nanoparticles.

1. Introduction

Seaweeds are the natural and renewable living resources in the marine ecosystem and they are consumed for food, feed, and medicine. Seaweeds contain more than 60 elements, macro- and micronutrients, proteins, carbohydrates, vitamins, and aminoacids [1]. Seaweeds are the sources for extracting industrial products such as phycocolloids: carrageenan, alginates, and agar [2, 3]. Sargassum is a big family of marine brown algae and it has a broad application field. Most of the seaweeds have the antibacterial activity against pathogenic bacteria like *Vibrio parahaemolyticus*, *Salmonella* sp., *Shewanella* sp., *Escherichia coli*, *Klebsiella pneumoniae*, *Streptococcus pyogenes*, *Staphylococcus aureus*, *Enterococcus faecalis*, *Pseudomonas aeruginosa*, and *Proteus mirabilis* [4],

antibiotic resistant postoperative infectious pathogens [5], and also used as antitumor compounds [6]. Moreover, seaweeds play an important role in adsorption of heavy metals like lead, copper, zinc, and manganese [7].

In the 21st century, nanotechnology is the newly emerging multidisciplinary research area with synthesis of nanosized materials [8]. Nanotechnology is the manipulation and production of materials ranging in size from 1 to 100 nanometer scale [9]. The nanoparticles can play a topmost role in the field of nanomedicines such as health care and medicine diagnostic and screening purposes, drug delivery systems, antisense and gene therapy applications, and tissue engineering and expectations of nanorobots configuration [10]. Many methods adopted in the field of synthesis of nanoparticles are chemical and physical. Nowadays, biological method of nanoparticles synthesis is a vast growing technique in the field of nanotechnology [11]. The biological sources had the more quantity of trouble-free protocols and when applied for the human health associated field, it is easy to approach for maintain aseptic environment during the synthesis process of nanoparticles [12]. Recently, biological materials such as bacteria, fungi, plant, and algae were used to synthesis of nanoparticles. Some are *Enterobacteria* [13], *Aspergillus fumigatus* [14], Coriander leaf [15], and *Sargassum wightii* [16] were used to synthesise of nanoparticles. Among the biological materials, algae is called as "bionanofactories" because the live and dead dried biomass was used for synthesis of metallic nanoparticles. It is low cost and environmentally effective, macroscopic structured material, and has the distinct advantage due to its high metal uptake capacity [17]. The rate and size of the nanoparticles were controlled by optimizing the parameters such as pH, temperature, substrate concentration, and incubation time [18].

This study reported that the exposure time and pH play an important role in the controlling of nanoparticles synthesis by using the *S. longifolium* algae extract. Nanoparticles synthesis was characterized by UV-vis spectroscopy, crystalline, and morphological structure which were characterized by XRD, SEM, and TEM. The antifungal activity of silver nanoparticles against pathogenic fungus was studied as well.

2. Materials and Methods

2.1. Chemicals. Analytical grade chemicals are silver nitrate, sodium hydroxide, and hydrochloric acid used for preparation of silver nanoparticles (NPs) and role of pH on NPs synthesis. Agar agar, Rose Bengal agar, and Sabouraud Dextrose agar were used for assessment of antifungal activity. All the chemicals and media were purchased from HiMedia (Mumbai, India).

2.2. Collection and Preparation of Algal Extract. The brown algae *Sargassum longifolium* was collected from the Tuticorin coastal area, Tamilnadu, India. The marine brown seaweed was thoroughly washed with fresh water and distilled water to remove the salt minerals and metallic compounds on the surface of the seaweed. Clean seaweed was dried at a shady place for ten days. The dried leaves were ground into fine powder. 1 gm of algal powder was mixed with 100 mL of distilled water in the 250 mL Erlenmeyer flask and boiled at 60°C for 10 min. The boiled extract was filtered through Whatman No. 1 filter paper, collected the supernatant, and stored at 4°C for nanoparticles synthesis.

2.3. Green Synthesis of Silver Nanoparticles. Typically, 10 mL of pure algal extract solution was mixed with aqueous solution of 90 mL of 1 mM silver nitrate ($AgNO_3$) solution and kept in room temperature with constant stirring at 120 rpm. A color change of the solution was noted by visual inspection and UV-vis spectroscopy at different time and wavelength confirming the synthesis of silver nanoparticles.

2.4. Effect of pH. The role of pH in the synthesis of silver nanoparticles was carried out by altering the pH of algal extract. The pH range was varied from 6.2, 6.8, 7.8 and 8.4 by using analytical graded 0.1 N sodium hydroxide and 0.1 N hydrochloric acid standard solutions. The influence of pH on the synthesis process was analyzed by UV-vis spectrophotometer in the wavelength range of 380–600 nm.

2.5. Purification and Characterization of Synthesized Silver Nanoparticles. The bioreduction of silver ions in aqueous solution using algae extract was monitored by double beam UV-vis spectrophotometer at different wavelengths from 320 to 700 nm (Perkin Elmer, Singapore). Green synthesized silver nanoparticles were purified by distilled water by repeated centrifugation at 10,000 rpm for 15 min. Crystalline nature of the purified silver nanoparticles was analyzed by XRD (Bruker, Germany, model: D8Advance) and particle morphology was characterized by Scanning Electron Microscope (Hitachi, Model: S-3400N). The functional biomolecules such as carboxyl groups present in the seaweed responsible for the silver nanoparticles formation were characterized by FT-IR (BrukerOptik GmbH Model No.—Tensor 27). The dried silver nanoparticles were compressed with KBr into thin pellets and measured at the wavelength range from 4000 to 400 cm^{-1}.

2.6. Antifungal Assay of Silver Nanoparticles

2.6.1. Clinical Fungal Pathogens. The three fungal pathogenic strains used in the present study were isolated from clinical samples and identified from Microlabs, Vellore District, India, which were *Aspergillus fumigatus*, *Candida albicans*, and *Fusarium* sp.

2.6.2. Assay of Antifungal Activity. The antifungal activity of green synthesized silver nanoparticles against various fungal strains was assayed by Agar well diffusion method. The fungicidal effect of the silver nanoparticles could be assessed by the formation of zone around the well. 100 mL of sterilized Sabouraud Dextrose Agar medium was poured into three sterilized Petri dishes. The fungal strains were grown in Rose Bengal agar and their spores were mixed into the 10 mL sterile distilled water and swapped on the agar. Three wells of 5 mm diameter were prepared and loaded with silver nanoparticles at different concentrations (30, 60, and 90 μL). The plates loaded with the fungal and silver nanoparticles were incubated at 37°C. The antifungal activities against the fungal strains were confirmed by forming the zone around the wells and measured after 24 hrs of incubation. The zone of inhibition was expressed in mm in diameter. The experiments were repeated three times to find the standard deviation and standard error.

3. Results and Discussion

3.1. Visual and UV-Vis Spectrophotometer Analysis. Reduction of silver ions to silver nanoparticles was visually identified by color change from yellow to brown in the aqueous

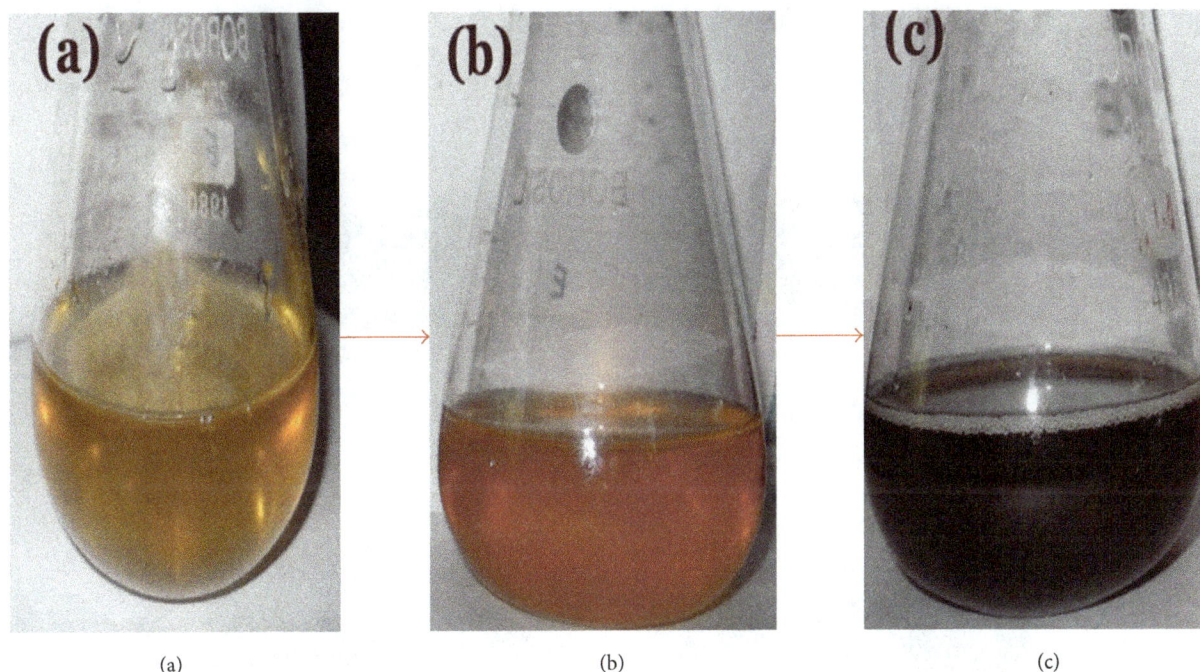

FIGURE 1: Color change after the addition of algae extract with 1 mM AgNO$_3$ indicates the formation of silver nanoparticles. (a) Initial color change, (b) 1 hr, (c) 32 hr.

solution of reaction mixture at 1 hr incubation time (Figure 1). Brown formation occurred due to the oscillation of free electrons in the reaction mixture. The color change depended on the incubation time. The deep brown color for silver nanoparticles was attained at 32 hr indicating that the increasing of color intensity is directly proportion to the time of incubation. Furthermore, the nanoparticles formation by the algal extract was confirmed by UV-vis spectroscopy at different wavelengths. Similarly, Chandran et al. [19] synthesized silver nanoparticles using *Aloe vera* extract taking 24 hr of reaction time in the presence of ammonia which enhances the nanoparticles formation.

UV-vis spectroscopy analysis depends on the arising of color in the reaction due to the excitation of surface Plasmon resonance band in a reaction mixture and was recorded as different functional time. Figure 2 shows that, there is no peak were formed at the initial stage indicates that there is no synthesis of silver nanoparticles was observed. After 1 hr of incubation time the surface Plasmon resonance band for silver nanoparticles was positioned at around 460 nm and the synthesis was steadily increasing with the increasing in time of reaction without change in the peak position. It shows no peak variation in their position with broad band indicates presence of polydispersed nanoparticles. The nanoparticles synthesis was completed at 64 h time of incubation.

3.2. Effect of pH. The pH of the reaction mixture plays an important role in the nanoparticles synthesis. At low pH 6.2, the color change of the reaction mixture was slower than that at high pH 8.4. The color intensity of the reduction process increased with the increase of the pH (Figure 3). Figure 4 shows the UV-vis spectra to analyze the pH impact on the

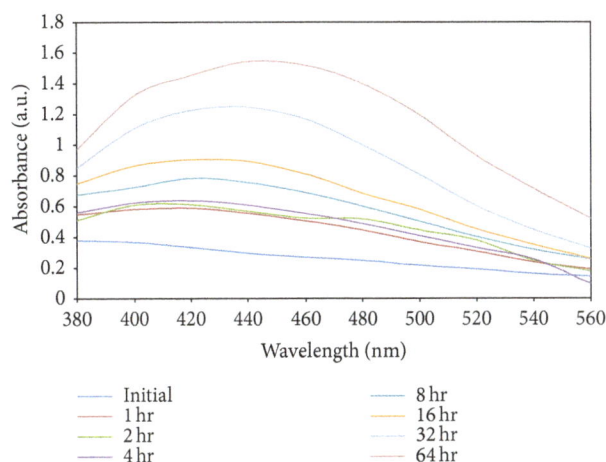

FIGURE 2: UV-vis spectra recorded the formation of nanoparticles in the reaction mixture of algae extract and AgNO$_3$ at different time intervals showing the peak at 460 nm.

nanoparticles synthesis. At lower pH 6.2 and 6.8, the broadened SPR band was shown at 460 nm indicating polydispersed nanoparticles formation. Broadening peaks occurred at lower pH due to the excitations of longitudinal Plasmon vibrations [20]. Broad band at low pH is due to the formation of anisotropic nanoparticles [21]. The narrow peak at higher pH that occurred is due to the formation of monodispersed and small-sized silver nanoparticles [22]. Lower pH suppresses nanoparticles formation and higher pH enhances the nanoparticles synthesis process. The high absorbance and narrow band was formed at 440 nm in the higher pH. Similarly, some reports explained that pH plays an important role

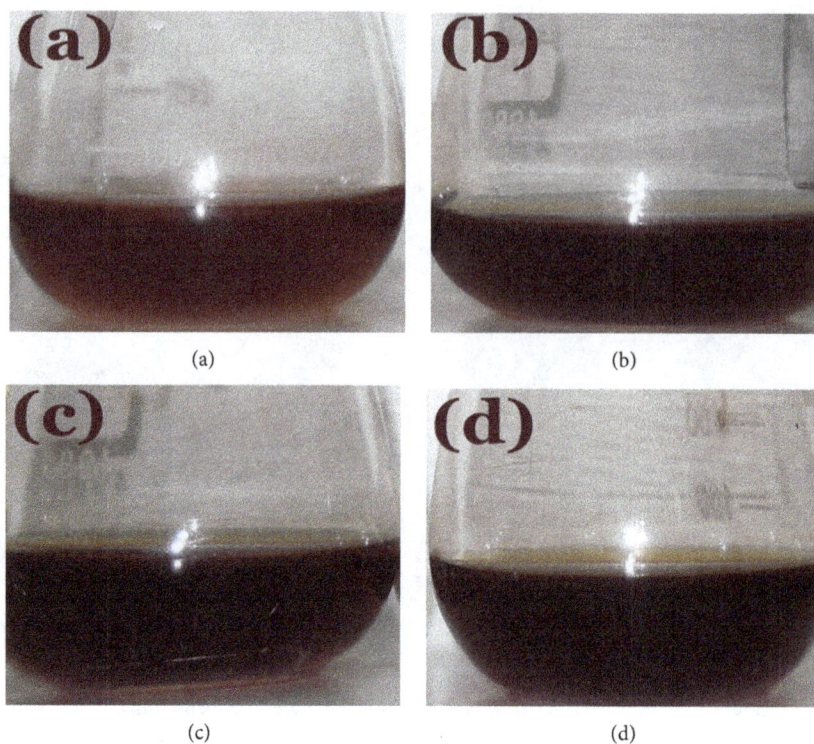

FIGURE 3: Color change variation at different pH levels (a) pH 6.2 (b) 6.8 (c) 7.8 (d) 8.4.

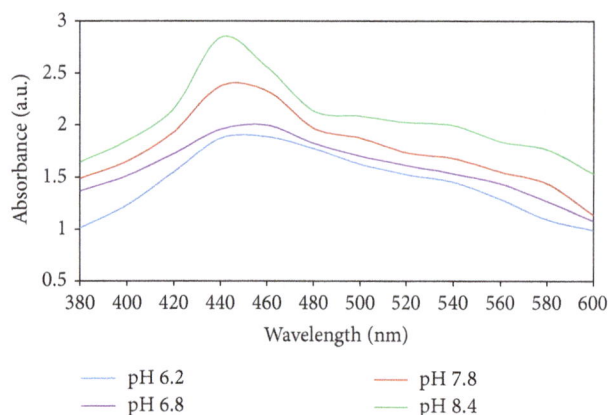

FIGURE 4: Effect of pH on the synthesis of silver nanoparticles shows that maximum synthesis occurred at high pH.

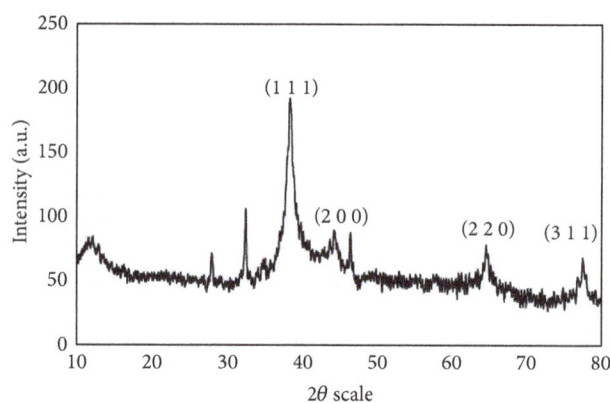

FIGURE 5: XRD spectrum of silver nanoparticles synthesized from *S. longifolium* extract shows crystalline structure of nanoparticles.

in shape and size control synthesis process of silver nanoparticles. This is also in agreement with earlier reports that the addition of an alkaline ion in the reaction mixture is necessary to carry out the reduction of metal ions [23].

3.3. XRD Analysis. The crystalline structure of the silver nanoparticles was determined by XRD in the whole spectrum of 2θ values ranging from 10 to 80. The green synthesized silver nanostructure by employing algae extract of *S. longifolium* was confirmed by the characteristic peaks observed in the XRD image (Figure 5). The four distinct diffraction peaks of the 2θ values of 38.27°, 44.16°, 65.54°, and 77.42° could be

assigned the plane of (1 1 1), (2 0 0), (2 2 0), and (3 1 1), respectively, indicating that the silver nanoparticles are fcc (face centered cubic) and crystalline in nature. The synthesized silver nanoparticles are compared with standard silver nitrate and pure silver particles which are published by Joint Committee on Powder Diffraction Standards (File nos. 04-0783 and 84-0713). The XRD clearly shows that the four distinct planes for the silver nanoparticles synthesized from algal extract of *S. longifolium* are highly crystallized and purified.

3.4. SEM Analysis. SEM image shows the morphologies of the algae mediated synthesized bionanoparticles at different

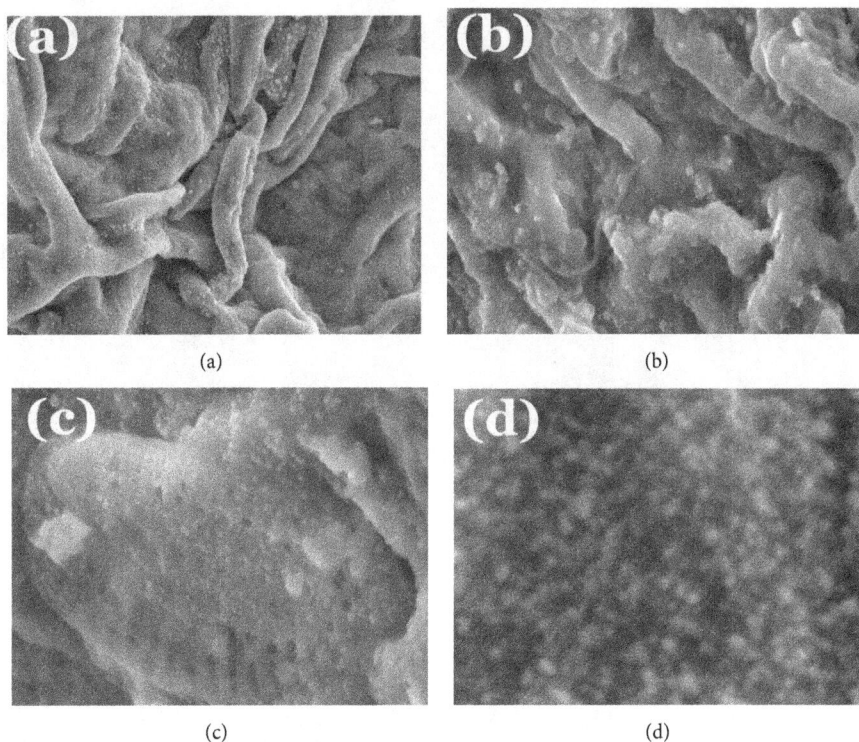

FIGURE 6: Scanning electron microscope image of silver nanoparticles synthesized from *S. longifolium* extract showing predominantly spherical shape nanoparticles on the surface of the cell at different magnifications: (a) 5 μm, (b) 3 μm, (c) 1 μm, and (d) 500 nm.

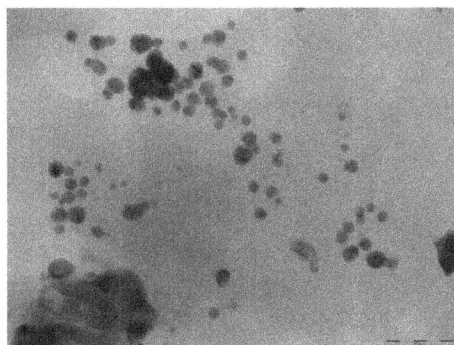

FIGURE 7: TEM image of silver nanoparticles shows polydispersed synthesized silver nanoparticles.

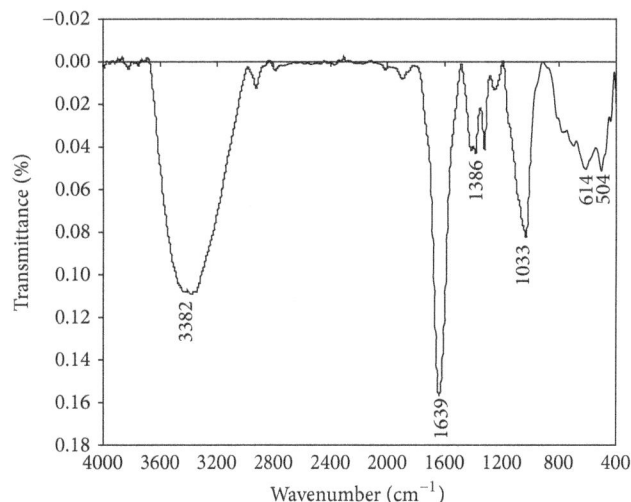

FIGURE 8: FTIR shows the functional groups associated with silver nanoparticles synthesized from extract of *S. longifolium*.

magnifications (Figure 6). High density of silver nanoparticles is shown in the SEM image which is uniformly distributed on the surface of the cells of algae. The polydispersed silver nanoparticles were attached on the surface of the biomolecules of the algal cells. Silver nanoparticles adhered on the cells due to weakly dislodging of bound silver nanoparticles from the biomass and silver nitrate reaction mixture during preparation [24]. Similar results were obtained in the gold nanoparticles using algal extract of *S. marginatum* with the size ranging from 40 to 85 nm [25] and also by using the *T. conoides* algae [26]. It reveals the involvement of the polyphenolic or secondary capping agent in the nanoparticles formation.

3.5. TEM Image of Silver Nanoparticles. The morphology of *S. longifolium* algae extract mediated synthesized silver nanoparticles was determined by TEM. The synthesized silver nanoparticles were formed predominantly spherical and some are truncated and ellipsoidal in the form of agglomerates. Some of the nanoparticles noted that the edges of the particles were smoother than the centers (marked by arrows) in Figure 7. The edges of nanoparticles were shown smoother

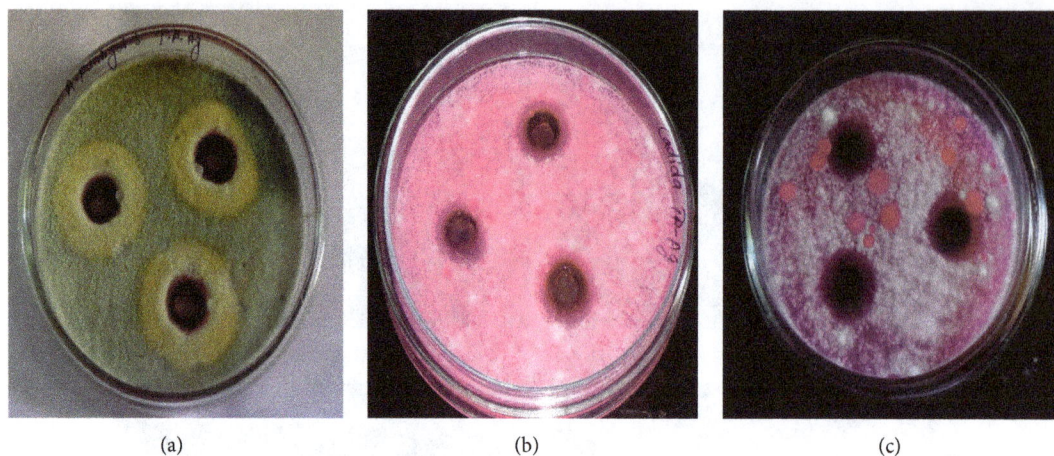

(a)   (b)   (c)

FIGURE 9: Antifungal activity of SNPs synthesized by using marine algae *S. longifolium*.

than the center indicating that the proteins in algal extract may cap the silver nanoparticles. Similarly, Ahmad et al. [27] suggested that proteins in basil plant are present among the particles and they adhered to their surfaces.

3.6. FTIR.

FTIR was used to identify the biomolecules in *S. longifolium* responsible for the silver ions reduction and stabilization of reduced silver ions (Figure 8). The broad spectrum at $3382 \, cm^{-1}$ shows the O–H bond stretching of alcohols and phenols. The narrow band at $1639 \, cm^{-1}$ (–NH–C=O) indicates the presence of N–H bended primary amines. A weak band at $1386 \, cm^{-1}$ corresponds to asymmetrical stretching for nitro compounds. The absorption peak at $1033 \, cm^{-1}$ illustrates the presence of carboxylic acid functional group. The very weak bands were formed at 613 and $504 \, cm^{-1}$ due to the occurrence of vibrations of alkyl halides. FTIR revealed that soluble organic compounds or proteins in the extract may bind with silver ions and reduce the silver ions to nanoparticles. The carboxylic groups in the extract may enhance the stability of silver nanoparticles.

3.7. Antifungal Activity of Algal Synthesized Silver Nanoparticles.

Nanostructured silver materials can be initialized in some potential applications in the field of bactericidal activities and they vigorously involve the fight against the pathogenic microbes causing dangerous disease. The green synthesized silver nanoparticles exhibited excellent antibacterial activity against pathogenic fungi such as *A. fumigatus*, *Candida albicans*, and *Fusarium* sp. Silver could be used as an antibacterial agent for many infectious diseases at ancient time and before the emergence of antibiotics [28]. In this study, the antifungal activity of *S. longifolium* mediated synthesized silver nanoparticles against harmful pathogenic fungi at different concentrations (50, 100, and $150 \, \mu L$) was carried out. Table 1 shows that with the increase in the concentration of silver nanoparticles, the zone of inhibition increased. So the zone formation around the well is directly proportional to the concentration of silver nanoparticles (Figure 9).

TABLE 1: Antifungal activity of silver bionanoparticles against pathogenic fungi.

Pathogenic fungi	Zone of inhibition (mm in diameter)		
	$50 \, \mu L$	$100 \, \mu L$	$150 \, \mu L$
Aspergillus fumigatus	11.17 ± 0.152	12.07 ± 0.120	13.10 ± 0.130
Candida albicans	11.20 ± 0.256	13.87 ± 0.134	14.20 ± 0.152
Fusarium sp.	18.10 ± 0.208	20.23 ± 0.186	22.03 ± 0.033

±Standard deviation.

Silver nanoparticles highly form the zone against *Fusarium* sp. (22.03 ± 0.033) mm at the concentration of $150 \, \mu L$. Minimum zone inhibition was observed against *A. fumigates* (13.10 ± 0.130) mm.

The mechanism of antifungal activity of silver nanoparticles was not fully known. A few of the literature illustrated the possible mechanism of antifungal activity. Antifungal activity depends on the size and shape of the silver nanoparticles. Small size nanoparticles have large surface area ensuring the inhibition of microbial growth. Spherical shape with size-reduced silver ions has the increased contact area so that it can eliminate the bacterial growth. Activity of silver nanoparticles has similar effects as silver ions [29]. Positively charged silver ions may attach with negatively charged cell membranes of microbes by electrostatic attraction [30]. Silver nanoparticles form the pits in the cell wall and damage the cell permeability [31] and induce the proton leakage caused by ROS in the membrane [32, 33] resulting in cell death. Kim et al. [34] demonstrated that silver nanoparticles inhibit the conidial germination on fungi. Finally, the silver nanoparticles have a great potential to control the spore producing fungi.

4. Conclusion

In conclusion, this present study stated the green mediated synthesis of silver nanoparticles under optimized parameters using the extract of enormously available algae *S. longifolium*. The broad peak was observed under UV-vis spectra at 460 nm for silver nanoparticles. Highly synthesis and spherical shape nanoparticles were obtained at higher pH confirmed by TEM. Crystalline structure of nanoparticles was identified by XRD. Purity and component of silver nanoparticles were confirmed by EDX. *S. longifolium* extract mediated synthesized silver nanoparticles show high antifungal activity that can be used therapeutically in biomedical applications.

Conflict of Interests

The authors declare that there is no conflict of interests regarding the publication of this paper.

Acknowledgments

The authors gratefully acknowledge STIC, Cochin for providing XRD and EDX, IIT Bombay for TEM facility, and DST for FIST grant (Ref. no. S/FST/ESI-101/2010).

References

[1] A. Jensen, "Present and future needs for algae and algal products," *Hydrobiologia*, vol. 260-261, no. 1, pp. 15–23, 1993.

[2] A. K. Semesi, "Coastal resource of Bagamoyo District, Tanzania," *Trends in Plant Science*, vol. 11, pp. 517–533, 2000.

[3] M. M. Chandraprabha, R. Seenivasan, H. Indu, and S. Geetha, "Biochemical and nanotechnological studies in selected seaweeds of Chennai Coast," *Journal of Applied Pharmaceutical Science*, vol. 2, no. 11, pp. 100–107, 2012.

[4] R. Lavanya and N. Veerappan, "Antibacterial potential of six seaweeds collected from gulf of mannar of Southeast Coast of India," *Advances in Biological Research*, vol. 5, no. 1, pp. 38–44, 2011.

[5] S. Ravikumar, L. Anburajan, G. Ramanathan, and N. Kaliaperumal, "Screening of seaweed extracts against antibiotic resistant post operative infectious pathogens," *Seaweed Research and Utilization*, vol. 24, no. 1, pp. 95–99, 2002.

[6] A. Ayesha, H. Hira, V. Sultana, J. Ara, and S. Ehteshamul-Haque, "In vitro cytotoxicity of seaweeds from Karachi coast on brine shrimp," *Pakistan Journal of Botany*, vol. 42, no. 5, pp. 3555–3560, 2010.

[7] K. Vijayaraghavan, T. T. Teo, R. Balasubramanian, and U. M. Joshi, "Application of Sargassum biomass to remove heavy metal ions from synthetic multi-metal solutions and urban storm water runoff," *Journal of Hazardous Materials*, vol. 164, no. 2-3, pp. 1019–1023, 2009.

[8] M. Amin, F. Anwar, M. R. S. A. Janjua, M. A. Iqbal, and U. Rashid, "Green synthesis of silver nanoparticles through reduction with *Solanum xanthocarpum* L. berry extract: characterization, antimicrobial and urease inhibitory activities against *Helicobacter pylori*," *International Journal of Molecular Sciences*, vol. 13, pp. 9923–9941, 2012.

[9] P. Mohanpuria, N. K. Rana, and S. K. Yadav, "Biosynthesis of nanoparticles: technological concepts and future applications," *Journal of Nanoparticle Research*, vol. 10, no. 3, pp. 507–517, 2008.

[10] T. Kubik, K. Bogunia-Kubik, and M. Sugisaka, "Nanotechnology on duty in medical applications," *Current Pharmaceutical Biotechnology*, vol. 6, no. 1, pp. 17–33, 2005.

[11] S. Sinha, I. Pan, P. Chanda, and S. K. Sen, "Nanoparticles fabrication using ambient biological resources," *Journal of Applied Biosciences*, vol. 19, pp. 1113–1130, 2009.

[12] L. Rastogi and J. Arunachalam, "Sunlight based irradiation strategy for rapid green synthesis of highly stable silver nanoparticles using aqueous garlic (*Allium sativum*) extract and their antibacterial potential," *Materials Chemistry and Physics*, vol. 129, no. 1-2, pp. 558–563, 2011.

[13] A. R. Shahverdi, S. Minaeian, H. R. Shahverdi, H. Jamalifar, and A.-A. Nohi, "Rapid synthesis of silver nanoparticles using culture supernatants of *Enterobacteria*: a novel biological approach," *Process Biochemistry*, vol. 42, no. 5, pp. 919–923, 2007.

[14] K. C. Bhainsa and S. F. D'Souza, "Extracellular biosynthesis of silver nanoparticles using the fungus *Aspergillus fumigatus*," *Colloids and Surfaces B*, vol. 47, no. 2, pp. 160–164, 2006.

[15] K. B. Narayanan and N. Sakthivel, "Coriander leaf mediated biosynthesis of gold nanoparticles," *Materials Letters*, vol. 62, no. 30, pp. 4588–4590, 2008.

[16] G. Singaravelu, J. S. Arockiamary, V. G. Kumar, and K. Govindaraju, "A novel extracellular synthesis of monodisperse gold nanoparticles using marine alga, *Sargassum wightii* Greville," *Colloids and Surfaces B*, vol. 57, no. 1, pp. 97–101, 2007.

[17] T. A. Davis, B. Volesky, and A. Mucci, "A review of the biochemistry of heavy metal biosorption by brown algae," *Water Research*, vol. 37, no. 18, pp. 4311–4330, 2003.

[18] M. Gericke and A. Pinches, "Microbial production of gold nanoparticles," *Gold Bulletin*, vol. 39, no. 1, pp. 22–28, 2006.

[19] S. P. Chandran, M. Chaudhary, R. Pasricha, A. Ahmad, and M. Sastry, "Synthesis of gold nanotriangles and silver nanoparticles using Aloe vera plant extract," *Biotechnology Progress*, vol. 22, no. 2, pp. 577–583, 2006.

[20] P. V. Kamat, M. Flumiani, and G. V. Hartland, "Picosecond dynamics of silver nanoclusters: photoejection of electrons and fragmentation," *Journal of Physical Chemistry B*, vol. 102, no. 17, pp. 3123–3128, 1998.

[21] D. Philip, "Green synthesis of gold and silver nanoparticles using *Hibiscus rosasinensis*," *Physica E*, vol. 42, no. 5, pp. 1417–1424, 2010.

[22] T. P. Amaladhas, S. Sivagami, T. A. Devi, N. Ananthi, and S. P. Velammal, "Biogenic synthesis of silver nanoparticles by leaf extract of *Cassia angustifolia*," *Advances in Natural Sciences*, vol. 3, no. 7, Article ID 045006, 2012.

[23] R. Sanghi and P. Verma, "Biomimetic synthesis and characterisation of protein capped silver nanoparticles," *Bioresource Technology*, vol. 100, no. 1, pp. 501–504, 2009.

[24] M. Sastry, A. Ahmad, M. Islam Khan, and R. Kumar, "Biosynthesis of metal nanoparticles using fungi and actinomycete," *Current Science*, vol. 85, no. 2, pp. 162–170, 2003.

[25] F. A. A. Rajathi, C. Parthiban, V. Ganesh Kumar, and P. Anantharaman, "Biosynthesis of antibacterial gold nanoparticles using brown alga, *Stoechospermum marginatum* (kützing)," *Spectrochimica Acta A*, vol. 99, pp. 166–173, 2012.

[26] K. Vijayaraghavan, A. Mahadevan, M. Sathishkumar, S. Pavagadhi, and R. Balasubramanian, "Biosynthesis of Au(0) from Au(III) via biosorption and bioreduction using brown marine alga *Turbinaria conoides*," *Chemical Engineering Journal*, vol. 167, no. 1, pp. 223–227, 2011.

[27] N. Ahmad, S. Sharma, M. K. Alam et al., "Rapid synthesis of silver nanoparticles using dried medicinal plant of basil," *Colloids and Surfaces B*, vol. 81, no. 1, pp. 81–86, 2010.

[28] H. J. Klasen, "Historical review of the use of silver in the treatment of burns. I: early uses," *Burns*, vol. 26, no. 2, pp. 117–130, 2000.

[29] S. Pal, Y. K. Tak, and J. M. Song, "Does the antibacterial activity of silver nanoparticles depend on the shape of the nanoparticle? A study of the gram-negative bacterium *Escherichia coli*," *Applied and Environmental Microbiology*, vol. 73, no. 6, pp. 1712–1720, 2007.

[30] I. Sondi and B. Salopek-Sondi, "Silver nanoparticles as antimicrobial agent: a case study on *E. coli* as a model for Gram-negative bacteria," *Journal of Colloid and Interface Science*, vol. 275, no. 1, pp. 177–182, 2004.

[31] M. Raffi, F. Hussain, T. M. Bhatti, J. I. Akhter, A. Hameed, and M. M. Hasan, "Antibacterial characterization of silver nanoparticles against *E. coli* ATCC-15224," *Journal of Materials Science and Technology*, vol. 24, no. 2, pp. 192–196, 2008.

[32] P. Dibrov, J. Dzioba, K. K. Gosink, and C. C. Häse, "Chemiosmotic mechanism of antimicrobial activity of Ag+ in *Vibrio cholerae*," *Antimicrobial Agents and Chemotherapy*, vol. 46, no. 8, pp. 2668–2670, 2002.

[33] S. H. Dehkordi, F. Hosseinpour, and A. E. Kahrizangi, "An in vitro evaluation of antibacterial effect of silver nanoparticles on *Staphylococcus aureus* isolated from bovine subclinical mastitis," *African Journal of Biotechnology*, vol. 10, no. 52, pp. 10795–10797, 2011.

[34] S. W. Kim, K. S. Kim, K. Lamsal et al., "An in vitro study of the antifungal effect of silver nanoparticles on oak wilt pathogen *Raffaelea* sp," *Journal of Microbiology and Biotechnology*, vol. 19, no. 8, pp. 760–764, 2009.

23

Response of Functionally Graded Material Plate under Thermomechanical Load Subjected to Various Boundary Conditions

Manish Bhandari[1] and Kamlesh Purohit[2]

[1]Jodhpur Institute of Engineering and Technology, Jodhpur, Rajasthan, India
[2]Jai Narain Vyas University, Jodhpur, Rajasthan, India

Correspondence should be addressed to Manish Bhandari; manish.bhandari@jietjodhpur.com

Academic Editor: Massimo Pellizzari

Functionally graded materials (FGMs) are one of the advanced materials capable of withstanding the high temperature environments. The FGMs consist of the continuously varying composition of two different materials. One is an engineering ceramic to resist the thermal loading from the high-temperature environment, and the other is a light metal to maintain the structural rigidity. In the present study, the properties of the FGM plate are assumed to vary along the thickness direction according to the power law distribution, sigmoid distribution, and exponential distribution. The fundamental equations are obtained using the first order shear deformation theory and the finite element formulation is done using minimum potential energy approach. The numerical results are obtained for different distributions of FGM, volume fractions, and boundary conditions. The FGM plate is subjected to thermal environment and transverse UDL under thermal environment and the response is analysed. Numerical results are provided in nondimensional form.

1. Introduction

Composite materials are widely in use due to their intrinsic mechanical property such as high strength, modulus of elasticity, and lower specific gravity. Further, as a result of intensive studies into metallurgical aspects of material and better understanding of structural property, it has become possible to develop new composite materials with improved physical and mechanical properties. The functionally graded material (FGM) is one such material whose property can be useful to accomplish the specific demands in various engineering applications to achieve the advantage of the properties of individual material. This is possible due to the material composition of the FGM which changes according to a law in a preferred direction. The thermomechanical analysis of FGM structures is one dimension which has attracted the attention of many researchers in the past few years. The applications of FGMs include design of aerospace structures, heat engine components, and nuclear power plants.

A large number of research papers have been published to evaluate the behaviour of FGM using both experimental and numerical techniques which include both linearity and nonlinearity in various areas. A few of published literatures highlight the importance of the present work. The FGM can be produced by gradually and continuously varying the constituents of multiphase materials in a predetermined profile. Most researchers use the power law function (P-FGM), sigmoid function (S-FGM), or exponential function (E-FGM) to describe the effective material properties. Delale and Erdogan [1] indicated that the effect of Poisson's ratio on the deformation is much less as compared to that of Young's modulus. Praveen and Reddy [2] examined the thermoelastostatic response of simply supported square FG plates subjected to pressure loading and thickness varying temperature fields. They used the first order shear deformation plate theory (FSDT) to develop the governing equations. They reported that the basic response of the plates which corresponds to properties intermediate to those of the metal and the ceramic does not necessarily lie in between those of the ceramic and metal. Reddy [3] found that the nondimensional deflection reached a minimum value at a particular volume fraction index. They used the power law function to calculate the material gradient. Cheng and Batra [4] computed deformations due to thermal and mechanical loads applied to the top and bottom surfaces of the rigidly clamped elliptic FG plate separately. It was found that the through-thickness distributions of the in-plane displacements and transverse shear stresses in a functionally graded plate do not agree with those assumed in classical and shear deformation plate theories. Reddy and Zhen [5] solved the problem using a higher order shear and normal deformable plate theory (HONSDPT) since, in the HOSNDPT, the transverse normal and shear stresses are computed from equation of the plate theory rather than by integrating the balance of linear momentum with respect to the thickness coordinate. It was reported that the assumption of constant deflection is not true for thermal load but it was found to be true for mechanical load. Qian and Batra [6] found that the centroidal deflection for a clamped plate is nearly one-third of that for simply supported plate, and the maximum magnitude of the axial stress induced at the centroid of the top surface is nearly 40% larger than that for a simply supported plate. Dai et al. [7] analyzed the plate under the mechanical loading as well as thermal gradient and found that the relations between the deflection and the volume fraction exponent are quite different under the two loadings. Ferreira et al. [8] used collocation method third order shear order deformation theory and presented the effect of aspect ratio of the plate and the volume fraction of the constituents on the transverse deflection. Chi and Chung [9, 10] studied the effect of loading conditions on the mechanical behaviour of a simply supported rectangular FGM plate. They assumed that Young's moduli vary continuously throughout the thickness direction according to the volume fraction of constituents defined by sigmoid function. The maximum tensile stress of the FGM plate was found to be at the bottom surface of the plate. Wang and Qin [11] concluded that the appropriate graded parameter can lead to low stress concentration and little change in the distribution of stress fields. They assumed that the thermoelastic constants and the temperature vary exponentially through the thickness. Mahdavian [12] obtained the equilibrium and stability equations based on the classical plate theory (CPT) and Fourier series expansion. They found that the critical buckling coefficients for FGM plates are considerably higher than isotropic plates. Ashraf and Daoud [13] derived the equilibrium and stability equations using sinusoidal shear deformation plate theory (SPT). It was concluded that the critical buckling temperature differences of functionally graded plates are generally lower than the corresponding ones for homogeneous ceramic plates. Alieldin et al. [14] proposed three approaches to determine the property details of an FG plate equivalent to the original laminated composite plate. They developed the equations of motion based on the combination of the first order plate theory and the Von Karman strains. Kyung-Su and Ji-Hwan [15] compared numerical results for three types of materials. It was found that the minimum compressive stress ratio is

observed for the fully FGM plate with largest volume fraction index. Suresh et al. [16] studied the effect of shear deformation and nonlinearity response of functionally graded material plate and concluded that the effect of nonlinearity in functionally graded composite plates is more predominant in decreasing the deflections in thin plates for side to thickness ratio of 10. Mohammad and Singh [17] obtained the numerical results for different thickness ratios, aspect ratios, volume fraction index, and temperature rise with different loading and boundary conditions. They employed the boundary conditions; for example, all edges are simply supported (SSSS), all edges are clamped (CCCC), two edges are simply supported and two are clamped (SCSC), two edges are clamped and two are free (CFCF), all edges are hinged (HHHH), and two edges are clamped and two are hinged (CHCH). It was noticed that the maximum centre deflection was found for simply supported boundary conditions and least central deflection was found for clamped (CCCC) boundary conditions. Nguyen-Xuan et al. [18] applied the method for static, free vibration and mechanical/thermal buckling problems of functionally graded material (FGM) plates. They analysed the behaviour of FGM plates under mechanical and thermal loads numerically in detail through a list of benchmark problems. Alshorbagy et al. [19] concluded that FG plates provide a high ability to withstand thermal stresses, which reflects its ability to operate at elevated temperatures. Bhandari and Purohit [20] presented the response of FG plates under mechanical load for various boundary conditions, for example, SSSS, CCCC, SCSC, CFCF, CCSS, SSFF, SSSC, SSSF, and SSCF. The power law, sigmoid law, and exponential law were used for the calculations of the properties through the thickness. They also compared the behaviour of isotropic plates (ceramic and metal) with that of the FGM plates. It was concluded that the isotropic ceramic plate has the lowest tensile stress for all the boundary conditions. It was also reported that the maximum tensile stress occurs for CCFF boundary condition and the minimum tensile stress was observed for SCSC boundary condition. Bhandari and Purohit [21] presented the response of FG plates under mechanical load for varying aspect ratios. The power law, sigmoid law, and exponential law were used for the calculations of the properties through the thickness. They also compared the behaviour of isotropic plates (ceramic and metal) with that of the FGM plates.

With the increasing applications of functionally graded materials, it is vital to understand the behaviour of thermomechanical response of FG plates under various boundary conditions. It is also important to study the response of the FG materials following various material gradient laws, for example, power law function, sigmoid function, and exponential function for properties. In both power law and exponential functions, the stress concentrations appear in one of the interfaces in which the material is continuous but is rapidly changing. In sigmoid FGM, which is composed of two power law functions, there is a gradual change in volume fraction as compared to power law and exponential function. Power law function has been applied to many of the FGMs' but application of sigmoid function is sparse in literature. Power law function, sigmoid function, and exponential function have been used in various research works separately but the comparisons of the three FGM laws for various volume fraction exponents, ceramic and metal, have been sparse. In most of the research works, FGM plates with edges simply supported and clamped have been considered. Plate subjected to other boundary conditions, for example, clamped, free, hinged, and combined, is also useful but they have been rarely reported. Thermomechanical loaded FGM structures have been researched but thermally loaded plates have been rarely reported.

Keeping this in consideration, the objective of the present work is to examine the thermomechanical behaviour for various boundary conditions, various material gradient laws, and various volume fraction exponents. The results are presented in the form of nondimensional parameters. The comparison of isotropic ceramic, metal, and FGM plates is also presented.

2. Material Gradient of FGM Plate

The material properties in the thickness direction of the FGM plates vary with power law functions (P-FGM), exponential functions (E-FGM), or sigmoid functions (S-FGM). A mixture of the two materials composes the through-thickness characteristics. The FGM plate of thickness "h" is modeled usually with one side of the material being ceramic and the other side being metal as shown in Figure 1.

2.1. Power Law Distribution.
Material properties of P-FGM are dependent on the volume fraction (V_f) which obeys power law as defined in

$$V_f = \left(\frac{z}{h} + \frac{1}{2} \right)^n,\tag{1}$$

where n is a parameter that dictates the material variation profile through the thickness known as the volume fraction exponent, h is thickness of plate, and z is depth measured from the neutral axis of the plate.

The material properties of a P-FGM can be determined by the rule of mixture as given in

$$P(z) = (P_t - P_b) V_f + P_b,\tag{2}$$

where $P(z)$ is generic material property, for example, elastic modulus, at a particular depth z, P_t and P_b are generic properties at top and bottom surface of the plate, and $n = 0$ and ∞ denotes fully ceramic plate and fully metal plate, respectively.

Consequently, at bottom surface, $(z/h) = -1/2$ and $V_f = 0$; hence, $P(z) = P_b$ and, at top surface, $(z/h) = 1/2$ and so $V_f = 1$; hence, $P(z) = P_t$.

2.2. Sigmoid Law Distribution.
Material properties vary continuously throughout the thickness direction according to the volume fraction of constituents defined by sigmoid function [8]. The volume fraction is calculated using two power law

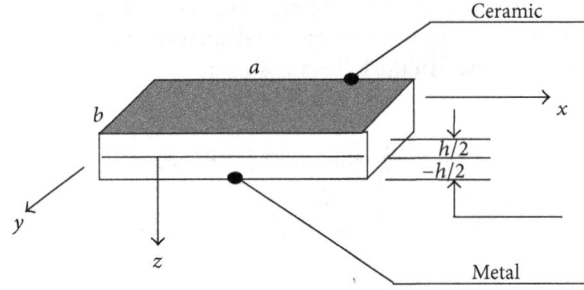

FIGURE 1: FGM plate.

functions to ensure smooth distribution of stresses among all the interfaces. The two power law functions are defined by

$$g_1(z) = 1 - \frac{1}{2} \left(\frac{h/2 - z}{h/2} \right)^P \quad \text{for } 0 \leq z \leq \frac{h}{2},\tag{3a}$$

$$g_2(z) = \frac{1}{2} \left(\frac{h/2 + z}{h/2} \right)^P \quad \text{for } -\frac{h}{2} \leq z \leq 0.\tag{3b}$$

By using the rule of mixture, Young's modulus of the S-FGM can be calculated by

$$E(z) = E_1 g_1(z) + [1 - g_1(z)] E_2 \quad \text{for } 0 \leq z \leq \frac{h}{2},\tag{4a}$$

$$E(z) = E_1 g_2(z) + [1 - g_2(z)] E_2 \quad \text{for } -\frac{h}{2} \leq z \leq 0,\tag{4b}$$

where E_1 is Young's modulus at the top surface and E_2 is Young's modulus at the bottom surface.

2.3. Exponential Law.
Material properties vary continuously throughout the thickness direction according to the volume fraction of constituents defined exponentially through the thickness. Accordingly, the exponential law is defined as

$$E(z) = E_2 e^{(1/h)\ln(E_1/E_2)(z+h/2)},\tag{5}$$

where E_1 is Young's modulus at the top surface and E_2 is Young's modulus at the bottom surface.

3. Governing Equations

3.1. Displacement and Strain Field.
The FSDT theory [3] takes into account transverse shear strain in the formulation with the following assumptions.

(1) The transverse normals remain straight after deformation but may not be orthogonal to the midsurface of the plate.

(2) The out-of-plane normal stress $\sigma_z = 0$.

(3) The layers of the composite plate are perfectly bonded.

(4) The material of each layer is linear elastic and isotropic.

Under the same assumptions and restrictions, the first order plate theory is based on the displacement field which can be expressed in the following form:

$$u(x, y, z) = u_0(x, y) + z\Phi_x(x, y),$$

$$v(x, y, z) = v_0(x, y) + z\Phi y(x, y), \quad (6)$$

$$w(x, y, z) = w_0(x, y),$$

where $(u_0, v_0, w_0, \Phi x, \Phi y)$ are unknown functions to be determined.

The Von Karman nonlinear strains associated with the displacement field are $(\varepsilon_{zz} = 0)$ given by

$$
\begin{bmatrix}
\varepsilon_{xx}(x, y, z) \\
\varepsilon_{yy}(x, y, z) \\
\gamma_{yz}(x, y, z) \\
\gamma_{xz}(x, y, z) \\
\gamma_{xy}(x, y, z)
\end{bmatrix}
=
\begin{bmatrix}
\varepsilon_{xx}^0(x, y, z) \\
\varepsilon_{yy}^0(x, y, z) \\
\gamma_{yz}^0(x, y, z) \\
\gamma_{xz}^0(x, y, z) \\
\gamma_{xy}^0(x, y, z)
\end{bmatrix}
+ z
\begin{bmatrix}
\varepsilon_{xx}^1(x, y, z) \\
\varepsilon_{yy}^1(x, y, z) \\
0 \\
0 \\
\gamma_{xy}^1(x, y, z)
\end{bmatrix}
$$

$$
=
\begin{bmatrix}
\dfrac{\partial u_0}{\partial x} + \dfrac{1}{2}\left(\dfrac{\partial w_0}{\partial x}\right)^2 \\[2mm]
\dfrac{\partial u_0}{\partial x} + \dfrac{1}{2}\left(\dfrac{\partial w_0}{\partial y}\right)^2 \\[2mm]
\dfrac{\partial w_0}{\partial y} + \phi_y \\[2mm]
\dfrac{\partial w_0}{\partial x} + \phi_x \\[2mm]
\dfrac{\partial u_0}{\partial y} + \dfrac{\partial v_0}{\partial x} + \dfrac{\partial w_0}{\partial x}\dfrac{\partial w_0}{\partial x}
\end{bmatrix}
\quad (7)
$$

$$
+ z
\begin{bmatrix}
\dfrac{\partial \phi_x}{\partial x} \\[2mm]
\dfrac{\partial \phi_y}{\partial y} \\[2mm]
0 \\
0 \\
\dfrac{\partial \phi_x}{\partial y} + \dfrac{\partial \phi_y}{\partial x}
\end{bmatrix}.
$$

ε_{ij} are the total strain components. The total strain components are the sum of the elastic strains ε_{ijm} (due to the applied mechanical loads) and thermal strains ε_{Ti} (due to temperature change). So the total strains are given by

$$\varepsilon_{ij} = \varepsilon_{ijm} + \varepsilon_{Ti}. \quad (8)$$

The total strain components can be divided into tensile strain (ε_{xx}) and shear strain (ε_{xy}) components given by

$$
\begin{bmatrix}
\varepsilon_{yy}(x, y, z) \\
\gamma_{xy}(x, y, z)
\end{bmatrix}
=
\begin{bmatrix}
1 & 0 & 0 & z & 0 & 0 \\
0 & 1 & 0 & 0 & z & 0 \\
0 & 0 & 1 & 0 & 0 & z
\end{bmatrix}
$$

$$
\cdot
\begin{bmatrix}
u_{0,x}(x, y) \\
v_{0,x}(x, y) \\
u_{0,x}(x, y) + v_{0,x}(x, y) \\
\phi_{0,x}(x, y) \\
\phi_{0,y}(x, y) \\
\phi_{0,x}(x, y) + \phi_{0,y}(x, y)
\end{bmatrix}
\quad (9)
$$

Or $\quad \{\varepsilon_b\} = [Z_s]\{\varepsilon_b^0\} \quad (10)$

$$
\begin{bmatrix}
\gamma_{xz}(x, y, z) \\
\gamma_{yz}(x, y, z)
\end{bmatrix}
=
\begin{bmatrix}
1 & 0 \\
0 & 1
\end{bmatrix}
\begin{bmatrix}
w_{0,x}(x, y) + \phi_x(x, y) \\
w_{0,y}(x, y) + \phi_y(x, y)
\end{bmatrix}
\quad (11)
$$

Or $\quad \{\varepsilon_s\} = [Z_s]\{\varepsilon_s^0\}, \quad (12)$

where $\{\varepsilon_b^0\}$ and $\{\varepsilon_s^0\}$ are the nodal bending strains and the nodal shear strains, respectively.

3.2. Minimum Total Potential Energy Formulation. For all the conditions of equilibrium, the potential energy is minimal. The minimum total potential energy formulation is a common approach in generating finite element models in solid mechanics. External loads applied to a body will cause the body to deform. During the deformation, the work done by the external forces is stored in the material in the form of elastic energy, called strain energy. The governing equations for the plate equilibrium are derived based on the principle of minimum total potential energy. So, the total potential energy takes the form as in

$$
\Pi = \left(0.5\int_A \{\varepsilon_b^0\}^T [DE_b]\{\varepsilon_b^0\}\,dA - \int_A \{\varepsilon_b^0\}^T [DT_b]\,dA\right)
$$
$$
+ \left(0.5\int_A \{\varepsilon_s^0\}^T [DE_s]\{\varepsilon_s^0\}\,dA - \int_A \{\varepsilon_s^0\}^T [DT_s]\,dA\right)
$$
$$
- \Sigma\{P\}\{u^o\},
$$
$$(13)$$

where $[DT_b]$ and $[DT_s]$ are given by

$$
[DT_b] = \int_z \{Z_b\}^T [D_b]\,\varepsilon_{Tb}\,dz,
$$
$$
[DT_s] = \int_z \{Z_s\}^T [D_s]\,\varepsilon_{Ts}\,dz,
$$
$$(14)$$

where $\{\varepsilon_{Tb}\} = \begin{bmatrix} \alpha(z)\delta T(z) \\ \alpha(z)\delta T(z) \\ 0 \end{bmatrix}$, $\{\varepsilon_{Ts}\} = \begin{bmatrix} 0 \\ 0 \end{bmatrix}$, $\alpha(z)$ is the thermal coefficient of expansion, and $\delta T(z)$ is the continuum temperature change through the plate thickness. Based on the principal of the equivalent single-layer theories, a heterogeneous plate is treated as a statically equivalent, single layer having a complex constitutive behaviour, reducing the 3D continuum problem to 2D problem. The equivalent layer of the FG plate can be obtained by integrating the plate material properties through the plate thickness as in

$$[DE_b] = \int_{-h/2}^{h/2} \{Z_b\}^T [D_b]^T \{Z_b\} \, dz, \qquad (15a)$$

$$[DE_s] = \int_{-h/2}^{h/2} \{Z_s\}^T [D_s]^T \{Z_s\} \, dz, \qquad (15b)$$

where $[D_b]$ and $[D_s]$ are the bending and shear material matrices, respectively. These material matrices provide the stress-strain relations for FG plates as in

$$[D_b] = \begin{bmatrix} \overline{Q_{11}} & \overline{Q_{12}} & \overline{Q_{16}} \\ \overline{Q_{12}} & \overline{Q_{22}} & \overline{Q_{26}} \\ \overline{Q_{16}} & \overline{Q_{26}} & \overline{Q_{66}} \end{bmatrix},$$

$$[D_s] = \begin{bmatrix} \overline{Q_{44}} & \overline{Q_{45}} \\ \overline{Q_{45}} & \overline{Q_{55}} \end{bmatrix}. \qquad (16)$$

The $\overline{Q_{ij}}(z)$ are the equivalent material property types of stiffness as a function of the plate thickness direction (z) which follows the power law function (2), sigmoid function ((3a) and (3b)), and exponential function (5). The equivalent material types of stiffness of isotropic FG plate are as in

$$\overline{Q_{11}}(z) = \overline{Q_{22}}(z) = \frac{E(z)}{1 - v^2},$$

$$\overline{Q_{12}}(z) = v\overline{Q_{11}}(z),$$

$$\overline{Q_{66}}(z) = \frac{1 - v}{2}\overline{Q_{11}}(z), \qquad (17)$$

$$\overline{Q_{44}}(z) = \overline{Q_{55}}(z) = k\frac{1 - v}{2}\overline{Q_{11}}(z),$$

$$\overline{Q_{16}}(z) = \overline{Q_{26}}(z) = \overline{Q_{45}}(z) = 0,$$

where $Q(z)$ is the effective young's modulus, $k \; (= 5/6)$ is shear correction factor, and v is the effective Poisson's ratio of the material through the plate thickness.

3.3. Finite Element Model.
The displacements and normal rotations at any point into a finite element "e" may be expressed, in terms of the n nodes of the element, as in

$$\begin{bmatrix} u_0(x, y) \\ v_0(x, y) \\ w_0(x, y) \\ \phi_x(x, y) \\ \phi_y(x, y) \end{bmatrix} = \sum_{i=1}^{n} \begin{bmatrix} \varphi_i^e & 0 & 0 & 0 & 0 & 0 \\ 0 & \varphi_i^e & 0 & 0 & 0 & 0 \\ 0 & 0 & \varphi_i^e & 0 & 0 & 0 \\ 0 & 0 & 0 & \varphi_i^e & 0 & 0 \\ 0 & 0 & 0 & 0 & 0 & \varphi_i^e \end{bmatrix} \begin{bmatrix} u_j \\ v_j \\ w_j \\ s_{j1} \\ s_{j2} \end{bmatrix}, \qquad (18)$$

where φ_i^e is the Lagrange interpolation function at node i.

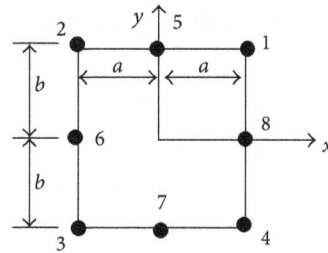

FIGURE 2: Eight-node quadratic Lagrange element.

The Lagrange interpolation functions for eight-node rectangular element (Figure 2) are given by (19) in terms of the natural coordinates:

$$\{L_e\} = \begin{bmatrix} \varphi_1 \\ \varphi_2 \\ \varphi_3 \\ \varphi_4 \\ \varphi_5 \\ \varphi_6 \\ \varphi_7 \\ \varphi_8 \end{bmatrix} = \frac{1}{4} \begin{bmatrix} (1 - \xi)(1 - \eta)(-\xi - \eta - 1) \\ (1 + \xi)(1 - \eta)(\xi - \eta - 1) \\ (1 + \xi)(1 + \eta)(\xi + \eta - 1) \\ (1 - \xi)(1 + \eta)(-\xi + \eta - 1) \\ 2(1 - \xi^2)(1 - \eta) \\ 2(1 + \xi)(1 - \eta^2) \\ 2(1 - \xi^2)(1 + \eta) \\ 2(1 - \xi)(1 - \eta^2) \end{bmatrix}. \qquad (19)$$

The nodal bending strain can be written as in

$$\{\varepsilon_b^0\} = \begin{bmatrix} u_{0,x}(x, y) \\ v_{0,x}(x, y) \\ u_{0,x}(x, y) + v_{0,x}(x, y) \\ \phi_{x,x}(x, y) \\ \phi_{y,y}(x, y) \\ \phi_{x,y}(x, y) + \phi_{y,x}(x, y) \end{bmatrix}$$

$$= \sum_{i=1}^{8} \begin{bmatrix} \varphi_i^e & 0 & 0 & 0 & 0 \\ 0 & \varphi_i^e & 0 & 0 & 0 \\ \varphi_i^e & \varphi_i^e & 0 & 0 & 0 \\ 0 & 0 & 0 & \varphi_i^e & 0 \\ 0 & 0 & 0 & 0 & \varphi_i^e \\ 0 & 0 & 0 & \varphi_i^e & \varphi_i^e \end{bmatrix} \begin{bmatrix} u_j \\ v_j \\ w_j \\ s_{j1} \\ s_{j2} \end{bmatrix} \qquad (20)$$

Or $\quad \{\varepsilon_b^0\} = \sum_{i=1}^{8} [B_{bj}]\{u_j^0\}.$

The nodal shearing strain can be written as in

$$\{\varepsilon_s^0\} = \begin{bmatrix} w_{0,x}(x,y) + \phi_x(x,y) \\ w_{0,y}(x,y) + \phi_y(x,y) \end{bmatrix}$$

$$= \sum_{i=1}^{8} \begin{bmatrix} 0 & 0 & \varphi_i^e & \varphi_i^e & 0 \\ 0 & 0 & \varphi_i^e & 0 & \varphi_i^e \end{bmatrix} \begin{bmatrix} u_j \\ v_j \\ w_j \\ s_{j1} \\ s_{j2} \end{bmatrix} \qquad (21)$$

Or $\quad \{\varepsilon_s^0\} = \sum_{i=1}^{8} [B_{sj}] \{u_j^0\},$

where $[B_{bj}]$ is the curvature-displacement matrix, $[B_{sj}]$ is the shear strain-displacement matrix, and $\{u_j^0\}$ are the nodal degrees of freedom.

The total potential energy can be given in

$$\Pi = \left(0.5 \int_A \{u^o\}^T [B_b]^T [DE_b] \{u^o\} [B_b] \, dA \right.$$

$$- \int_A \{u^o\}^T [B_b]^T [DT_b] \, dA \right)$$

$$+ 0.5 \int_A \{u^o\}^T [B_s]^T [DE_s] \{u^o\} [B_s] \, dA \qquad (22)$$

$$- \int_A \{u^o\}^T [B_s]^T [DT_s] \, dA$$

$$- \Sigma \{P\} \{u^o\}^T.$$

The minimum potential energy principle states that

$$\delta\Pi = \left(\int_A [B_b]^T [DE_b] [B_b] \, dA \right) \{u^o\}$$

$$+ \left(\int_A [B_s]^T [DE_s] [B_s] \, dA \right) \{u^o\}$$

$$- \int_A [B_b]^T [DT_b] \, dA \qquad (23)$$

$$- \int_A [B_s]^T [DT_s] \, dA - \Sigma \{P\} [\varphi_i^e]^T = 0.$$

In another form,

$$[[K_b] + [K_s]] \{u^o\} = \{F_T\} + \{P\}, \qquad (24)$$

where $[K_b]$, $[K_s]$ are the element bending and shear stiffness matrices, respectively, defined as in (25a) and (25b). $\{F_T\}$,

$\{P\}$ are the element thermal and mechanical load vectors, respectively, defined as in (26a), (26b), and (26c). Consider

$$[K_b] = \int_A [B_b]^T [DE_b] [B_b] \, dA, \qquad (25a)$$

$$[K_s] = \int_A [B_s]^T [DE_s] [B_s] \, dA, \qquad (25b)$$

$$\{P\} = \int_A [\varphi_i^e]^T \{\sigma\} \, dA, \qquad (26a)$$

$$\{F_{Tb}\} = \int_A [B_b]^T [DT_b] \, dA, \qquad (26b)$$

$$\{F_{Ts}\} = \int_A [B_s]^T [DT_s] \, dA. \qquad (26c)$$

Substituting (20) and (21) into (9) and (11), the bending and shear strain vectors can be obtained. The normal stress components can be determined as in

$$\begin{bmatrix} \sigma_{xx} \\ \sigma_{yy} \\ \sigma_{xy} \end{bmatrix} = \begin{bmatrix} \overline{Q_{11}} & \overline{Q_{12}} & 0 \\ \overline{Q_{12}} & \overline{Q_{22}} & 0 \\ 0 & 0 & \overline{Q_{66}} \end{bmatrix}$$

$$\times \left[\begin{bmatrix} \varepsilon_{xx} \\ \varepsilon_{yy} \\ \gamma_{xy} \end{bmatrix} - \begin{bmatrix} \alpha(z)\delta T(z) \\ \alpha(z)\delta T(z) \\ 0 \end{bmatrix} \right] \qquad (27)$$

Or $\quad \{\sigma_b\} = [D_b] \{\varepsilon_b\}.$

The shear stress components are determined as in

$$\begin{bmatrix} \sigma_{yz} \\ \sigma_{xz} \end{bmatrix} = \begin{bmatrix} \overline{Q_{44}} & 0 \\ 0 & \overline{Q_{55}} \end{bmatrix} \begin{bmatrix} \gamma_{xz}(x,y,z) \\ \gamma_{yz}(x,y,z) \end{bmatrix} \left[\begin{bmatrix} \gamma_{yz} \\ \gamma_{xz} \end{bmatrix} - \begin{bmatrix} 0 \\ 0 \end{bmatrix} \right] \qquad (28)$$

Or $\quad \{\sigma_s\} = [D_s] \{\varepsilon_s\}.$

4. Numerical Examples

To ascertain the accuracy and proficiency of the present finite element formulation, two examples have been analysed for thermomechanical deformations of the FGM plates. In this section, the finite element formulation of FGM plate described in previous section has been applied to a few problems to test its validity, versatility, and accuracy.

Example 1. The results are compared with those given by Ferreira et al. [8] in which meshless collocation method and first order shear deformation theory have been used. A square (1 m × 1 m) simply supported FGM plate is considered for the investigation. The plate is made of a ceramic material (ZrO_2 : E = 151 GPa, ν = 0.3) at the top surface and metallic material (Al : E = 70 Gpa, ν = 0.3) at the bottom surface. The length to thickness ratio (a/h) was taken to be 20. Volume fraction exponent studied was n = 0; pure ceramic 0.5, 1.0, 2, and ∞; pure metal. Where E is modulus of elasticity and ν is Poisson's ratio, the FGM plate problem with

TABLE 1: Transverse displacement of a simply supported (SSSS) square FGM plate subjected to UDL.

Volume fraction exponent (n)	Transverse displacement $\overline{u_z} = u_z/h$		% Difference
	Ferreira et al. [8]	Present results	
Ceramic (0)	0.0205	0.0203	0.97
0.5	0.0262	0.0264	0.76
1	0.0294	0.0297	1.02
2	0.0323	0.0326	0.93
Metal ($\propto$)	0.0443	0.0447	0.90

abovementioned inputs is solved by proposed finite element formulation, implemented through ANSYS Software. The properties were calculated using power law function and external mechanical load in the form of uniformly distributed load which have been applied with intensity of 10 kN/m^2. The numerical solution of this FGM plate for nondimensional transverse displacement $\overline{u_z} = u_z/h$ is computed for different values of volume fraction exponent "n." These results are reported in Table 1.

The comparison of present results with those of Ferreira et al.'s [8] shows that the results obtained by proposed finite element methodology is in good agreement with published results. The difference between the two results is nearly 1%.

Example 2. In this example, the results are compared with those given by Chi and Chung [10] in which classical plate theory and Fourier series expansion have been used. A simply supported FGM plate is considered for the investigation. The aspect ratio (a/b) of the plate in which $a = 0.1$ m and "b" is kept varying. Also the length to thickness ratio a/h is taken as 50. The properties at the top surface are $E = 21$ GPa, $\nu = 0.3$ and those at the bottom surface are $E = 210$ GPa, $\nu = 0.3$. Volume fraction exponent studied was $n = 2$ where E is modulus of elasticity and ν is Poisson's ratio. The FGM plate problem with abovementioned inputs is solved by proposed finite element formulation, implemented through ANSYS Software. The properties were calculated using sigmoid law function and external mechanical load in the form of uniformly distributed load which have been applied with intensity of 10 kN/m^2. In order to study the convergence, nondimensional transverse displacement $\overline{u_z} = u_z/h$ and nondimensional tensile stress $\overline{S_y} = S_y/p$ are compared with those of the published results. The comparison of present results with Chi and Chung's [10] are presented in Figures 3 and 4.

Figures 3 and 4 show the variation of maximum transverse displacement and maximum tensile stress for varying aspect ratios (a/b). An excellent agreement between the present and published results can be observed. The results show that the performance of the present formulation is very good in terms of solution accuracy.

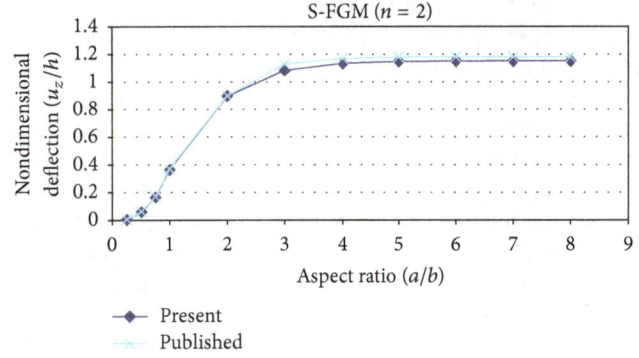

FIGURE 3: Maximum transverse displacement of an S-FGM plate versus aspect ratio (a/b).

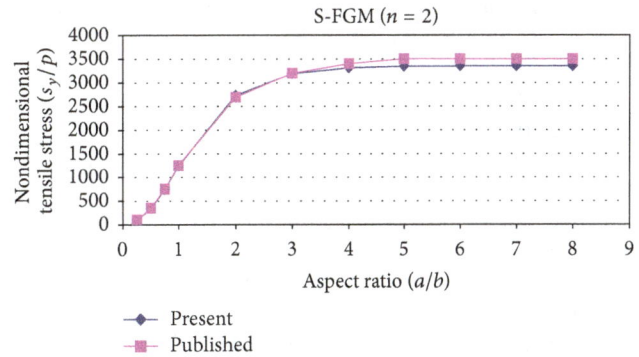

FIGURE 4: Maximum tensile stress of an S-FGM plate versus aspect ratio (a/b).

5. Thermal and Thermomechanical Analysis

The thermal and thermomechanical analysis is conducted for FGM made of combination of metal and ceramic. The metal and ceramic chosen are aluminium and zirconia, respectively. Young's modulus for aluminium is 70 GPa and that for zirconia is 151 GPa. The coefficient of thermal expansion for aluminium is $23 \times 10^{-6}/°$C and that for zirconia is $10 \times 10^{-6}/°$C. Poisson's ratio for both the materials was chosen to be 0.3. The effect of Poisson's ratio on the deformation is much less as compared to that of Young's modulus [1]. Thermal analysis was performed by applying thermal load on the FGM plate. The ceramic top surface is exposed to a temperature of 100°C. The lower metallic surface and all the edges are kept at a temperature of 0°C. The thermomechanical analysis has been performed by applying uniformly distributed load ($1E6$ N/m^2) along with thermal load on FGM plate for various boundary conditions. Various boundary conditions of plate used for the analysis are as follows: all edges are simply supported (SSSS), all edges are clamped (CCCC), alternate edges are simply supported and clamped (SCSC), alternate edges are clamped and free (CFCF), two edges are clamped and two are free (CCFF), two edges are clamped and two are simply supported (CCSS), two edges are simply supported and two are free (SSFF), three edges are simply supported and one is clamped (SSSC), three edges are simply supported and one is free (SSSF), and two edges are simply supported, one edge is clamped, and one is kept free (SSCF). A square FGM plate is considered here. The thickness of the plate (h) is taken as 0.02 m and the side lengths are taken as 1 m; that is, aspect ratio is taken unity and the length to thickness ratio is 50. The uniformly distributed load (udl) was equal to 1×10^6 N/m^2. The analysis is performed for E-FGM and for various values of the volume fraction exponent (n) in P-FGM and S-FGM. The results are presented in terms of nondimensional parameters, that is, nondimensional deflection ($\overline{u_z}$), nondimensional tensile stress ($\overline{\sigma_x}$), and nondimensional shear stress ($\overline{\sigma_{xy}}$).

The various nondimensional parameters used are as follows.

Nondimensional deflection $\overline{u_z} = u_z/h$, nondimensional tensile stress ($\overline{\sigma_x}$) = σ_x/p_o, and nondimensional shear stress ($\overline{\sigma_{xy}}$) = σ_{xy}/p_o.

"u_z" is deflection, "σ" is stress, "h" is plate thickness, "a" and "b" are side lengths of plate, and "p_o" is applied load ($1E6$ N/m^2).

The material properties of the FGM vary throughout the thickness; the numerical model is to be broken up into number of "layers" in order to capture the change in properties. These "layers" are of finite thickness and are treated like isotropic materials. Material properties are calculated using various volume fraction distribution laws. The "layers" and their associated properties are then layered together to establish the through-thickness variation of material properties. Although the layered structure does not reflect the gradual change in material properties, a sufficient number of "layers" can reasonably approximate the material gradation.

In this paper, the modeling and analysis of FGM plate is carried out using ANSYS-APDL Software. ANSYS offers a number of elements to choose from for the modeling of gradient materials. An eight-node quadratic Lagrange element with six degrees of freedom at each node for the present model is used. The FGM characteristics under thermal and thermomechanical loads are studied on a flat plate. Based on the established approach and analysis 100 × 100 mesh has been used for the analysis. These have been used for computing results unless it is stated otherwise.

5.1. Variation of Boundary Conditions in Thermal Environment. This section discusses the results of the analyses performed on FGM plate with various boundary conditions subject to constant thermal environment. The results are presented in terms of nondimensional parameters, that is, nondimensional deflection ($\overline{u_z}$), nondimensional tensile stress ($\overline{\sigma_x}$), and nondimensional shear stress ($\overline{\sigma_{xy}}$).

5.1.1. Nondimensional Deflection ($\overline{u_z}$). Tables 2 and 3 show the nondimensional deflection $\overline{u_z}$ for various boundary conditions of a square plate in constant thermal environment for P-FGM, S-FGM, and E-FGM, respectively. In case of P-FGM and S-FGM, the comparison of various values of volume fraction exponent (n) has been presented. The following can be observed from Tables 2 and 3.

(a) The metal plate has the largest deflection for all the boundary conditions considered here as compared to

the other FGM plate. The deflection values of FGM plate are much lower than those of fully metal plate. This clearly shows that the FGM plate can resist high-temperature conditions very well.

(b) The nondimensional deflection in the ceramic rich portion may be comparable to that in the metal rich region, because the ceramic has a lower coefficient of thermal expansion than that of the metal. Hence, the nondimensional deflection depends on the product of the temperature and the thermal expansion coefficient. Therefore, the response of the graded plate is not intermediate to the metal and ceramic plate.

(c) The maximum deflection occurs for clamped free (CCFF) boundary conditions and minimum deflection occurs for clamped (CCCC) boundary condition among all the cases considered here.

5.1.2. Nondimensional Tensile Stress ($\overline{\sigma_x}$). Tables 4 and 5 show the variation of nondimensional tensile stress ($\overline{\sigma_x}$) for various boundary conditions of a square plate in thermal environment for P-FGM, S-FGM, and E-FGM, respectively. In case of P-FGM and S-FGM, the comparison of various values of volume fraction exponent (n) has been presented.

A close study of Tables 4 and 5 reveals the following.

(a) The nondimensional tensile stress in the ceramic rich portion may be comparable to that in the metal rich region, because the ceramic has a lower coefficient of thermal expansion than the metal and, at the same time, ceramic has more stiffness than that of the metal. Hence, the nondimensional tensile stress depends on the product of the modulus of elasticity and the thermal expansion coefficient. Therefore, the response of the graded plate is not intermediate to the metal and ceramic plate.

(b) The tensile stress increases with increasing volume fraction exponent "n" for the FGM plate.

(c) The maximum tensile stress occurs for simply supported clamped free (SSCF) boundary conditions and minimum tensile stress occurs for clamped free (CFCF) boundary condition among all the cases considered here.

5.1.3. Nondimensional Shear Stress ($\overline{\sigma_{xy}}$). Tables 6 and 7 show the variation of nondimensional shear stress ($\overline{\sigma_{xy}}$) for various boundary conditions of a square plate in thermal environment for P-FGM, S-FGM, and E-FGM, respectively. In case of P-FGM and S-FGM, the comparison of various values of volume fraction exponent (n) has been presented.

The following can be observed from Tables 6 and 7.

(a) The isotropic ceramic and metal plate has the lowest shear stress for all the boundary conditions considered here.

(b) The shear stress becomes higher with increasing n for the FGM plates.

(c) The maximum shear stress occurs for simply supported clamped (SSSC) boundary conditions and

TABLE 2: Nondimensional deflection $(\overline{u_z})$ for various boundary conditions of a square plate in thermal environment for P-FGM and E-FGM.

| BC | P-FGM | | | | | | | | | | E-FGM |
	$n = 0$	0.1	0.2	0.5	1	2	5	10	100	$\propto$	
SSSS	0.46	0.10	0.15	0.23	0.29	0.33	0.35	0.41	0.46	0.52	0.30
CCCC	0.01	0.00	0.00	0.01	0.01	0.01	0.01	0.01	0.01	0.01	0.01
SCSC	0.10	0.02	0.03	0.05	0.06	0.07	0.08	0.09	0.10	0.11	0.07
CFCF	0.22	0.05	0.07	0.11	0.14	0.16	0.17	0.20	0.22	0.25	0.15
CCFF	1.98	0.44	0.65	1.01	1.25	1.40	1.52	1.75	1.97	2.22	1.29
CCSS	0.18	0.04	0.06	0.09	0.11	0.13	0.14	0.16	0.18	0.20	0.12
SSFF	1.77	0.40	0.58	0.90	1.11	1.25	1.35	1.56	1.76	1.98	1.15
SSSC	0.27	0.06	0.09	0.14	0.17	0.19	0.21	0.24	0.27	0.30	0.18
SSSF	0.63	0.14	0.21	0.32	0.39	0.44	0.48	0.55	0.62	0.70	0.41
SSCF	0.27	0.06	0.09	0.14	0.17	0.19	0.21	0.24	0.27	0.30	0.18

TABLE 3: Nondimensional deflection $(\overline{u_z})$ for various boundary conditions of a square plate in thermal environment for S-FGM.

| BC | S-FGM | | | | | | | | | |
	$n = 0$	0.1	0.2	0.5	1	2	5	10	100	$\propto$
SSSS	0.46	0.10	0.12	0.14	0.19	0.21	0.25	0.30	0.35	0.52
CCCC	0.01	0.00	0.00	0.00	0.00	0.00	0.01	0.01	0.01	0.01
SCSC	0.10	0.02	0.03	0.03	0.04	0.05	0.05	0.07	0.08	0.11
CFCF	0.22	0.05	0.06	0.07	0.09	0.10	0.12	0.15	0.17	0.25
CCFF	1.98	0.42	0.50	0.62	0.82	0.89	1.06	1.28	1.50	2.22
CCSS	0.18	0.04	0.05	0.06	0.07	0.08	0.10	0.12	0.14	0.20
SSFF	1.77	0.38	0.45	0.55	0.73	0.79	0.95	1.15	1.34	1.98
SSSC	0.27	0.06	0.07	0.08	0.11	0.12	0.15	0.18	0.21	0.30
SSSF	0.63	0.13	0.16	0.20	0.26	0.28	0.34	0.41	0.48	0.70
SSCF	0.27	0.06	0.07	0.08	0.11	0.12	0.15	0.18	0.20	0.30

TABLE 4: Nondimensional tensile stress $(\overline{\sigma_x})$ for various boundary conditions (BC) of a square plate in thermal environment for P-FGM and E-FGM.

| BC | P-FGM | | | | | | | | | | E-FGM |
	$n = 0$	0.1	0.2	0.5	1	2	5	10	100	$\propto$	
SSSS	8.03	19.10	22.04	27.87	37.40	43.25	68.56	86.49	106.73	12.34	37.90
CCCC	41.82	99.43	114.71	145.09	194.67	225.14	356.86	450.21	555.57	64.25	197.27
SCSC	14.42	34.29	39.57	50.04	67.14	77.65	123.08	155.28	191.62	22.16	68.04
CFCF	4.66	11.08	12.79	16.17	21.70	25.10	39.78	50.18	61.93	7.16	21.99
CCFF	29.22	69.47	80.15	101.38	136.02	157.31	249.34	314.57	388.19	44.90	137.84
CCSS	9.08	21.59	24.91	31.51	42.28	48.89	77.50	97.77	120.65	13.95	42.84
SSFF	23.65	56.22	64.87	82.05	110.08	127.31	201.79	254.58	314.16	36.33	111.55
SSSC	5.73	13.62	15.71	19.87	26.66	30.83	48.87	61.65	76.08	8.80	27.01
SSSF	50.08	119.06	137.37	173.75	233.11	269.60	427.33	539.12	665.29	76.94	236.23
SSCF	52.03	123.71	142.73	180.52	242.21	280.12	444.00	560.16	691.25	79.95	245.45

minimum shear stress occurs for clamped (CCCC) boundary condition among all the cases considered here.

The nondimensional deflection, tensile stress, and shear stress for S-FGM remain closer for various values of "n" as compared to those of P-FGM since material gradation is more uniform in S-FGM as compared to P-FGM.

5.2. Comparison of P-FGM, S-FGM, E-FGM, Ceramic, and Metal. It is also interesting to see the comparison of various parameters like nondimensional deflection, tensile stress,

TABLE 5: Nondimensional tensile stress $(\overline{\sigma_x})$ for various boundary conditions (BC) of a square plate in thermal environment for S-FGM.

BC	S-FGM									
	$n = 0$	0.1	0.2	0.5	1	2	5	10	100	∞
SSSS	8.03	18.34	25.00	31.04	37.40	40.04	43.60	45.60	47.44	12.34
CCCC	41.82	95.49	130.16	161.59	194.67	208.44	226.95	237.36	246.93	64.25
SCSC	14.42	32.93	44.89	55.73	67.14	71.89	78.28	81.87	85.17	22.16
CFCF	4.66	10.64	14.51	18.01	21.70	23.23	25.30	26.46	27.52	7.16
CCFF	29.22	66.72	90.94	112.91	136.02	145.64	158.58	165.85	172.53	44.90
CCSS	9.08	20.74	28.27	35.09	42.28	45.27	49.29	51.55	53.63	13.95
SSFF	23.65	54.00	73.60	91.38	110.08	117.87	128.33	134.22	139.63	36.33
SSSC	5.73	13.08	17.82	22.13	26.66	28.54	31.08	32.50	33.81	8.80
SSSF	50.08	114.35	155.86	193.50	233.11	249.60	271.77	284.23	295.69	76.94
SSCF	52.03	118.81	161.94	201.06	242.21	259.34	282.37	295.32	307.23	79.95

TABLE 6: Nondimensional shear stress $(\overline{\sigma_{xy}})$ for various boundary conditions (BC) of a square plate in thermal environment for P-FGM and E-FGM.

BC	P-FGM										E-FGM
	$n = 0$	0.1	0.2	0.5	1	2	5	10	100	∞	
SSSS	378.1	423.4	426.5	429.6	465.0	515.7	580.6	642.4	644.9	400.2	487.0
CCCC	97.9	109.6	110.4	111.2	120.4	133.5	150.3	166.3	166.9	103.6	126.1
SCSC	324.5	363.4	366.1	368.6	399.1	442.6	498.3	551.3	553.5	343.4	418.0
CFCF	237.3	265.7	267.7	269.5	291.8	323.6	364.3	403.1	404.7	251.1	305.6
CCFF	238.7	267.3	269.3	271.2	293.5	325.6	366.5	405.5	407.1	252.6	307.4
CCSS	331.2	370.9	373.6	376.3	407.3	451.8	508.6	562.7	564.9	350.5	426.6
SSFF	224.1	250.9	252.7	254.5	275.5	305.6	344.0	380.6	382.1	237.1	288.6
SSSC	397.6	445.2	448.5	451.7	488.9	542.3	610.5	675.5	678.2	420.8	512.1
SSSF	351.9	394.0	396.9	399.7	432.7	479.9	540.3	597.8	600.2	372.4	453.2
SSCF	258.2	289.1	291.3	293.3	317.5	352.2	396.4	438.7	440.4	273.2	332.5

TABLE 7: Nondimensional shear stress $(\overline{\sigma_{xy}})$ for various boundary conditions (BC) of a square plate in thermal environment for S-FGM.

BC	S-FGM									
	$n = 0$	0.1	0.2	0.5	1	2	5	10	100	∞
SSSS	378.1	386.2	410.9	437.1	465.0	477.6	537.6	594.9	597.2	400.2
CCCC	97.9	100.0	106.3	113.1	120.4	123.6	139.2	154.0	154.6	103.6
SCSC	324.5	331.5	352.6	375.1	399.1	409.9	461.4	510.5	512.5	343.4
CFCF	237.3	242.3	257.8	274.3	291.8	299.7	337.4	373.3	374.8	251.1
CCFF	238.7	243.8	259.4	275.9	293.5	301.5	339.4	375.5	377.0	252.6
CCSS	331.2	338.3	359.9	382.9	407.3	418.3	470.9	521.1	523.1	350.5
SSFF	224.1	228.8	243.4	259.0	275.5	283.0	318.6	352.5	353.9	237.1
SSSC	397.6	406.1	432.0	459.6	488.9	502.2	565.3	625.5	628.0	420.8
SSSF	351.9	359.4	382.3	406.7	432.7	444.4	500.3	553.6	555.8	372.4
SSCF	258.2	263.7	280.5	298.4	317.5	326.1	367.1	406.2	407.8	273.2

shear stress, transverse strain and shear strain for ceramic, metal, and FGMs following power law, sigmoid, and exponential distribution. Figures 5, 6, and 7 show the comparison graphs for pure ceramic ($n = 0$), pure metal ($n = \infty$), P-FGM ($n = 2$), P-FGM ($n = 0.5$), S-FGM ($n = 2$), S-FGM ($n = 0.5$), and E-FGM.

5.2.1. Nondimensional Deflection $(\overline{u_z})$. The following is observed from Figures 5, 6, and 7.

(a) The nondimensional deflection for the three FGMs is more than that of the ceramic and metal.

(b) The nondimensional parameters, for example, tensile stress and shear stress, for the three FGMs are less than those of the ceramic and metal.

(c) It is evident from the above comparison that P-FGM ($n = 0.5$) plate has the smallest deflection and stress among all kinds of FGM plates. The reason is

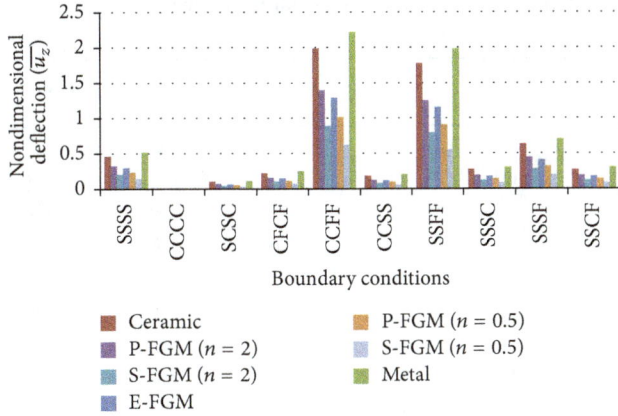

FIGURE 5: Nondimensional deflection ($\overline{u_z}$) for various boundary conditions of a square plate for various FGMs in thermal environment.

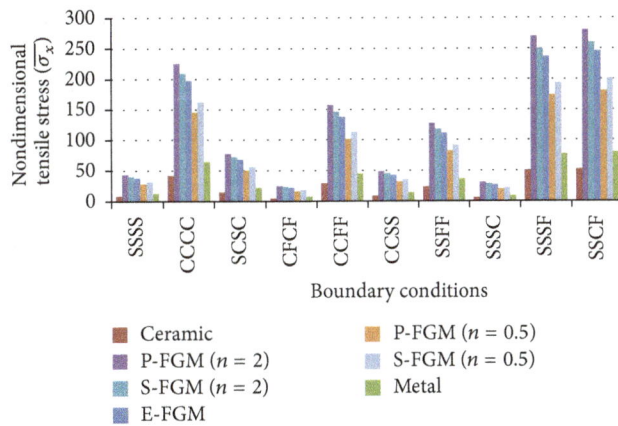

FIGURE 7: Nondimensional shear stress ($\overline{\sigma_{xy}}$) for various boundary conditions of a square plate for various FGMs in thermal environment.

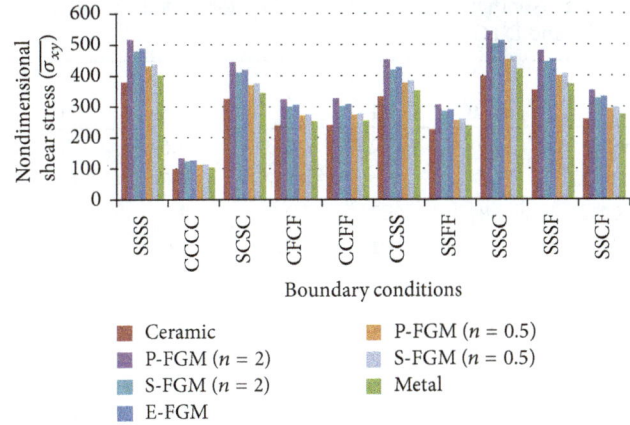

FIGURE 6: Nondimensional tensile stress ($\overline{\sigma_x}$) for various boundary conditions of a square plate for various FGMs in thermal environment.

observable that the coefficient of thermal expansion of the P-FGM ($n = 0.5$) plate is more than that of E-FGM plate and stiffness of the E-FGM plate is more than that of P-FGM ($n = 2$).

5.2.2. Nondimensional Tensile Stress ($\overline{\sigma_x}$). See Figure 6.

5.2.3. Nondimensional Shear Stress ($\overline{\sigma_{xy}}$). See Figure 7.

5.3. Variation of Boundary Condition in under UDL in Thermal Environment. This section discusses the results of the analyses performed on FGM plate with various boundary conditions subject to constant UDL in thermal environment. The results are presented in terms of nondimensional parameters, that is, nondimensional deflection ($\overline{u_z}$), nondimensional tensile stress ($\overline{\sigma_x}$), and nondimensional shear stress ($\overline{\sigma_{xy}}$).

5.3.1. Nondimensional Deflection ($\overline{u_z}$). Tables 8 and 9 show nondimensional deflection ($\overline{u_z}$) for various boundary conditions of a square plate under uniformly distributed load

in thermal environment for P-FGM, S-FGM, and E-FGM, respectively.

A study of Tables 8 and 9 reveals the following information.

(a) The nondimensional deflection in the ceramic rich portion may be comparable to that in the metal rich region. The nondimensional deflection is maximum for the case of pure metal ($n = \infty$) and pure ceramic ($n = 0$). The nondimensional deflection of both the metallic and the ceramic plates is higher in magnitude than the graded plates. The deflection therefore depends on the product of the temperature and the thermal expansion coefficient. Therefore, the response of the graded plates is not intermediate to the metal and ceramic plates.

(b) The deflections become higher with increasing n. It is observed that when thermal effect is induced, the bending response of the functionally graded plate is not necessarily intermediate to those of the metal and the ceramic plate.

(c) It is also found that the maximum deflection occurs for simply supported free (SSFF) boundary conditions and minimum deflection occurs for clamped (CCCC) boundary condition for all the cases considered here.

5.3.2. Nondimensional Tensile Stress ($\overline{\sigma_x}$). Tables 10 and 11 show the variation of nondimensional tensile stress ($\overline{\sigma_x}$) for various boundary conditions of a square plate under uniformly distributed load in thermal environment for P-FGM, S-FGM, and E-FGM, respectively. In case of P-FGM and S-FGM, the comparison of various values of volume fraction exponent (n) have been presented.

The following can be observed from Tables 10 and 11.

(a) The isotropic ceramic and metallic plates have the lowest tensile stress for all the boundary conditions

TABLE 8: Nondimensional deflection ($\overline{u_z}$) for various boundary conditions (BC) of a square plate under udl in thermal environment for P-FGM and E-FGM.

BC	P-FGM										E-FGM
	$n = 0$	0.1	0.2	0.5	1	2	5	10	100	∞	
SSSS	3.9	2.2	2.3	2.6	3.0	3.2	3.4	3.5	3.8	4.0	3.1
CCCC	1.2	0.6	0.7	0.7	0.8	0.9	1.0	1.0	1.1	1.2	0.9
SCSC	1.8	1.0	1.0	1.2	1.3	1.4	1.5	1.6	1.7	1.9	1.4
CFCF	2.7	1.4	1.5	1.6	1.7	1.9	2.1	2.3	2.6	2.9	1.8
CCFF	40.7	21.5	22.3	24.5	27.1	29.6	32.4	34.7	38.9	42.9	28.0
CCSS	2.1	1.1	1.2	1.4	1.5	1.7	1.8	1.9	2.0	2.2	1.6
SSFF	169.3	91.0	95.3	111.3	125.4	136.4	139.2	148.1	158.6	178.2	121.8
SSSC	2.7	1.5	1.6	1.8	2.0	2.2	2.4	2.4	2.6	2.8	2.1
SSSF	12.1	6.6	6.9	7.8	8.7	9.5	10.2	10.7	11.7	12.7	9.0
SSCF	5.6	3.0	3.2	3.5	3.9	4.3	4.6	4.9	5.4	5.9	4.0

TABLE 9: Nondimensional deflection ($\overline{u_z}$) for various boundary conditions (BC) of a square plate under udl in thermal environment for S-FGM.

BC	S-FGM									
	$n = 0$	0.1	0.2	0.5	1	2	5	10	100	∞
SSSS	3.9	2.7	2.7	2.8	3.0	3.1	3.2	3.3	3.3	4.0
CCCC	1.2	0.8	0.8	0.8	0.8	0.8	0.9	0.9	0.9	1.2
SCSC	1.8	1.2	1.2	1.3	1.3	1.4	1.4	1.4	1.4	1.9
CFCF	2.7	1.8	1.8	1.7	1.7	1.8	1.8	1.8	1.8	2.9
CCFF	40.7	26.9	26.8	26.9	27.1	27.5	28.0	28.1	28.2	42.9
CCSS	2.1	1.4	1.4	1.5	1.5	1.6	1.7	1.7	1.7	2.2
SSFF	169.3	113.4	115.5	120.8	125.4	131.6	136.5	137.8	138.2	178.2
SSSC	2.7	1.9	1.9	2.0	2.0	2.1	2.2	2.2	2.2	2.8
SSSF	12.1	8.2	8.3	8.5	8.7	9.0	9.3	9.3	9.4	12.7
SSCF	5.6	3.8	3.8	3.8	3.9	4.0	4.1	4.1	4.1	5.9

TABLE 10: Nondimensional tensile stress ($\overline{\sigma_x}$) for various boundary conditions (BC) of a square plate under udl in thermal environment for P-FGM and E-FGM.

BC	P-FGM										E-FGM
	$n = 0$	0.1	0.2	0.5	1	2	5	10	100	∞	
SSSS	400.8	418.9	421.0	438.7	479.6	534.5	570.4	584.2	613.1	380.7	488.8
CCCC	520.5	566.0	586.8	637.2	688.9	732.8	778.9	807.4	769.4	494.5	710.9
SCSC	148.0	155.8	155.8	185.1	212.4	232.8	417.6	277.4	303.8	140.6	220.7
CFCF	1130.4	1205.6	1244.5	1342.6	1445.5	1536.1	1632.3	1690.8	1628.1	1073.9	1472.3
CCFF	4170.4	4381.9	4482.4	4744.8	5041.1	5330.3	5699.6	5962.2	5857.1	3961.9	5143.8
CCSS	893.0	990.7	1037.2	1141.8	1241.5	1321.3	1388.6	1416.8	1327.7	848.4	1256.9
SSFF	1249.9	1315.7	1307.7	1377.9	1506.4	1678.8	1609.6	1834.7	1925.7	1187.4	1416.5
SSSC	256.0	269.5	269.9	282.7	318.1	358.9	393.1	408.7	436.1	243.2	326.3
SSSF	1231.7	1296.5	1288.7	1300.9	1374.3	1495.6	1591.7	1636.7	1720.4	1170.1	1390.8
SSCF	1937.0	2078.3	2148.8	2317.8	2492.4	2646.8	2808.6	2901.4	2787.5	1840.2	2525.5

considered here. In the presence of the above temperature field, compression occurs at the top surface while tension is at the bottom surface. Excepting fully ceramic or fully metal plates, the stress distribution of FGM plates has a similar trend. The nondimensional tensile stress is minimum for the case of pure metal ($n = \infty$) and pure ceramic ($n = 0$).

(b) The tensile stress becomes higher with increasing n. This is due to the fact that the bending stiffness is the maximum for ceramic plate, while being minimal for metallic plate, and degrades continuously as n increases.

(c) The nondimensional tensile stress therefore depends on the product of the temperature and the thermal

TABLE 11: Nondimensional tensile stress ($\overline{\sigma}_x$) for various boundary conditions (BC) of a square plate under udl in thermal environment for S-FGM.

| BC | S-FGM | | | | | | | | | |
	$n=0$	0.1	0.2	0.5	1	2	5	10	100	∞
SSSS	400.8	490.2	483.6	475.2	479.6	500.6	528.0	535.8	538.4	380.7
CCCC	520.5	623.0	636.9	666.2	688.9	698.9	698.8	698.5	698.5	494.5
SCSC	148.0	217.9	209.7	196.1	212.4	222.0	223.1	222.8	222.4	140.6
CFCF	1130.4	1346.1	1366.8	1411.1	1445.5	1459.6	1457.0	1455.2	1454.1	1073.9
CCFF	4170.4	4883.3	4923.0	4997.3	5041.1	5038.1	5006.9	4998.4	4995.4	3961.9
CCSS	893.0	1109.6	1137.4	1195.6	1241.5	1264.5	1267.0	1265.8	1264.5	848.4
SSFF	1249.9	1469.0	1449.3	1424.1	1506.4	1572.3	1658.4	1682.8	1690.8	1187.4
SSSC	256.0	334.0	326.8	315.8	318.1	330.8	343.2	347.8	349.2	243.2
SSSF	1231.7	1456.7	1430.6	1386.7	1374.3	1405.3	1458.2	1473.7	1478.8	1170.1
SSCF	1937.0	2332.1	2367.8	2439.2	2492.4	2511.2	2504.2	2499.9	2497.4	1840.2

TABLE 12: Nondimensional shear stress ($\overline{\sigma}_{xy}$) for various boundary conditions (BC) of a square plate under udl in thermal environment for P-FGM and E-FGM.

| BC | P-FGM | | | | | | | | | | E-FGM |
	$n=0$	0.1	0.2	0.5	1	2	5	10	100	∞	
SSSS	544.6	605.2	614.8	635.1	668.7	713.7	734.5	735.9	745.2	517.4	657.7
CCCC	129.2	135.5	139.2	148.2	157.6	165.9	175.6	182.4	178.3	122.7	160.6
SCSC	294.5	310.0	322.6	349.7	379.8	420.4	294.9	499.9	468.0	279.8	351.6
CFCF	365.6	286.2	287.8	291.8	304.7	329.0	351.9	360.0	369.0	347.4	296.6
CCFF	750.7	780.3	800.0	849.8	904.3	957.3	1023.0	1065.9	1041.6	713.1	917.4
CCSS	339.4	315.5	328.4	356.1	386.8	428.2	486.6	508.9	476.3	322.5	358.1
SSFF	2745.2	2889.7	2979.2	3032.5	3193.1	3408.0	3445.8	3513.8	3558.5	2608.0	3270.0
SSSC	530.3	558.2	566.8	584.6	615.0	658.0	681.8	685.3	694.3	503.8	601.7
SSSF	688.4	724.6	733.1	753.6	791.9	842.9	864.4	867.4	885.5	653.9	785.2
SSCF	309.2	286.2	295.3	315.3	329.9	332.3	373.1	401.1	409.5	293.7	352.3

expansion coefficient. Therefore, the response of the graded plates is not intermediate to the metal and ceramic plates.

(d) It is also found that the maximum tensile stress occurs for simply supported free (CCFF) boundary conditions and minimum tensile stress occurs for clamped (SCSC) boundary condition for all the cases considered here.

5.3.3. Nondimensional Shear Stress ($\overline{\sigma}_{xy}$).

Tables 12 and 13 show the variation of nondimensional shear stress ($\overline{\sigma}_{xy}$) for various boundary conditions of a square plate under uniformly distributed load in thermal environment for P-FGM, S-FGM, and E-FGM, respectively. In case of P-FGM and S-FGM, the comparison of various values of volume fraction exponent (n) has been presented.

The following can be observed from Tables 12 and 13.

(a) The isotropic ceramic and metallic plate has the lowest shear stress for all the boundary conditions considered here. In the presence of the above temperature field, compression occurs at the top surface while tension is at the bottom surface. Excepting fully ceramic or fully metal plates, the stress distribution of FGM plates has a similar trend. The nondimensional

tensile stress is minima for the case of pure metal ($n = \infty$) and pure ceramic ($n = 0$).

(b) The shear stress becomes higher with increasing n. This is due to the fact that the bending stiffness is maximum for ceramic plate, while being minimal for metallic plate, and degrades continuously as n increases.

(c) The response of the graded plates is not intermediate to the metal and ceramic plates.

(d) It is also found that the maximum shear stress occurs for simply supported free (SSFF) boundary conditions and minimum shear stress occurs for clamped (CCCC) boundary condition for all the cases considered here.

The nondimensional deflection, tensile stress, and shear stress for S-FGM remain closer for various values of "n" as compared to those of the P-FGM since material gradation is more uniform in S-FGM as compared to P-FGM.

5.4. Comparison of P-FGM, S-FGM, E-FGM, Ceramic, and Metal.

It is also interesting to see the comparison of various parameters like nondimensional deflection, tensile stress, shear stress, transverse strain, and shear strain for ceramic,

TABLE 13: Nondimensional shear stress ($\overline{\sigma_{xy}}$) for various boundary conditions (BC) of a square plate under udl in thermal environment for S-FGM.

BC	S-FGM									
	$n = 0$	0.1	0.2	0.5	1	2	5	10	100	$\propto$
SSSS	544.6	675.0	670.6	664.5	668.7	688.6	713.4	718.8	719.5	517.4
CCCC	129.2	150.0	151.7	155.1	157.6	158.5	158.4	158.4	158.4	122.7
SCSC	294.5	410.5	408.6	398.5	379.8	354.4	328.4	318.6	313.4	279.8
CFCF	365.6	330.9	324.6	312.2	304.7	304.9	309.4	310.2	309.9	347.4
CCFF	750.7	875.7	883.1	896.7	904.3	903.3	897.3	895.4	894.6	713.1
CCSS	339.4	417.9	416.0	405.8	386.8	361.0	334.5	324.6	319.2	322.5
SSFF	2745.2	3169.0	3148.4	3119.6	3193.1	3288.1	3406.6	3432.4	3435.7	2608.0
SSSC	530.3	629.7	624.2	614.9	615.0	629.5	649.1	653.1	653.2	503.8
SSSF	688.4	797.3	792.0	785.3	791.9	817.8	850.0	857.6	859.0	653.9
SSCF	309.2	337.7	331.4	315.2	329.9	350.1	366.4	372.2	375.1	293.7

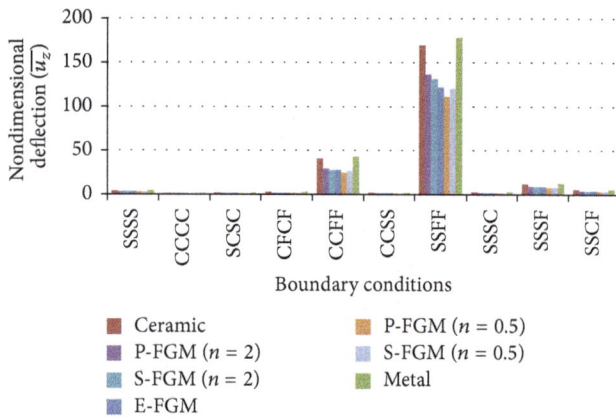

FIGURE 8: Nondimensional deflection ($\overline{u_z}$) for various boundary conditions of a square plate under uniformly distributed load for various FGMs in thermal environment.

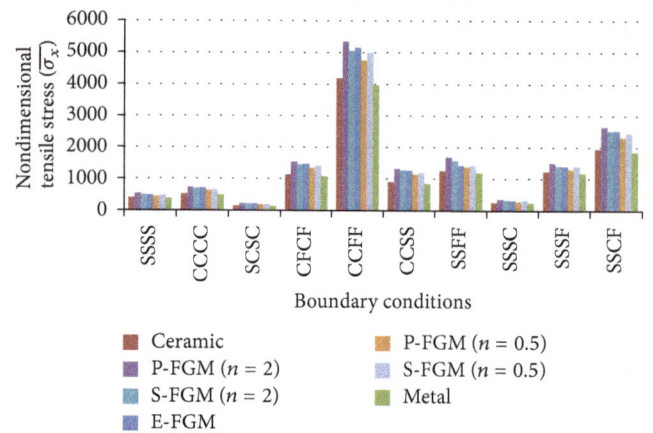

FIGURE 9: Nondimensional tensile stress ($\overline{\sigma_x}$) for various boundary conditions of a square plate under uniformly distributed load for various FGMs in thermal environment.

metal, and FGMs following power law, sigmoid, and exponential distribution. Figures 8, 9, and 10 show the comparison graphs for pure ceramic ($n = 0$), pure metal ($n = \infty$), P-FGM ($n = 2$), P-FGM ($n = 0.5$), S-FGM ($n = 2$), S-FGM ($n = 0.5$), and E-FGM.

5.4.1. Nondimensional Deflection ($\overline{u_z}$). See Figure 8.

5.4.2. Nondimensional Tensile Stress ($\overline{\sigma_x}$). See Figure 9.
   The following is observed from Figures 8, 9, and 10:

(a) The nondimensional parameters deflection, strain, and shear strain for the three FGMs are maximum for the ceramic and metal.

(b) The nondimensional parameters tensile stress and shear stress for the three FGMs are minimum for the ceramic and metal.

(c) P-FGM ($n = 0.5$) plate has the smallest deflection and stress among all kinds of FGM plate. The reason is observable in which the stiffness of the P-FGM ($n = 0.5$) plate is more than that of E-FGM plate and

stiffness of the E-FGM plate is more than that of P-FGM ($n = 2$).

5.4.3. Nondimensional Shear Stress ($\overline{\sigma_{xy}}$). See Figure 10.

6. Conclusion and Future Scope

(a) It is seen that the intermediate response of graded plates under thermal and thermomechanical loads is quite different from the pure mechanical load [20]. The deflections become higher with increasing n. It is observed that when thermal effect is induced, the bending response of the functionally graded plate is not necessarily intermediate to those of the metal and the ceramic plate.

(b) The nondimensional deflection of isotropic plates (pure metal and pure ceramic) is higher in magnitude than the graded plates. The deflection therefore depends on the product of the temperature and the thermal expansion coefficient. Therefore, the response of the graded plates is not intermediate to the metal and ceramic plates. It is clear that the FGM plates can resist high-temperature conditions very well.

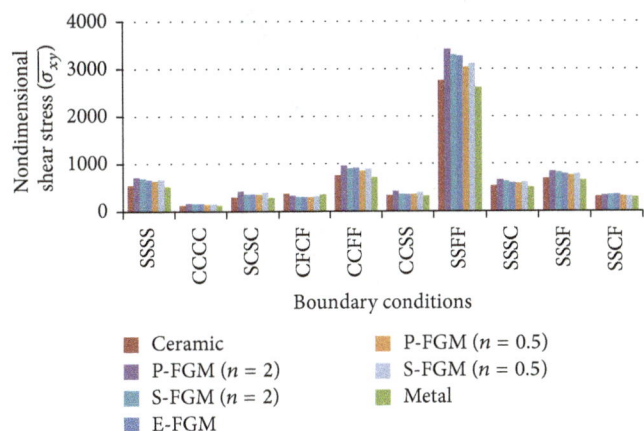

FIGURE 10: Nondimensional shear stress $(\overline{\sigma_{xy}})$ for various boundary conditions of a square plate under uniformly distributed load for various FGMs in thermal environment.

(c) The maximum deflection occurs for clamped free (CCFF) boundary conditions and minimum deflection occurs for clamped (CCCC) boundary condition under thermal load while, under thermomechanical load, the maximum deflection occurs for simply supported free (SSFF) boundary conditions and minimum deflection occurs for clamped (CCCC) boundary condition.

(d) The maximum tensile stress under thermal load occurs for simply supported clamped free (SSCF) boundary conditions and minimum tensile stress occurs for clamped free (CFCF) boundary condition while, under thermomechanical load, the maximum tensile stress occurs for simply supported free (CCFF) boundary conditions and minimum tensile stress occurs for clamped (SCSC) boundary condition.

(e) The isotropic ceramic and metallic plates have the lowest tensile and shear stress for all the boundary conditions considered here. In the presence of the above temperature field, compression occurs at the top surface while tension is at the bottom surface. Excepting fully ceramic or fully metal plates, the stress distribution of FGM plates has a similar trend.

(f) The maximum shear stress under thermal load occurs for simply supported clamped (SSSC) boundary conditions and minimum shear stress occurs for clamped (CCCC) boundary conditions while, under thermomechanical load, maximum shear stress occurs for simply supported free (SSFF) boundary conditions and minimum shear stress occurs for clamped (CCCC) boundary condition.

(g) The nondimensional deflection, nondimensional tensile stress, and nondimensional shear stress for S-FGM remain closer for various values of "n" as compared to those of the P-FGM.

The efforts taken in this work are to solve and analyze FGM plate with various loadings and boundary conditions should pave a way for more research in the future.

(1) More complex geometries can be taken for analysis. The geometries to be analyzed can be selected in such a way that they could be used in real-time engineering in the future.

(2) A further investigation of functionally graded plate structures with material properties varying in directions other than through the thickness is recommended.

(3) Since the prediction of the thermomechanical properties is not a simple task, the techniques for estimating effective material properties of functionally graded material structure are required.

Conflict of Interests

The authors declare that there is no conflict of interests regarding the publication of this paper.

References

[1] F. Delale and F. Erdogan, "The crack problem for a nonhomogeneous plane," NASA Contractor Report, Lehigh University, 1982.

[2] G. N. Praveen and J. N. Reddy, "Nonlinear transient thermoelastic analysis of functionally graded ceramic-metal plates," International Journal of Solids and Structures, vol. 35, no. 33, pp. 4457–4476, 1998.

[3] J. N. Reddy, "Thermomechanical behavior of functionally graded materials," Final Report for Afosr Grant F49620-95-1-0342, Cml Report 98-01, 1998.

[4] Z.-Q. Cheng and R. C. Batra, "Three-dimensional thermoelastic deformations of a functionally graded elliptic plate," Composites Part B: Engineering, vol. 31, no. 2, pp. 97–106, 2000.

[5] J. N. Reddy and Q. C. Zhen, "Three-dimensional thermo mechanical deformations of functionally graded rectangular plates," European Journal of Mechanics A/Solids, vol. 20, no. 5, pp. 841–855, 2001.

[6] L. F. Qian and R. C. Batra, "Transient thermoelastic deformations of a thick functionally graded plate," Journal of Thermal Stresses, vol. 27, no. 8, pp. 705–740, 2004.

[7] K. Y. Dai, G. R. Liu, X. Han, and K. M. Lim, "Thermomechanical analysis of functionally graded material (FGM) plates using element-free Galerkin method," Computers and Structures, vol. 83, no. 17-18, pp. 1487–1502, 2005.

[8] A. J. M. Ferreira, R. C. Batra, C. M. C. Roque, L. F. Qian, and P. A. L. S. Martins, "Static analysis of functionally graded plates using third-order shear deformation theory and a meshless method," Composite Structures, vol. 69, no. 4, pp. 449–457, 2005.

[9] S. H. Chi and Y. L. Chung, "Mechanical behavior of functionally graded material plates under transverse load-Part I: analysis," International Journal of Solids and Structures, vol. 43, no. 13, pp. 3657–3674, 2006.

[10] S.-H. Chi and Y.-L. Chung, "Mechanical behavior of functionally graded material plates under transverse load-part II: numerical results," International Journal of Solids and Structures, vol. 43, no. 13, pp. 3675–3691, 2006.

[11] H. Wang and Q.-H. Qin, "Meshless approach for thermomechanical analysis of functionally graded materials," Engineering Analysis with Boundary Elements, vol. 32, no. 9, pp. 704–712, 2008.

[12] M. Mahdavian, "Buckling analysis of simply-supported functionally graded rectangular plates under non-uniform in-plane compressive loading," Journal of Solid Mechanics, vol. 1, no. 3, pp. 213–225, 2009.

[13] M. Z. Ashraf and S. M. Daoud, "Thermal buckling analysis of ceramic-metal functionally graded plates," *Natural Science*, vol. 2, no. 9, pp. 968–978, 2010.

[14] S. S. Alieldin, A. E. Alshorbagy, and M. Shaat, "A first-order shear deformation finite element model for elastostatic analysis of laminated composite plates and the equivalent functionally graded plates," *Ain Shams Engineering Journal*, vol. 2, no. 1, pp. 53–62, 2011.

[15] N. Kyung-Su and K. Ji-Hwan, "Comprehensive studies on mechanical stress analysis of functionally graded plates," *World Academy of Science, Engineering and Technology*, vol. 60, pp. 768–773, 2011.

[16] K. Suresh, S. R. Jyothula, E. R. C. Bathini, and K. R. K. Vijaya, "Nonlinear thermal analysis of functionally graded plates using higher order theory," *Innovative Systems Design and Engineering*, vol. 2, no. 5, pp. 1–13, 2011.

[17] T. Mohammad and B. N. Singh, "Thermo-mechanical deformation behavior of functionally graded rectangular plates subjected to various boundary conditions and loadings," *International Journal of Aerospace and Mechanical Engineering*, vol. 6, no. 1, pp. 14–25, 2012.

[18] H. Nguyen-Xuan, L. V. Tran, C. H. Thai, and T. Nguyen-Thoi, "Analysis of functionally graded plates by an efficient finite element method with node-based strain smoothing," *Thin-Walled Structures*, vol. 54, pp. 1–18, 2012.

[19] E. Alshorbagy, S. S. Alieldin, M. Shaat, and F. F. Mahmoud, "Finite element analysis of the deformation of functionally graded plates under thermomechanical loads," *Mathematical Problems in Engineering*, vol. 2013, Article ID 569781, 14 pages, 2013.

[20] M. Bhandari and K. Purohit, "Analysis of functionally graded material plate under transverse load for various boundary conditions," *IOSR Journal of Mechanical and Civil Engineering*, vol. 10, no. 5, pp. 46–55, 2014.

[21] M. Bhandari and K. Purohit, "Static response of functionally graded material plate under transverse load for varying aspect ratio," *International Journal of Metals*, vol. 2014, Article ID 980563, 11 pages, 2014.

24

Synthesis of MgO Nanoparticles by Solvent Mixed Spray Pyrolysis Technique for Optical Investigation

K. R. Nemade and S. A. Waghuley

Department of Physics, Sant Gadge Baba Amravati University, Amravati 444 602, India

Correspondence should be addressed to S. A. Waghuley; sandeepwaghuley@sgbau.ac.in

Academic Editor: Yuanshi Li

Solvent mixed spray pyrolysis technique has attracted a global interest in the synthesis of nanomaterials since reactions can be run in liquid state without further heating. Magnesium oxide (MgO) is a category of the practical semiconductor metal oxides, which is extensively used as catalyst and optical material. In the present study, MgO nanoparticles were successfully synthesized using a solvent mixed spray pyrolysis. The X-ray diffraction pattern confirmed the formation of MgO phase with an excellent crystalline structure. Debye-Scherrer equation is used for the determination of particle size, which was found to be 9.2 nm. Tunneling electron microscope analysis indicated that the as-synthesized particles are nanoparticles with an average particle size of 9 nm. Meanwhile, the ultraviolet-visible spectroscopy of the resulting product was evaluated to study its optical property via measurement of the band gap energy value.

1. Introduction

In last two decades, synthesis of metal oxide nanoparticles has attracted considerable attention [1, 2]. Abundant techniques have been also developed to prepare nanoparticles of MgO. This nanoparticle has attracted much attention due to its wide band gap [3]. However, most of the techniques need high temperatures and perform under a costly inert atmosphere. Kaviyarasu and Devarajan reported a versatile route to synthesize MgO nanoparticles by combustion technique [4]. Lange and Obendorf studied the effect of plasma etching on destructive adsorption properties of polypropylene fibres containing magnesium oxide nanoparticles [5]. Jin and He demonstrated antibacterial activities of MgO nanoparticles against food borne pathogens [6]. Mirzaei and Davoodnia reported the microwave assisted sol-gel synthesis of MgO nanoparticles and their catalytic activity in the synthesis of Hantzsch 1,4-dihydropyridines [7]. Camtakan et al. studied the uranium sorption properties of MgO [8].

For the present work, solvent mixed spray pyrolysis method was undertaken to prepare MgO nanoparticles.

To the best of our knowledge, no report is present in the literature of material science on optical study of MgO nanoparticles synthesized by solvent mixed spray pyrolysis method. As-synthesized MgO nanoparticles were characterized by X-ray diffraction (XRD), transmission electron microscopy (TEM), and ultraviolet-visible (UV-VIS) spectroscopy.

2. Experimental

All the reagents were of analytical grade and they were used without further purification. MgO nanoparticles were prepared via quick precipitation route using magnesium nitrate $(Mg(NO_3)_2 \cdot 6H_2O)$ and hexamethylenetetramine $(C_6H_{12}N_4)$. In the typical procedure, a stock solution of 1M solution of $C_6H_{12}N_4$ was prepared by dissolving suitable quantity in distilled water. Similarly 1 M $Mg(NO_3)_2 \cdot 6H_2O$ solution was prepared by dissolving in distilled water. Both solutions were mixed under magnetic stirring for 10 min at room temperature. After this procedure, prepared solution was loaded in chamber of spray pyrolysis. The reaction unit

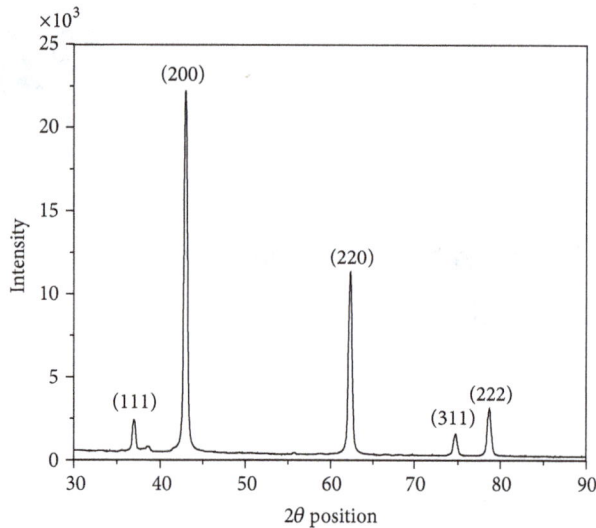

FIGURE 1: XRD pattern of the resulting MgO nanoparticles.

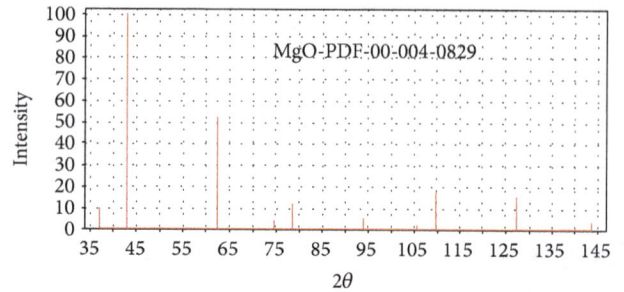

FIGURE 2: PDF card of MgO.

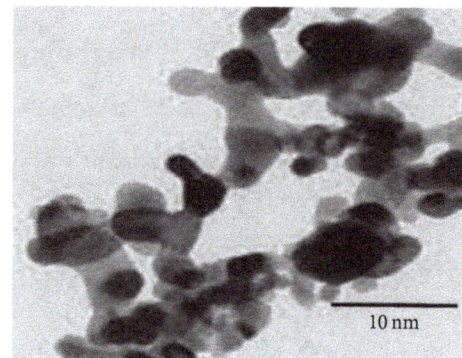

FIGURE 3: TEM images of the resulting MgO nanoparticles.

was a flame-spray apparatus consisting of high pressure gas assisted nozzle, which is made of a capillary tube with an outer diameter of 1 mm (inner diameter 0.6 mm) and an opening of 1.2 mm in diameter. The spray was evaporated by a supporting flamelets maintained at 573 K. The flow rate of the dispersion of solution was controlled by a flow controller. The product was collected on a SiO_2 substrate. The prepared sample was characterized by X-ray diffraction (XRD), transmission electron microscopy, (TEM) and ultraviolet-visible (UV-VIS) spectroscopy. X-ray diffraction pattern was recorded using a Rigaku miniflex-II diffractometer with CuKα radiation in the range 30°–90°. The morphology and grain size of the sample was observed by using TEM (JEOL-1200ex). UV-VIS spectrum was recorded on Perkin Elmer UV spectrophotometer in the range 200–1100 nm in solution of MgO nanoparticles dispersed in double distilled water.

3. Results and Discussion

The XRD pattern of the final product is shown in Figure 1. This pattern clearly confirmed the presence of the MgO cubic phase with a lattice parameter of $a = b = c = 4.213$ Å and space group (Fm-3m (225)). The diffraction peaks at 2θ values of 36.94°, 42.90°, 62.30°, 74.67°, and 78.61° matching the cubic MgO (PDF- 00-004-0829) indicated the formation of this compound (Figure 2). No other peaks for impurities were detected. The average crystallite size was calculated from diffraction peaks using the Debye-Scherrer equation, which was found to be 9.2 nm [9].

Figure 3 reveals the TEM images of the obtained MgO nanoparticles. This image illustrates that small amount of agglomeration is present in the as-synthesized sample. The average crystallite size seen in the micrograph is of the order of 9 nm. Both XRD and TEM analyses of samples are in concurrence with each other.

Figure 4 shows the UV-VIS spectrum of MgO nanoparticles, reflecting variation of % absorbance of MgO as a function of wavelength. Broad peak around 203 nm shows that the particle possesses quantum confinement [10]. This shows that charge particles are confined about three spatial dimensions. The band gap of MgO nanoparticles was estimated by plotting the $(\alpha h\nu)^2$ versus hν(eV) as shown in Figure 5. The band gap energy of MgO nanoparticle was found to be 4.2 eV. The value of particle size and band gap of MgO clearly shows that this nanomaterial is applied in photocatalytic activities and optical devices.

The variation of extinction coefficient with wavelength is shown in Figure 6. The extinction coefficient (K) is a measure of the fraction of light lost due to scattering and absorption per unit distance of the penetration medium. The extinction coefficient is computed in the sample during the exposure of UV spectra by using the relation between % absorption and wavelength [11]. Extinction coefficient (K) is as shown in (1):

$$K = \frac{\alpha\lambda}{4\pi}, \qquad (1)$$

where α is % absorption and λ is wavelength. The curve of extinction coefficient clearly shows that scattering increases gradually from 400 nm up to 1100 nm for constant distance of the penetration medium.

The optical reaction of a material is mainly studied in terms of the optical conductivity (σ) which is given by the relation (2) [12]

$$\sigma = \frac{\alpha n c}{4\pi}, \qquad (2)$$

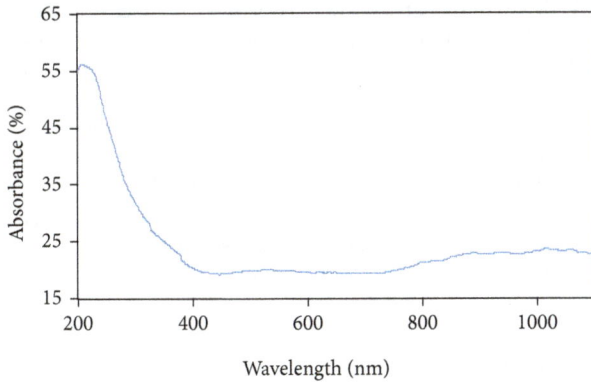

FIGURE 4: UV-VIS of MgO nanoparticles.

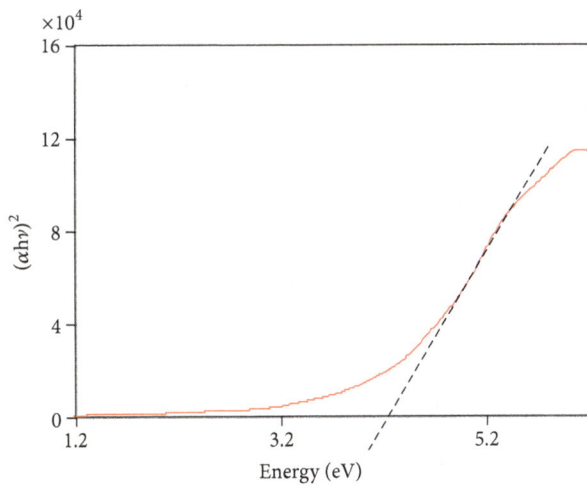

FIGURE 5: Plot of $(\alpha h\nu)^2$ versus $(h\nu)$ of MgO nanoparticles.

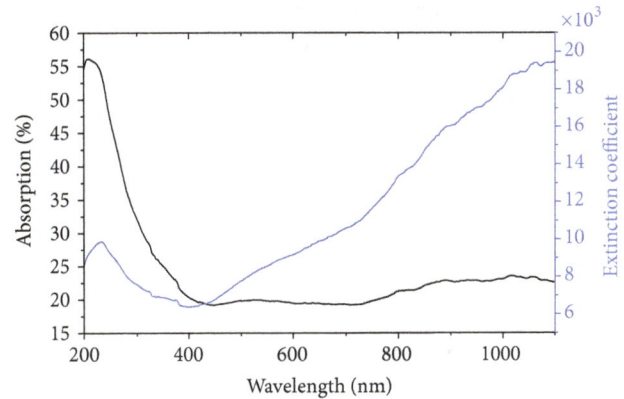

FIGURE 6: Variation of % absorbance and extinction coefficient of MgO nanoparticles as a function of wavelength.

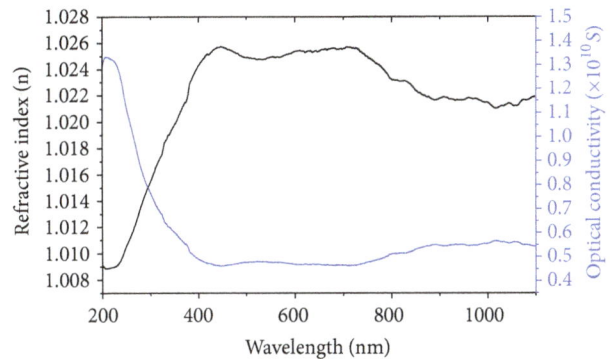

FIGURE 7: Variation of refractive index and optical conductivity of MgO nanoparticles as a function of wavelength.

where c is the velocity of light, α is the absorption coefficient, and n is the refractive index. It can be seen clearly from Figure 7 that optical conductivity directly depends on the absorption coefficient and the refractive index of the material. It can be noticed that optical conductivity increases abruptly between 200 and 400 nm. The sudden increase in optical conductivity can be attributed to the decrease in absorption coefficient.

The real and imaginary dielectric constant is a fundamental property of the material. The real and the imaginary parts of the dielectric constant can be estimated using the relations (3) [12]

$$\varepsilon_r = n^2 - K^2, \qquad \varepsilon_i = 2nK. \qquad (3)$$

The real part of the dielectric constant is to measure how much it will slow down the speed of light in the material. Figure 8 shows that variation of real dielectric constant as a function of photon energy. While the imaginary dielectric constant part shows how a dielectric material absorbs energy from an electric field due to dipole motion, Figure 9 shows the variation of imaginary dielectric constant as a function of photon energy.

4. Conclusions

In summary, nanoparticles of MgO were synthesized using a solvent mixed spray pyrolysis technique. This product was obtained by using $Mg(NO_3)_2 \cdot 6H_2O$ and $C_6H_{12}N_4$ as starting materials. An average particle size of the resulting nanoparticles was found to be 9 nm computed using XRD and TEM analyses. The optical property of the produced nanoparticles was studied by measuring the % absorbance and band gap energy. The estimated optical band gap energy is an accepted value for the photocatalytic activities in visible light and also for application in the solar cells and optical devices. The solvent mixed spray pyrolysis technique used in this study is a simple, useful, and an economic technique to prepare MgO nanoparticles.

Conflict of Interests

The authors declare that there is no conflict of interests regarding the publication of this paper.

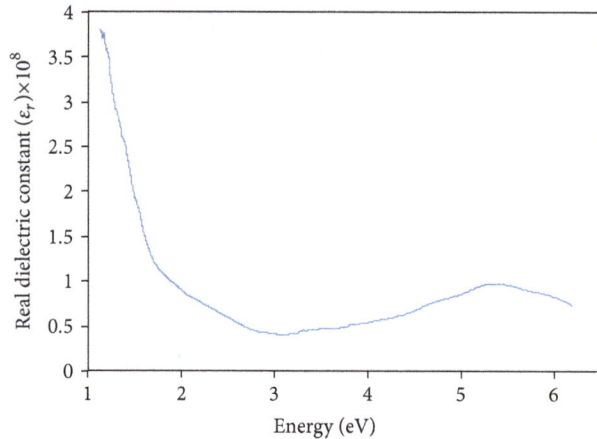

FIGURE 8: Variation of real dielectric constant as a function of photon energy.

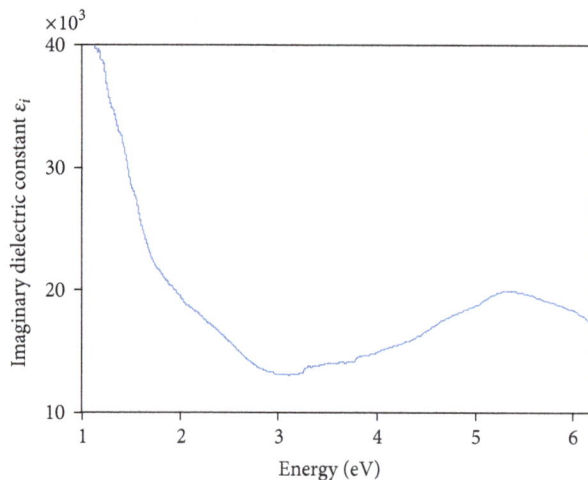

FIGURE 9: Variation of imaginary dielectric constant as a function of photon energy.

Acknowledgment

The authors are thankful to the Head of the Department of Physics, Sant Gadge Baba Amravati University, Amravati, for providing necessary facilities.

References

[1] Y. Q. Huang, L. Meidong, Z. Yike, L. Churong, X. Donglin, and L. Shaobo, "Preparation and properties of ZnO-based ceramic films for low-voltage varistors by novel sol-gel process," *Materials Science and Engineering B: Solid-State Materials for Advanced Technology*, vol. 86, no. 3, pp. 232–236, 2001.

[2] F. Nastase, I. Stamatin, C. Nastase, D. Mihaiescu, and A. Moldovan, "Synthesis and characterization of PAni-SiO$_2$ and PTh-SiO$_2$ nanocomposites' thin films by plasma polymerization," *Progress in Solid State Chemistry*, vol. 34, no. 2–4, pp. 191–199, 2006.

[3] L. de Matteis, L. Custardoy, R. Fernandez-Pacheco et al., "Ultrathin MgO coating of superparamagnetic magnetite nanoparticles by combined coprecipitation and sol-gel synthesis," *Chemistry of Materials*, vol. 24, no. 3, pp. 451–456, 2012.

[4] K. Kaviyarasu and P. A. Devarajan, "A versatile route to synthesize MgO nanocrystals by combustion technique," *Der Pharma Chemica*, vol. 3, no. 5, pp. 248–254, 2011.

[5] L. E. Lange and S. K. Obendorf, "Effect of plasma etching on destructive adsorption properties of polypropylene fibers containing magnesium oxide nanoparticles," *Archives of Environmental Contamination and Toxicology*, vol. 62, no. 2, pp. 185–194, 2012.

[6] T. Jin and Y. He, "Antibacterial activities of magnesium oxide (MgO) nanoparticles against foodborne pathogens," *Journal of Nanoparticle Research*, vol. 13, no. 12, pp. 6877–6885, 2011.

[7] H. Mirzaei and A. Davoodnia, "Microwave assisted sol-gel synthesis of MgO nanoparticles and their catalytic activity in the synthesis of hantzsch 1,4-dihydropyridines," *Chinese Journal of Catalysis*, vol. 33, no. 9-10, pp. 1502–1507, 2012.

[8] Z. Camtakan, S. Erenturk, and S. Yusan, "Magnesium oxide nanoparticles: preparation, characterization, and uranium sorption properties," *Environmental Progress and Sustainable Energy*, vol. 31, no. 4, pp. 536–543, 2012.

[9] K. R. Nemade and S. A. Waghuley, "LPG sensing application of graphene/CeO$_2$ quantum dots composite," in *Proceedings of the AIP International Conference on Recent Trends in Applied Physics and Material Science (RAM '13)*, vol. 1536, pp. 1258–1259, Rajasthan, India, 2013.

[10] K. R. Nemade and S. A. Waghuley, "UV–VIS spectroscopic study of one pot synthesized strontium oxide quantum dots," *Results in Physics*, vol. 3, pp. 52–54, 2013.

[11] N. A. Bakr, A. M. Funde, V. S. Waman et al., "Determination of the optical parameters of a-Si:H thin films deposited by hot wire–chemical vapour deposition technique using transmission spectrum only," *Pramana: Journal of Physics*, vol. 76, no. 3, pp. 519–531, 2011.

[12] P. Sharma and S. C. Katyal, "Determination of optical parameters of a-(As$_2$Se$_3$)$_{90}$Ge$_{10}$ thin film," *Journal of Physics D: Applied Physics*, vol. 40, no. 7, article 038, pp. 2115–2120, 2007.

25

Optimization of Conversion Treatment on Austenitic Stainless Steel Using Experimental Designs

S. El Hajjaji,[1] C. Cros,[2] and L. Aries[2]

[1] *Laboratoire de Spectroscopie, Modélisation Moléculaire, Matériaux et Environnement (LS3ME), Faculté des Sciences, Université Med V-Agdal, Avenu Ibn Battouta, BP 1014, Rabat, Morocco*
[2] *CIRIMAT-LCMIE, Université Paul Sabatier, 118 route de Narbonne, 31064 Toulouse Cedex 4, France*

Correspondence should be addressed to S. El Hajjaji; selhajjaji@hotmail.com

Academic Editor: Chi Tat Kwok

Conversion coating is commonly used as treatment to improve the adherence of ceramics films. The conversion coating properties depend on the structure of alloy as well as on the treatment parameters. These conversion coatings must be characterized by strong interfacial adhesion, high roughness, and high real surface area, which were measured by an electrochemical method. The influence of all the elaboration factors (temperature, time, and bath composition: sulphuric acid, thiosulphate as accelerator, propargyl alcohol as inhibitor, and surface state) and also the interactions between these factors were evaluated, using statistical experimental design. The specific surface area and optical factor (α) correspond to the quantitative responses. The evaluation showed, by using a designed experimental procedure, that the most important factor was "surface state." Sanded surface allows the formation of conversion coating with high real surface area. A further aim was to optimise two parameters: treatment time and temperature using Doehlert shell design and simplex method. The growth of the conversion coating is also influenced by treatment time and temperature. With such optimized conditions, the real surface area of conversion coating obtained was about $235\,\mathrm{m^2/m^2}$.

1. Introduction

Coatings have been developed from various materials using several deposition methods [1–3]. Electrochemical deposition is an interesting technique to obtain corrosion protection coatings, but the problem for such coatings is adhesion. In previous papers [4–6], we described an original method to strengthen the interface between ceramic layer and stainless steel or super alloy substrate. This method involves three steps. In the first, the metal surface is modified by a conversion treatment in an acid bath with S^{2-} and acetylenic alcohol as additions, allowing the control of the conversion coating growth [6, 7]. This pretreatment of the surface leads to a conversion coating which is very adherent, with a particular morphology, with micropores that allow deposition during the second step and contribute to the "anchoring" of the ceramic layer. In the second step, a refractory character is conferred to the surface by a cathodic treatment in a suitable bath, which induces the deposition of oxides or hydroxides

with varying degrees of hydration. In the third step, a thermal treatment leads to ceramic oxides and stabilized the coating.

So, to strengthen the interface between ceramic and substrate, a specific pretreatment of the metal surface is proposed so as to form a conversion coating. The morphology of the surface is important and must present a very porous structure and a high specific area to facilitate the anchoring of the ceramic layer [5–7]. Many authors have studied the influence of different parameters in conversion treatment for different metal substrates and different applications [5, 8–10].

This study was undertaken to elucidate the role of the different parameters and to optimise conversion coating on austenitic stainless steel. The parameters of conversion treatment have been studied using statistical experimental designs. The treatment process and the statistical designs are briefly reviewed before the experimental results are presented.

TABLE 1: Chemical composition of austenitic stainless steel (wt%).

C	Si	Cu	Mo	S	Cr	Ni	Fe
0.031	0.77	0.06	0.10	0.007	18.2	10.3	70.5

2. Materials and Methods

2.1. Conversion Treatment. Conversion coatings were prepared on an austenitic stainless steel, its composition is given in Table 1. Samples were cleaned with tetrahydrofurane (ACROS ORGANIC, purity Z99%), washed with distilled water, and then dried in air at room temperature. Austenitic stainless steel conversion coating was obtained by chemical treatment in acid bath containing suitable additives and particularly substances containing chalcogenides such as sulphur (sulphides, thiosulphates) [4–10]. Corrosion inhibitors like acetylenic alcohols are also required to facilitate the control of film growth in order to obtain coats with specific properties [4–10].

In order to homogenize the surface hardness, samples have undergone to a surface treatment of sanding or of microball tests.

After treatment, the samples were rinsed in demineralised water, and then dried at 70°C for 10 minutes.

2.2. Electrochemical Study. The electrochemical measurement was performed using a Tacussel model PRT 20-02 potentiostat. A saturated calomel electrode (SCE) was used as the reference electrode and a platinum electrode was used as the counter electrode.

2.3. Methodology of Experimental Research. The objective of the methodology of experimental research (MER) is to search for an optimal strategy which allows obtaining the largest number of good quality information concerning a studied phenomenon, while carrying a limited number of experiments. These are informationally optimal mathematical schemes in which all important factors are changed simultaneously, thereby facilitating the identification of process relations as well as the location of the real process optimum.

2.3.1. Screening (Design I). The main purpose of a screening study is to identify the most influential factors and those that may be regarded as inert. Fractional factorial designs [11] were chosen to evaluate the factors that significantly influence conversion coating morphology. For each problem formulated, the first problem is the choice of the factors which are the parameters that we can control. We must choose the variation limit of these factors which determines the experimental domain. These variations may have very different orders of magnitude, so that, to be able to compare the factor effects, it is necessary to work with the code levels of variation of each factor. For the present work, we have to evaluate the influence of fix factors, each at two levels (high (+1) and low (−1)). The selected parameters are listed in Table 2.

A factor is an assigned variable and the levels of the factor are the values assigned to the factor. Each experiment represents a particular point of the experimental domain and provides a measurement with one or several responses of the phenomenon in this point.

In first step we used a 2^{6-1} experiment; six factors each at two levels (+1, −1) were investigated; 32 trials were necessary for this fractional factorial design (Table 3). A factor is an assigned variable and the levels of the factor are the values assigned to the factor. The fractional factorial design consists in expressing the estimated effects in contrast. All experiments were performed in random order and the calculation was obtained by the NEMROD program [12].

2.3.2. Optimisation. In the second step, once the most significant factors have been identified, the next step is to optimise the process with respect to these factors. In this work, we used a "Doehlert uniform shell design" [13–15] and a simplex method [15].

In the present work, we studied two factors: treatment time (X_1) and temperature (X_2), requiring that six coefficients be determined as follows:

$$y = b_0 + b_1 X_1 + b_2 X_2 + b_{11} X_1^2 + b_{22} X_2^2 + b_{12} X_1 X_2. \quad (1)$$

The experimental design is presented in Table 4. Their variation domains were determined in preliminary experiments (Table 4). To minimize the effect of uncontrolled factors and time variations, all experiments were performed in random order.

2.4. Measurement of Responses. In order to show the effect of each factor, the studied responses are the real surface area or specific surface area (SS) of the conversion coating expressed in m^2/m^2 and its optical properties (α).

2.4.1. Measurement of Specific Area of the Coating. The main characteristic of conversion coatings is their high porosity. The porous character was evaluated using cyclic voltammetry to obtain the real surface area. This method involves application of a potential E, which varies with time to an electrode, between −0.2 and −1.5 V/SCE in a 1 M sodium sulphate medium. The scanning rate was 20 mV/s. The measurements were performed with the three-electrode technique. This measurement assumes the formation of a monolayer of adsorbed hydrogen and one atom of hydrogen is taken as occupying 10Å^2 [16, 17]. The surface area for 1 cm^2 was given below:

$$SS = \frac{QN10\text{Å}^2}{nF}, \quad (2)$$

where Q is the quantity of electricity (coulomb) corresponding to the anodic peak area. N is Avogadro's number. N is the number of electron ($H^+ + 1e^- \rightarrow H_{ads}$, $n = 1$) and $F = 96500 \, C \, mol^{-1}$.

2.4.2. Measurement of Optical Property (α). The total hemispheric solar adsorption factor α (ration of the energy adsorbed by the surface to the incident solar energy) was measured with an EL510 alpha meter (Elan Informatique).

TABLE 2: Factors and their levels for the experiments.

H_2SO_4 % (X_1)		T (°C) (X_2)		$Na_2S_2O_3$, $5H_2O$ g/L (X_3)		C_3H_4O % (X_4)		Treatment time (min) (X_5)		Surface state (X_6)	
−1	+1	−1	+1	−1	+1	−1	+1	−1	+1	−1	+1
0.2	2	40	70	0.5	1.5	0	0.25	10	30	Sanded	Microball

TABLE 3: Fractional factorial design 2^{6-1}: theoretical values of coded variables.

No.	X_1	X_2	X_3	X_4	X_5	X_6
1	−1	−1	−1	−1	−1	1
2	1	−1	−1	−1	−1	−1
3	−1	1	−1	−1	−1	−1
4	1	1	−1	−1	−1	1
5	−1	−1	1	−1	−1	−1
6	1	−1	1	−1	−1	1
7	−1	1	1	−1	−1	1
8	1	1	1	−1	−1	−1
9	−1	−1	−1	1	−1	1
10	1	−1	−1	1	−1	−1
11	−1	1	−1	1	−1	−1
12	1	1	−1	1	−1	1
13	−1	−1	1	1	−1	−1
14	1	−1	1	1	−1	1
15	−1	1	1	1	−1	1
16	1	1	1	1	−1	−1
17	−1	−1	−1	−1	1	−1
18	1	−1	−1	−1	1	1
19	−1	1	−1	−1	1	1
20	1	1	−1	−1	1	−1
21	−1	−1	1	−1	1	1
22	1	−1	1	−1	1	−1
23	−1	1	1	−1	1	−1
24	1	1	1	−1	1	1
25	−1	−1	−1	1	1	−1
26	1	−1	−1	1	1	1
27	−1	1	−1	1	1	1
28	1	1	−1	1	1	−1
29	−1	−1	1	1	1	1
30	1	−1	1	1	1	−1
31	−1	1	1	1	1	−1
32	1	1	1	1	1	1

TABLE 4: Doehlert design: theoretical values of coded variables and their levels for the experiments.

No.	X_1	X_2	Treatment time (s)	T (°C)
1	1	0	300	55.00
2	−1	0	180	55.00
3	0.5	0.866	270	60.02
4	−0.5	−0.866	210	49.98
5	0.5	−0.866	270	49.98
6	−0.5	0.866	210	60.02
7	0	0	240	55.00

3. Results and Discussion

3.1. Determination of the Significant Factors (FFDs 2^{6-1}). Experimental treatment conditions and the specific surface area and α of the as-prepared conversion coatings are shown in Table 5. Values of the 11 contrasts were computed with NEMROD^W software and are given in Table 6. The study of these results indicates that the process can be explained by the strong effects corresponding to sulphuric acid, alcohol concentrations, and to the effect of the interaction of alcohol concentration with acid concentration and temperature. Figure 1 serves as an illustration for studying these interactions.

The effect of alcohol depends on the level of temperature. The results show that the increase of propargyl alcohol concentration for a low temperature has no effect on the real surface. But, at high temperature, the best result corresponds to a low alcohol concentration (SS = 115 m^2/m^2).

The effect of alcohol depends on the level of acid. At low or high alcohol concentration, the responses are highly influenced by the variation of acid concentration and the best result corresponds to a low concentrations of sulfuric acid and propargyl alcohol (SS = 129 m^2/m^2). One other factor also has a significant main effect and does not display any interaction: surface state ($b_6 = -30.4$). We can therefore state that the value of SS is higher if the surface is at level −1. In this case, sand surface is the surface that serves to obtain the highest SS.

A great real surface area of the conversion coating is achieved under the following experimental conditions.

Sulphuric acid concentration 0.2%.

Thiosulphate concentration: 0.5 or 1.5 g/L.

Alcohol concentration: 0%.

Temperature: 70°C.

Treatment time: 10 min or 30 min.

2.5. Measurement of Electrochemical Impedance. EIS measurements were performed using EGG PAR apparatus model 16310. Impedance spectra were obtained in the frequency range of 10 KHz to 10 MHz. AC amplitude was 5 mV. Experiments were performed in aqueous aerated solution of 1 M Na_2SO_4 at 25°C.

TABLE 5: Fractional factorial design (2^{6-1}): variables, their levels, and data of the responses (SS and α).

No.	H_2SO_4 % (X_1)	T (°C) (X_2)	$Na_2S_2O_8$ g/L (X_3)	C_3H_4O % (X_4)	Treatment time (min) (X_5)	Surface state (X_6)	SS (m^2/m^2)	α
1	0.2	40	0.5	0	10	Microball	40.20	0.76
2	2	40	0.5	0	10	Sanded	15.10	0.45
3	0.2	70	0.5	0	10	Sanded	226.00	0.94
4	2	70	0.5	0	10	Microball	8.00	0.83
5	0.2	40	1.5	0	10	Sanded	131.20	0.90
6	2	40	1.5	0	10	Microball	6.40	0.27
7	0.2	70	1.5	0	10	Microball	160.10	0.92
8	2	70	1.5	0	10	Sanded	97.00	0.89
9	0.2	40	0.5	0.25	10	Microball	19.70	0.54
10	2	40	0.5	0.25	10	Sanded	83.70	0.31
11	0.2	70	0.5	0.25	10	Sanded	65.70	0.75
12	2	70	0.5	0.25	10	Microball	44.90	0.66
13	0.2	40	1.5	0.25	10	Sanded	95.30	0.71
14	2	40	1.5	0.25	10	Microball	44.70	0.46
15	0.2	70	1.5	0.25	10	Microball	29.20	0.76
16	2	70	1.5	0.25	10	Sanded	102.40	0.61
17	0.2	40	0.5	0	30	Sanded	164.70	0.91
18	2	40	0.5	0	30	Microball	5.30	0.75
19	0.2	70	0.5	0	30	Microball	142.80	0.94
20	2	70	0.5	0	30	Sanded	136.10	0.96
21	0.2	40	1.5	0	30	Microball	46.70	0.94
22	2	40	1.5	0	30	Sanded	139.60	0.94
23	0.2	70	1.5	0	30	Sanded	119.50	0.91
24	2	70	1.5	0	30	Microball	37.10	0.92
25	0.2	40	0.5	0.25	30	Sanded	47.40	0.77
26	2	40	0.5	0.25	30	Microball	54.20	0.73
27	0.2	70	0.5	0.25	30	Microball	10.20	0.49
28	2	70	0.5	0.25	30	Sanded	107.80	0.82
29	0.2	40	1.5	0.25	30	Microball	60.50	0.79
30	2	40	1.5	0.25	30	Sanded	146.40	0.84
31	0.2	70	1.5	0.25	30	Sanded	25.60	0.70
32	2	70	1.5	0.25	30	Microball	22.40	0.88

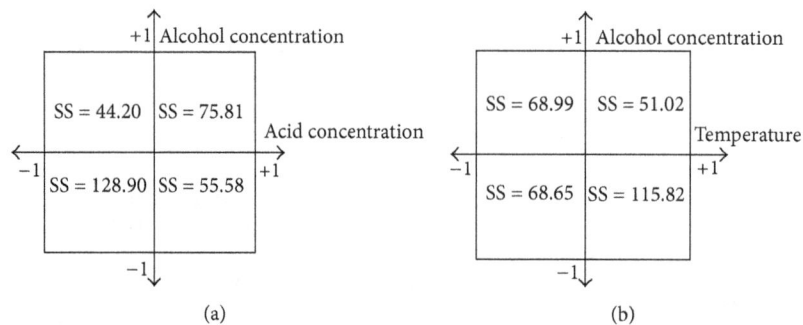

FIGURE 1: Illustration of interactions b_{14} and b_{24} between (a) sulphuric acid concentration (X_1)—alcohol concentration (X_4) and (b) temperature (X_2)—alcohol concentration (X_4).

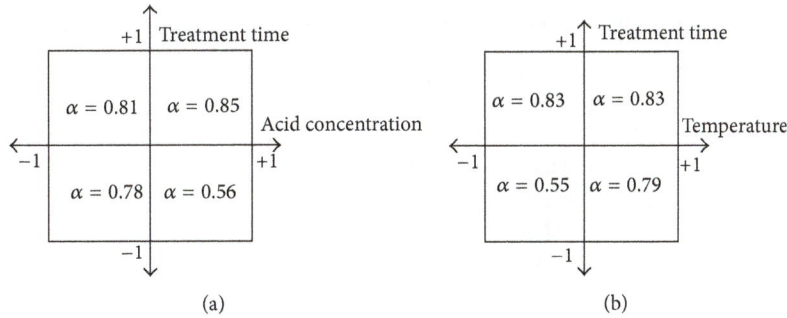

FIGURE 2: Illustration of interactions b_{15} and b_{25} between (a) sulphuric acid concentration (X_1)—treatment time (X_5) and (b) temperature (X_2)—treatment time (X_4).

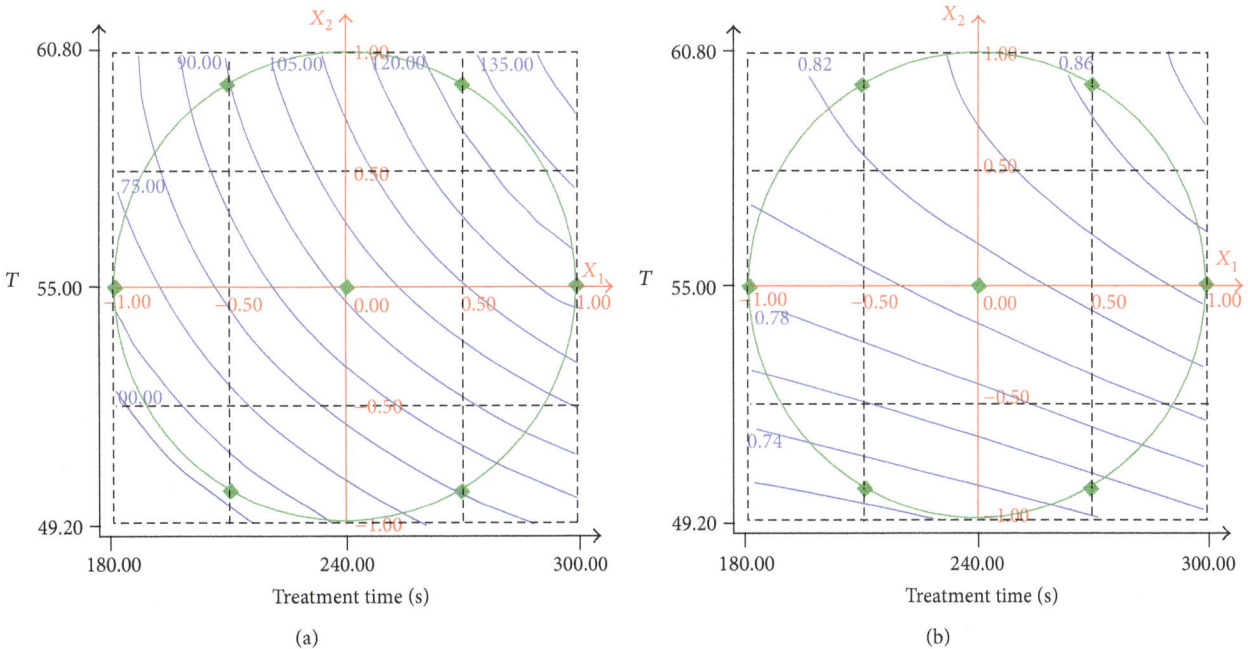

FIGURE 3: Response surface of the real surface area (a) and α (b) versus X_1 (treatment time) and X_2 (temperature).

For α response, the significant one here is also alcohol concentration that must be at level (−1). Two interactions (b_{15} and b_{25}), corresponding to the interaction acid-treatment time and temperature-treatment time, respectively, have a significant effect and can be illustrated as in Figure 2. It emerges from these two interactions that the acid concentration, treatment time, and temperature must be at level (−1).

We therefore decided to work in the following conditions:

Acid concentration: 0.2% (−1).

Thiosulphate concentration: 0.5 g/L (−1).

Alcohol concentration: 0 (−1).

Surface sanded surface: (−1).

And in these conditions, we focused essentially on the effect of the treatment time and temperature using a Doehlert uniform shell design for two parameters, and in which a second order response surface is fitted to the experimental result by least squares multiple regressions by NEMROD software.

3.2. Doehlert Shell Design and Simplex Matrix. The results obtained according to Doehlert's matrix are given in Table 7. The experimental domain was determined from their best levels.

Processing of the data led to the estimation of six coefficients for the polynomial equation for each response as follows:

$$SS = 99 + 27.2X_1 + 24X_2 - 3X_1^2 - 2.7X_2^2 + 7.5X_1X_2,$$

$$\alpha = 0.81 + 0.03X_1 + 0.06X_2 + 0.005X_1^2$$

$$- 0.03X_2^2 + 0.006X_1X_2$$

$$(3)$$

$$(X_1: \text{treatment time}, X_2: \text{temperature}).$$

TABLE 6: Main and interactions effects calculated from factorial fractional design 2^{6-1}.

b_i	Estimates (for responses SS)	Estimates (for responses α)
b_0	76.1	0.75
b_1	−10.4	−0.04
b_2	7.3	0.06
b_3	2.9	0.03
b_4	**−16.1**	**−0.08**
b_5	3.0	**0.08**
b_6	**−30.4**	0.02
b_{14}	**26.2**	0.03
b_{24}	**−16.3**	−0.03
b_{15}	12.4	**0.07**
b_{25}	−11.3	−0.06

The bold font refers to the most important values according to their levels as it is explain in the text.

TABLE 7: Doehlert design: results for each experiment.

No.	Time (s)	T(°C)	SS (m²/m²)	α
1	300	55.00	120	0.85
2	180	55.00	72	0.78
3	270	60.02	134	0.86
4	210	49.98	59	0.73
5	270	49.98	86	0.75
6	210	60.02	94	0.83
7	240	55.00	99	0.81

Figure 3 represents the variation of responses SS (Figure 3(a)) and α (Figure 3(b)) according to temperature and treatment time. The high effects of treatment time and temperature appear clearly and the coefficient values corresponding to these factors are very important. The results show that the increase of treatment time and temperature increases the real surface area and α as can be seen in Figure 3. To find the optimum via an alternative, the simplex sequential optimisation method was used. Experiment 8 was carried out at a point symmetrical to experiment 7 with respect to the midpoint between points 1 and 3 (Figure 4). Here, the response is high (SS = 327 m² · m⁻², α = 0.91), which shows that the increase of a treatment time and temperature allows the formation of conversion coating with high surface area.

3.3. Determination of Fractal Dimension of the Coating by Impedance Measurements. The electrochemical impedance diagrams of the conversion coating (Figure 5) show a capacitive arc characteristic of the charge transfer process at the electrode-solution interface. At very high frequencies, a process of diffusion in the pores is observed.

It has been shown that the transfer semicircle, which is centered for a flat smooth interface, becomes rotated around its high frequency when the surface is porous and/or rough. This difference from a smooth interface is due to the distribution of the system response time constant. The

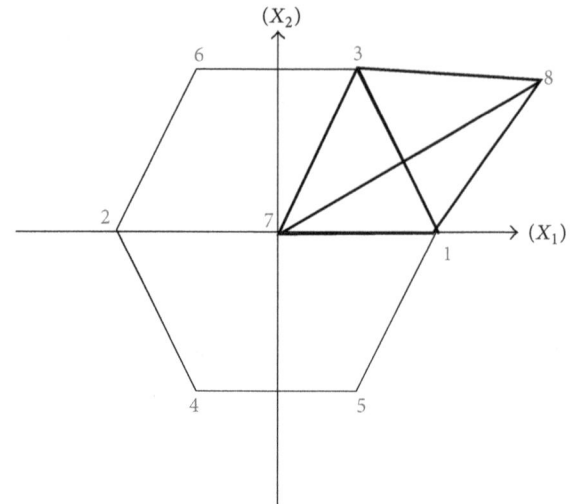

FIGURE 4: Illustration of simplex defined on X_1 (treatment time)—X_2 (temperature).

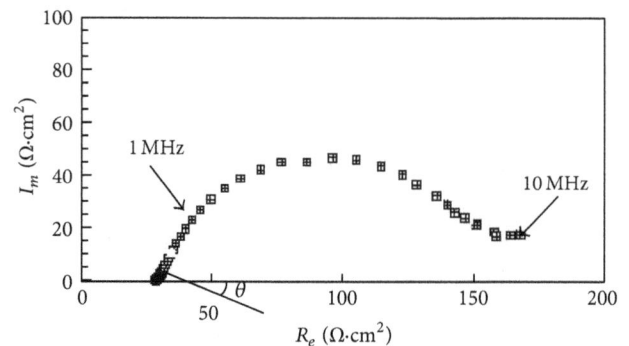

FIGURE 5: EIS Nyquist plot in aerated 1 M Na_2SO_4 aqueous solution of the austenitic stainless steel with conversion coating.

angle of rotation of capacitive loop around its high frequency is noted θ (Figure 5). In order to interpret any correlation that might exist between the particular texture of certain interfaces and the angle of rotation θ, a nondimensional parameter d_f, representing the difference from an ideal surface (perfectly smooth and homogeneous), is often introduced. Several tentative relationships have been proposed to determine d_f from the angle θ [16–21]. In this work, we used the relationship proposed by Le Mehauté and Crepy [16]:

$$d_f = \frac{180}{180 - 2\theta} + 1. \tag{4}$$

For our optimal conversion coating, the obtained value of d_f is about 2.15 (θ = 12.2°).

The double layer capacitance C_{dl} relevant to the EIS diagram in Figure 5 is 11775.3 μF · cm⁻², a high value attributed to the presence of a porous layer on the surface, while a double layer capacitance for a smooth surface is considered to be about 50 μF · cm⁻². The real surface area was estimated to be 235 m²/m².

The calculated value of SS is in good agreement with what has been determined by cyclic voltammetry method.

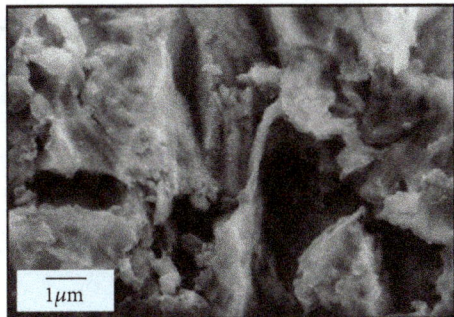

FIGURE 6: Micrograph of conversion coating obtained in an optimum bath.

Examination of the optimal coating by scanning electron microscopy showed that the surface is rough and porous (Figure 6).

4. Conclusion

The aim of this study is the optimisation of conversion coating on iron-chromium-nickel alloy. This conversion coating must have a high specific area. The present work has demonstrated that the experimental domain that we defined appears suitable for the optimisation of conversion coating. The fractional factorial design, Doehlert design, and simplex design allow a rapid overall study of conversion coating growth in sulphuric acid medium, under relatively strong experimental constraints. The primary conclusion of this study is that interactions between bath compounds and surface state have an important influence on the formation of conversion coatings. The experimental designs used for these experiments led to the optimum conditions being obtained. The real surface area of optimal conversion coating is very high ($235 \, m^2/m^2$). The fractal dimension was determined by impedance measurement. The measurements of specific area by impedance and cyclic voltammetry present a good agreement and indicate high porosity.

References

[1] P. Álvarez, A. Collazo, A. Covelo, X. R. Nóvoa, and C. Pérez, "The electrochemical behaviour of sol-gel hybrid coatings applied on AA2024-T3 alloy: effect of the metallic surface treatment," *Progress in Organic Coatings*, vol. 69, no. 2, pp. 175–183, 2010.

[2] Y. Song, D. Shan, R. Chen, F. Zhang, and E. Han, "A novel phosphate conversion film on Mg-8.8Li alloy," *Surface and Coatings Technology*, vol. 203, no. 9, pp. 1107–1113, 2009.

[3] A. A. Zuleta, E. Correa, C. Villada, M. Sepúlveda, J. G. Castaño, and F. Echeverría, "Comparative study of different environmentally friendly (Chromium-free) methods for surface modification of pure magnesium," *Surface and Coatings Technology*, vol. 205, no. 23-24, pp. 5254–5259, 2011.

[4] L. Bamoulid, M. T. Maurette, D. De Caro et al., "Investigations on composition and morphology of electrochemical conversion layer/titanium dioxide deposit on stainless steel," *Surface and Coatings Technology*, vol. 201, no. 6, pp. 2791–2795, 2006.

[5] L. Bamoulid, M.-T. Maurette, D. De Caro et al., "An efficient protection of stainless steel against corrosion: combination of a conversion layer and titanium dioxide deposit," *Surface and Coatings Technology*, vol. 202, no. 20, pp. 5020–5026, 2008.

[6] S. El Hajjaji, M. El Alaoui, P. Simon et al., "Preparation and characterization of electrolytic alumina deposit on austenitic stainless steel," *Science and Technology of Advanced Materials*, vol. 6, no. 5, pp. 519–524, 2005.

[7] A. Lgamri, A. Guenbour, A. Ben Bachir, S. El Hajjaji, and L. Aries, "Characterisation of electrolytically deposited alumina and yttrium modified alumina coatings on steel," *Surface and Coatings Technology*, vol. 162, no. 2-3, pp. 154–160, 2003.

[8] A. Komla, L. Aries, B. Naboulsi, and J. P. Traverse, "Texture of selective surfaces for photothermal conversion," *Solar Energy Materials*, vol. 22, no. 4, pp. 281–292, 1991.

[9] S. El Hajjaji, A. Lgamri, E. Puech-Costes, A. Guenbour, A. Ben Bachir, and L. Aries, "Optimization of conversion coatings: study of the influence of parameters with experimental designs," *Applied Surface Science*, vol. 165, no. 2, pp. 184–192, 2000.

[10] S. El Hajjaji, A. Guenbour, A. Ben Bachir, and L. Aries, "Effect of treatment baths nature on the characteristics of conversion coatings modified by electrolytic alumina deposits," *Corrosion Science*, vol. 42, no. 6, pp. 941–956, 2000.

[11] G. E. P. Box, W. G. Hunter, and J. S. Hunter, *Statistics for Experimenters: An Introduction to Design, Data Analysis and Model Building*, Wiley, New York, NY, USA, 1978.

[12] D. Mathieu and R. Phan-Tan-Luu, *NEMROD Software*, LPRAI, Marseille, France, 1995.

[13] D. H. Doehler, "Uniform shell designs," *Journal of the Royal Statistical Society C*, vol. 19, pp. 231–239, 1970.

[14] D. H. Doehlert and V. L. Klee, "Experimental designs through level reduction of the d-dimensional cuboctahedron," *Discrete Mathematics*, vol. 2, no. 4, pp. 309–334, 1972.

[15] W. Splendley, G. R. Hext, and F. R. Himsworth, "Sequential application of simplex design of optimization and evolutionary operations," *Technometrics*, vol. 4, pp. 441–461, 1962.

[16] A. Le Mehauté and G. Crepy, "Introduction to transfer and motion in fractal media: the geometry of kinetics," *Solid State Ionics*, vol. 9-10, pp. 17–30, 1983.

[17] M. Keddam and H. Takenouti, "Impedance of fractal interfaces : new data on the Von Koch model," *Electrochimica Acta*, vol. 33, pp. 445–448, 1986.

[18] A. J. Bard and L. R. Faulkner, *Electrochimie*, Masson, Paris, France, 1983.

[19] L. Nyikos and T. Pajkossy, "Fractal dimension and fractional power frequency-dependent impedance of blocking electrodes," *Electrochimica Acta*, vol. 30, pp. 1533–1540, 1985.

[20] T. Pajkossy and L. Nyikos, "Impedance of fractal blocking electrodes," *Journal of Electrochemical Society*, vol. 133, no. 10, pp. 2061–2064, 1986.

[21] S. H. Liu, "Fractal model for the ac response of a rough interface," *Physical Review Letters*, vol. 55, no. 5, pp. 529–532, 1985.

26

First Principles Study of Electronic Structure and Magnetic Properties of TMH (TM = Cr, Mn, Fe, Co)

S. Kanagaprabha,[1] R. Rajeswarapalanichamy,[2] and K. Iyakutti[3]

[1] *Kamaraj College, Tuticorin, Tamilnadu 628003, India*
[2] *Department of Physics, N.M.S.S. Vellaichamy Nadar College, Madurai, Tamilnadu 625019, India*
[3] *SRM University, Chennai, Tamilnadu 600030, India*

Correspondence should be addressed to R. Rajeswarapalanichamy; rrpalanichamy@gmail.com

Academic Editor: Dong B. Lee

First principles calculations are performed using a tight-binding linear muffin-tin orbital (TB-LMTO) method with local density approximation (LDA) and atomic sphere approximation (ASA) to understand the electronic properties of transition metal hydrides (TMH) (TM = Cr, Mn, Fe, Co). The structural property, electronic structure, and magnetic properties are investigated. A pressure induced structural phase transition from cubic to hexagonal phase is predicted at the pressures of 50 GPa for CrH and 23 GPa for CoH. Also, magnetic phase transition is observed in FeH and CoH at the pressures of 10 GPa and 180 GPa, respectively.

1. Introduction

Metal hydrides have been attracting attention of scientists for decades. Their physical properties are interesting from both fundamental and practical points of view. Many transition metals react readily with hydrogen to form stable metal hydrides [1]. Metal hydrides are of intense scientific and technological interest in the view of their potential application, for example, for hydrogen storage, in fuel cells and internal combustion engines, as electrodes for rechargeable batteries, and in energy conversion devices. Hydrides for hydrogen storage need to be able to form hydrides with a high hydrogen to metal mass ratio but should not be too stable, so that the hydrogen can easily be released without excessive heating [2]. Among those, the metal hydrides are of particular interest due to their application in hydrogen storage for fuel cells [3, 4]. The physical properties of metal hydrides are quite interesting from the practical point of view. For example, the density of hydrogen in metal hydrides is larger than that in liquid hydrogen [5]. Previously the first principles calculations were performed for metal dihydrides with the fluorite structure [6, 7]. The stability and electronic structure of the transition-metal hydrides were reported [8]. Phase

transformations, crystal and magnetic structures of high-pressure hydrides of d-metals were analyzed by Antonov [9]. γ-CrH and ε-CrH hydrides were prepared electrolytically and studied by neutron diffraction and inelastic neutron scattering at liquid helium temperatures [10]. Previously we have investigated the pressure induced magnetic phase transition in hexagonal FeH and CoH [11].

In this work, the structural, electronic, and magnetic properties of 3d transition metal hydrides (CrH, MnH, FeH, CoH) are investigated using TB-LMTO method. The weight percentage of hydrogen in Cr, Mn, Fe, and Co is also calculated.

2. Computational Details

The TB-LMTO (tight-binding linear muffin-tin orbital) method has been used [12, 13] to compute the electronic structure and the basic ground state properties of TMH. In the LMTO scheme, the crystal potential is approximated by a series of nonoverlapping atomic like spherical potentials and a constant potential between the spheres. The Schrödinger equation can be solved in both regions. These solutions are then matched at the sphere boundaries to produce muffin-tin orbitals. These muffin-tin orbitals are used to construct a basis which is energy independent, linear order in energy and rapidly convergent. Each orbital must satisfy Schrödinger's differential equation in the region between the atoms. Here the potential is flat on a scale of 1 Ry and, since the energy range of interest begins near the point where the electron can pass between the atoms and extends upwards by about 1 Ry, it seems natural to choose orbital, which have zero kinetic energy, that is, satisfy the Laplace equation, in the interstitial region. In the tight-binding muffin-tin orbital the solution of Schrödinger's equation is written as follows:

$$|\chi^{\alpha}(\varepsilon)\rangle = \begin{cases} |\phi(E)\rangle N^{\alpha}(E) + |J^{\alpha}\rangle P^{\alpha}(E) & r \le w, \\ |K^{\alpha}\rangle & r \ge w, \end{cases} \quad (1)$$

where r is any distance from the centre of the muffin-tin sphere, w is the average Wigner-Seitz radius, and α are dimensionless screening constants. Inside the MT sphere, the base field or unscreened field $|K\rangle$ is defined by

$$|K\rangle = |\phi(E)\rangle N^{\alpha}(E) + |J^{\alpha}\rangle P^{\alpha}(E). \quad (2)$$

In the interstitial region, the screened field $|k^{\alpha}\rangle$ is defined by

$$|k^{\alpha}\rangle = |\phi(E)\rangle N^{\alpha}(E) + |J^{\alpha}\rangle P^{\alpha}(E) - |J^{\alpha}\rangle S^{\alpha}, \quad (3)$$

where $\phi(E)$ is normalized to unity in its sphere. J^{α} is the screen field radial function and S^{α} is the screened structure matrix. The elements of the diagonal matrices P and N are

$$P^{\alpha}(E) = \frac{\{\phi(E), k\}}{\{\phi(E), J^{\alpha}\}} = \frac{P^{o}(E)}{1 - \alpha P^{o}(E)},$$

$$N^{\alpha}(E) = \frac{\{J^{\alpha}, k\}}{\{J^{\alpha}, \phi(E)\}} = \left(\frac{w}{2}\right)^{1/2} \dot{P}^{\alpha}(E)^{1/2}. \quad (4)$$

The set of energy dependent MTO's $|\chi^{\alpha}(E)\rangle$ thus equals $|k^{\alpha}\rangle$ and $|k\rangle$; the linear combination $|\chi^{\alpha}(E)\rangle u^{\alpha}$, specified by a column vector u^{α}, is seen to be a solution of Schrödinger's equation at energy E for the MT potential if it equals the one center expansions $|\phi(E)\rangle > N^{\alpha}(E) u^{\alpha}$ in the spheres, that is, if the set of linear homogeneous equations $[P^{\alpha}(E) - S^{\alpha}] u^{\alpha} = 0$ has a proper solution. This is the generalization of the so-called tail cancellation or Kerringa-Kohn-Rostoker (KKR) condition.

The secular matrix $[P^{\alpha}(E) - S^{\alpha}]$ depends on the potential only through the potential functions along the diagonal and for the most localized set it has the TB two-center form with S^{α} playing the role of the transfer integrals.

The screened field $|k^{0}\rangle$ is given by the superposition of bare fields $|k^{0}\rangle$ and that the relationship between the bare and screened structure matrices is as follows:

$$S^{\alpha} = S^{o}(1 - S^{o})^{-1}. \quad (5)$$

The expansion coefficient S^{o} forms a Hermitian matrix which is dimensionless and independent of the scale of the structure. This is the so-called (bare) canonical structure matrix.

The KKR equations have the form of Eigen value problem if P^{α} is a linear function of E. This is true, if $\alpha = \gamma$ (potential parameter) in this case the effective two-center Hamiltonian is as follows:

$$H_{ij}^{\gamma} = C_{i} S_{ij} + \left(\sqrt{\Delta_{i}}\right) S_{ij}^{\gamma} \left(\sqrt{\Delta_{i}}\right). \quad (6)$$

For crystals, where the matrix inversion in (6) can be performed, one may obtain from S^{o} (or) S^{α}.

To obtain Eigen value when $\alpha \ne \gamma$, energy independent orbitals are needed. Now, $|\chi^{\alpha}(E)\rangle$ is independent of energy; in the interstitial region and in the spheres, its first energy derivative at E_{γ} will vanish. Therefore, the orbital base is as follows:

$$|\chi^{\alpha}\rangle = |\phi\rangle + |\dot{\phi}^{\alpha}\rangle h^{\alpha}. \quad (7)$$

In this base, the Hamiltonian matrices are

$$\langle \chi |H - E_{\gamma}| \chi \rangle = h(1 + oh), \quad (8)$$

$$(H - E) |\phi(E)\rangle = 0 \quad (9)$$

$H - E_{\gamma}$ is the effective two-center TB Hamiltonian. This has a shorter range. Therefore, this set of equations is used in self-consistent calculations. In the interstitial region (8) becomes

$$\langle \chi^{\alpha} |H - E_{\gamma}| \chi^{\alpha} \rangle = h^{\alpha}(1 + o^{\alpha} h^{\alpha}). \quad (10)$$

The set of $|\chi^{\alpha}\rangle$ is thus complete to first order in $(E - E_{\gamma})$ and it can yield energy estimates correct to the third order. A Von-Barth and Hedin [12] parameterization scheme has been used for the exchange correlation potential within the local density approximation (LDA). The accuracy of the total energies obtained within the density functional theory, often using LDA, is in many cases sufficient to predict which structure at a given pressure has the lowest free energy [14]. The atomic sphere approximation (ASA) has been used in the present work. The Wigner-Seitz sphere is chosen in such a way that the sphere boundary potential is at its minimum and the charge flow is in accordance with the electronegativity criteria. s, p, and d partial waves are included. The tetrahedron method [15] of the Brillouin zone (k-space) integration has been used to calculate the density of states.

3. Results and Discussion

3.1. Stability of TMH and Formation Energy. At ambient condition, all the metal hydrides TMH (TM = Cr, Mn, Fe, Co) crystallize in fcc structure with the space group Fm3m (225). The Wyckoff positions for each metal and hydrogen are 4a:(0,0,0) and 4b:(0.5,0.5,0.5), respectively, and contain four formula units per unit cell. The stability of these metal hydrides is analyzed by computing the formation energy using the following relation:

$$\Delta H = E_{\text{tot}}(\text{TMH}) - E_{\text{tot}}(\text{TM}) - E_{\text{tot}}(\text{H}), \quad (11)$$

where $E_{\text{tot}}(\text{TMH})$ is the energy of the primitive cell of TMH. $E_{\text{tot}}(\text{TM})$ and $E_{\text{tot}}(\text{H})$ are the energies of a transition metal atom and a hydrogen atom. The variation of heat of formation for chromium, manganese, iron, and cobalt monohydride with rocksalt structure is shown in Figure 1. It is observed that the stability of the hydride rapidly increases as we move from chromium to iron and the heat of formation of CoH is less than FeH.

Figure 2 offers a further illustration of the importance of the valence electron count for the chemical hydrogen insertion energy. It shows the chemical component of the hydrogen insertion energy as function of the number of valence electrons (s + d) for the hydrides (TMH) in rock salt structure.

3.2. Structural Phase Transition.

The total energy calculation is carried out for cubic and hexagonal structures of CrH and CoH for various pressures and the total energy versus reduced volume is plotted in Figure 3. Hexagonal structure with the space group $P6_3/mmc$ (194) is used for total energy calculation. The Wyckoff positions for each metal and hydrogen are 2a:(0,0,0) and 2b:(0.333,0.667,0.25), respectively, and contain two formula units per unit cell. In order to calculate the transition pressure, the Gibbs free energy is calculated for the two phases using the expression

$$G = E_{tot} + PV - TS. \tag{12}$$

Since the theoretical calculations are performed at 0 K, the Gibbs free energy will become equal to the enthalpy (H)

$$H = E_{tot} + PV. \tag{13}$$

At a given pressure, a stable structure is one in which the enthalpy has its lowest value. The transition pressures are calculated at which the enthalpies of the two phases are equal. The enthalpy versus pressure curves corresponding to the cubic and hexagonal phases of CrH and CoH are shown in Figure 4. It is observed that, at normal and low pressures, the thermodynamically stable phase is cubic structure. A pressure induced structural phase transition from cubic to hexagonal phase is observed at the pressure of 50 GPa for CrH and 23 GPa for CoH.

Metal hydrides represent an exciting way of hydrogen storage which is inherently safer than the compressed-gas or liquid storing. Also, some intermetallics (including metals and alloys) store hydrogen at a higher volume density than liquid hydrogen. Elements, especially those in groups I–IV and some transition metals, have their hydride and amide/imide forms. That has, therefore, still plenty of scope for further exploring metal-H systems for hydrogen storage. In our study, the storage capacity, formation energy per unit volume (Ry.), and weight percentage of hydrogen for chromium, manganese, iron, and cobalt are computed and are given in Table 1. Our results conclude that the hydrogen storage capacity decreases as we move from chromium to cobalt.

3.3. Band Structure.

The configurations of Cr $4s^2\ 3d^4$, Mn $4s^2\ 3d^5$, Fe $4s^2\ 3d^6$, Co $4s^2\ 3d^7$, and H $1s^1$ are treated as the valence electrons. The band structure of all above-mentioned

TABLE 1: Weight % of hydrogen in Cr, Mn, Fe, and Co.

	Formation energy/unit volume (Ry.)	Density of H/unit cell	Weight % of hydrogen
CrH	−0.1148	8.7450	1.90
MnH	−0.1033	9.4799	1.80
FeH	−0.0953	9.7968	1.77
CoH	−0.0999	9.7163	1.68

metal hydrides at normal pressure is given in Figure 5. The band structures of CrH and MnH have 4 valence bands corresponding to 7 and 8 valence electrons. FeH and CoH have 5 valence bands corresponding to 9 and 10 valence electrons, respectively. At the bottom, the band structure shows first deep narrow band with low dispersion, which is mainly due to the nonmetal H-1s state. The valence states are separated by a wide gap from the occupied states, indicating covalent behavior. The empty conduction bands are highly overlapping with the valence bands. Above the Fermi level the empty conduction bands are present with a mixed s, p, and d characters. The optimized lattice constants, the equilibrium volume V_0 (Å^3), valence electron density ρ (electrons/Å^3), Fermi energy, density of states at Fermi energy, Wigner-Seitz radius, and band energy of cubic metal hydrides are given in Table 2. Valence electron density (VED) is defined as the total number of valence electrons divided by volume per unit cell which is an important factor for analyzing the super hard materials. It is worth noting that the calculated VEDs for CrH are higher than those of Cr (0.5187) and for MnH, FeH, and CoH are slightly less than those of Mn (0.6644), Fe (0.78), and Co (0.89) metals, respectively, and are comparable to 0.70 electrons/Å^3 for diamond [16].

3.4. Density of States.

To understand the correlation between the electronic and mechanical properties, we have computed the density of states under equilibrium geometry. The total DOS of all metal hydrides at normal pressure is shown in Figure 6 and it is observed that the spike near −0.5 Ry. is due to metals hydrogen bonding which is clearly shown in partial DOS (Figure 7). It is observed that the H-1s states are near −0.5 Ry. and they are hybridized with the metal s and p states. The dotted line indicates the Fermi level. The d-state electrons of the metal atoms contribute to majority of the DOS near the Fermi level. This indicates that this 3d state intensively hybridize with the H-2p states. The DOS of M-d and H-p states are energetically degenerated from the top of the valence band, indicating the possibility of covalent bonding between the metal (TM-Cr, Mn, Fe, and Co) and H atom. However, the main hybridization in this energy window concerns with 3d state of metal atom and s state of hydrogen atom.

3.5. Charge Density Analysis.

An understanding of the nature of the chemical bond can be aided by the studies of the distribution of charges in real space. The real space charge density can also be used to understand features of

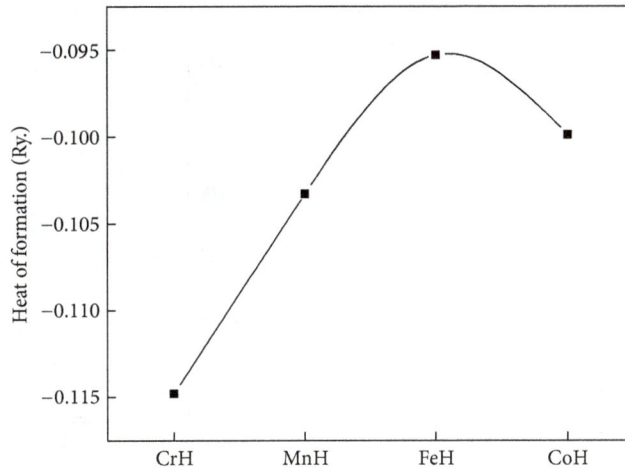

FIGURE 1: Formation energy of TMH.

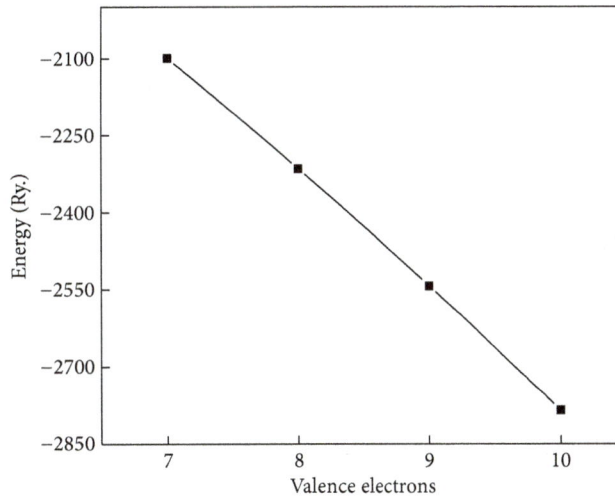

FIGURE 2: Chemical contributions to the hydride formation energy as function of the number of valence electrons (energy given per H_2 molecule).

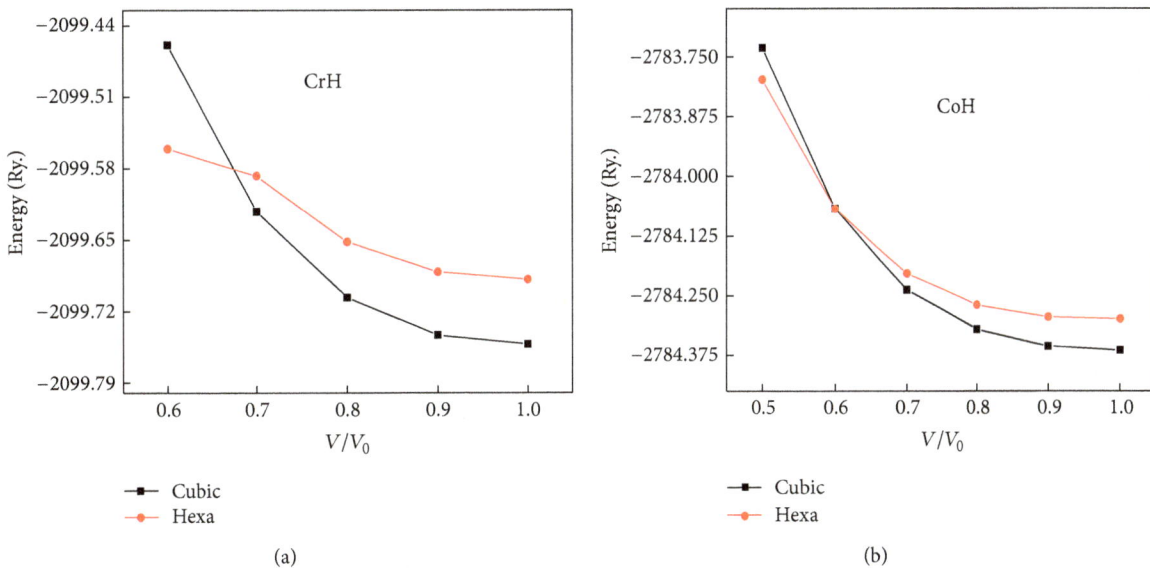

(a)   (b)

FIGURE 3: Total energy versus reduced volume for TMH.

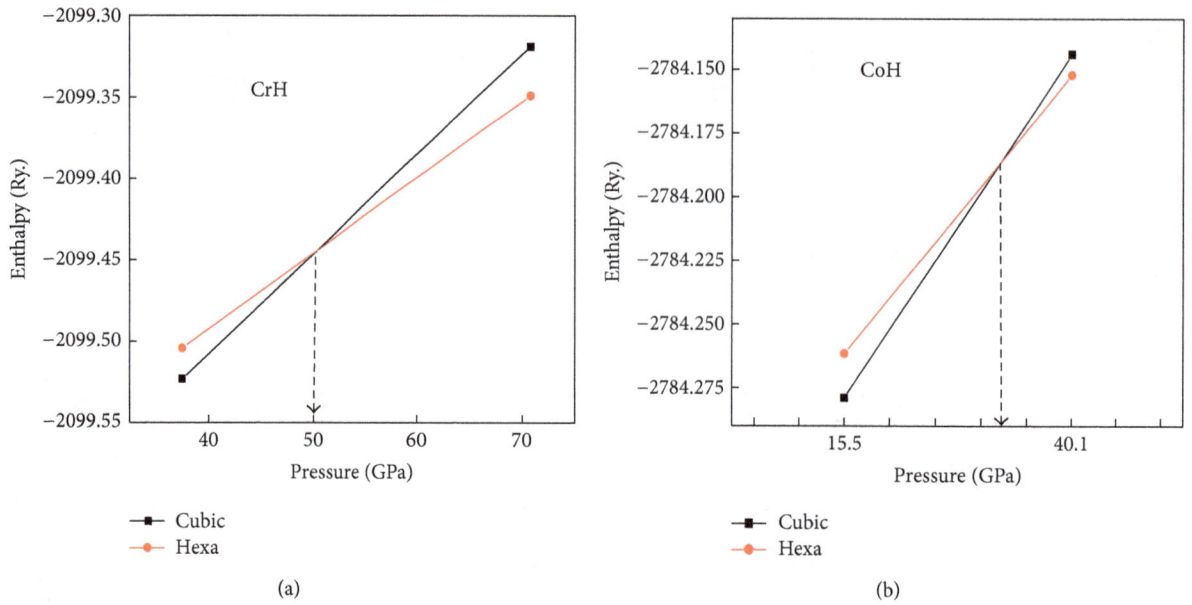

FIGURE 4: Enthalpy as a function of pressure.

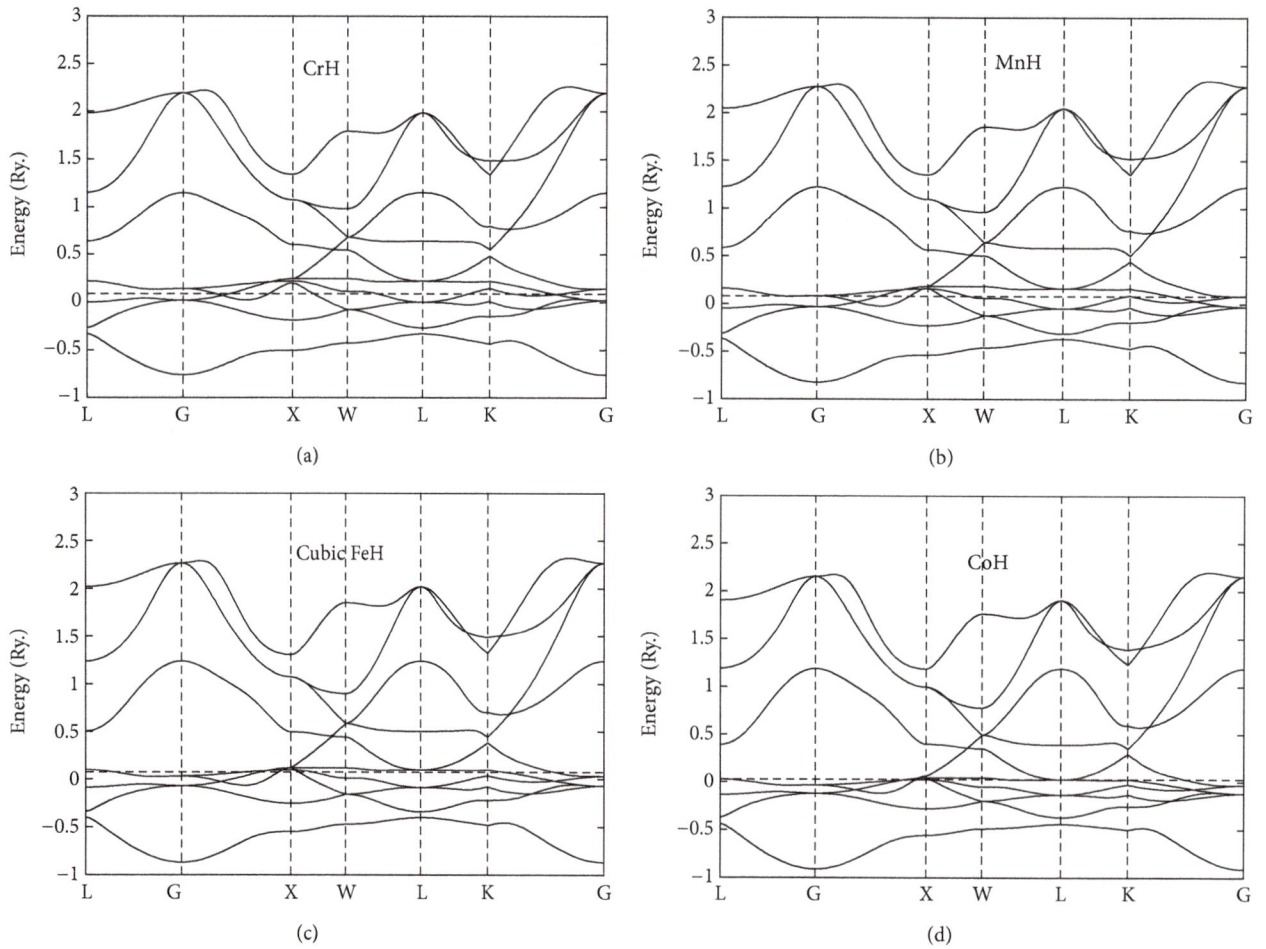

FIGURE 5: Band structures of cubic TMHs at normal pressure.

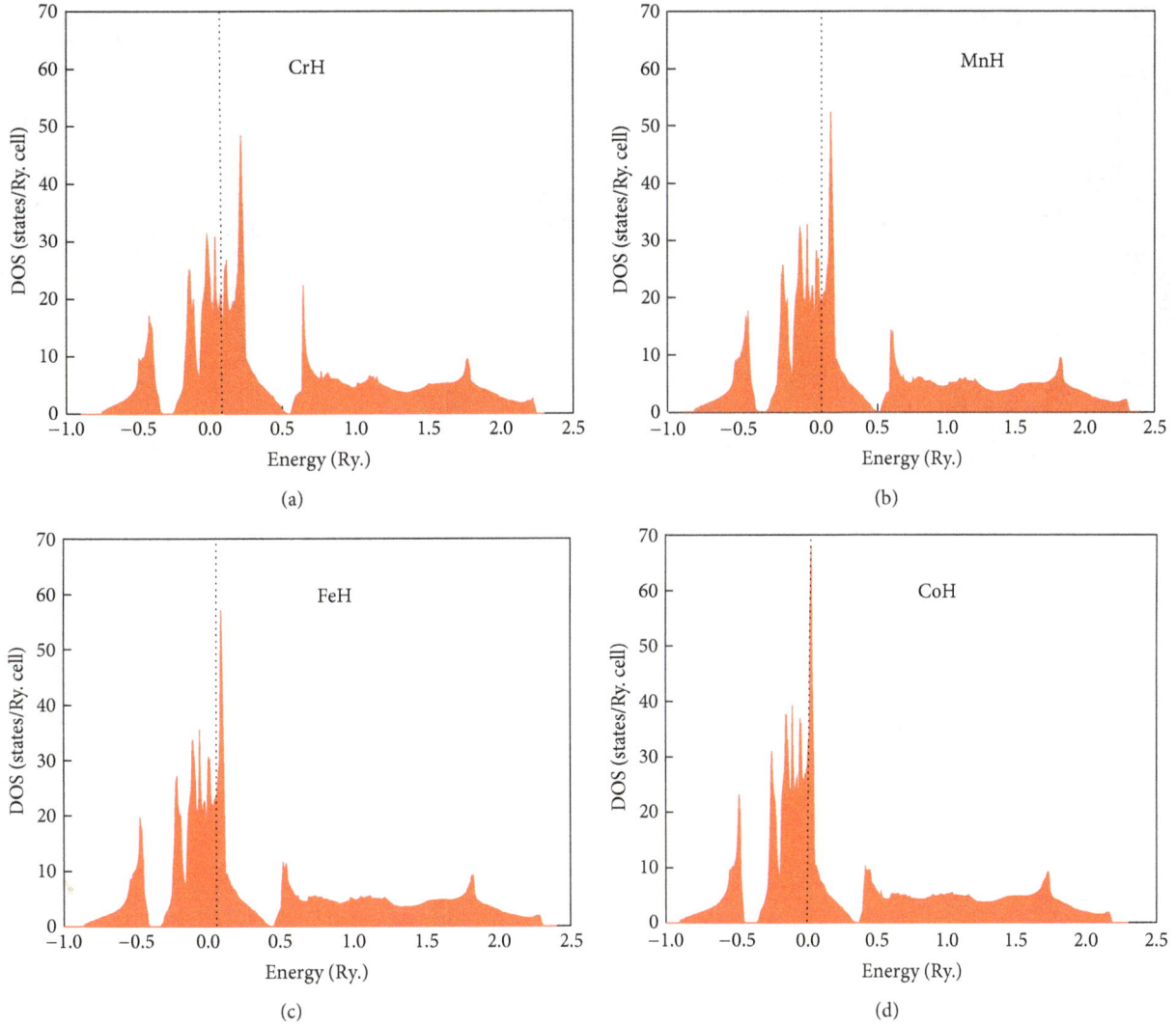

FIGURE 6: Total density of states of cubic TMHs at normal pressure.

TABLE 2: Lattice constant a_o (Å), the equilibrium volume V_0 (Å^3), valence electron density ρ (electrons/Å^3), Fermi energy E_F (Ry.), density of states at Fermi energy $N(E_F)$, Wigner-Seitz radius R_{WZ} (a.u), and band energy E_B (Ry).

	CrH	MnH	FeH	CoH
	3.77	3.67	3.63	3.64
a_o (Å)	3.71[a]	3.62[a]	3.59[a]	3.60[a]
	3.854[b]			
V_0	13.395	12.358	11.958	12.057
ρ	0.5226	0.6474	0.7526	0.8294
E_F (Ry)	0.0881	0.0817	0.0743	0.0256
$N(E_F)$	15.8469	18.8391	26.8858	63.6115
R_{WZ}	2.2105	2.1519	2.1284	2.1343
E_B	−1.1722	−1.4232	−1.65767	−2.0608

[a]Ref. [8].
[b]Ref. [10], expt.

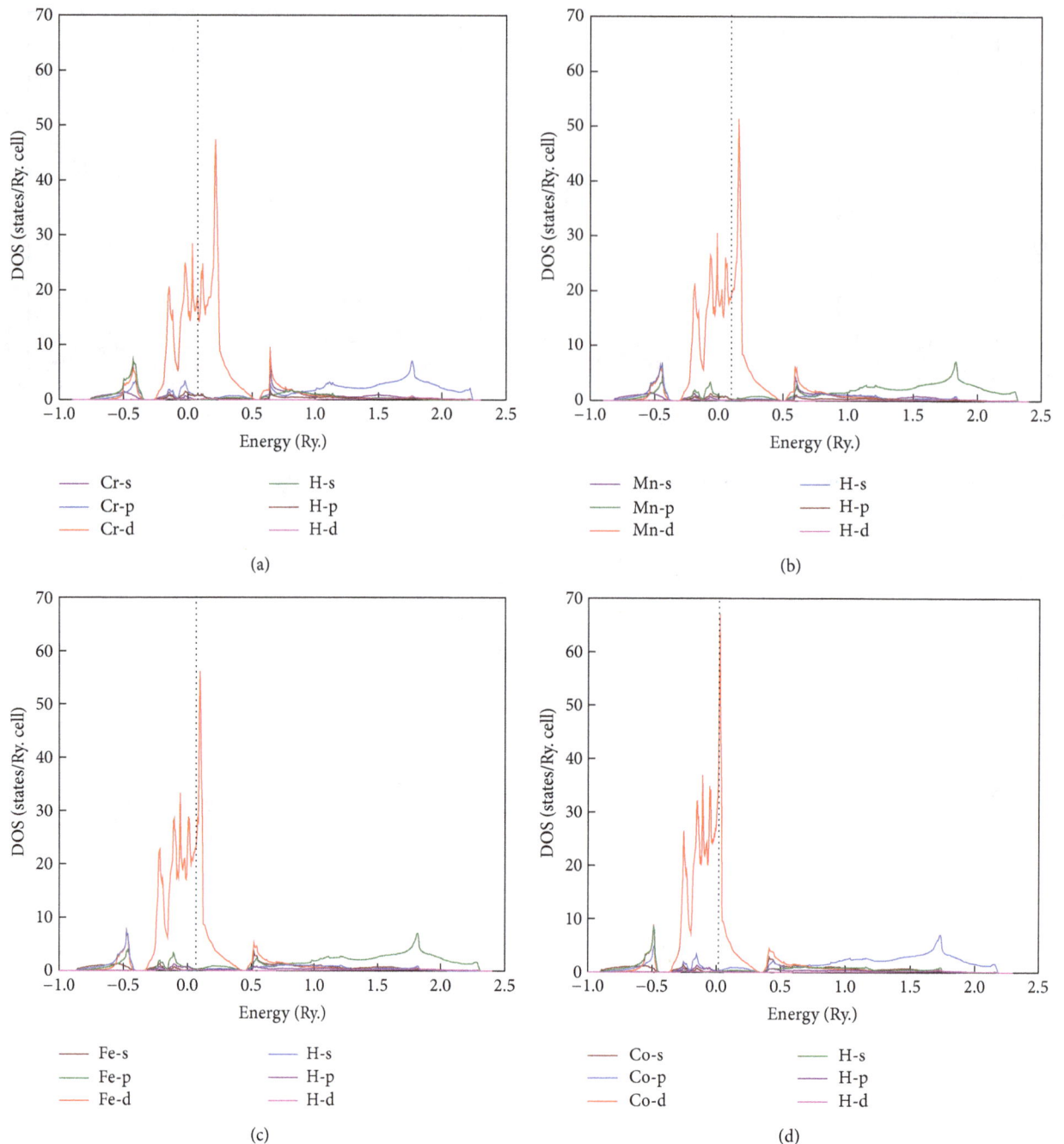

FIGURE 7: Partial density of states of cubic TMHs at normal pressure.

the electronic properties in a real material. One has to remember that the charge density very much depends on the crystal structure and different characteristics may be formed for one and the same compound in different structural arrangements. The covalent characteristics between metal and H atoms can be confirmed by the charge density distribution. The charge density distribution of TMHs is shown in Figure 8. It is clearly seen that the charge strongly accumulate between the transition metals and hydrogen atoms, which means that a strong directional bonding exist between them. Thus, the covalent characteristics between them can be

confirmed by the charge density distribution. The total charge transfer from the transition metals to hydrogen atom has some ionic character. Thus, the bonding may be a mixture of metallic and covalent and ionic attribution.

3.6. Magnetic Phase Transition of FeH and CoH. The magnetic moments of FeH and CoH at normal and various pressures are investigated and are given in Figure 9. It is observed that magnetic moment decreases as the pressure increases and it will become zero at $V/V_0 = 0.5$ in CoH and at $V/V_0 = 0.9$ in FeH. It indicates that as pressure increases

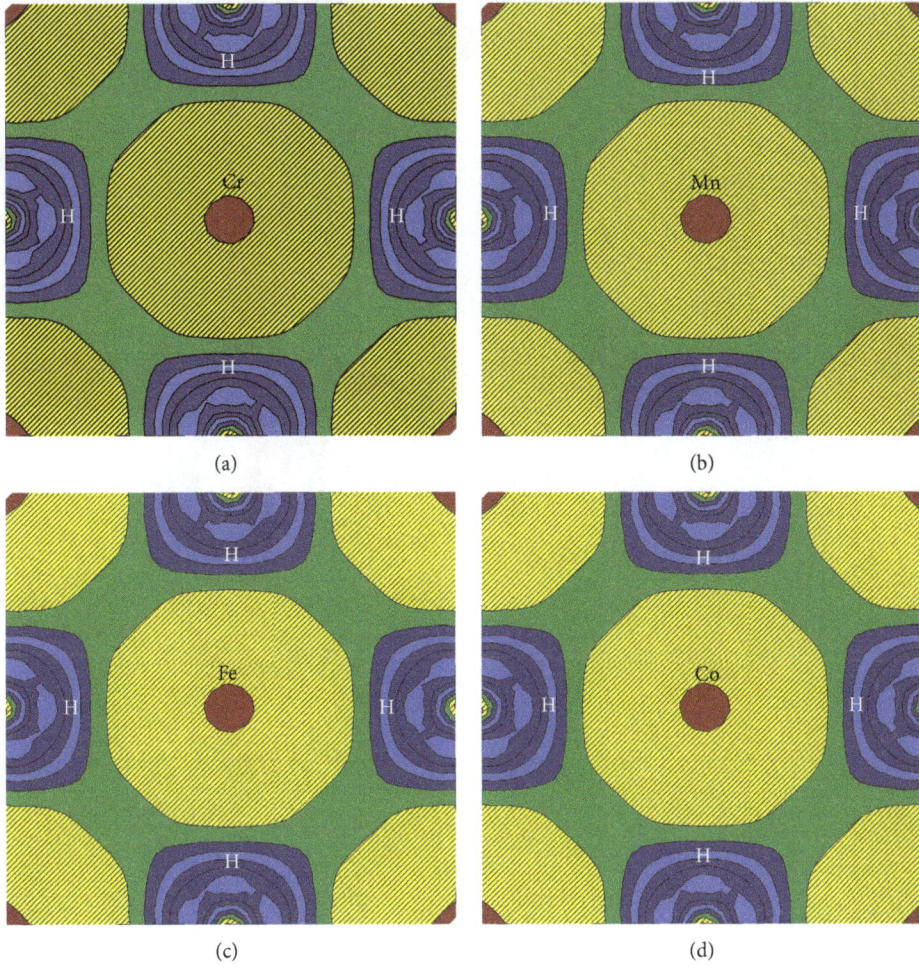

FIGURE 8: Charge density plot.

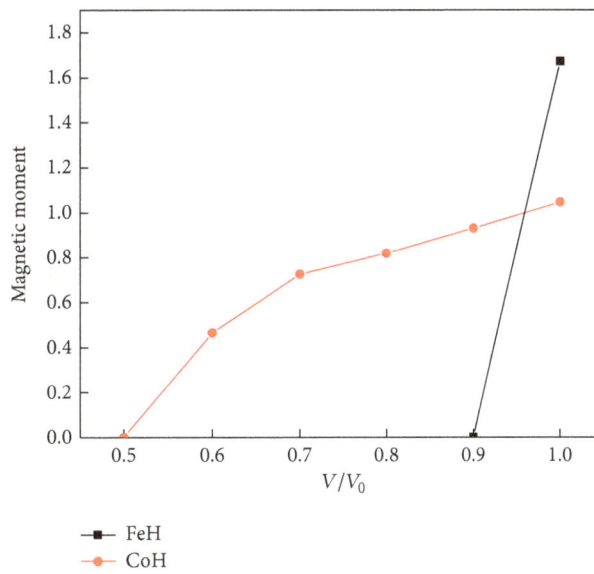

FIGURE 9: Variation of magnetic moment with reduced volume.

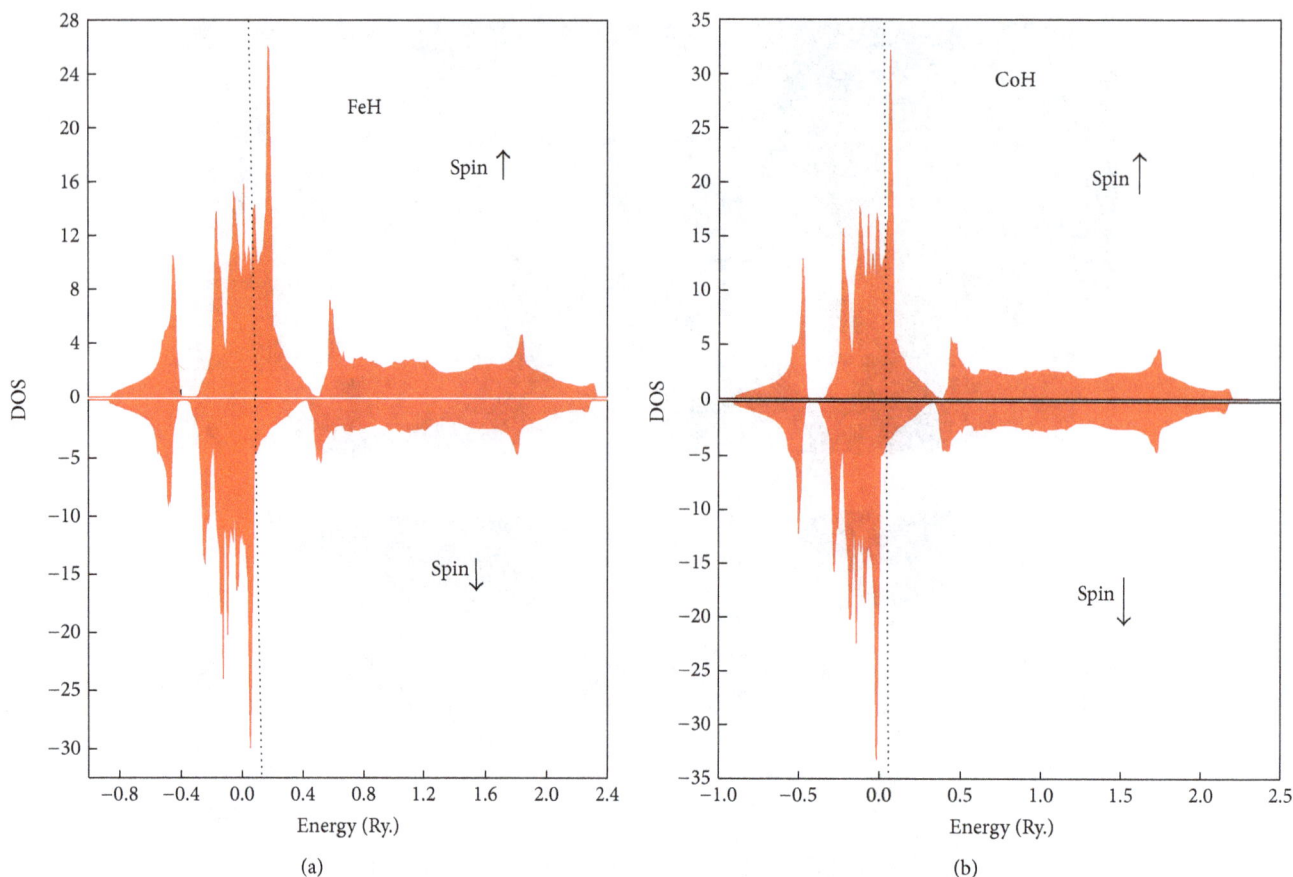

FIGURE 10: Spin dependent total density of states of FeH and CoH.

the magnetic property of the metal hydrides decreases. The magnetic to nonmagnetic transition occurs at a pressure of 10 GPa in FeH and 180 GPa in CoH. The main reason behind this magnetic transition is magnetic collapse due to band widening in transition metal ions under pressure. It has been found that the contribution to the magnetic moment is entirely from Fe and Co atoms rather than H atom. The spin dependent total density of states is given in Figure 10. It is found that the highest spike due to 3d state lies slightly above the Fermi level in the minority spin (spin up) whereas in the majority spin (spin down), it lies slightly below the Fermi level.

4. Conclusions

The band structure, density of states, magnetic phase transition, and charge density under various pressures are investigated based on first principles calculation under the frame work of tight-binding theory within local density approximation. It is found that a pressure induced structural phase transition from cubic to hexagonal phase is observed at the pressure of 50 GPa for CrH and 23 GPa for CoH. The magnetic to nonmagnetic transition occurs at the pressure of 10 GPa in FeH and 180 GPa in CoH.

Acknowledgment

The authors thank the college management for their constant encouragement.

References

[1] P. Vajeeston, P. Ravindran, A. Kjekshus, and H. Fjellvåg, "Structural stability of BeH_2 at high pressures," *Applied Physics Letters*, vol. 84, no. 1, pp. 34–36, 2004.

[2] Y. Fukai, *Metal Hydrogen System: Basics Bulk Properties*, Springer, Berlin, Germany, 2005.

[3] N. L. Rosi, J. Eckert, M. Eddaoudi et al., "Hydrogen storage in microporous metal-organic frameworks," *Science*, vol. 300, no. 5622, pp. 1127–1129, 2003.

[4] I. P. Jain, C. Lal, and A. Jain, "Hydrogen storage in Mg: a most promising material," *International Journal of Hydrogen Energy*, vol. 35, no. 10, pp. 5133–5144, 2010.

[5] M. Latroche, "Structural and thermodynamic properties of metallic hydrides used for energy storage," *Journal of Physics and Chemistry of Solids*, vol. 65, no. 2-3, pp. 517–522, 2004.

[6] R. Quijano, R. de Coss, and D. J. Singh, "Electronic structure and energetics of the tetragonal distortion for TiH_2, ZrH_2, and HfH_2: a first-principles study," *Physical Review B*, vol. 80, no. 18, Article ID 184103, 8 pages, 2009.

[7] K. Miwa and A. Fukumoto, "First-principles study on 3d transition-metal dihydrides," *Physical Review B*, vol. 65, no. 15, Article ID 155114, 7 pages, 2002.

[8] H. Smithson, C. A. Marianetti, D. Morgan, A. Van der Ven, A. Predith, and G. Ceder, "First-principles study of the stability and electronic structure of metal hydrides," *Physical Review B*, vol. 66, no. 14, Article ID 144107, 10 pages, 2002.

[9] V. E. Antonov, "Phase transformations, crystal and magnetic structures of high-pressure hydrides of d-metals," *Journal of Alloys and Compounds*, vol. 330–332, pp. 110–116, 2002.

[10] V. E. Antonov, A. I. Beskrovnyy, V. K. Fedotov et al., "Crystal structure and lattice dynamics of chromium hydrides," *Journal of Alloys and Compounds*, vol. 430, no. 1-2, pp. 22–28, 2007.

[11] S. Kangaprabha, A. T. Asvini Meenaatci, R. Rajeswara-palanichamy, and K. Iyakutti, "Electronic structure and magnetic properties of FeH and CoH," in *Proceedings of the AIP Conference Proceedings*, vol. 1447, pp. 71–72, 2012.

[12] O. K. Andersen, "Linear methods in band theory," *Physical Review B*, vol. 12, no. 8, pp. 3060–3083, 1975.

[13] H. L. Skriver, *The LMTO Method*, Springer, Heidelberg, Germany, 1984.

[14] N. E. Christensen, D. L. Novikov, R. E. Alonso, and C. O. Rodriguez, "Solids under pressure. Ab initio theory," *Physica Status Solidi B*, vol. 211, no. 1, pp. 5–16, 1999.

[15] O. Jepson and O. K. Anderson, "The electronic structure of h.c.p. Ytterbium," *Solid State Communications*, vol. 9, no. 20, pp. 1763–1767, 1971.

[16] E. Sjöstedt, L. Nordström, and D. J. Singh, "An alternative way of linearizing the augmented plane-wave method," *Solid State Communications*, vol. 114, no. 1, pp. 15–20, 2000.

27

Effect of Alternating Bending on Texture, Structure, and Elastic Properties of Sheets of Magnesium Lithium Alloy

N. M. Shkatulyak, S. V. Smirnova, and V. V. Usov

South Ukrainian National Pedagogical University Named after K. D. Ushinsky, 26 Staroportofrankovskaya Street, Odessa 65020, Ukraine

Correspondence should be addressed to N. M. Shkatulyak; shkatulyak@mail.ru

Academic Editor: Manoj Gupta

The effect of low-cycle alternating bending at room temperature on the crystallographic texture, metallographic structure, and elastic properties of sheets of MgLi5 (mass) magnesium alloy after warm cross-rolling has been studied. Texture of alloy is differed from the texture of pure magnesium. The initial texture of alloy is characterized by a wide scatter of basal poles in the transverse direction. In the process of alternating bending, the changes in the initial texture and structure (which is represented by equiaxed grains containing twins) lead to regular changes in the anisotropy of elastic properties.

1. Introduction

Alloying of magnesium with lithium (Li) with a density of $530 \, \text{kg/m}^3$ not only significantly reduces the density of Mg but also significantly increases the ductility and toughness of magnesium alloys [1]. Despite the fact that the effect of lithium on the microstructure and mechanical properties of magnesium alloys is known [2], many questions regarding texture formation and anisotropy of properties (in particular, elastic) for various kinds of heat treatment and deformation of magnesium alloys with lithium are yet sufficiently studied. For example, the behavior of binary alloys Mg-Li with a hexagonal structure at tension and compression is unknown, which is important for straightening sheet metal. Typically, the sheets or rolled metals are subjected to straightening by roller straightening machines. Such processing that consists of the alternating bending (AB) reduces the internal stresses of the metal and obtaining a flat sheet [3]. There undergo substantial changes in the structure of the metal and its characteristics despite the relatively low plastic deformation during AB. The research of these changes has important applied significance.

Effect of low-cycle alternating bending (AB) on crystallographic texture, microstructure and anisotropic properties of hexagonal magnesium alloys with aluminum and zinc (AZ31), zinc, zirconium and rare earth metals (ZE10), and titanium has been previously studied [4–9]. However, data on complex studies of the effects of the AB on the texture, structure, and anisotropy of elastic properties of magnesium alloys with lithium are absent.

The aim of this article is to study the effect of alternating bending on the crystallographic texture, metallographic structure and elastic properties of sheets of alloy Mg-5% Li (mass) with a hexagonal lattice.

2. Materials and Methods

Material used for the study are the cylindrical ingots of alloy MgLi5 (mass) with a length of 120 mm. Bars of 6 mm of thickness and 60 mm of width were received after turning and pressing of cast billets at 350°C. Then the bars were rolled along longitudinal direction of work pieces to a thickness of 4.5 mm in two passes. Next rolling was carried out in the transverse direction with a reduction of 10% per pass to a thickness of 2 mm for 10 passes. Heating to 350°C was performed after each pass. Then the direction of rolling is changed in the 90° one pass with a reduction of 10%.

Then again the rolling direction is changed to 90° to a final thickness of 1 mm.

The process of straightening the sheets was simulated by AB on a three-roll bending device. The diameter of the bending roller was 50 mm. The speed of the motion of metal upon bending was ~150 mm/s. One cycle of bending consisted of bending in one direction (0.25 cycle), returning to the flat state (0.5 cycle), bending in the direction (0.75 cycle), and straightening (1.0 cycle). Rectangular samples of size 100 × 10 mm were cut out from the initial sheet and deformed by AB with 0.5, 1, 3, and 5 cycles through every 15° from the last rolling direction (RD) up to transverse direction (TD) for measuring the anisotropy in Young's modulus, as well as samples for investigating the structure and texture. Samples for measurement of the elastic properties in the package were handled by a milling machine to reduce the influence on the accuracy of measurements of geometrical dimensions.

The elastic modulus was measured by dynamic method via frequency of forced flexural vibrations of a flat sample [10]. The measurement error did not exceed 1%.

Before the study of the texture, the samples were chemically polished to the depth of 0.1 mm to remove the distorted surface layer. The crystallographic texture was investigated on two surfaces of the samples after the above indicated number of cycles by constructing inverse pole figures (IPFs) for the normal direction (ND) and rolling direction (RD) as well as by means of incomplete direct pole figures {0002} ($0° \leq \alpha \leq 70°$) of rolling plane using a DRON-3M diffractometer in filtered Mo Kα radiation. A sample without texture measurements was prepared from fine recrystallized sawdust of the alloy investigated. To plot IPFs that correspond to the RD (IPF RD), composite samples were prepared. When constructing IPFs, the normalization according to Morris was used [11].

Metallographic structure was investigated in the rolling plane by means of metallographic microscope MIM-7 using the camera Veb-E-TREK DEM 200 to output the image structure on the computer monitor.

3. Results and Discussions

The IPFs of the samples of the Mg-5% Li (mass) magnesium alloy investigated are shown in Figure 1. The corresponding direct pole figures are represented in Figure 2. Microstructures of the alloy are shown in Figure 3.

In the ND IPF (Figure 1(a)), the absolute maximum of the pole density (4.50) coincides with the pole $\langle 21\bar{3}2 \rangle$ with the scattering up to pole $\langle 30\bar{3}2 \rangle$. In the pole $\langle 0002 \rangle$ a relatively high pole density of 3.00 is observed. On the corresponding direct PF (Figure 2(a)) it can be seen that the basal planes are aligned with the sheet surface, but there are splits in the texture peaks where the basal poles are rotated approximately 15° towards the TD (recall that this TD coincides with the initial RD). The difference in the numerical value of the pole density on IPF and direct PF is due to their different normalization.

Thus, the texture of the initial sheet of a magnesium-lithium alloy may be described as a complicated texture of the deflected basal type with deviation angles of basal

plane toward TD approximately 15 and 70°. In the IPF RD, the region of the heightened pole density occupies an area bounded by the poles of $\langle 30\bar{3}2 \rangle$, $\langle 11\bar{2}0 \rangle$, and $\langle 10\bar{1}0 \rangle$ with an absolute maximum at the pole $\langle 21\bar{3}1 \rangle$. Thus, the rolling direction coincides mainly with the crystallographic direction $\langle 11\bar{2}0 \rangle$ with the scattering up to $\langle 10\bar{1}0 \rangle$. The above-described texture of the initial sheet is different from the basal central type texture of pure magnesium.

We have previously observed a similar rotation of basal poles in TD direction in the study of the texture of magnesium alloy with zinc, zirconium, and rare earth metals (REM) [9]. This was to the influence of additives of REMs to the magnesium alloy, which promote activation of nonbasal deformation mechanisms, particularly of prismatic and pyramidal slip of dislocations [9].

It is known that the alloying magnesium by lithium promotes the scattering of the basal planes in the TD [12]. Alloying of magnesium by lithium eases the glide of $\langle a \rangle$ dislocations on prismatic planes because lithium reduces the hcp c/a ratio [13]. Quimby et al. [14] also reported that the critical resolved shear stress (CRSS) for basal slip in alpha solid solution (12.5 at. % Li in Mg) is 10 times higher than that for pure magnesium. Even when there is evidence of $\langle c + a \rangle$ dislocation glide during deformation of Mg-Li alloys the significant role played by prismatic slip during deformation has been acknowledged [15]. Thus, the texture of the initial sheet of the alloy is caused by the activation of nonbasal mechanisms of sliding.

Twinning processes can also play some roles in the formation of the observed texture. In pure magnesium, the twinning at room temperature usually occurs on planes {$10\bar{1}2$} although in the early work [16] needle like twins {$30\bar{3}4$} were also observed. In [17], after the hot rolling of pure-magnesium samples made from an ingot, tension-induced {$10\bar{1}2$} $\langle 10\bar{1}1 \rangle$ twins and compression-related twins {$10\bar{1}1$} $\langle 10\bar{1}2 \rangle$ have been revealed. The alloying and deformation performed at elevated temperatures can change the mechanisms of deformation and twinning in the alloys of magnesium [18]. For example, in [17] the formation of double-twinning structures {$10\bar{1}1$}–{$10\bar{1}2$} was revealed. In the RD IPFs of the initial sample, an absolute maximum equal to 3.79 is observed at the pole $\langle 21\bar{3}1 \rangle$ and the enhanced pole density occupies a wide region, which includes poles $\langle 30\bar{3}2 \rangle$, $\langle 20\bar{2}1 \rangle$, $\langle 10\bar{1}0 \rangle$, $\langle 21\bar{3}0 \rangle$, and $\langle 11\bar{2}0 \rangle$ (hatched region in Figure 1(b)). It is interesting to note that, according to [17], this region corresponds to the orientations of secondary twins that arise after primary twinning on the planes {$10\bar{1}1$}.

In the photographs of the microstructure (Figure 3), almost equiaxed grains are seen (which is characteristic of recrystallization), the average size of which in two mutually perpendicular directions is almost identical (approximately 21 × 26 μm) both in the initial state and after deformation using different numbers of AB cycles. Thus, the contribution of dynamic recrystallization in the texture formation under effect of the warm rolling of initial alloy sheet cannot be ruled out.

Twins also can be seen in the appropriate photographs of the microstructure of initial sample as well as after AB

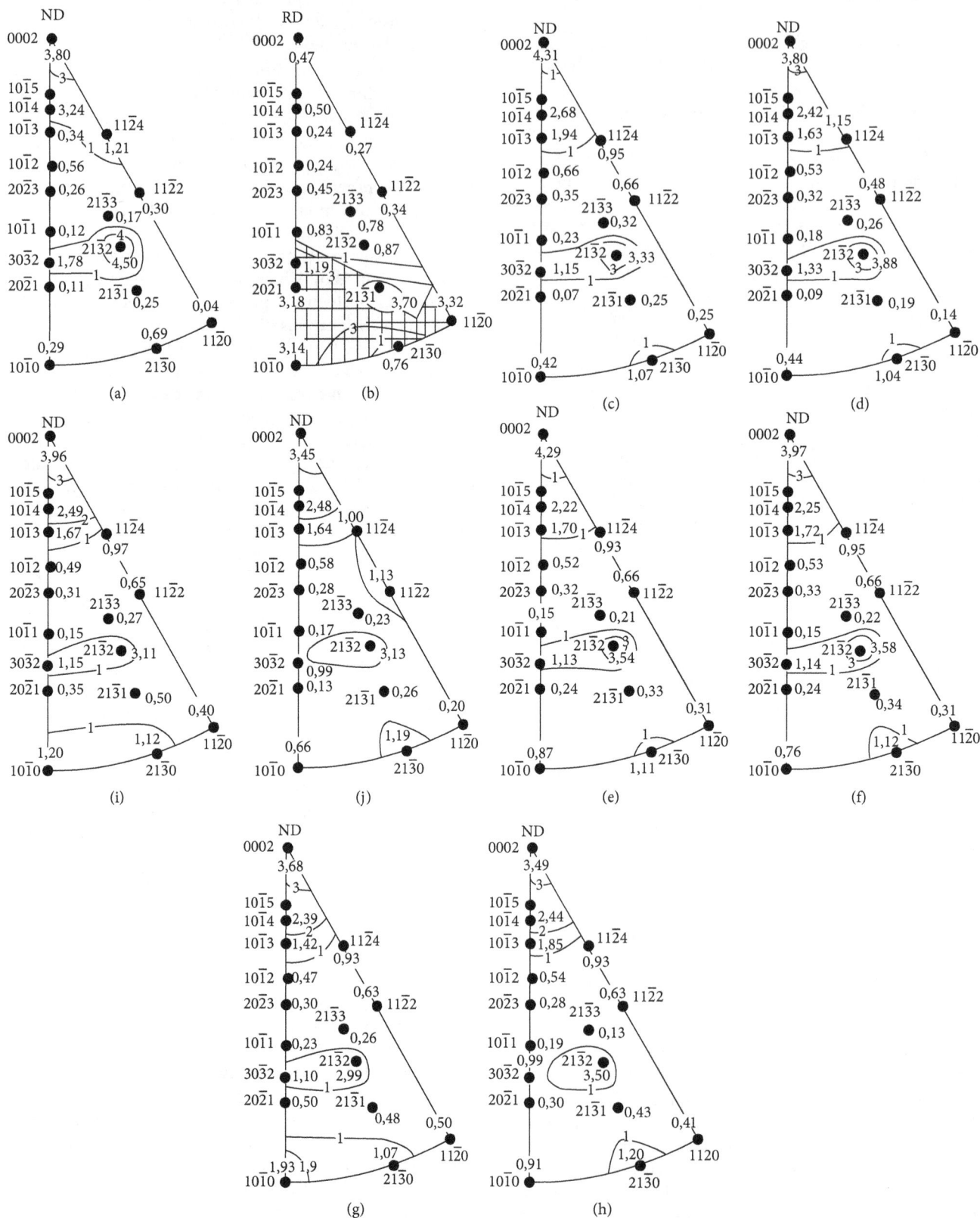

FIGURE 1: Experimental inverse pole figures of alloy Mg-5% Li (mass): (a, b) initial state; (c–j) after alternating bending using different number of cycles: (c, d) 0.5, (e, f) 1.0, (g, h) 3.0, and (i, j) 5.0 cycles. Hatched regions correspond to twin reorientations of crystals on the twinning planes {10$\bar{1}$0}; (d, f, g, i) correspond to the stretched side of the sheet; (c, e, h, j) correspond to the compressed side of sheets of alloy.

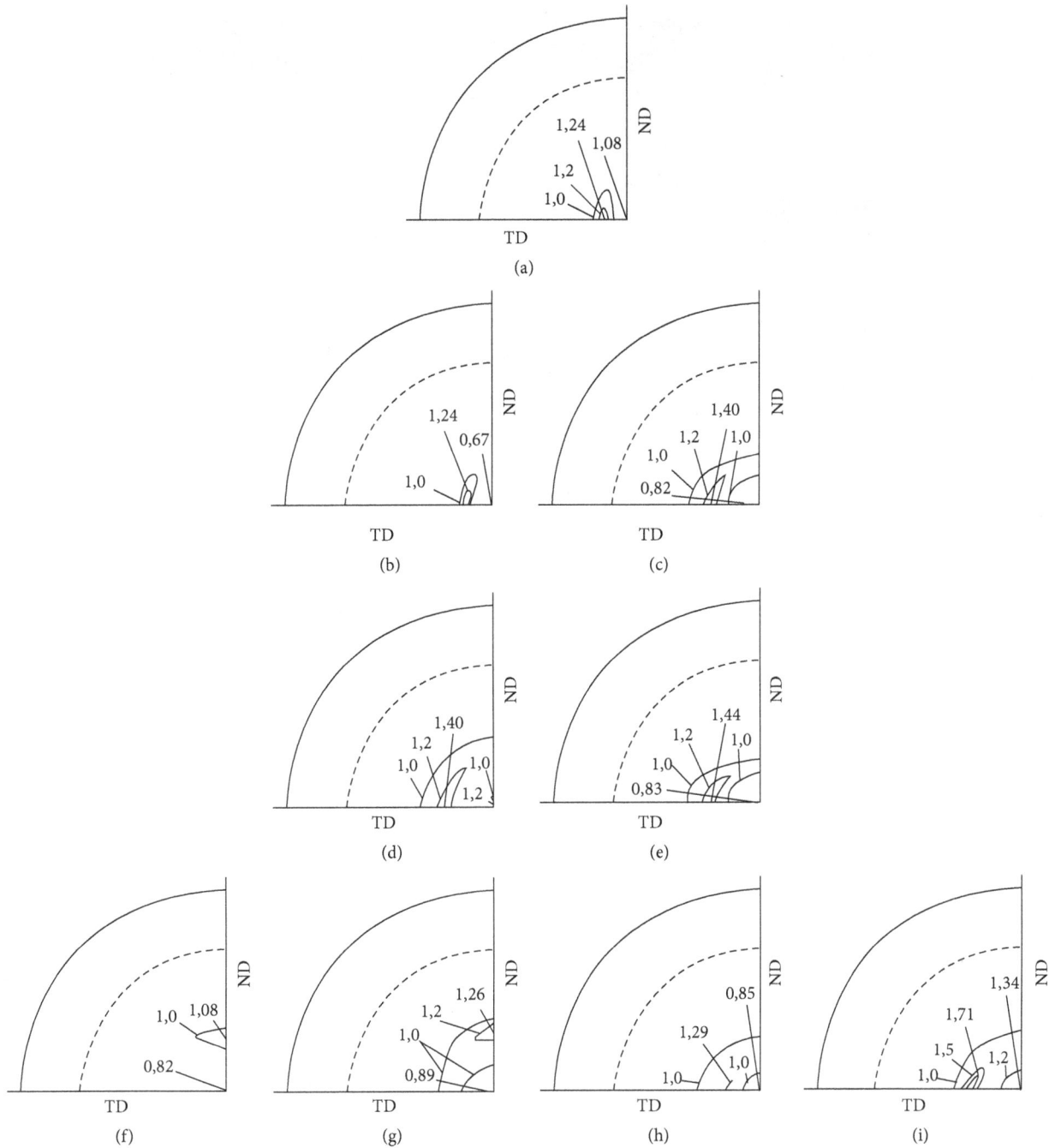

FIGURE 2: Experimental incomplete direct pole figures {0002} of alloy Mg-5% Li (mass): (a) initial state; (b–i) after alternating bending using different number of cycles: (b, c) 0.5, (d, e) 1.0, (f, g) 3.0, and (h, i) 5.0, cycles. (c, d, g, i) Correspond to the stretched side of the sheet; (b, e, f, h) correspond to the compressed side.

(Figure 3). Moreover, with the number of cycles of AB, the number of twins in micrographs increases. In pictures there are both broad and sharp at ends twins ($\langle 10\bar{1}2 \rangle$) and paired thin twins $\langle 10\bar{1}1 \rangle$ [19].

The character of texture scattering and value of pole density on pole figures depend on the number of cycles of alternating bending (Figures 1 and 2). The IPFs (Figures 1(c), 1(d), 1(e), and 1(f)) vary a little after 0.5 and 1.0 cycles of AB. For example, the pole density in the pole $\langle 0002 \rangle$ of IPF ND is

increased slightly to 4.31 on the compressed side of sheet alloy after 0.5 cycles of AB and on a stretched side of the sheet has not changed. Pole density in $\langle 21\bar{3}2 \rangle$ decreased slightly after 0.5 cycles of AB.

The pole density $\langle 0002 \rangle$ on IPF ND has increased to 4.29 on the compressed side of the sheet compared to the original value of 3.80 after 1.0 cycle of AB (Figure 1(e)). At the same time the pole density $\langle 0002 \rangle$ has slightly increased to a value of 3.97 on the stretched side of the sheet (Figure 1(f)). The pole

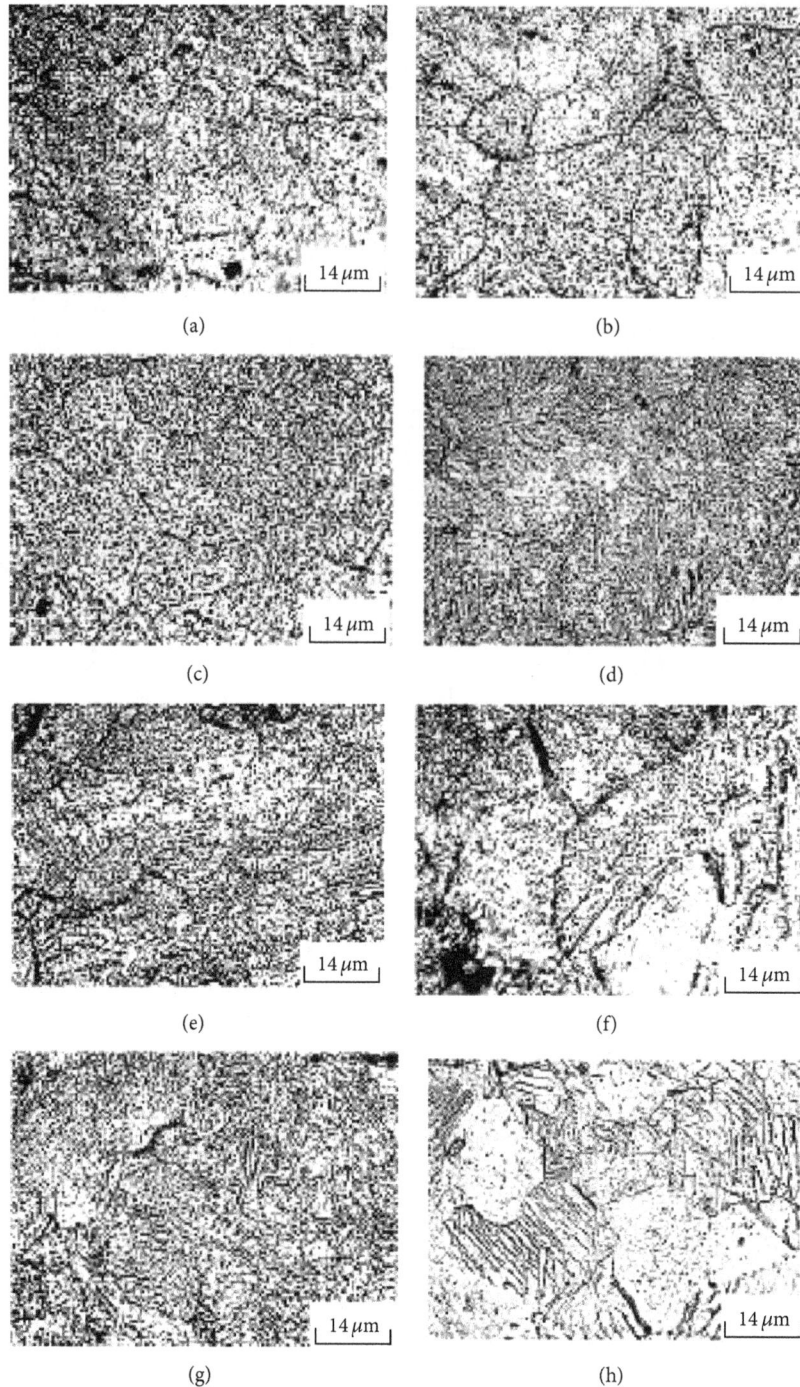

FIGURE 3: Microstructures of MgLi5 alloy after alternating bending using different numbers of cycles: (a, b) 0.5, (c, d) 1.0, (e, f) 3.0, and (f, h) 5 cycles; (b, c, e, g) correspond to the stretched side of the sheet; (a, d, f, h) correspond to the compressed side.

density $\langle 21\bar{3}2 \rangle$ has decreased slightly to values 3.54 and 3.58 after 1.0 cycle of AB on compressed and stretched side sheet, respectively (Figures 1(e) and 1(f)).

More significant texture changes are observed after 3 and 5 cycles of AB. The pole density $\langle 10\bar{1}0 \rangle$ on stretched sides of sheets reached 1.93 and 1.20 after 3 and 5 cycles of AB, respectively. The formation of such orientations can be caused by the basal and pyramidal slip under stretching

[19]. Values of the pole density $\langle 21\bar{3}2 \rangle$ have decreased on the stretched side (up to 2.99 and 3.11) as well as on the compressed sides of the sheets (up to 3.50 and 3.13) (Figures 1(g), 1(h), 1(i), and 1(j)). The local maximum of 1.13 is observed at the pole $\langle 1132 \rangle$ on the IPF ND of compressed sheet side after 5 cycles of AB (Figure 1(j)). This maximum corresponds to a twin orientation with the twinning plane of $\{11\bar{2}2\}$ [20].

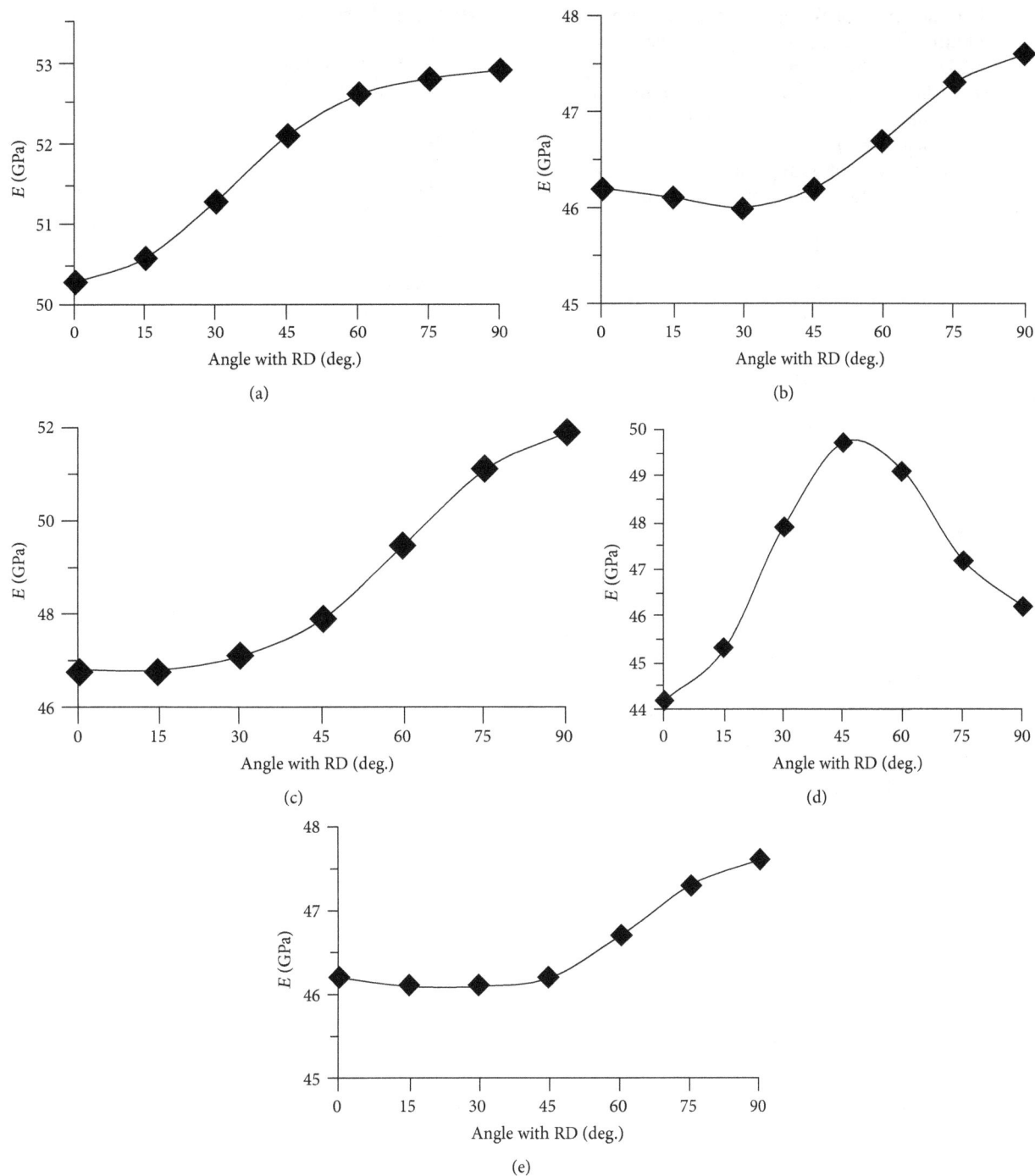

FIGURE 4: Elastic properties of alloy sheet (a), (b), (d), (c), and (e) after 0, 0.5, 1.0, 3.0, and 5.0 cycles of AB, respectively.

The pole density $\langle 21\bar{3}0 \rangle$ has become more than 1 (Figure 1) on all IPFs RD after AB. This indicates that scattering of basal plane reached 90° in the TD.

Direct PFs (Figure 2) show that the texture scattering and its average sharpness have increased after 0.5 cycles of the AB. Average sharpness of texture and its scattering have increased also after 1 cycle of AB. Texture has changed after 3 cycles of AB. Now the pole density maximum is observed in RD, but not in the TD. The location of maximums after 5 cycles of

AB again becomes similar to what was after 0.5 and 1 cycles of AB. The texture scattering has increased, as well as the average value of the basal pole density on the centre of direct PF {0002} has increased.

The above-described texture changes are reflected in the behavior of the curves of the anisotropy of the elastic properties (Figure 4). The maximum value of Young's modulus is observed in the TD in initial sheet as well as after 0.5, 1, and 5 cycles of AB (Figures 4(a), 4(b), 4(c), and 4(e)). But after 3 cycles of AB a maximum of Young's modulus is observed in RD + 45° (Figure 4(d)). Young's modulus has the minimum value in the RD + 30° after 0.5 and 5 cycles of AB (Figures 4(b) and 4(e)). The minimal value in the RD shows Young's modulus in the initial sheet and after 1 and 3 cycles of AB (Figures 4(a), 4(c), and 4(d)).

The anisotropy degree of Young's modulus can be represented quantitatively by the anisotropy coefficient:

$$\eta = \frac{E_{\max} - E_{\min}}{E_{\min}} \times 100\%, \qquad (1)$$

where E is Young's modulus. The initial material anisotropy coefficient η is 4.5%. After 0.5, 1, 3, and 5 cycles of AB η is 3.4%, 10.7%, 12.5, and 3.3%, respectively.

The minimum anisotropy of the elastic properties after 5 cycles of AB is explained by increasing of texture scattering and density of basal poles in the centre of PF {0002} (Figures 2(h) and 2(i)).

4. Conclusion

(1) Of the initial sheet MgLi5 (mass) alloy received by warm cross-rolling is different from the basal central type texture of pure magnesium. The texture of the initial sheet of MgLi5 (mass) alloy may be described as a complicated texture of the deflected basal type with deviation angles of basal plane toward transverse direction of approximately 15 and 70°. The rolling direction coincides mainly with the crystallographic direction $\langle 11\bar{2}0 \rangle$ with the scattering up to $\langle 10\bar{1}0 \rangle$. Such texture may be caused by the activation of nonbasal mechanisms of sliding as well as twinning due to the alloying by lithium. The contribution of dynamic recrystallization in the texture formation under effect of the warm rolling of initial alloy sheet cannot be ruled out.

(2) The character of texture scattering and value of pole density on pole figures depend on the number cycles of alternating bending. More significant texture changes are observed after 3 and 5 cycles of alternating bending. The texture scattering is increased, and the average value of the basal pole density in the centre of direct PF {0002} increased as well with the increasing of number cycles of alternating bending. The number of twins on respective images of the microstructure is also increased.

(3) The texture changes are reflected in the anisotropy of the elastic properties. Young's modulus has maximum in the transverse direction in the initial sheet, as well as after 0.5, 1, and 5 cycles of alternating bending. After 3 cycles of alternating bending a maximum of Young's modulus is observed in RD + 45°. The minimal Young's modulus takes place in the rolling direction in the initial sheet, as well as after 1 and 3 cycles of alternating bending. Young's modulus has a minimum in the RD + 30° after 0.5 and 5 cycles of alternating bending.

Conflict of Interests

The authors declare that there is no conflict of interests regarding the publication of this paper.

References

[1] Z. Yang, J. P. Li, J. X. Zhang, G. W. Lorimer, and J. Robson, "Review on research and development of magnesium alloys," *Acta Metallurgica Sinica*, vol. 21, no. 5, pp. 313–328, 2008.

[2] H. Haferkamp, R. Boehm, U. Holzkamp, C. Jaschik, V. Kaese, and M. Niemeyer, "Alloy development, processing and applications in magnesium lithium alloys," *Materials Transactions*, vol. 42, no. 7, pp. 1160–1166, 2001.

[3] Technology of sraightening in ARKU rollers, http://www.tkzentrum.ru/equipment/arku/item17/.

[4] Y. V. Zilberg, F. Bach, D. Bormann, M. Rodman, M. Sharper, and M. Hepke, "Effect of alternating bending on the structure and properties of strips from AZ31 magnesium alloy," *Metal Science and Heat Treatment*, vol. 51, no. 3-4, pp. 170–175, 2009.

[5] A. A. Bryukhanov, P. P. Stoyanov, Yu. V. Zilberg, and D. Rodman, "Anisotropy of mechanical properties of magnesium alloy AZ31 sheets as a result of sign-variable bending deformation," *Metallurgical and Mining Industry*, vol. 2, no. 3, pp. 215–219, 2010.

[6] A. A. Bryukhanov, M. Rodman, A. F. Tarasov, P. P. Stoyanov, M. Shaper, and D. Bormann, "Mechanism of the plastic deformation of the AZ31 alloy upon low-cycle reverse bending," *Physics of Metals and Metallography*, vol. 111, no. 6, pp. 623–629, 2011.

[7] T. Uota, T. Suzu, S. Fukumoto, and A. Yamamoto, "EBSD observation for reversible behavior of deformation twins in AZ31B magnesium alloy," *Materials Transactions*, vol. 50, no. 8, pp. 2118–2120, 2009.

[8] V. V. Usov, P. A. Bryukhanov, M. Rodman et al., "Influence of reversed bending on texture, structure and mechanical properties of α-titanium sheets," *Deformation and Fracture of Materials*, no. 9, pp. 32–37, 2012.

[9] N. M. Shkatulyak, V. V. Usov, N. A. Volchok et al., "Effect of reverse bending on texture, structure, and mechanical properties of sheets of magnesium alloys with zinc and zirconium," *The Physics of Metals and Metallography*, vol. 115, no. 6, pp. 609–616, 2014.

[10] ATCP, "Elastic moduli: overview and characterization methods," Technical Review ITC-ME/ATCP, 2010, http://www.atcp-ndt.com/images/products/sonelastic/articles/RT03-ATCP.pdf.

[11] P. R. Morris, "Reducing the effects of nonuniform pole distribution in inverse pole figure studies," *Journal of Applied Physics*, vol. 30, no. 4, pp. 595–596, 1959.

[12] L. W. F. Mackenzie and M. Pekguleryuz, "The influences of alloying additions and processing parameters on the rolling microstructures and textures of magnesium alloys," *Materials Science and Engineering A*, vol. 480, no. 1-2, pp. 189–197, 2008.

[13] F. E. Hauser, P. R. Landon, and J. E. Dorn, "Deformation and fracture of alpha solid. Solutions of lithium in magnesium," *Transactions of American Society for Metals*, vol. 50, pp. 856–883, 1958.

[14] R. M. Quimby, J. D. Mote, and J. E. Dorn, "Yield point phenomena in magnesium-lithium single crystals," *Transactions of American Society for Metals*, vol. 55, pp. 149–157, 1962.

[15] S. R. Agnew, P. Mehrotra, T. M. Lillo, G. M. Stoica, and P. K. Liaw, "Texture evolution of five wrought magnesium alloys

during route a equal channel angular extrusion: experiments and simulations," *Acta Materialia*, vol. 53, no. 11, pp. 3135–3146, 2005.

[16] S. L. Couling and C. S. Roberts, "New twinning systems in magnesium," *Acta Crystallographica*, vol. 9, no. 11, pp. 972–973, 1956.

[17] M. R. Barnett, Z. Keshavarz, A. G. Beer, and X. Ma, "Non-Schmid behaviour during secondary twinning in a polycrystalline magnesium alloy," *Acta Materialia*, vol. 56, no. 1, pp. 5–15, 2008.

[18] J. Koike and D. Ando, "Strain accommodation twins and fracture initiation twins in magnesium alloys," http://www.magnet .ubc.ca/news/images/MagNET%20Workshop%20Abstracts.pdf.

[19] M. R. Barnett, "Twinning and the ductility of magnesium alloys. Part II. "Contraction" twins," *Materials Science and Engineering A*, vol. 464, no. 1-2, pp. 8–16, 2007.

[20] Ya. D. Vishnyakov, A. A. Babareko, S. A. Vladimirov, and I. V. Egiz, *Theory of Texture Formation in Metals and Alloys*, Nauka, Moscow, Russia, 1979 (Russian).

28

Thermodynamic Properties of $La_{1-x}Sm_xCoO_3$

Rasna Thakur and N. K. Gaur

Superconductivity Research Lab., Department of Physics, Barkatullah University, Bhopal 462026, India

Correspondence should be addressed to Rasna Thakur; rasnathakur@yahoo.com

Academic Editor: Franz Demmel

We have investigated the bulk modulus and thermal properties of $La_{1-x}Sm_xCoO_3$ ($0 \leq x \leq 0.2$) at temperatures $1\,K \leq T \leq 300\,K$ probably for the first time by incorporating the effect of lattice distortions using the modified rigid ion model (MRIM). The calculated specific heat, thermal expansion, bulk modulus, and other thermal properties reproduce well with the available experimental data, implying that MRIM represents properly the nature of the pure and doped cobaltate. The specific heats are found to increase with temperature and decrease with concentration (x) for the present. The increase in Debye temperature (θ_D) indicates an anomalous softening of the lattice specific heat because increase in T^3-term in the specific heat occurs with the decrease of concentration (x).

1. Introduction

Cobaltites of rare-earth elements with the chemical formula $LnCoO_3$ are important agile and multifunctional materials, which are very promising for high temperature oxygen separation membranes and cathodes in solid oxide fuel cells (SOFCs), heterogeneous catalysts, and gas sensors [1, 2]. The $LaCoO_3$ ceramic is at present the best studied representative of the rare-earth cobaltite family. $LaCoO_3$ exhibits two spin state transitions as the temperature increases. The first transition is from low temperature low spin (LS) to intermediate spin (IS) state near 100 K characterized by a steep jump of magnetization at the transition [3–5] and the second one is from IS to high spin (HS) state leading to an insulator-metal (I-M) transition around 500 K [3, 5]. Throughout the $ACoO_3$ series, only $LaCoO_3$ has been analyzed with rhombohedral symmetry [6]; the rest of the members of the family with the ionic radius of the rare-earth smaller than the ionic radius of La exhibit an orthorhombic crystallographic structure [7]. A structural transition from rhombohedral ($R\bar{3}C$) to orthorhombic ($Pbnm$) symmetry with Sm doping level $x > 0.08$ is found from the X-ray diffraction data in $La_{1-x}Sm_xCoO_3$ [8].

Recently, we have applied the modified rigid ion model (MRIM) to study the specific heat of cobaltates and manganites [9, 10]. Motivated from the applicability and versatility of MRIM, we have applied MRIM to investigate the temperature dependence of the specific heat, thermal expansion, and elastic and thermal properties of $La_{1-x}Sm_xCoO_3$ ($0 \leq x \leq 0.2$). It is found that the model is successful in describing temperature dependent ($1\,K < T < 300\,K$) specific heat (C), cohesive energy (ϕ), molecular force constant (f), Reststrahlen frequency (v), Debye temperature (θ_D), and Gruneisen parameter (γ) of $La_{1-x}Sm_xCoO_3$ ($0 \leq x \leq 0.2$). The various properties in $La_{1-x}Sm_xCoO_3$ ($0 \leq x \leq 0.2$) are affected by the exchange of the A ions due to their different ionic radii. The paper is organized in the following way.

The computational details of model formalism are given in Section 2. In Section 3, we discuss the elastic and thermal properties of $La_{1-x}Sm_xCoO_3$ ($0 \leq x \leq 0.2$). The calculated results are compared with the available experimental results in the same section. Finally, in Section 4, the findings of the present study are concluded.

2. Formalism of MRIM

The effective interionic potential corresponding to the modified rigid ion model (MRIM) frame work is expressed as [9, 10]

$$\phi(r) = -\frac{e^2}{2}\sum_{kk'}Z_kZ_{k'}r_{kk'}^{-1} - \sum_{kk'}C_{kk'}r_{kk'}^{-6}$$
$$+ \left[n_ib_i\beta_{kk'}\exp\left\{\frac{(r_k + r_{k'} - r_{kk'})}{\rho_i}\right\}\right] + \frac{n_i'}{2}b_i$$

TABLE 1: Values of average cation radius at A-site, tolerance factor, critical radius, and model parameters of $La_{1-x}Sm_xCoO_3$ ($0.0 \leq x \leq 0.2$).

Compound	r_A (Å) (A-site)	Tolerance factor (t)	r_{cr} (Å)	Model parameters			
				$b_1 \times 10^{-19}$ (J)	ρ_1 (Å)	$b_2 \times 10^{-19}$ (J)	ρ_2 (Å)
$LaCoO_3$	1.360	0.9958	0.9148	1.662	1.114	0.225	0.404
$La_{0.98}Sm_{0.02}CoO_3$	1.357	0.9953	0.9159	1.636	1.105	0.222	0.399
$La_{0.96}Sm_{0.04}CoO_3$	1.355	0.9947	0.9164	1.605	1.095	0.219	0.394
$La_{0.94}Sm_{0.06}CoO_3$	1.352	0.9941	0.9168	1.596	1.093	0.218	0.392
$La_{0.92}Sm_{0.08}CoO_3$	1.350	0.9936	0.9168	1.551	1.071	0.213	0.383
$La_{0.88}Sm_{0.12}CoO_3$	1.205	0.9420	1.0102	1.317	1.757	0.197	0.364
$La_{0.84}Sm_{0.16}CoO_3$	1.202	0.9413	1.0143	1.288	1.751	0.192	0.355
$La_{0.80}Sm_{0.20}CoO_3$	1.199	0.9407	1.0169	1.269	1.751	0.190	0.351

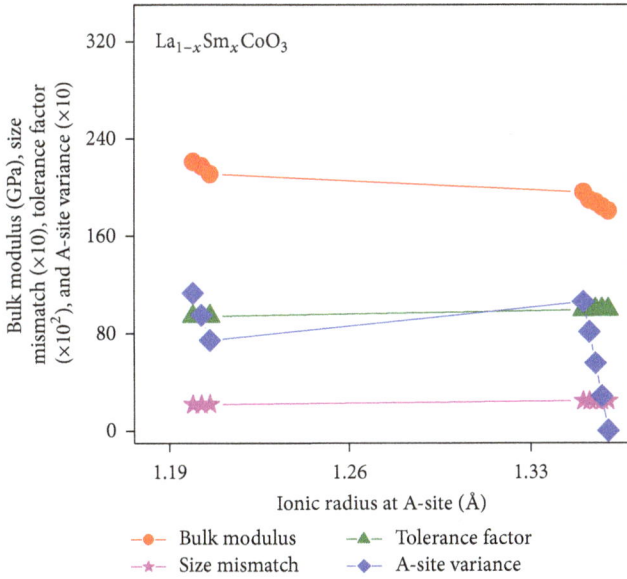

FIGURE 1: The variation of bulk modulus, size mismatch, tolerance factor, and A-site variance of $La_{1-x}Sm_xCoO_3$ ($0.0 \leq x \leq 2.0$) with the radius of A-site cation.

$$\times \left[\beta_{kk} \exp \left\{ \frac{(2r_k - r_{kk})}{\rho_i} \right\} \right.$$

$$\left. + \beta_{k'k'} \exp \left\{ \frac{(2r_{k'} - r_{k'k'})}{\rho_i} \right\} \right] \right].$$

$$(1)$$

Here, first term is attractive long range (LR) coulomb interactions energy. The second term represents the contributions of van der Waals (vdW) attraction for the dipole-dipole interaction and is determined by using the Slater-Kirkwood Variational (SKV) method [11]. The third term is short range (SR) overlap repulsive energy represented by the Hafemeister-Flygare-type (HF) interaction extended up to the second neighbour. In (1), $r_{kk'}$ represents separation between the nearest neighbours, while r_{kk} and $r_{k'k'}$ appearing in the next terms are the second neighbour separation. r_k ($r_{k'}$) is the ionic radii of k (k') ion. n (n') is the number of nearest (next nearest neighbour) ions. The summation is performed over all the kk' ions. b_i and ρ_i are the hardness

and range parameters for the ith cation-anion pair ($i = 1, 2$), respectively, and $\beta_{kk'}$ is the Pauling coefficient [12] expressed as

$$\beta_{kk'} = 1 + \left(\frac{Z_k}{N_k} \right) + \left(\frac{Z_{k'}}{N_{k'}} \right) \tag{2}$$

with Z_k ($Z_{k'}$) and N_k ($N_{k'}$) as the valence and number of electrons in the outermost orbit of k (k') ions, respectively. The model parameters (hardness and range) are determined from the equilibrium condition

$$\left[\frac{d\phi(r)}{dr} \right]_{r=r_0} = 0 \tag{3}$$

and the bulk modulus

$$B = \frac{1}{9Kr_0} \left[\frac{d^2\phi(r)}{dr^2} \right]_{r=r_0}. \tag{4}$$

The symbol K is the crystal structure constant, r_0 is the equilibrium nearest neighbor distance of the basic perovskite cell, and B is the bulk modulus. The cohesive energy for $La_{1-x}Sm_xCoO_3$ ($0 \leq x \leq 0.2$) is calculated using (1) and other thermal properties such as the Debye temperature (θ_D), Reststrahlen frequency (ν), molecular force constant (f), Gruneisen parameter (γ), specific heat (C), and thermal expansion (α) are computed using the expressions given in [9, 10]. The results are thus obtained and discussed below.

3. Results and Discussion

3.1. Model Parameters. The values of input data like unit cell parameters and interionic distances for Sm doped $LaCoO_3$ are taken from [8], for the evaluation of model parameters (b_1, ρ_1) and (b_2, ρ_2) corresponding to the ion pairs Co^{3+}-O^{2-} and La^{3+}/Sm^{3+}-O^2. The values of model parameters (b_1, b_2, ρ_1 and ρ_2) are listed in Table 1. The tolerance factor t ($t = (r_A + r_O)/2(r_B + r_O)$) for these compounds is reported in Table 1, which satisfies the condition that t for a stable perovskite phase is above 0.84 [16]. Depending on the composition of the perovskite, critical radius (r_{cr}) can be calculated by using (5) (Table 1) which describes the maximum size of the mobile ion to pass through. Consider

$$r_{cr} = \frac{a_0 \left((3/4) a_0 - \sqrt{2}r_B \right) + r_B^2 - r_A^2}{2 \left(r_A - r_B \right) + \sqrt{2}a_0}, \tag{5}$$

TABLE 2: Values of bulk modulus, cohesive and thermal properties of $La_{1-x}Sm_xCoO_3$ ($0.0 \leq x \leq 0.2$).

Compound	B_T (GPa)	Cohesive property		Thermal property			
		ϕ (eV) MRIM	ϕ (eV) Kapustinskii Eqn.	f (N/m)	v (THz)	θ_D (K)	γ
$LaCoO_3$	180.6	−148.5	−147.5	35.20	9.40	451.9	3.08
$La_{0.98}Sm_{0.02}CoO_3$	184.2	−148.8	−147.9	35.87	9.49	456.0	3.12
$La_{0.96}Sm_{0.04}CoO_3$	187.8	−149.2	−148.2	36.56	9.57	460.2	3.15
$La_{0.94}Sm_{0.06}CoO_3$	189.5	−149.4	−148.4	36.85	9.61	461.9	3.17
$La_{0.92}Sm_{0.08}CoO_3$	195.9	−150.0	−148.9	38.05	9.76	469.1	3.23
$La_{0.88}Sm_{0.12}CoO_3$	211.1	−151.4	−150.5	41.23	10.2	487.9	3.40
$La_{0.84}Sm_{0.16}CoO_3$	217.5	−151.7	−151.1	42.50	10.3	495.1	3.47
$La_{0.80}Sm_{0.20}CoO_3$	221.3	−152.1	−151.5	43.25	10.4	499.1	3.52
Others	180[a]	−144.54[b]				480[c]	

[a]reference [13], [b]reference [14], and [c]reference [15].

where r_A and r_B are the radius of the A ion and B ion, respectively, and a_0 corresponds to the lattice parameter ($V^{1/3}$). The condition that critical radius does not exceed 1.05 Å [17] for typical perovskite material is satisfied in our compounds. The values of ionic radii and atomic compressibility for La^{3+}, Sm^{3+}, Co^{3+}, and $O^{2−}$ are taken from [18, 19]. We defined the variance of the La/Sm ionic radii as a function of x, $\sigma^2 = \sum(x_i r_i^2 - r_A^2)$. Here, x_i, r_i, and r_A are the fractional occupancies, the effective ionic radii of cation of La and Sm, and the averaged ionic radius ($r_A = (1 − x)r_{R^{3+}} + x r_{A^{3+}}$), respectively. The effect of ionic radii on tolerance factor and A-site variance is shown in Figure 1. In $La_{1-x}Sm_xCoO_3$, the A-site cation radius reduces with decrease in x, and the buckling of Co-O-Co angle progressively increases with x, which leads to increased distortions of the lattice. Here, we have computed the bulk modulus (B) on the basis of Atoms in Molecules (AIM) theory [20] which emphasizes the partitioning of static thermophysical properties in condensed systems into atomic or group distributions. The computed values of bulk modulus for $La_{1-x}Sm_xCoO_3$ ($0.02 \leq x \leq 0.2$) are in good agreement with value predicted by Cornelius et al. [13].

3.2. Cohesive Energy.

The cohesive energy (ϕ) is a measure of the strength of forces that bind atoms together in the solid states and is descriptive in studying the phase stability. The cohesive energy of Sm doped $LaCoO_3$ is computed using (1) and is reported in Table 2. The negative value of the ϕ indicates that these compounds are stable at ambient temperature. The calculated cohesive energy of $La_{1-x}Sm_xCoO_3$ is quite comparable with the reported value of the lattice energy for $SmCoO_3$, −144.54 eV [14]. The variation of the cohesive energy (ϕ) of $La_{1-x}Sm_xCoO_3$ ($0 \leq x \leq 0.2$) as a function of A-site ionic radius is displayed Figure 2. Further, to ascertain the validity of the MRIM, we have used the generalized Kapustinskii equation [21]. The lattice energy obtained using the Kapustinskii equation is close to the MRIM. According to generalized Kapustinskii equation, the lattice energies of crystals with multiple ions are given as

$$U/KJ\,mol^{-1} = -\frac{1213.9}{\langle r \rangle}\left(1 - \frac{\rho}{\langle r \rangle}\right)\sum n_k z_k^2, \quad (6)$$

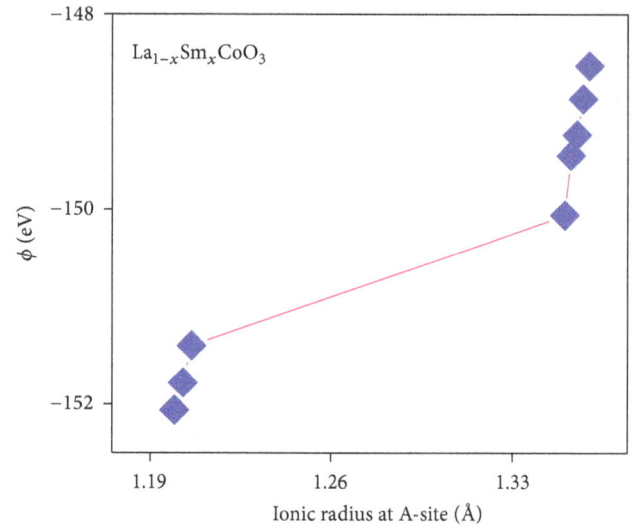

FIGURE 2: Variation of cohesive energy with radius of A-site cation for $La_{1-x}Sm_xCoO_3$ ($0.0 \leq x \leq 2.0$).

where $\langle r \rangle$ = weighted mean cation-anion radius sum and ρ is the average value of model parameters ρ_1 and ρ_2.

3.3. Thermal Properties.

We have also predicted the molecular force constant (f), Reststrahlen frequency (v), and Gruneisen parameter (γ) for Sm doped $LaCoO_3$ (Table 2). In the Debye approach, we consider the vibrations of the collective positive ion lattice with respect to the negative ion lattice. The frequency of vibration obtained by this model is also reported here as Reststrahlen frequency. This is clear from Table 2 that as the Sm doping increases the Reststrahlen frequency increases. The Debye temperature estimated for the analysis of specific heat is also reported in Table 2. Our calculated values of Debye temperature for $La_{1-x}Sm_xCoO_3$ ($0.02 \leq x \leq 0.2$) are close to reported value of $LaCoO_3$ which is 480 K [15]. Since the ionic radius of Sm is larger than that of La, the bulk modulus and other thermal properties systematically increase with increasing x (Table 2). We have

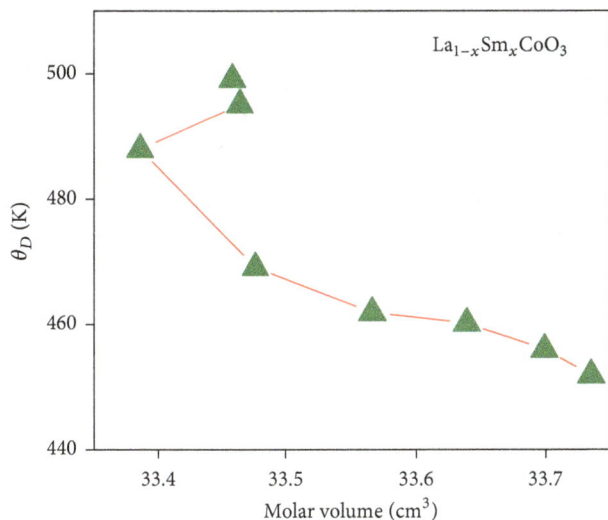

FIGURE 3: Variation of Debye temperature (θ_D) for La$_{1-x}$Sm$_x$CoO$_3$ ($0.0 \leq x \leq 2.0$) with molar volume (cm^3) of the basic perovskite cell.

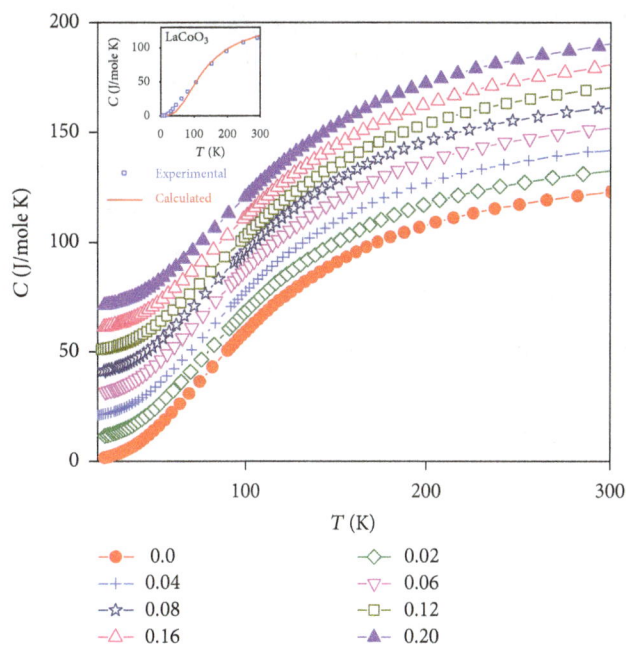

FIGURE 4: The variation of calculated specific heat of La$_{1-x}$Sm$_x$CoO$_3$ ($0.0 \leq x \leq 0.2$) as a function of temperature ($10 \leq T \leq 300$ K). The curves of higher cation doping are shifted upward by 10 J each from the preceding curve for clarity. Inset shows the specific heat of LaCoO$_3$ in the temperature interval (10 K $\leq T \leq 300$ K) and its comparison with the experimental values [22]. The solid line (—) and open square (□) represent the model calculation and experimental curve, respectively.

displayed the variation of the Debye temperature (θ_D) of Sm doped LaCoO$_3$ as a function of molar volume in Figure 3 and the θ_D is observed to be increasing with increasing basic perovskite molar cell volume.

3.4. *Specific Heat and Thermal Expansion.* The specific heat in the normal state of the material is usually approximated by the contribution of the lattice specific heat. The temperature dependence (10 K $\leq T \leq 300$ K) of specific heat for La$_{1-x}$Sm$_x$CoO$_3$ ($0 \leq x \leq 0.2$) is computed as displayed in Figure 4. We find that when the temperature is below 300 K the specific heat (Figure 4) is strongly dependent on temperature which is due to the anharmonic approximation. However, at higher temperatures, the anharmonic effect on C is suppressed, and C is almost constant at high temperature. The computed results on specific heat for LaCoO$_3$ compared with the experimental work of Tsubouchi et al. [22] in temperature range (10 K $\leq T \leq 300$ K) are displayed in the inset of Figure 4. The calculated dependence of C on temperature does not show any anomalous behavior and is in good agreement with the other experimental data [22].

Encouraged by the specific heat results, we tried to compute the thermal expansion coefficient (α) as a function of temperature using the well-known relation, $\alpha = \gamma C / B_T V$, where B_T, V, and C are the isothermal bulk modulus, unit formula volume, and specific heat at constant volume, respectively, and γ is the Gruneisen parameter. It is important to note that in Figure 5 (for 10 K) that thermal expansion shows a significant decrease due to the doping of Sm in LaCoO$_3$. Our results on volume thermal expansion coefficients will certainly serve as a guide to experimental workers in future.

4. Conclusion

On the basis of an overall discussion, it may be concluded that the description of the thermodynamic properties of Sm

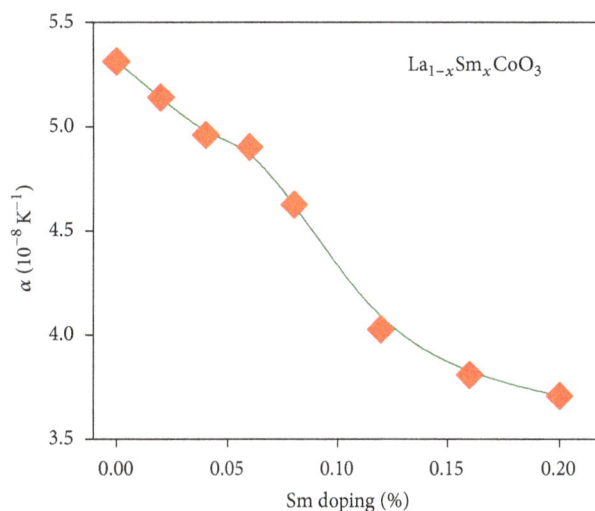

FIGURE 5: The variation of thermal expansion (α) with composition (x) in La$_{1-x}$Sm$_x$CoO$_3$ ($0.0 \leq x \leq 0.2$) at 10 K.

doped LaCoO$_3$ given by us is remarkable in view of the inherent simplicity and less parametric nature of the modified rigid ion model (MRIM). Our results are probably the first reports of specific heat and thermal expansion at these temperatures for La$_{1-x}$Sm$_x$CoO$_3$ ($0.02 \leq x \leq 0.2$). To the best of our knowledge, the values on bulk modulus, cohesive and lattice thermal properties for $La_{1-x}Sm_xCoO_3$ $(0.0 \leq x \leq 0.2)$ have not yet been measured or calculated, hence our results can serve as a prediction for future investigations.

Conflict of Interests

The authors declare that there is no conflict of interests regarding the publication of this paper.

Acknowledgment

The authors are thankful to the University Grant Commission (UGC), New Delhi, for providing the financial support.

References

[1] T. Inagaki, K. Miura, H. Yoshida et al., "High-performance electrodes for reduced temperature solid oxide fuel cells with doped lanthanum gallate electrolyte: II. La(Sr)CoO$_3$ cathode," *Journal of Power Sources*, vol. 86, no. 1-2, pp. 347–351, 2000.

[2] C. H. Chen, H. J. M. Bouwmeester, R. H. E. Van Doom, H. Kruidhof, and A. J. Burggraaf, "Oxygen permeation of La$_{0.3}$Sr$_{0.7}$CoO$_{3-\delta}$," *Solid State Ionics*, vol. 98, no. 1-2, pp. 7–13, 1997.

[3] S. Yamaguchi, Y. Okimoto, H. Taniguchi, and Y. Tokura, "Spin-state transition and high-spin polarons in LaCoO$_3$," *Physical Review B: Condensed Matter and Materials Physics*, vol. 53, no. 6, pp. R2926–R2929, 1996.

[4] Y. Kobayashi, N. Fujiwara, S. Murata, K. Asai, and H. Yasuoka, "Nuclear-spin relaxation of ^{59}Co correlated with the spin-state transitions in LaCoO$_3$," *Physical Review B—Condensed Matter and Materials Physics*, vol. 62, no. 1, pp. 410–414, 2000.

[5] M. Señarís-Rodríguez and J. Goodenough, "LaCoO$_3$ revisited," *Journal of Solid State Chemistry*, vol. 116, no. 2, pp. 224–231, 1995.

[6] P. G. Radaelli and S. W. Cheong, "Structural phenomena associated with the spin-state transition in LaCoO$_3$," *Physical Review B*, vol. 66, Article ID 094408, 2002.

[7] J. A. Alonso, M. J. Martínez-Lope, C. de La Calle, and V. Pomjakushin, "Preparation and structural study from neutron diffraction data of RCoO$_3$ (R = Pr, Tb, Dy, Ho, Er, Tm, Yb, Lu) perovskites," *Journal of Materials Chemistry*, vol. 16, pp. 1555–1560, 2006.

[8] J. R. Sun, R. W. Li, and B. G. Shen, "Spin-state transition in La$_{1-x}$Sm$_x$CoO$_3$ perovskites," *Journal Of Applied Physics*, vol. 89, p. 1331, 2001.

[9] R. Thakur, R. K. Thakur, and N. K. Gaur, "Thermophysical properties of Pr$_{1-x}$Ca$_x$CoO$_3$," *Thermochimica Acta*, vol. 550, pp. 53–58, 2012.

[10] R. K. Thakur, S. Samatham, N. Kaurav, V. Ganesan, N. K. Gaur, and G. S. Okram, "Dielectric, magnetic, and thermodynamic properties of Y$_{1-x}$Sr$_x$MnO$_3$ (x = 0.1 and 0.2)," *Journal of Applied Physics*, vol. 112, no. 10, Article ID 104115, 2012.

[11] J. C. Slater and J. G. Kirkwood, "The van der waals forces in gases," *Physical Review*, vol. 37, no. 6, pp. 682–697, 1931.

[12] L. Pauling, *Nature of the Chemical Bond*, Cornell University Press, Ithaca, NY, USA, 1945.

[13] A. L. Cornelius, S. Kletz, and J. S. Schilling, "Simple model for estimating the anisotropic compressibility of high temperature superconductors," *Physica C*, vol. 197, no. 3-4, pp. 209–223, 1992.

[14] M. A. Farhan and M. J. Akhtar, "Negative pressure driven phase transformation in Sr doped SmCoO$_3$," *Journal of Physics: Condensed Matter*, vol. 22, no. 7, Article ID 075402, 2010.

[15] C. He, H. Zheng, J. F. Mitchell, M. L. Foo, R. J. Cava, and C. Leighton, "Low temperature Schottky anomalies in the specific heat of LaCoO$_3$: defect-stabilized finite spin states," *Applied Physics Letters*, vol. 94, no. 10, Article ID 102514, 2009.

[16] C.-M. Fang and R. Ahuja, "Structures and stability of ABO$_3$ orthorhombic perovskites at the Earth's mantle conditions from first-principles theory," *Physics of the Earth and Planetary Interiors*, vol. 157, no. 1-2, pp. 1–7, 2006.

[17] M. Mogensen, D. Lybye, N. Bonanos, P. V. Hendriksen, and F. W. Poulsen, "Factors controlling the oxide ion conductivity of fluorite and perovskite structured oxides," *Solid State Ionics*, vol. 174, no. 1–4, pp. 279–286, 2004.

[18] R. D. Shannon, "Revised effective ionic radii and systematic studies of interatomic distances in halides and chalcogenides," *Acta Crystallographica Section A*, vol. 32, pp. 751–767, 1976.

[19] C. Kittel, *Introduction to Solid State Physics*, Wiley, New York, NY, USA, 5th edition, 1976.

[20] A. M. Pendás, A. Costales, M. A. Blanco, J. M. Recio, and V. Luaña, "Local compressibilities in crystals," *Physical Review B*, vol. 62, Article ID 13970, 2000.

[21] L. Glasser, "Lattice energies of crystals with multiple ions: a generalized Kapustinskii equation," *Inorganic Chemistry*, vol. 34, no. 20, pp. 4935–4936, 1995.

[22] S. Tsubouchi, T. Kyomen, M. Itoh, and M. Oguni, "Kinetics of the multiferroic switching in MnWO$_4$," *Physical Review B*, vol. 69, Article ID 144406, 2004.

29

Nonlinear Optical Properties of Novel Mono-O-Hydroxy Bidentate Schiff Base: Quantum Chemical Calculations

N. S. Labidi[1,2]

[1] Department of Chemistry, Faculty of Sciences, University of the Sciences and Technology of Oran (U.S.T.O.MB),
  BP 1505 El-M'naouer, 31000 Oran, Algeria
[2] Centre Universitaire de Tamanrasset, 11000 Tamanrasset, Algeria

Correspondence should be addressed to N. S. Labidi; labidi19722004@yahoo.fr

Academic Editor: Tao Zhang

The semiempirical AM1 SCF method is used to study the first static hyperpolarizabilities β of some novel mono-O-Hydroxy bidentate Schiff base in which electron donating (D) and electron accepting (A) groups were introduced on either side of the Schiff base ring system. Geometries of all molecules were optimized at the semiempirical AM1. The first static hyperpolarizabilities of these molecules were calculated using Hyperchem package. To understand this phenomenon in the context of molecular orbital picture, we examined the molecular HOMO and molecular LUMO generated via Hyperchem. The study reveals that the mono-O-Hydroxy bidentate Schiff bases have large β values and hence in general may have potential applications in the development of nonlinear optical materials.

1. Introduction

An intense research activity is currently associated with the synthesis and development of molecule-based second-order nonlinear optical (NLO) materials, involving organic chromophore and metal complexes [1–3]. The Schiff base compounds have been under investigation for several years because of their potential application to optical communications and because many of them have NLO behaviour [4, 5]. The design of efficient organic materials for the nonlinear effect is based on molecular units containing highly delocalized Π-electron moieties and extra electron donor and electron acceptor groups on opposite sides of the molecule at appropriate positions on the ring to enhance the conjugation. The effect of electron withdrawing and electron attracting substituents on the first hyperpolarizability β of conjugated systems has received a great deal of attention in recent years [6, 7]. It was shown that the type of substituent plays a major role in charge transfer through the molecule and therefore in nonlinear properties [8]. Prasad and Williams [9] explained that the certain classes of organic materials exhibit extremely larger NLO and electrooptic effect. The design of most efficient organic materials for the nonlinear effect is based on

molecular units containing highly delocalized pi-electron moieties and extra electron donor (D) and electron acceptor (A) groups on opposite sides of the molecule at appropriate positions on the ring to enhance the conjugation. The pi-electron cloud movement from donor to acceptor makes the molecule to be highly polarized. The chromophore design was mainly done by synthetic explorations which are time consuming and costly process. At the same time, measurement of the molecular nonlinear optical coefficients (β) requires the use of the well-known electric-field-induced second harmonic (EFISH) experiment, in which only the vector component of β_{tot} parallel to the molecule's ground state dipole moment can be determined [10]. To avoid these problems, the development of quantum procedures for the science of chromophore design and proprieties calculation has taken a different route especially due to the birth of quantum chemistry packages [11]. One of the best computer semiempirical program packages is Hyperchem 7.0 (molecular modelling system) [12]. This package can be used to study the electronic structure and energy of ground and excited states of atoms, molecules, ions, first hyperpolarizability, second hyperpolarizability, and so forth.

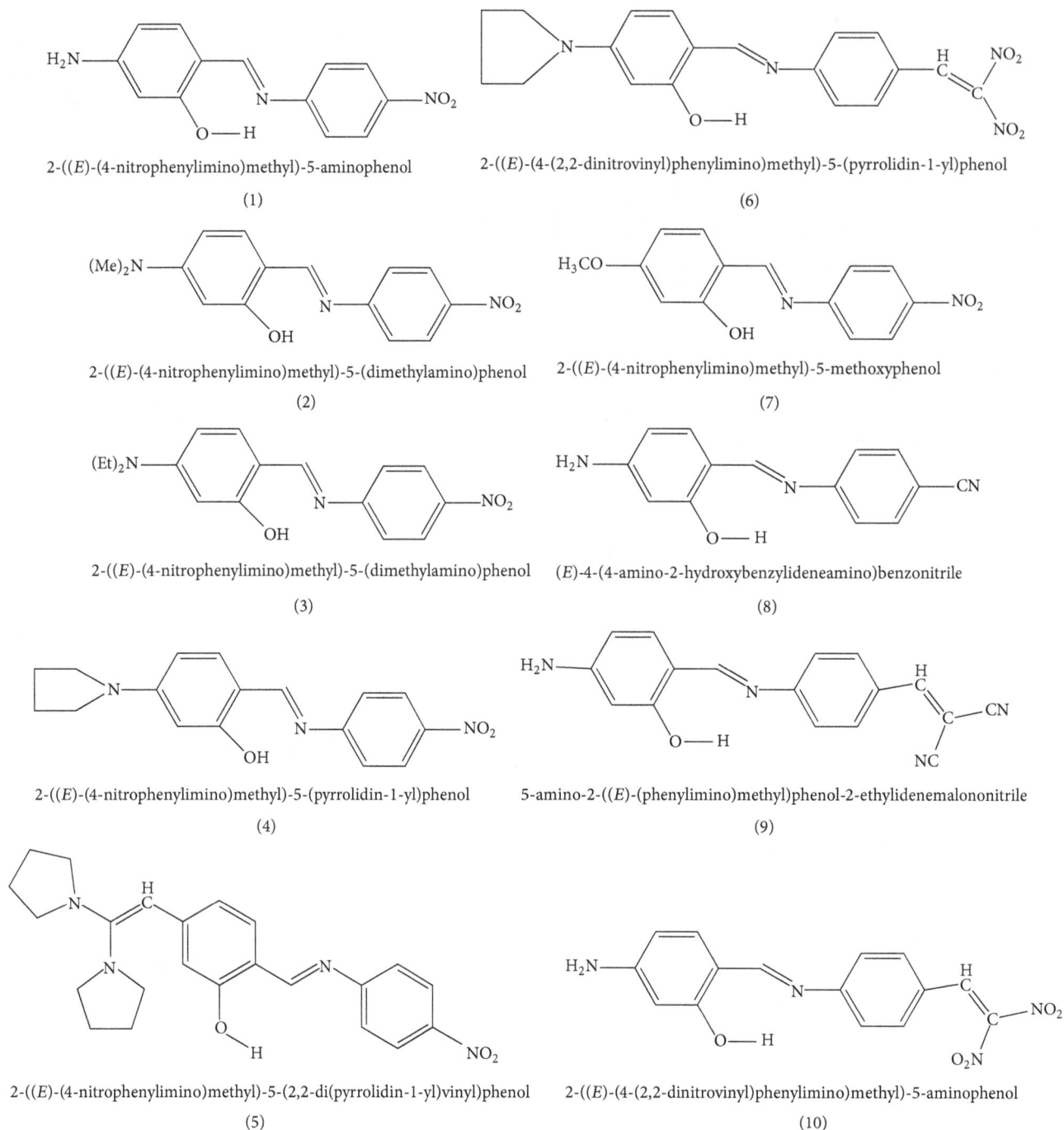

2-((E)-(4-nitrophenylimino)methyl)-5-aminophenol

(1)

2-((E)-(4-(2,2-dinitrovinyl)phenylimino)methyl)-5-(pyrrolidin-1-yl)phenol

(6)

2-((E)-(4-nitrophenylimino)methyl)-5-(dimethylamino)phenol

(2)

2-((E)-(4-nitrophenylimino)methyl)-5-methoxyphenol

(7)

2-((E)-(4-nitrophenylimino)methyl)-5-(dimethylamino)phenol

(3)

(E)-4-(4-amino-2-hydroxybenzylideneamino)benzonitrile

(8)

2-((E)-(4-nitrophenylimino)methyl)-5-(pyrrolidin-1-yl)phenol

(4)

5-amino-2-((E)-(phenylimino)methyl)phenol-2-ethylidenemalononitrile

(9)

2-((E)-(4-nitrophenylimino)methyl)-5-(2,2-di(pyrrolidin-1-yl)vinyl)phenol

(5)

2-((E)-(4-(2,2-dinitrovinyl)phenylimino)methyl)-5-aminophenol

(10)

FIGURE 1: The designed novel mono-O-Hydroxy Schiff base ligands.

Our objective is to design a range of novel asymmetric mono-O-Hydroxy Schiff base compounds as model to evaluate the electronic properties and NLO responses. The approach is based on the concept of charge transfer (CT) between donor and acceptor through the molecules. In this research work, first hyperpolarizabilities (β) are calculated using semiempirical AM1 method using Hyperchem program package.

2. Computational Procedures

All hyperpolarizability calculations of the O-Hydroxy bidentate Schiff bases (Figure 1) were performed using Hyperchem software [12], Intel Pentium P4 2002 XP with the Windows XP operating system. The AM1 Hamiltonian calculation was used for geometric optimization of the molecules and to calculate parameters like bond length, bond angle, core-core attraction, heat of formation, ionisation potential dipole moment, HOMO-LUMO gap, and Mulliken population analysis of the outermost MO for mono-O-Hydroxy Schiff base ligands 1–10.

3. Second-Order Nonlinearities

The existence of NLO phenomena is represented at the microscopic level as Taylor's expansion of the relationship between the spatial components of the induced dipole moment μ_i and the components of the electronic field E_j that create it [13, 14], as shown in

$$\mu_i = \mu_i^0 + \sum \alpha_{ij} E_j + \frac{1}{2} \sum \beta_{ijk} E_j E_k + \frac{1}{6} \sum \gamma_{ijk} E_j E_k E_l + \cdots,$$
(1)

where $\alpha_{ij} = (\partial \mu_i / \partial E_j)_{E0}$, $\beta_{ijk} = (\partial \mu i / \partial E_j \partial E_k)_{E0}$, and $\gamma_{ijkl} = (\partial \mu_i / \partial E_j \partial E_k \partial E_k \partial E_l)_{E0}$. These tensors can also be developed from Taylor's expansion of the Stark energy, as shown in

$$U(E) = U(0) - \sum \mu_i E_i - \frac{1}{2} \sum \alpha_{ij} E_i$$
$$- \frac{1}{6} \sum \beta_{ijk} E_i E_j E_k - \frac{1}{24} \sum \gamma_{ijkl} E_i E_j E_k E_l - \cdots,$$
(2)

where $U(0)$ is the energy in the absence of the electric field E, $\mu_i = (\partial \mu_i / \partial E_i)_{U0}$, $\alpha_{ij} = (\partial U_i / \partial E_i \partial E_j)_{U0}$, $\beta_{ijk} = (\partial U_i / \partial E_i \partial E_j \partial E_k)_{U0}$, and $\gamma_{ijkl} = (\partial U_i / \partial E_i \partial E_k \partial E_l)_{U0}$. In the above equations the subscripts i, j, k, and l refer to the molecular coordinate system and E_j, E_k, and so forth denote the components of the applied field. Note that these approximations are valid only for fields and polarizations that are small relative to atomic fields. When small electric fields are employed, the terms containing β and γ components can be ignored and the relation is assumed to be linear.

The values that are usually calculated are the molecular quantities: α, β, and γ, which are the polarizability, first-order hyperpolarizability, and the second-order hyperpolarizability tensors, respectively. A value that is useful in measuring second-order NLO properties is the mean first-order hyperpolarizability β, which is calculated as

$$\beta_{tot} = (\beta_{zxx} + \beta_{zyy} + \beta_{zzz}).$$
(3)

The indices indicate the projection of field components in the direction indicated by the second two indices (β_{zxx}) on the molecular axis indicated by the first index (β_{zxx}), and the vectorial nature of β for fixed molecular and field direction is clear. If we devise an experiment where the applied field is pointed in the direction of the molecular x-axis which is parallel to the dipolar axis of the molecule, β_{xxx} will be the dominant contributor to the nonlinear response. Since the values of the first hyperpolarizability tensors of the output file of Hyperchem are reported in atomic units (a.u.), the calculated values were converted into electrostatic units (1 a.u. = 8.6393× 10^{-33} esu).

TABLE 1: Nonlinear optical properties for mono-O-Hydroxy Schiff base ligands 1–10.

Molecule	β_{zxx}	β_{zyy}	β_{zzz}	β ($\times 10^{-30}$ esu)
1	204.219	852.595	1155.666	19.114
2	−240.801	1648.990	2784.197	36.219
3	−82.088	762.850	4224.402	42.377
4	1341.625	−1344.460	4442.795	38.358
5	1890.550	−31.837	16853.621	161.661
6	−7622.706	2416.968	14374.475	79.2114
7	−131.707	1555.308	2311.811	32.271
8	622.288	924.831	1427.131	25.69545
9	−5168.761	1480.576	5752.509	17.834
10	−443.862	3076.680	4611.84959	62.588

The calculated β values have been converted into electrostatic units (esu) (1 a.u. = 8.6393 · 10^{-33} esu).
$\beta_{tot} = (\beta_{zxx} + \beta_{zyy} + \beta_{zzz})$.

4. Results and Discussion

The molecules studied in this investigation employing semi-empirical AM1 calculation are shown in Figure 1. The study involves the calculation of first hyperpolarizability tensor for all 1–10 mono-O-Hydroxy Schiff bases ligands. It is intended to compare the electronic effect on the first hyperpolarizability of mono-O-Hydroxy Schiff bases ligands upon substitution of donor and acceptor groups at appropriate positions.

All first hyperpolarizabilities calculated for the molecules under investigation are given in Table 1. The magnitude of the first hyperpolarizability tensor of all N-salicylidene-aniline derivatives studied is dependent upon the availability of the lone pair of electrons on the nitrogen atom to conjugate with the N-salicylidene-aniline moiety.

The dramatic increase of first hyperpolarizability has been observed when the lone pair on the nitrogen atom of the donor group is forced to conjugate with the N-salicylidene-aniline ring system, upon substitution on nitrogen of NH_2 group with other groups, such as methyl, ethyl, and methylpyrrolidine. It is expected that molecule 1 with NH_2 as the donor group would give a higher value for first hyperpolarizability compared to molecule 7 where the donor group is OCH_3. The enhancement of first hyperpolarizability from molecule 1 to 6 is attributed to the enhanced availability of the lone pair for conjugation with the N-salicylidene-aniline system. The increase of β for molecules 1–6 is very significant.

Molecule 5 shows almost four times increase of first hyperpolarizability compared to molecule 4. Molecule 10 shows almost two times increase of first hyperpolarizability compared to molecule 7. The significant difference between molecule 4 and molecule 5 is the introduction of an extra double bond, which helped to introduce two pyrrolidine rings to the system which will enhance the conjugation. The magnitude of the first hyperpolarizability tensor of molecule 5 is the largest relative to the other molecules. The molecular hyperpolarizability value of this molecule is about 1000 times that of urea (0.14×10^{-30} esu).

TABLE 2: HOMO, LUMO, and band gap energies (ΔE) for Schiff base ligands 1–10.

Molecule	HOMO (eV)	LUMO (eV)	ΔE (eV) $(E_{LUMO} - E_{HOMO})$
1	−8.777	−1.377	7.400
2	−8.567	−1.325	7.242
3	−8.402	−1.320	7.082
4	−8.686	−1.485	7.201
5	−8.163	−1.368	6.795
6	−8.548	−1.854	6.694
7	−9.204	−1.449	7.755
8	−8.614	−0.932	7.682
9	−8.619	−1.356	7.263
10	−8.772	−1.878	6.894

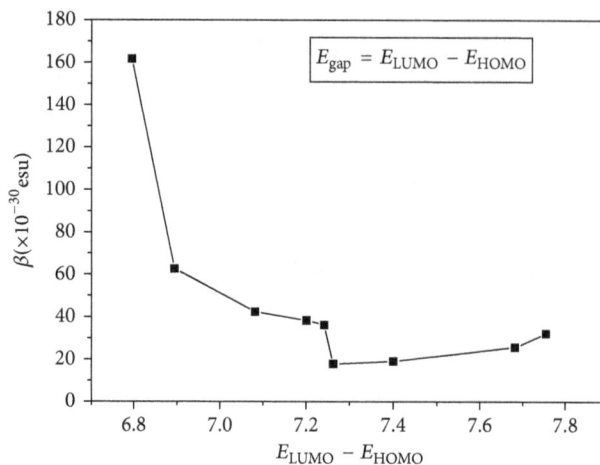

FIGURE 2: Variation of β_{tot} and E_{gap} values for Schiff base ligands.

In molecules 8–10 consisting of combination of NO_2 and CN as acceptor groups we observed a high value for β. The results obtained for molecule 9 when the R1 and R2 substituents were $-NH_2$ and $-CN$ groups, respectively, show a decrease for first hyperpolarizability compared to all molecules. This means that the presence of the $-NH_2$ group together with the $-CN$ in N-salicylidene-aniline based ligands decreases the nonlinear optic property of these types of ligands.

The AM1 calculated HOMO-LUMO gaps for all N-salicylidene-aniline derivatives are shown in Table 2 and summarized graphically in Figure 2.

As shown in Table 2 substitution of different electron pushing groups on nitrogen of NH_2 group in molecules 1–6 increases the energy of the molecular HOMO, while leaving the LUMO energy essentially unchanged due to the same acceptor group. Thus, the energy gap decreases with the substitution on nitrogen of NH_2 and produces a larger hyperpolarizability.

The substitution of $-OCH_3$ group for NH_2 reduces the energy of the HOMO while leaving the LUMO energy unchanged. This has led to a larger energy gap than that of molecules 1–6 and produces a decrease in beta value. As can be seen from Figure 2 ligands 1–6 differ only by R1 substituent. The significant differences of the band gaps of these ligands indicate that the role of this substituent in prediction of the band gap is very important.

The substitution of CN group decreased the LUMO energy while keeping the HOMO unchanged which has led to a larger energy gap and shows a further reduction of the beta. In molecules 8–10 the HOMO is comparable but a large decrease in LUMO of molecule 10 has been observed and this has resulted in the smallest gap and the largest beta.

Figure 2 shows also the inverse relationship between calculated β_{tot} and E_{gap}. The AM1 values for selected compounds show that it could be interesting to synthesize compounds with end parts in polyacetylene (NO_2/N(Et)$_2$, N(Me)$_2$, NH, NH_2, $NHNH_2$, NHOH, and OH) groups having the greatest and the lowest, respectively, β_{tot} and E_{gap} values.

The nonlinear optical (NLO) properties of conjugated molecules have been extensively studied as these compounds form a promising class of organic materials with interesting characteristics for photonic applications. Oudar and Zyss [15] found that the level of SHG response of a given material is inherently dependent upon its structural attributes. On a molecular scale, the extent of charge transfer (CT) across the NLO chromophore determines the level of SHG output. The greater the CT, the larger the SHG output. Chemla and Zyss [16] theoretical works have shown that the delocalization of Π-electrons in linear systems leads to large nonresonant optical molecular polarizabilities. Besides, geometric changes caused by incorporation of push-pull end groups can enhance the nonlinear polarizabilities of conjugated molecules.

Hayashi et al. [17] have calculated the linear and nonlinear polarizabilities in the side-chain direction of the polymer chains with all H atoms substituted by fluorine, hydroxyl, and cyano groups. Their results have shown that the coupling between electronic states of the side groups with those of the main chain increases the values of the perpendicular polarizabilities.

5. Conclusions

It is evident that the first hyperpolarizability β tensor of N-salicylidene-aniline derivatives strongly depends on the electronic structure of the molecule. In these molecules, where there is connectivity between two rings, system tends to rotate through the carbon-carbon sigma bond. This will increase the overlap of interacting orbitals, which will eventually increase the CT from donor to acceptor through the N-ethylidenemethanamine moiety.

The HOMO-LUMO calculations show that the first hyperpolarizability of these derivatives is directly related to the HOMO-LUMO energy gap. The highest is molecule 5 which has the smallest energy gap while the smallest is 7, which has the highest energy gap. The study reveals that these N-salicylidene-aniline derivatives have important first static hyperpolarizabilities and may have potential applications in the development of NLO materials. It is important to stress that, in these calculated β values, we do not take into account the effect of the field strength on the nuclear positions; we evaluate only the electronic component of β. The vibrational contributions which, for conjugated systems, can be important according to the NLO process are left for further investigations.

Acknowledgment

The author gratefully acknowledges the support of this work by the Research Laboratory of Chemical Materials, University of Batna in Algeria.

References

[1] A. Elmali, A. Karakaş, and H. Ünver, "Nonlinear optical properties of bis[(p-bromophenyl-salicylaldiminato)chloro]iron(III) and its ligand N-(4-bromo)-salicylaldimine," *Chemical Physics*, vol. 309, no. 2-3, pp. 251–257, 2005.

[2] A. Karakaş, H. Ünver, A. Elmali, and I. Svoboda, "Study on the second order optical properties of N-(2, 4-dichloro)-salicylaldimine," *Zeitschrift für Naturforschung A*, vol. 60, no. 5, pp. 376–382, 2005.

[3] J. F. Nicoud and R. J. Twieg, "Design and synthesis of organic molecular compounds for efficient second-harmonic generation," in *Nonlinear Optical Properties of Organic Molecules and Crystals*, D. S. Chemla and J. Zyss, Eds., vol. 1, pp. 227–296, Academic Press, New York, NY, USA, 1987.

[4] I. Ledoux, J. Zyss, F. Simoni, and C. Umeton, *Novel Optical Materials and Applications*, John Wiley & Sons, New York, NY, USA, 1997.

[5] S. Di Bella, I. Fragala, I. Ledoux, M. A. Díaz-García, and T. J. Marks, "Synthesis, characterization, optical. Spectroscopy, electronic structure and second order nonlinear optical (NLO) properties of a novel class of donor-acceptor bis(salicylaldiminato)nickel(II) Schiff base NLO chromophores," *Journal of the American Chemical Society*, vol. 119, no. 40, pp. 9550–9557, 1997.

[6] G. Raos and M. del Zoppo, "Substituent effects on the second-order hyperpolarisability of cyanine cations," *Journal of Molecular Structure*, vol. 589-590, pp. 439–445, 2002.

[7] A. Karakas, A. Elmali, H. Ünver, and I. Svoboda, "Nonlinear optical properties of some derivatives of salicylaldimine-based ligands," *Journal of Molecular Structure*, vol. 702, no. 1-2, pp. 103–110, 2004.

[8] M. Jalali-Heravi, A. A. Khandar, and I. Sheikshoaie, "Theoretical investigation of the structure, electronic properties and second-order nonlinearity of some azo Schiff base ligands and their monoanions," *Spectrochimica Acta A*, vol. 55, no. 12, pp. 2537–2544, 1999.

[9] P. N. Prasad and D. J. Williams, *Introduction to Nonlinear Optical Effects in Molecules and Polymers*, John Wiley & Sons, New York, NY, USA, 1990.

[10] B. F. Levine and C. G. Bethea, "Second and third order hyperpolarizabilities of organic molecules," *The Journal of Chemical Physics*, vol. 63, no. 6, pp. 2666–2682, 1975.

[11] V. J. Docherty, D. Pugh, and J. O. Morley, "Calculation of the second-order electronic polarizabilities of some organic molecules. Part 1," *Journal of the Chemical Society, Faraday Transactions II*, vol. 81, no. 8, pp. 1179–1192, 1985.

[12] Hyper Chem, *Molecular Modelling System*, Hypercube Inc., Gainesville, Fla, USA, 2000.

[13] H. A. Kurtz, J. J. P. Stewart, and K. M. Dieter, "Calculation of the nonlinear optical properties of molecules," *Journal of Computational Chemistry*, vol. 11, no. 1, pp. 82–87, 1990.

[14] S. P. Karna and M. J. Dupuis, "Frequency dependent nonlinear optical properties of molecules: formulation and implementation in the HONDO program," *Journal of Computational Chemistry*, vol. 12, no. 4, pp. 487–504, 1991.

[15] J. L. Oudar and J. Zyss, "Relations between microscopic and macroscopic lowest-order optical nonlinearities of molecular crystals with one- or two-dimensional units," *Physical Review A*, vol. 26, no. 4, pp. 2028–2048, 1982.

[16] D. S. Chemla and J. Zyss, *Nonlinear Optical Properties of Organic Molecules and Crystals*, vol. 1, Academic Press, New York, NY, USA, 1987.

[17] S.-I. Hayashi, S. Yabushita, and A. Imamura, "Ab initio calculations of linear and nonlinear polarizabilities in the side-chain direction on the conjugated polymers," *Chemical Physics Letters*, vol. 179, no. 4, pp. 405–409, 1991.

Permissions

All chapters in this book were first published in IJMET, by Hindawi Publishing Corporation; hereby published with permission under the Creative Commons Attribution License or equivalent. Every chapter published in this book has been scrutinized by our experts. Their significance has been extensively debated. The topics covered herein carry significant findings which will fuel the growth of the discipline. They may even be implemented as practical applications or may be referred to as a beginning point for another development.

The contributors of this book come from diverse backgrounds, making this book a truly international effort. This book will bring forth new frontiers with its revolutionizing research information and detailed analysis of the nascent developments around the world.

We would like to thank all the contributing authors for lending their expertise to make the book truly unique. They have played a crucial role in the development of this book. Without their invaluable contributions this book wouldn't have been possible. They have made vital efforts to compile up to date information on the varied aspects of this subject to make this book a valuable addition to the collection of many professionals and students.

This book was conceptualized with the vision of imparting up-to-date information and advanced data in this field. To ensure the same, a matchless editorial board was set up. Every individual on the board went through rigorous rounds of assessment to prove their worth. After which they invested a large part of their time researching and compiling the most relevant data for our readers.

The editorial board has been involved in producing this book since its inception. They have spent rigorous hours researching and exploring the diverse topics which have resulted in the successful publishing of this book. They have passed on their knowledge of decades through this book. To expedite this challenging task, the publisher supported the team at every step. A small team of assistant editors was also appointed to further simplify the editing procedure and attain best results for the readers.

Apart from the editorial board, the designing team has also invested a significant amount of their time in understanding the subject and creating the most relevant covers. They scrutinized every image to scout for the most suitable representation of the subject and create an appropriate cover for the book.

The publishing team has been an ardent support to the editorial, designing and production team. Their endless efforts to recruit the best for this project, has resulted in the accomplishment of this book. They are a veteran in the field of academics and their pool of knowledge is as vast as their experience in printing. Their expertise and guidance has proved useful at every step. Their uncompromising quality standards have made this book an exceptional effort. Their encouragement from time to time has been an inspiration for everyone.

The publisher and the editorial board hope that this book will prove to be a valuable piece of knowledge for researchers, students, practitioners and scholars across the globe.

List of Contributors

Michiko Yoshitake and Shinjiro Yagyu
National Institute for Materials Science, 3-13 Sakura, Tsukuba 305-0003, Japan

Toyohiro Chikyow
National Institute for Materials Science, 1-1 Namiki, Tsukuba 305-0044, Japan

Abdulganiyu Funsho Alabi, Ishaq Na'Allah Aremu and Segun Isaac Talabi
Department of Materials and Metallurgical Engineering, University of Ilorin, Ilorin, Nigeria

Samson Oluropo Adeosun
Department of Metallurgical and Materials Engineering, University of Lagos, Nigeria

Sulaiman Abdulkareem
Department of Mechanical Engineering, University of Ilorin, Ilorin, Nigeria

M. Y. Salunkhe
Department of Physics, Institute of Science, R. T. Road,Nagpur, Maharashtra 440001, India

D. S. Choudhary
Dhote Bandhu Science College, Gondia, Maharashtra 441614, India

S. B. Kondawar
Department of Physics, R. T. M. Nagpur University, Nagpur, Maharashtra 440033, India

Shujahadeen B. Aziz, Sarkawt Hussein, Ahang M. Hussein and Salah R. Saeed
Department of Physics, Faculty of Science, University of Sulaimani, Kurdistan Regional Government, Sulaimani City, Iraq

Olawale Olarewaju Ajibola
Metallurgical and Materials Engineering Department, Federal University of Technology Akure, Akure 340252, Nigeria
Materials and Metallurgical Engineering Department, Federal University Oye Ekiti, Oye Ekiti 371104, Nigeria

Daniel Toyin Oloruntoba and Benjamin O. Adewuyi
Metallurgical and Materials Engineering Department, Federal University of Technology Akure, Akure 340252, Nigeria

Oswaldo Antonio Hilders
Department of Physical Metallurgy, School of Metallurgical Engineering and Materials Science, Central University of Venezuela, Apartado 47514, Los Chaguaramos, Caracas 1041-A, Distrito Capital, Venezuela

Naddord Zambrano
Foundation for Professional Development,The Venezuelan College of Engineering, Caracas 1050, Venezuela

Ramón Caballero
Failure Analysis Laboratory, School of Metallurgical Engineering and Materials Science, Central University of Venezuela, Apartado 47514, Los Chaguaramos, Caracas 1041-A, Distrito Capital, Venezuela

A. Dhanapal, K. Chidambaram and A. R. Thoheer Zaman
Department of Mechanical Engineering, Sri Ramanujar Engineering College, Vandalur, Chennai, Tamil Nadu 600 048, India

S. Rajendra Boopathy
Department of Mechanical Engineering, College of Engineering, Anna University, Chennai 600 025, India

V. Balasubramanian
Center for Materials Joining & Research (CEMAJOR), Department of Manufacturing Engineering, Annamalai University, Annamalai Nagar, Chidambaram 608 002, India

Djillali Bensaid, Mohammed Ameri and Nour Eddine Bouzouira
Laboratory of Physical Chemistry of Advanced Materials, University of Djillali Liabes, BP 89, 22000 Sidi Bel Abbes, Algeria

Nadia Benseddik, Ali Mir and Fethi Benzoudji
Physics Department, Science Faculty, University of Sidi Bel Abbes, 22000 Sidi Bel Abbes, Algeria

M. Vanaja, K. Paulkumar, G. Gnanajobitha, C. Malarkodi and G. Annadurai
Environmental Nanotechnology Division, Sri Paramakalyani Centre for Environmental Sciences, Manonmaniam Sundaranar University, Alwarkurichi, Tamil Nadu 627412, India

S. Rajeshkumar
PG and Research Department of Biochemistry, Adhiparasakthi College of Arts and Science, Kalavai, Tamil Nadu 632506, India

Subir Paul, Anjan Pattanayak and Sujit K. Guchhait
Department of Metallurgical and Material Engineering, Jadavpur University, Kolkata 700032, India

J. U. Anaele, O. O. Onyemaobi, C. S. Nwobodo and C. C. Ugwuegbu
Department of Materials and Metallurgical Engineering, Federal University of Technology, PMB 1526, Owerri, Nigeria

Manish Bhandari
Jodhpur Institute of Technology, Jodhpur, Rajasthan 342003, India

Kamlesh Purohit
Jai Narain Vyas University, Jodhpur, Rajasthan 342005, India

G. Marinzuli, L. A. C. De Filippis and A. D. Ludovico
Politecnico di Bari, Dipartimento di Meccanica, Matematica e Management (DMMM), Viale Japigia 126, 70126 Bari, Italy

R. Surace
ITIA CNR, Institute of Industrial Technology and Automation, National Research Council, Via Paolo Lembo 38F, 70124 Bari, Italy

Paul Ocheje Ameh
Physical Chemistry Unit, Department of Chemistry, Nigeria Police Academy, PMB 3474,Wudil, Kano State, Nigeria

S. H. Gawande and A. A. Keste
Department of Mechanical Engineering, M. E. Society's College of Engineering, Pune, Maharashtra 411001, India

N. D. Pagar
Department of Mechanical & Materials Technology and Department of Technology, S.P. Pune University, Pune 411007, India

V. B. Wagh
Department of Mechanical Engineering, G.S.M. College of Engineering, Pune, Maharashtra 411045, India

Dirk J. Pons, Gareth Bayley, Christopher Tyree, Matthew Hunt and Reuben Laurenson
Department of Mechanical Engineering, University of Canterbury, Private Bag 4800, Christchurch 8020, New Zealand

S. V. San'kova, N. M. Shkatulyak, V. V. Usov and N. A. Volchok
South Ukrainian National Pedagogical University, 26 Staroportofrankovskaya Street, Odessa 65020, Ukraine

Rohit K.Mahadule
Smt. Radhikatai Pandav College of Engineering, Nagpur, Maharashtra 411204, India

Purushottam R. Arjunwadkar
Institute of Science, Nagpur, Maharashtra 440001, India

Megha P. Mahabole
School of Physical Sciences, S.R.T.M. University, Nanded, Maharashtra 431606, India

Nadeem Raza, Zafar Iqbal Zafar and Najam-ul-Haq
Institute of Chemical Sciences, Bahauddin Zakariya University, Multan 60800, Pakistan

Saeed Tamimi
Mining and Metallurgical Engineering Department, Amirkabir University of Technology, Tehran 15875-4413, Iran
Department of Mechanical Engineering, TEMA, University of Aveiro, 3810-193 Aveiro, Portugal

Mostafa Ketabchi and Nader Parvin
Mining and Metallurgical Engineering Department, Amirkabir University of Technology, Tehran 15875-4413, Iran

Mehdi Sanjari
Mining and Materials Engineering Department, McGill University, Montreal, QC, Canada H3A 0E8

Augusto Lopes
Departamento de Engenharia de Materiais e Ceramica, CICECO, University of Aveiro, 3810-193 Aveiro, Portugal

Urs Haßlinger, Christian Hartig and Robert Günther
Institute of Materials Physics and Technology, Hamburg University of Technology, Eißendorfer Straße 42, 21073 Hamburg, Germany

Norbert Hort
Helmholtz-Zentrum Geesthacht, Magnesium Innovation Centre (MagIC), Max-Planck-Straße 1, 21502 Geesthacht, Germany

Shanmugam Rajeshkumar, Chelladurai Malarkodi, Kanniah Paulkumar, Mahendran Vanaja, Gnanadas Gnanajobitha, and Gurusamy Annadurai
Environmental Nanotechnology Division, Sri Paramakalyani Centre for Environmental Sciences, Manonmaniam Sundaranar University, Alwarkurichi, Tamilnadu 627412, India

Manish Bhandari
Jodhpur Institute of Engineering and Technology, Jodhpur, Rajasthan, India

Kamlesh Purohit
Jai Narain Vyas University, Jodhpur, Rajasthan, India

K. R. Nemade and S. A.Waghuley
Department of Physics, Sant Gadge Baba Amravati University, Amravati 444 602, India

S. El Hajjaji
Laboratoire de Spectroscopie, Modélisation Moléculaire, Matériaux et Environnement (LS3ME), Faculté des Sciences, Université Med V-Agdal, Avenu Ibn Battouta, BP 1014, Rabat, Morocco

C. Cros and L. Aries
CIRIMAT-LCMIE, Universit´e Paul Sabatier, 118 route de Narbonne, 31064 Toulouse Cedex 4, France

S. Kanagaprabha
Kamaraj College, Tuticorin, Tamilnadu 628003, India

R. Rajeswarapalanichamy
Department of Physics, N.M.S.S. Vellaichamy Nadar College, Madurai, Tamilnadu 625019, India

K. Iyakutti
SRM University, Chennai, Tamilnadu 600030, India

N. M. Shkatulyak, S. V. Smirnova and V. V. Usov
South Ukrainian National Pedagogical University Named after K. D. Ushinsky, 26 Staroportofrankovskaya Street, Odessa 65020, Ukraine

Rasna Thakur and N.K.Gaur
Superconductivity Research Lab., Department of Physics, Barkatullah University, Bhopal 462026, India

N. S. Labidi
Department of Chemistry, Faculty of Sciences, University of the Sciences and Technology of Oran (U.S.T.O.MB), BP 1505 El-M'naouer, 31000 Oran, Algeria
Centre Universitaire de Tamanrasset, 11000 Tamanrasset, Algeria